NEW EVIDENCE FOR THE PLEISTOCENE PEOPLING OF THE AMERICAS

NEW EVIDENCE FOR THE PLEISTOCENE PEOPLING OF THE AMERICAS

Alan Lyle Bryan

Editor

PEOPLING OF THE AMERICAS

Symposia Series

CENTER FOR THE STUDY OF EARLY MAN
University of Maine
Orono, Maine

NEW EVIDENCE FOR THE PLEISTOCENE PEOPLING OF THE AMERICAS

Printed in U.S.A. by the University of Maine Printing Office, Orono, Maine.

ISBN: 0-912933-03-8

THE CENTER FOR THE STUDY OF EARLY MAN

The Center for the Study of Early Man is an affiliate of the Institute for Quaternary Studies and the Department of Anthropology at the University of Maine at Orono. It was established in July, 1981 by a seed grant from Mr. William Bingham's Trust for Charity. The Center's goals are to encourage research about Pleistocene peoples of the Americas, and to make this new knowledge available to both the scientific community and the interested public. Toward this end, the Center staff is developing research, public outreach, and publications programs.

"Peopling of the Americas," the Center's publication program, focuses on the earliest Americans and their environments. This program includes: (1) a monograph series presenting primary data on sites in North and South America which are more than 10,000 years old; (2) a process series presenting new methods and theories for interpreting early remains; (3) an edited volume series containing topical papers and symposia proceedings; (4) a popular book series making the most significant discoveries and research available to the general public; and (5) a bibliographic series.

In addition, the Center publishes an annual journal, *Current Research in the Pleistocene.* The journal publishes note-length articles about current research in the field of Pleistocene peopling of the Americas. The Center also publishes a quarterly newspaper called the *Mammoth Trumpet.* The newspaper is written for a general and a professional audience.

MANUSCRIPT SUBMISSIONS

BOOKS

The Center solicits high-quality original manuscripts in English (preferred), Spanish, Portuguese or French. For information write: Robson Bonnichsen, Director, Center for the Study of Early Man, 495 College Avenue, University of Maine, Orono, Maine 04473, or call (207) 581-2197.

CURRENT RESEARCH IN THE PLEISTOCENE

Researchers wishing to submit summaries for inclusion in the annual series should contact *CRP* Editor, Center for the Study of Early Man. The deadline for submissions is January 31 of each calendar year.

MAMMOTH TRUMPET

News of discoveries, reports on recent conferences and books, and news of current issues are invited. Contact editor Marcella H. Sorg at the Center.

BIBLIOGRAPHY

Authors are encouraged to submit reprints of published articles or copies of unpublished papers for inclusion in the bibliography. Please address contributions to the Center's librarian.

OTHER TITLES IN PEOPLING OF THE AMERICAS

UNDERSTANDING STONE TOOLS: A COGNITIVE APPROACH
David E. Young and Robson Bonnichsen

ARCHAEOLOGICAL SEDIMENTS IN CONTEXT
Julie K. Stein and William R. Farrand, Editors

ENVIRONMENTS AND EXTINCTIONS: MAN IN LATE GLACIAL NORTH AMERICA
Jim I. Mead and David J. Meltzer, Editors

CURRENT RESEARCH IN THE PLEISTOCENE

Contents

PREFACE

The Commission on the Peopling of the Americas convened at the X Congress of the International Union of Prehistoric and Protohistoric Sciences in Mexico City in October 1981. Papers presented at the Congress were prepublished in the Commission 10 volume, which was distributed to all participants. Most of those papers were revised and updated for publication in this volume. In addition, several other papers were solicited to extend the coverage.

The papers in this volume cover a wide spectrum of ongoing research in many countries. Some reports contain fairly complete analyses of data, while others are short summary reports of data available in more complete form elsewhere. A few are preliminary reports on important ongoing excavations. More detailed reports of excavations and laboratory analyses will be published in due course. Treatment of the data is therefore not uniform, but altogether this volume affords a comprehensive review of a great deal of important current research regarding the complex question of the peopling of the Americas.

The original intention was to translate all papers into English, but Miguel Bombin, who is absolutely fluent in Spanish and Portuguese as well as English, explained that it would be exceedingly difficult to translate because the method of presentation of data based upon authors' thought processes are so dissimilar. Bombin therefore agreed to compose an extended English abstract for each paper in Spanish and Portuguese so that the most important information is available for readers who do not know these languages. Anyone who needs the details can refer to the original versions. Shorter English abstracts are available for the papers in English.

This volume was scheduled for publication in late 1983; however, an unfortunate combination of circumstances delayed its completion. The unavoidable delay means that some authors may have changed their thinking about their materials; however, a few authors took advantage of the opportunity to update their research results and interpretations.

Production of this volume was handled in Edmonton through the paste-up stage. Lisa Bombin entered most of the foreign language papers into the mainframe computer. The remainder was entered by Russanne Low and Douglas Schnurrenberger, who also produced the type. Rick Will and Keary Walde helped redraft many illustrations, and Howard Kirkby did the paste-up. Ruth Gruhn assisted considerably with editing problems. Travel for conference participants to the X Congress, and development and publication of this book have been sponsored by the Center for the Study of Early Man, Orono, Maine, under the directorship of Robson Bonnichsen.

A.L.B.
Edmonton, Alberta
December 1984

Paleoamerican Prehistory as seen from South America

ALAN LYLE BRYAN
Department of Anthropology
University of Alberta
Edmonton, Alberta T6G 2H4
CANADA

Abstract

Recent archaeological research in South America supports the model that early Paleoamericans were general foragers who used a simple flaked stone technology. Between 12,000 and 13,000 B.P., people in central Colombia and southern Chile were collecting plants and hunting small game. There is no definite evidence that they hunted mastodonts, as did contemporary El Jobo people in northern Venezuela. In southern Patagonia people hunted horses and ground sloths about 11,000 B.P., but there is no evidence that people in central and northern Brazil ever predated on megamammals. Radiocarbon dates greater than 31,000 B.P. on hearths from undisturbed cultural contexts containing pebble and flake tools in southern Chile and northeastern Brazil indicate that people entered South America sometime before 35,000 B.P. Elsewhere in the Americas, controversial sites which have yielded simple stone and bone assemblages from less definite cultural contexts can be most reasonably evaluated by use of a model of early foraging rather than specialized hunting. It is necessary for American archaeologists to change their conception of the economy and technology of the earliest Americans.

THE NEED FOR A NEW PARADIGM IN AMERICAN PREHISTORY

During the last few years the controversy over evidence for Early Man in the Americas has been sharpened by new data. Although several sites with dated radiocarbon ages greater than 12,000 years have been reported during the last 20 years in North America, all of these reports have been subjected to stringent criticism by skeptics (e.g., Dincauze 1984; Owen 1984). Currently, the forefront of knowledge of Early Man is in South America, where archaeologists lack inhibiting preconceptions derived from the half century old discovery of fluted projectile points in undoubted association with mammoth and extinct bison at several kill sites in south-central North America. These Folsom and Clovis discoveries led to establishment of a paradigm that the earliest Americans were technologically specialized Upper Paleolithic big game hunters who migrated from the steppes of central Asia across the steppe tundra of Beringia and southward onto the Great Plains near the end of the Wisconsinan stadial (Haynes 1982; cf. Bryan 1978).

All who ever attempted to synthesize American prehistory have admitted the possibility that people could have lived in the Americas before the well-established time (11,500-9000 B.P.) when Paleoindians fashioned diagnostic bifacially flaked stone projectile points. Nevertheless, skeptics have consistently attacked all reported evidence for what Krieger (1964) termed the "pre-projectile point tradition" as due to incomplete assemblages, lack of proper stratigraphic contexts, or inaccurate radiometric dating. Such objections seemed reasonable at the time Krieger presented his thesis because much of the evidence he was able to marshal was from undatable surface sites. But the accelerating accumulation of new evidence, excavated from dated stratigraphic contexts, especially in South America, makes it increasingly difficult to explain everything away in order to maintain the accepted paradigm of a single late immigration of technologically specialized hunters. It should be recognized that the term "pre-projectile point" is usually misconstrued. Krieger meant to imply only that the early people did not make bifacially flaked stone projectile points. Undoubtedly they used projectile points made of perishable wood and bone, as well as sharp stone flakes. General foragers with a technologically less sophisticated (or at least less diagnostic) weaponry also used other animal procurement devices such as traps, nets, blowguns, throwing sticks, slings, and bolas. Only the latter device might be identifiable in a normal archaeological site, where the vast majority of weapons and other implements, such as those used by contemporary South American tropical forest Indians (cf. e.g., Albisetti and Venturelli 1962; Schmidt 1905) will not be preserved.

But archaeologists must base their interpretative models on concrete evidence, including clearly interpretable clues, not with what might have been but left no evidence. Normally perishable materials (other than bone) will preserve well in continually dry caves or continually wet open sites. The only other possibilities are carbonization or impregnation with mineral salts. Obviously, preservation under such ideal conditions is much less likely in Pleistocene than Holocene sites. In the Americas, the oldest perishable materials preserved by desiccation (string, wood, and hair) were excavated at Smith Creek Cave, Nevada, from occupation floors dated between 10,000 and 12,000 B.P. (Bryan 1979). Carbonized plaited basketry fragments from the oldest cultural zone in Meadowcroft Rockshelter, Pennsylvania, were dated at 19,600 B.P. (Stile 1982).

The most significant body of evidence giving us a remarkably complete picture of normally perishable material culture is being carefully excavated from the waterlogged Monte Verde village site, which was occupied 13,000 years ago in forested subantarctic Chile (Dillehay 1984). Dillehay and Collins and Dillehay present preliminary analyses in this volume on the artifacts and other materials recovered through the 1981 field season. Together with the 1983 excavations, summarized by Dillehay, a remarkable picture has emerged of wooden house structures and many simple but effective wooden artifacts, including bowls containing plant food remains, mortars, and handles hafted to naturally broken pebbles selected for their useful edges. Carefully observed cultural contexts suggest that several other naturally fractured and polished spherical stones were selected for use as tools; and microscopic analyses have demonstrated use on some of these natural objects. Little evidence has been found for intentionally flaked stone artifacts: a core, a split cobble, and three edge-battered stones were found, plus two large bifacially flaked objects fashioned of exotic stones, suggesting that they may have been traded from elsewhere. The only other examples of intentionally shaped stone artifacts are two bolas stones, carefully grooved by pecking.

The paleoenvironmental evidence recovered from the Monte Verde site is exceedingly important, but more significant for the question of understanding how Paleoamericans adapted to Pleistocene environments is the relatively complete inventory of the material culture used by a group of people who were able to live in a small village throughout the year with a generalized foraging economy based on plant collecting supplemented by hunting and fishing. Most stone artifacts, which would have constituted the entire artifact inventory in a normal open site, are very simple unifacially flaked tools which would be easily overlooked except for the context. Nevertheless, the people who made these simple tools were living quite successfully because they knew their environment well. The bones of mastodonts originally attracted archaeologists to the site, but there is no definite evidence that the people hunted and killed the proboscideans. Rather, broken bones and tusks appear to have been selected (possibly from natural kills) for their useful working surfaces and edges, in the same way that useful natural stones were selected for use.

Realization of the implications of these provocative interpretations of the evidence may be expected to revolutionize the quest for Paleoamericans. A new approach to the study of early Americans is essential. The restrictive search for predatory big game hunters who made technologically sophisticated hunting equipment (i.e., carefully flaked stone projectile points) must be expanded to include more elusive evidence for generalized foragers, who might occasionally kill a large animal if conditions were propitious, but who were always looking for sharp-edged pieces of bone, tusk or stone to use in the procurement and preparation of plant and animal materials. Any camper who has used a cobble to drive in tent pegs has realized that it is easier to use an object that is already available (whether or not it was fashioned for such use) than it is to carry around an artifact shaped for a specific task. Intentional retouch of useful natural objects might occur in order to resharpen an edge, but if sufficient natural objects were abundantly available, the worn piece may simply be discarded. The archaeologist faces a difficult challenge to find such elusive evidence at sites that lack the remarkable preservation afforded by waterlogging. Nevertheless, the problem must be tackled. Examination of likely working surfaces, whether or not flaked, by electron microscopy is one way to reveal evidence of use, even on unintentionally-shaped artifacts. We can no longer afford to assume that simple stone specimens constitute "background noise" that can be ignored.

The oldest directly dated artifacts in the Americas are from Old Crow, Yukon. The most famous of these bone artifacts is the serrated caribou tibia skin scraper, which yielded a date of 25,700 B.P. Seven other artifacts have yielded dates between 24,000 and 30,000 years, the span of time Morlan (this volume) thinks that people definitely occupied the Old Crow Basin. However, Jopling et al. (1981) and Irving et al. (this volume) have excavated altered bones, interpreted as artifacts, from deposits believed to be Penultimate (Illinoian) Glacial in age.

Most archaeologists believe that people could not have lived in the arctic that long ago, but actually there are no valid logical reasons for limiting the time when people might have lived in Beringia (cf. Bryan 1981, 1984). Only the actual archaeological evidence can resolve the question. But first, it is necessary to adopt a new paradigm so that available evidence can be seen to make sense. In the following paper Alsoszathai-Petheo suggests that entrenchment of the current paradigm that Paleoamericans were predatory hunters carrying sophisticated stone tools is a result of preconceptions inherent in our Western world-view. If we adopt the world-view of actual hunter-gatherers who conceive of themselves as an integral part of the environment and living in harmony with nature, the alternative paradigm would become more reasonable that Paleoamericans were foragers who made successful adaptations to various environments with simple but effective cultural equipment.

Worked bone and antler from the Old Crow localities have opened new possibilities, the pursuit of which has created many interpretive problems because all materials have been redeposited by natural action at least once. The lack of undisturbed occupation floors and associated stone artifacts has precipitated considerable controversy. Although it may not be possible at this time to prove beyond any doubt that Old Crow people deliberately flaked thick-walled mammoth bone as if it were stone, the presence of a few artifacts undisputably altered by cutting, scraping and grinding before fossilization cannot be denied.

THE RELEVANCE OF THE LOWER PALEOLITHIC TO PALEOAMERICAN PREHISTORY

One reason why skeptics have labelled the Old Crow evidence as equivocal is because no generally acceptable site earlier than about 35,000 years had been reported from eastern Siberia (Mochanov 1978); however, even the Dyuktai evidence from deposits older than about 18,000 B.P. has been questioned on the basis that only a few artifacts came from inconclusively dated terrace deposits subject to disturbance by cryoturbation (Yi and Clark 1983: 198). Mochanov is now excavating the Diring-Yurekh site on the 100 m high terrace of the Lena River, near the cold pole of the northern hemisphere. A pebble chopper industry has been recovered from beneath a Dyuktai occupation and above a pebble boulder conglomerate. The geological context for the pebble tools is claimed to date between 1 and 1.5 million years (Agenbroad 1984). Ulalinka, a similar site with a pebble tool industry has been reported from the Altai region of western Siberia. At Ulalinka, the sediments enclosing the quartzite pebble tools are reported to have reverse polarity, which suggests they may predate 690,000 years, the upper limit of the Bruhnes/Matuyama Polarity Epoch (Okladnikov and Pospelova 1982). If either of these claims for the early Pleistocene occupation of Siberia is correct, it means that some people were already adapted to subarctic and possibly arctic environments before the onset of the coldest Pleistocene climatic conditions. The further implication of that conclusion, that the genus *Homo* may have formed part of the Beringian faunal assemblage early in the Pleistocene, seems too revolutionary for serious contemplation at this time; so it is to be expected that the Siberian evidence will be subjected to the essential process of skeptical criticism.

The reality of the Lower Paleolithic would never have been established in the western Old World by sites such as Diring-Yurekh or Ulalinka, which evidently lack associated early Pleistocene faunal remains as well as bifacially flaked artifacts. It is instructive to review the history of the original establishment of the association of artifacts with Pleistocene mammals in the Old World. Six decades elapsed between Frere's (1800) reasonable report of artifacts associated with extinct animals at Hoxne, southern England, and general acceptance of such evidence. During that time the paradigm held sway that man had been specially created only recently. Operating with this paradigm, skeptics "wielded the authority of intellectual command" through their numerous publications (Gruber 1974:389). Most influential was the renowned geologist William Buckland, of Oxford University, who either ignored or explained away all claims that human bones or artifacts were associated with extinct animals. Father MacEnery, an amateur paleontologist digging in Kent's Cavern, Devon, informed Buckland that he had recovered flint tools associated with bones of extinct animals beneath a flowstone floor. Buckland refuted this association, stating that MacEnery must have overlooked holes in the stalagmite through which the artifacts had dropped (Roe 1981:21). In the face of opinion from such authority, MacEnery never published his evidence. Many other claims were examined with equally extreme criticism and rejected as, at best, "not proven" (Gruber 1974: 388). After his death, MacEnery's researches in Kent's Cavern were published and his work was continued by William Pengelly, a Torquay school-teacher. Pengelly was equally unsuccessful in convincing scientists that there were artifacts directly associated with megamammals; but eventually Hugh Falconer, a paleontologist trained as a physician, and John Prestwich, an avocational geologist, were convinced that it was important to excavate the newly-opened Windmill Hill Cave at Brixham, near Torquay. After obtaining a grant from the Geological Society of London simply for the purpose of understanding the faunal variation in undisturbed Pleistocene sediments, excavations commenced in Brixham Cave in 1858 (Gruber 1974). But as Pengelly had anticipated, Brixham Cave proved to have more than paleontological and geological significance. The discovery of flint implements in direct association with the bones of animals beneath a stalagmitic floor convinced Falconer and Prestwich that the artifacts were of the same age as the bones. Pengelly, the field director, reported his discoveries to the British Association, where they were received with excitement and incredulity. Rumors of forgery and intrusion of the artifacts circulated. To counter these rumors, Falconer impressed upon Pengelly the necessity of being able to demonstrate the exact position of the flint knives.

Looking for corroboration of the evidence, Falconer visited the avocational French archaeologist Boucher de Perthes in Abbeville. Falconer was greatly impressed by the mass of evidence for the association of handaxes and megamammal bones that had been recovered from the Somme gravels. Boucher de Perthes had been reporting his discoveries for years, but his publications were generally ignored and his claims had been relegated to the "not proven" category. Even Charles Darwin later confessed to Charles Lyell that he had dismissed de Perthes' book *Antiquities Celtiques et Diluviennes* as "rubbish" (Darwin 1887, v. 3:15). Falconer persuaded Prestwich and John Evans (an avocational archaeologist) to visit Abbeville in the spring of 1859. Immediately thereafter, Prestwich reported on the geology of the Somme gravels to the Royal Society, and Evans reported his conclusions to the Society of Antiquaries. Evans (1859) reasoned that Boucher de Perthes' "flint hachets" were artifacts by noting the repetition of form and the sharpness of the cutting edges. Evans also managed to replicate a handaxe, the form of which he stressed differed significantly from later stone tools. Also, the original artifacts were deeply patinated and therefore obviously had not been fabricated recently. Two months later, Sir Charles Lyell, a former skeptic who had been convinced by Falconer to visit Abbeville, admitted in his presidential address to the Geological Section of the British Association for the Advancement of Science that skepticism "had previously been pushed to an extreme" (Lyell 1859: 93). Some skeptics continued to argue that the reported evidence was somehow flawed; however, the reasoned conclusions presented by Evans, Prestwich, and Lyell resulted in general acceptance of a new paradigm — that man had lived with extinct Pleistocene megamammals. Acceptance of Darwin's *Origin of the Species* later the same year was certainly eased by the quite unrelated revolution that had already occurred.

The above historical sketch of occurrences that led to the acceptance of a new paradigm is analogous to the current controversy about the acceptability of evidence of Pleistocene man in America. The observed and recorded evidence had been ignored or explained away by overzealous scientific caution, perhaps partly because the advocates themselves had been too enthusiastic. The roles that actors played in the nineteenth century scene can be categorized. Conservative skeptics were in the majority, although some (notably Prestwich and Lyell) were willing to change their opinions in the face of good evidence that they undertook to examine themselves. Other diehard skeptics never changed their position, which was that all the reported evidence was somehow flawed. Some field workers (e.g., MacEnery) accepted the opinion of authority and thereby doubted their own evidence. Thus they could not bring themselves to publish their evidence, let alone their interpretations. Others (e.g., Pengelly and Boucher de Perthes) persisted with their convictions until they had the opportunity to allow open-minded scientists (e.g., Falconer and Prestwich) to discover the evidence themselves. Finally, the paradigm switch occurred when respected conservative scientists (Prestwich, Evans, and Lyell) presented their analyses of the evidence in a positive but not an affirmative manner. When the current controversy over Early Man in America is finally relegated to history, individual actors on the current scene will be seen to fit similar roles.

The intellectual climate in mid-nineteenth century Europe resulted in prolonged controversy and resistance to acceptance of the actual evidence because the evidence was so unexpected, given the mindset of the intellectual authorities of the time. A paradigm shift was necessary before the unexpected evidence was seen to make sense. In the New World, the analogy is the completely unexpected nature of the earliest attested human remains: a simple stone and/or bone industry supporting a foraging lifestyle, with even remote northeastern Brazil and southern Chile populated by at least 32,000 years ago (cf. Dillehay; Guidon, this volume). American archaeologists, with the long-standing image of earliest New World man as a specialized big game hunter using a sophisticated lithic technology, simply cannot believe the contradictory evidence of a much earlier entry and a considerably different technology.

The present controversy over Early Man in America is analogous to that in Europe more than a century ago because the intellectual climate has been dominated for over 50 years by a particular paradigm which has seemed to fit most of the evidence, but which fails to explain an increasing body of data. Rather than considering a new paradigm which might make the evidence sensible, skeptics have demanded that all evidence for "pre-Clovis" be judged by more rigid standards of evidence and argument than are applied to later sites. Insistence by skeptics on rigorous methodology is an essential part of the scientific process to balance the intuitive insight of the "possibilists"; however, obsessive preoccupation with laying down special conditions for the acceptance of evidence smothers scientific advance. In this case, for example, if any skeptic has ever expressed any doubt about the interpretation of the evidence from a site, whether or not they have ever visited the locality, the evidence is rendered "equivocal." Arbitrary application of such rigid criteria to later sites, including Clovis sites, would relegate nearly all archaeological evidence to the "not proven" category. Because acceptance of a new paradigm will require a major shift in thinking about the history of the American Indians, it is essential to be certain that the evidence is secure. However, continued appeal to the authority of respected conservative skeptics in order to discredit all reported evidence for "pre-Clovis" man (e.g., Dincauze 1984; Owen 1984) means that new interpretations and new models can never be considered, and the field will not advance. Indisputable evidence for the presence of man in datable stratigraphic contexts has been excavated at several sites, especially in South America. The evidence from Monte Verde cannot possibly be denied because of a combination of unusual preservation with lack of subsequent geological disturbance, meticulous excavation, and insightful analyses. In my opinion we are at a critical turning point. In order for the field to advance, skeptics cannot insist that this unique set of circumstances at Monte Verde is required for all early sites. Instead, acceptance of Monte Verde requires that the evidence from all other sites be examined with a new framework so that the abundantly available evidence can be seen to make sense.

Admission that previous skepticism had been pushed to an extreme was necessary before the essential paradigm shift could occur in 1859. The same admission is necessary now, or the evidence will continue to be ignored and explained away because the current conceptualization of early American man does not allow for unexpected evidence. The association of American man with Pleistocene mammals was confirmed nearly sixty years ago; however, the paradigm that emerged from the discovery of mammoth and giant bison kill sites has gradually solidified into the notion that all early Americans had to be aggressive megamammal hunters using sophisticated technological equipment in order to survive.

This idea is based on the argument that people adapted to the Beringian steppe-tundra environment must have hunted mammals, not only because people living in a cold climate must have a high protein diet, but because there was very little else to eat except the partially digested stomach contents of herbivores (Schweger et al. 1982). The presence of 30,000 year old cut and butchered bones from Old Crow (Morlan, this volume) supports the conclusion that eastern Beringians hunted megamammals during the Wisconsinan. Available evidence, therefore, indicates that Beringians adapted to the continental interior were not general foragers (i.e., hunters, fishers and gatherers). However, contemporary people adapted to the maritime littoral environments of the North Pacific rim would have had access to much more varied resources, including fish, shellfish, birds, sea mammals and edible plants. The economic adaptation to the interior steppe-tundra ecosystem could have been made by groups of maritime-adapted foragers as part of the gradual process of population expansion into adjacent ecosystems. In other words, the earliest Beringians may have been foragers adapted to the maritime coast and not to the continental interior.

Each ecosystem requires its own adaptive strategy for economic survival. The strategy of big game hunting, which was necessary for survival on the High Plains as well as on the Beringean steppe-tundra, should not be extrapolated to other ecosystems. The actual evidence for economic adaptation to each environmental region must be judged on its own merits without preconceptions. Therefore, if the archaeological evidence suggests that a foraging way of life without intensive predation of megamammals was a successful adaptation to most Pleistocene ecosystems, then the model that all Paleoamericans were specialized big game hunters should be abandoned.

The fundamental need, then, is to evaluate the actual archaeological evidence for economic adaptations to various Pleistocene environments. Much of this evidence will have to be inferred from careful analysis of stone and bone tools; but the best evidence will come from careful analyses of materials recovered from sites with unusual preservation such as at Monte Verde, where waterlogging has permitted preservation of wood and food remains. The evidence from Monte Verde indicates that these people were foragers who supplemented their plant food diet by hunting with bolas and wooden spears. There is no evidence that they actually killed the mastodonts, whose remains originally attracted attention to the site.

In fact, there is only one definite Pleistocene kill site in South America, at Taima-taima, near the Caribbean coast of Venezuela, dated repeatedly at 13,000 B.P. At Taima-taima, the El Jobo contemporaries of the villagers at Monte Verde had developed special equipment, including bifacial projectile points, for hunting mastodonts (Bryan et al. 1978; Gruhn and Bryan 1984; Ochsenius and Gruhn 1979). But there is no evidence that contemporary people living on the Sabana de Bogotá in the northern Andes were specialized big mammal hunters (Correal, this volume). Like their contemporaries at Monte Verde, they did not flake bifacial projectile points. In fact, bifacial retouch is essentially absent in the Abriense industry, which has been recovered from several sites on the Sabana de Bogotá dated between 13,000 and 10,000 B.P. The carefully excavated Tibitó site yielded the bones of mastodonts, horses, and deer; many, if not most of which appear to have been modified by use. Although Tibitó might have been a butchering site, the evidence suggests that other activities were also engaged in. As at Monte Verde, animal and perhaps plant materials were prepared with the useful edges of broken bones and tusks as well as simple unifacial flakes.

Sites like Monte Verde and Tibitó are more similar to the early occupation floors in Bed I at Olduvai Gorge, Tanzania, than they are to the Devon caves or the Somme terraces. In mid-nineteenth century England, sites like Olduvai, Monte Verde, and Tibitó probably would not have caused a paradigm shift because the lack of undisputably shaped bifacial tools probably would have delayed their recognition as archaeological sites. The definite cultural context on living floors, not the presence of definitely shaped artifacts, demonstrated human activity at these sites. Although the presence of mammal bones may have led to their discovery a century earlier, the lack of realization of the significance of meticulous excavation of living floors may have precluded recognition of the full potential of Olduvai, Monte Verde, and Tibitó.

If only simple unifacial flake and pebble tools had been found in the Devon caves and Somme gravels, establishment of the Lower Paleolithic may have been delayed for several more decades. The reality of the British Lower Paleolithic has never been seriously questioned since 1859 because there could be no doubt whatever that bifacially flaked hand-axes, even if badly rolled by water action, must have been fabricated by man and not by nature. Although Lower Paleolithic living floors were discovered (e.g., Smith 1894: 103-157), until recently it was not considered essential to record carefully everything on undisturbed living floors. The significance of the presence of bifaces to the original acceptance of the European Lower Paleolithic was emphasized by the subsequent failure to establish a "pre-Paleolithic" stage lacking hand-axes even though it was argued that repetition of form ("rostrocarinates") and useful working edges suggested that people used these objects (Moir 1927, n.d.).

The debate over whether or not such steeply unifacially flaked objects were flaked by man has been resolved to the satisfaction of British archaeologists; however, the fundamental question of distinguishing naturefacts from artifacts continues to resurface in other parts of the world where very early flaked stone assemblages lacking bifacial tools are recovered. Thus, claims for the early Pleistocene occupation of Vallonet Cave in France or the three-million-year-old locality at Omo in Ethiopia where australopithecine remains were recovered in association with sharp pieces of quartz are generally relegated to the "not proven" category.

The question of how to distinguish naturefacts from artifacts is far from being resolved and demands more research. The way the problem was resolved in England, by application of the Barnes' (1939) statistical method of measuring the angles of platform scar, is not generally applicable to all problems of differentiating naturefacts from artifacts (Schnurrenberger and Bryan 1985). For instance, strict application of Barnes' method would exclude blades removed with a punch or pebbles split bipolarly on an anvil, as well as nearly all stone artifacts recovered from activity areas in and around the house structures at Monte Verde. But the undisputable cultural context at Monte Verde makes it unnecessary to apply the Barnes test. The essential paradigm switch will occur in the Americas because of the reporting of meticulously excavated and well-dated cultural contexts. Monte Verde presents indisputable evidence that Paleoamericans with a simple technology were making a good living in a subantarctic environment. It seems reasonable to extrapolate from this evidence that earlier Paleoamericans had been making successful adaptations to other Pleistocene environments, including subarctic Beringia. Monte Verde should break the log jam that has held back acceptance of dozens of other sites now floating in the "equivocal" pool. The evidence recovered from less definite cultural contexts will be seen to make sense if the paradigm is accepted that the earliest Paleoamericans were foragers and not predatory hunters carrying sophisticated stone tools.

There can be little doubt that the earliest people moved into America from northeast Asia. All other possibilities are exceedingly remote. Therefore, the Lower Paleolithic materials of Europe are not directly relevant to the question of the early peopling of the Americas. The Early Paleolithic of East Asia differs significantly from the western Lower Paleolithic because bifacially flaked artifacts, including hand-axes, are uncommon and usually absent in early or even late assemblages. For example, less than 2% of the 100,000 artifacts recovered from the living floors at Choukoutien Locality I exhibit bifacial edge retouch (Li 1983). Some early Asians occasionally flaked bifacially but most did not, and this is true for Siberia as well. The persistent search for bifacial projectile points in Siberian Pleistocene sites has been successful; but the total number of excavated points is probably less than 100, and most of these are from final Pleistocene contexts. Most East Asians, including Siberians, preferred to use unifacially flaked tools. Therefore, it should be expected that early Americans also preferred to flake unifacial and not bifacial tools. The fact that North Americans at the onset of the Holocene developed several bifacial projectile point traditions which spread rapidly and became ubiquitous throughout the continent was a phenomenon which distinguishes North American prehistory from that of all other continents. Even in South America, where bifacial projectile points were innovated in pre-Clovis times in northern Venezuela and evidently in southern Brazil (v. Beltrão; Danon et al.; Hurt, this volume), bifacial points were never adopted in large parts of the continent. Projectile points of bone and wood continue to be preferred by tropical forest foragers. The best known bifacial projectile point tradition, distributed from Patagonia to Panama, is examined in detail by Mayer-Oakes in this volume.

Bifacially thinned artifacts are notably sparse in Australia, whose aborigines also immigrated ultimately from East Asia. Although Australian aborigines are often depicted as hunters, they were actually foragers who used bifacial projectile points only recently and in a limited part of northwestern Australia. Nevertheless, Australian archaeologists have managed to deal with the assemblages of simple unifacial flake tools that lack conventional sharply demarcated forms. Instead, classification has been established from analysis of useful edges (e.g., White and O'Connell 1982:116).

But in America there persists the insistence that acceptable criteria for early man must include definitely shaped ("diagnostic") artifacts (preferably bifaces) found in undisturbed Pleistocene contexts. Most archaeologists realize the difficulty of locating such evidence by themselves, so they either ignore the problem or are content to pursue the location of diagnostic bifaces (like Clovis points) in the hope that they may discover an undisturbed Paleoindian site. This situation is circular, leading to the recovery and recording of many fluted points and the occasional discovery of a new Paleoindian occupation or kill site, but it is not intended to lead to the discovery of evidence for "pre-Clovis." So far the only evidence for "pre-Clovis" has come from sites yielding either a simple flaked stone industry and/or modified bones of extinct animals. Obviously, any claims for great antiquity of such sites must be supported by independent means of dating, either geological or, preferably radiometric. But, for one reason or another, all such claims for "pre-Clovis" man have been disputed. Either the cultural context has been disturbed, the method of dating has not been adequately verified, or no one except the advocates of the site saw the original context. But such objections, which sometimes seem quite desperate, really stem from the belief that some definitely shaped tool (preferably something "diagnostic") must be present in order to have acceptable "proof" for the presence of Early Man. Anything less is now being labelled a "myth," and believers of myths cannot be scientific archaeologists.

But if the Australian archaeologists had adhered to such strict criteria they would not have searched for and thereby recovered evidence for Pleistocene man on that continent. Because diagnostic bifaces were rare and late in Southeast Asia and Indonesia as well as Australia, it was realized that the only "diagnostic" artifact categories may be simple flakes and cores. It was realized that simple retouched flakes are adequate to demonstrate the presence of early man, if they are recovered from datable stratigraphic contexts.

Simple stone flakes and modified bones associated with occupation debris, including hearths, should be acceptable as evidence for Pleistocene man in America. Currently, such evidence is acceptable in Australia. Early Australians and early Americans all came ultimately from East Asia, so there should be standard criteria for acceptance of their respective early cultural remains. It is illogical to require the presence of diagnostic shaped tools in America and not to require their presence in Australia in order to prove that that continent was populated at least by 40,000 years ago.

But if there are no "diagnostic" shaped tools of either stone or bone in early American sites, how does the archaeologist locate such early sites? Almost all occupation sites with undisturbed Pleistocene living floors have been located because they were stratigraphically beneath later occupations which contain "diagnostic" artifacts.

The extreme importance of the inception of radiocarbon dating to the Pleistocene prehistory of America and Australia now becomes obvious. An assemblage lacking "diagnostic" artifacts or any means of dating the cultural context would have no recognizable significance if found on the surface because there is nothing that could not have been made at a later time. The stratigraphic context must be definite and must be datable. Before 1950 the only way to demonstrate great antiquity was by the stratigraphic association of definite artifacts with extinct Pleistocene fauna. Since then the hesitancy in accepting simple flakes or modified bones from living floors lacking "diagnostic" tools, even if they are associated with datable carbon, is because American archaeologists have been particularly cautious about accepting radiocarbon dates, not only because many dates have been proven to be wrong, but more often because the dates have not fit the archaeologist's preconceptions, most of which are based upon the situation before 1950, the year of the inception of the radiocarbon method of dating. It is much easier to argue that a date is wrong than it is to change one's preconceptions. But logically, particularly if the date is verified by other confirmatory dates, those dates are more likely to be correct than one's preconceptions. In fact, in my experience, except for those dates that are obviously impossible (like 1800 year old ground sloths!), any date should tentatively be considered as a reasonable possibility which should lead to consideration of alternative working hypotheses rather than adhering to those based upon one's preconceptions.

Although it is essential for the scientific process to be critical of all claims which reach beyond the bounds of current models, it is equally essential for the maintenance of a healthy progressive outlook in the discipline to balance skepticism with reasonable alternative interpretations of the available evidence. In this way an alternative model can emerge by which to evaluate new evidence, rather than continually being forced to explain away most evidence because it does not fit the current model.

The evidence from Old Crow and Ulalinka for early human occupation of continental subarctic environments cannot be ignored or simply explained away because the original cultural contexts have been destroyed, or because alternative possibilities for flaking of stone and bone must first be examined by experimentation. Although such sites may not themselves contain conclusive proof for the presence of man, the evidence should be used as clues or suggestions as to what archaeologists concerned about the problem of Early Man should be looking for in less disturbed contexts.

Elsewhere, I have suggested that the apparent absence of serrated bone scrapers, antler wedges, and deliberate flaking of proboscidean bone in the American tropics suggests that these bone and antler tools and techniques, characteristic of northern North America between at least 30,000 B.P. in the Yukon and about 11,000 B.P. in the northern United States, have a limited distribution within America because the technological tradition developed in Beringia after the earliest people had passed through the region (Bryan 1978, 1983). Whether or not this refutable working hypothesis is verified, it demonstrates how redeposited materials from disturbed sites can be used as clues for suggesting what archaeologists should be looking for in areas south of the continental glaciers.

The small quartzite pebble tools from the Ulalinka River terraces in the Altai Mountains about 51° north latitude, can also be used to formulate a similar working hypothesis. It happens that the flaked pebbles from Ulalinka, dated to greater than 690,000 years by detection of the Matuyama paleomagnetic reversal in deposits containing the pebble tools, are very similar to the flaked quartz and quartzite pebbles from undisturbed occupation floors (dated between 17,000 and 31,500 B.P.) in Toca do Boqueirão da Pedra Furada, Brazil (Guidon, this volume). The similarity in pebble tools from the oldest dated site in South America to those from the oldest dated site in Siberia suggests the hypothesis that the people who occupied the rock shelter in northeastern Brazil about 31,000 years ago were descendants of the earliest people to expand into subarctic portions of Siberia earlier in the Pleistocene. If this hypothesis has any validity, similar pebble tools should be found in early contexts in intervening areas. Skeptics will properly scrutinize any such claims for great antiquity. Among other things, the possibility of flaking of quartzite and quartz pebbles by natural processes must be tested by experimental replication. However, the presence of pebble tools in datably early undisturbed occupation contexts will provide the only secure demonstration of the hypothesis.

The first general recognition of the Lower Paleolithic in East Asia, the source area for both Australians and Americans, occurred when simple flake and core tools, including pebble tools, were recovered from primary occupation contexts in undoubted association with remnants of hearths, actual human bones, and a Middle Pleistocene fauna at Choukoutien Locality I during the 1920s and 1930s. This important cave is still being excavated. The simple flake and core tools, 88% of which are pieces of vein quartz (Li 1983), often bipolarly flaked, probably would not have been recognized as artifacts if they had been redeposited in river terrace gravels.

THE RELEVANCE OF SOUTH AMERICA TO EARLY MAN IN NORTH AMERICA

Obviously, pebble tools will be found only in areas where suitable pebbles occur. Early people might also be expected to use naturally sharp rocks, including vein quartz fragments, and to break up any breakable (including flakeable) stone in order to get the sharp edges desired for working other materials. In areas where flakeable stone is scarce or lacking, other more perishable materials would necessarily have sufficed, and the ability to get along without flaked stone could be continued even in regions containing flakeable stone (cf. discussion in Bryan 1977a). That some prehistoric people did not flake stone is demonstrated by the lack of flaked stone in certain shell midden occupation sites in areas such as Puget Sound (Bryan 1963) and south coastal Brazil (Bryan 1977b), where stones lacking evidence of modification nevertheless must obviously have been used because almost anything found in a shell midden must have been carried there by man. The extreme limitation of possibilities for natural deposition makes the excavation of shell mounds and sand dune sites invaluable for providing insights as to what artifacts may be overlooked in a normal occupation site. In turn, this information gives valuable clues as to what kinds of things should be looked for in sites containing early cultural contexts. Various kinds of clues may also be found by careful examination of bones, particularly those of extinct megafauna (Bryan 1983).

A glimpse at what most archaeologists miss in the way of an artifact repertoire at an early site is provided by the Monte Verde site. Monte Verde also gives us many important insights into the variety of lifeways early Americans must have developed as they gradually adapted themselves to the many environmental regions of the New World with their simple basic technology, composed of flakes and simple core tools which could be used to make useful objects from fiber, wood, sinew, skin, and bone (cf. discussion in Bryan 1977a). The "Paleoindian lifeway" that developed on the High Plains of North America between 11,500 and 10,000 B.P. was one of the most highly specialized ways of making a living that developed anywhere in the Americas. Therefore, it is not legitimate to attempt to extrapolate into environmentally dissimilar areas that particular adaptation to rather extreme environmental conditions on the Plains, which always (until the advent of the steel plow) required occupants to emphasize the hunting of large game as the primary means of survival.

Major differences are evident between "Paleoindian" sites in North America and their counterparts in South America. With the exception of Taima-taima, there are no "kill" sites comparable to the well-known Llano, Plano, and later sites on the High Plains of North America. This difference should not be particularly surprising, because kill sites are quite rare in the Old World as well (e.g., the Tomsk mammoth is the only reported "kill" site in Siberia).

In other words, the situation in the Plains area of North America is unusual. For that reason alone, it should be obvious that a model based upon the Paleoindian "kill" sites on the High Plains is logically not extrapolatable to other ecosystems in North America, and certainly is not extendable to South America.

Occupants of early South American sites, including Monte Verde and Tibitó, undoubtedly ate meat obtained from the hunt, but actually there is very little direct evidence that early South Americans deliberately hunted Pleistocene megafauna. Most of the evidence for processing large game animals could as well be interpreted as products of scavenging (for raw materials, not necessarily food) rather than hunting. Certainly all people hunted smaller animals, but it is possible that only a few social groups in Venezuela and in Patagonia, where sloths in Buitreras Cave (Borrero, this volume) and horses in Fell's Cave (Bird 1938) evidently were butchered and their bones processed, were really big game hunters. This is not to say that other groups did not take the occasional large animal that was discovered in a vulnerable state, but rather that very few cultures, notably those adapted to environmentally marginal areas where other kinds of foods may have been scarce (at least seasonally), actually developed a specialized flaked stone technology for more efficient procurement of the big beasts. Thus, Bird's basally thinned points with expanding stems (Mayer-Oakes, this volume), presumably for attachment to split stick hafts, and Cruxent's cylindrical El Jobo points for insertion into sockets are the apparent result of technological specialization at opposite ends of the continent in Late Pleistocene times.

The only other early projectile point style is the tanged or contracting stemmed form in southern Brazil (Hurt; Beltrão et al., this volume) which may have been the earliest bifacial projectile point tradition in the Americas. Although these tanged points may have been developed by 14,000 B.P., they were not commonly used until much later; and they have never been found associated with extinct megafauna. In fact, the only reported south Brazilian sites with direct association of artifacts and megafauna are on the Pampas in Rio Grande do Sul (Bombin and Bryan 1978). Farther south on the Argentine Pampas, there is also apparent association of man with megafauna (Fidalgo et al., this volume).

But these southern sites are not as early as the majority of the excavated cave and rockshelter sites farther north on the Planalto of Brazil, which are dated between 11,000 and more than 31,500 B.P. (cf. Guidon; Prous; Schmitz, this volume). Direct association of man and megafauna has recently been found in two limestone caves in Bahia (Bryan and Gruhn 1985), but there is no direct evidence that people killed the megamammals. Rather, they appear to have used the edges of naturally broken bones as scrapers. Only one horse carpal exhibits a definite butchering cut. In other caves people occasionally, for some unknown reason, chopped green megafaunal bones that had probably been scavenged (Bryan 1978, 1983; Prous, this volume), there is as yet no evidence that these early people ever developed a specialized flaked stone technology for hunting Pleistocene megamammals. Even the tanged points are found rarely, and only after 10,000 B.P. in northeast Brazil. In many parts of northeastern Brazil, including interior Bahia and southern Piaui (Guidon, this volume), not only were projectile points never made, but there is no evidence that local people practiced bifacial flaking at all, even after the introduction of pottery. Although it is possible that the megafauna had essentially become extinct earlier on the Planalto than on the Pampas, the abundance of megamammal bones in the caves of northeastern Brazil has led to the opposite opinion by paleontologists — that the large beasts persisted quite late in this area. Obviously, it is essential to obtain better dating control in order to resolve this question, but meanwhile it looks as if early people in this large region did not often intentionally hunt the local megafauna. Even in Patagonia, an environment more similar to the High Plains, there is little conclusive evidence that the early occupants spent much time hunting megamammals (cf. Borrero; Caviglia et al., this volume). Actually, the evidence indicates that small and medium-sized mammals and birds were always the most important game in Patagonia from the time of the first known occupation by man about 13,500 B.P. (Mengoni, this volume).

Another interesting difference between North and South America is that there have been few claims on the latter continent for really early sites such as Calico (cf. Simpson et al., this volume), Old Crow (Jopling et al. 1981), or Hueyatlaco (Steen-MacIntyre et al. 1981). Long ago, Ameghino (1880) claimed association of man with Pampean megafauna and even argued that *Homo sapiens* had evolved on the Pampas. Hrdlicka (1912), with the help of the American geologist Bailey Willis, successfully refuted these claims; but unfortunately they practiced overkill. Only now is credible evidence emerging (Fidalgo et al., this volume) that man was indeed contemporary with Pampean megamammals. Some of the Pampean sites recorded by Ameghino should be reexamined with a new perspective and modern techniques.

The Itaboraí site in the state of Rio de Janeiro, Brazil, may prove to be early or mid-Wisconsinan in age (Beltrão et al., this volume). Flaked angular vein quartz pebbles have been excavated from colluvial sediments now situated on the apex of a high hill. The only radiometric date, 8100 B.P., is on a hearth found at the base of silts stratigraphically overlying the artifact-bearing gravels. Attempts are being made to date the gravels, which have been deeply weathered.

Two important sites, still being excavated, have yielded evidence that people entered South America before 35,000 years ago. Preliminary notices of these early occupations are presented in this volume by Dillehay and Guidon. Dillehay has found pebble artifacts from a level at Monte Verde two meters beneath the 13,000 year old village. Ongoing excavations are designed to determine the extent of what appears to be an earlier excavation floor. Charcoal plastered on the surface of a pebble hammerstone and a piece of wood have yielded confirmatory dates of 33,000 B.P.

Hearths from different parts of an extensive occupation stratum (Camada XIX) at Toca do Boqueirão da Pedra Furada, a large rockshelter in southern Piaui, northeastern Brazil, have yielded dates of 26,000, 26,400 and 31,500 B.P. An abundant quartz pebble and flake industry has been recovered in primary association with these hearths, as well as with hearths from deeper levels which are undergoing radiocarbon analysis at the time of writing. Higher levels have yielded stratigraphically consistent dates on more than a dozen hearths extending from 2000 to 25,000 B.P.

In Lapa Vermelha, in the Lagoa Santa region of Minas Gerais, Brazil, quartz cores and flakes and a limestone sidescraper were associated with hearths dated between 25,000 and 15,300 B.P. (Prous, this volume). Ongoing excavations in northeastern Brazilian caves by several archaeologists, including Schmitz (this volume) are designed to confirm these early dated occupations of South America.

If the initial entry of South America occurred sometime before 35,000 B.P., the implication is clear that people had been living in North America long before that time because there appears to be no reasonable way that people could have reached South America except via Beringia and North America. It is unfruitful to speculate on how long ago that may have been because the new evidence from South America means that we no longer have any logical basis for proclaiming an arbitrary limit beyond which people could not have entered America. If, at the other end of the long route to South America, the suggestive evidence from Siberia is confirmed that people lived in subarctic latitudes at the beginning of the Pleistocene, then it should not be seen as foolish for American archaeologists to examine reports of cultural materials from any Pleistocene deposits. Only a great deal more work by archaeologists with an unbridled perspective will eventually reveal when man the ingenious toolmaker first saw what is now known as America.

REFERENCES CITED

Agenbroad, L.
1984 Opening doors. *Mammoth Trumpet* 1(3):6.

Albisetti, C., y A.J. Venturelli
1962 *Enciclopêdia Bororo.* Vol I. Publicão no. 1 do Museu Regional Dom Bosco, Campo Grande, Mato Grosso.

Ameghino, F.
1880(1947)
La Antigüedad del Hombre en el Plata. Tomo II. Editorial Intermundo, Buenos Aires.

Barnes, A.S.
1939 The difference between natural and human flaking on prehistoric flint implements. *American Anthropologist* 41: 99-112.

Bird, J.
1938 Antiquity and migrations of the early inhabitants of Patagonia. *Geographical Review* 28:250-275.

Bombin, M., and A.L. Bryan
1978 New perspectives on early man in southwestern Rio Grande do Sul, Brazil. In Early man in America from a circum-Pacific perspective, edited by A.L. Bryan, pp. 301-302. *Department of Anthropology, University of Alberta, Occasional Papers* No. 1. Edmonton.

Bryan, A.L.
1963 An archaeological survey of northern Puget Sound. *Idaho State University Museum, Occasional Papers* No. 11, Pocatello.

1977a Developmental stages and technological traditions. *New York Academy of Sciences, Annals* 288:355-368.

1977b Resumo da arqueologia do sambaqui de Forte Marechal Luz. *UFMG Museu de Historia Natural, Archivos,* vol. II:9-30. Belo Horizonte.

1978 An overview of Paleo-American prehistory from a circum-Pacific perspective. In Early man in America from a circum-Pacific perspective, edited by A.L. Bryan, pp. 306-327. *Department of Anthropology, University of Alberta, Occasional Papers* No. 1. Edmonton.

1979 Smith Creek Cave. In The archaeology of Smith Creek Canyon, edited by D.R. Tuohy and D.L. Randall, pp. 163-251. *Nevada State Museum Anthropological Papers* 17. Carson City.

1981 The archaeological evidence for human adaptation to cold climates on the way to America. *Union Internacional de Ciencias Prehistôricas y Protohistôricas, X Congreso, Miscelanea,* pp. 44-62. Mexico, D.F.

1983 Bone alteration patterns as clues for the identification of early man sites: or, an attempt to demythify the search for early man. In *Carnivores, human scavengers and predators: A question of bone technology,* pp. 193-217. Department of Archaeology, University of Calgary, Calgary.

1984 Human adaptation to cold climate: archaeological evidence for migration to America. *Centro Camuno di Studi Preistorici, Bollettino* 21: 95-106. Capo de Ponti.

Bryan, A.L., and R. Gruhn
1985 Stone and bone artifacts with Pleistocene fauna in two cave sites in interior Bahia, northeast Brazil. *Current Research in the Pleistocene 2:7-9.*

Bryan, A.L., R.M. Casamiquela, J.M. Cruxent, R. Gruhn, and C. Ochsenius
1978 An El Jobo mastodon kill at Taima-taima, Venezuela. *Science* 200:1275-1277.

Darwin, F. (editor)
1887 *The life and letters of Charles Darwin.* 3 vols., Murray, London.

Dillehay, T.D.
1984 A Late Ice-Age settlement in southern Chile. *Scientific American* 251: 106-117.

Dincauze, D.F.
1984 An archaeological evaluation of the case for pre-Clovis occupations. *Advances in World Archaeology* 3: 275-323.

Evans, J.
1859 On the occurrence of flint implements in undisturbed beds of gravel, sands and clay. *Archaeologia* 38: 280-307.

Frere, J.
1800 Account of flint weapons discovered at Hoxne in Suffolk. *Archaeologia* 13: 204-205.

Gruber, J.
1974 Brixham Cave and the antiquity of man. In *Readings in the History of Anthropology,* edited by R. Darnell, pp. 380-400. Harper & Row, N.Y.

Gruhn, R., and A.L. Bryan
1984 The record of Pleistocene megafaunal extinction at Taima-taima, northern Venezuela. In *Quaternary Extinctions: A Prehistoric Revolution,* edited by P.S. Martin and R.G. Klein, pp. 128-137. University of Arizona Press, Tucson.

Haynes, C.V.
1982 Were Clovis progenitors in Beringia? In *Paleoecology of Beringia,* edited by D.M. Hopkins, J.V. Matthews Jr., C.E. Schweger and S.B. Young, pp. 282-398. Academic Press, New York.

Hrdlička, A.
1912 Early man in South America. *Bureau of American Ethnology, Bulletin* 52. Washington D.C.

Jopling, A.V., W.N. Irving, and B.F. Beebe
1981 Stratigraphic, sedimentological and faunal evidence for the occurrence of Pre-Sangamonian artefacts in northern Yukon. *Arctic* 34(1):3-33.

Krieger, A.D.
1964 Early Man in the New World. In *Prehistoric man in the New World,* edited by J.D. Jennings and E. Norbeck, pp. 23-81. University of Chicago Press, Chicago.

Li Yanxian
1983 Le Paleolithique Inférior en Chine du nord. *L'Anthropologie* 87: 185-199.

Lyell, C.
1859 On the occurrence of works of human art in post-Pliocene deposits. *British Association for the Advancement of Science, Report* 29: 93-95.

Mochanov, I.A.
1978 Stratigraphy and absolute chronology of the Paleolithic of Northeast Asia. In Early man in America from a circum-Pacific perspective, edited by A.L. Bryan, pp. 54-66. *Department of Anthropology, University of Alberta, Occasional Papers* No. 1. Edmonton.

Moir, J.R.
1927 The antiquity of man in East Anglia. Cambridge University Press, Cambridge.

n.d. *Pre-Paleolithic man.* W.E. Harrison, Ipswich.

Ochsenius, C., and R. Gruhn (editors)
1979 *Taima-taima, A Late Pleistocene Paleo-Indian kill site in northernmost South America — Final reports of 1976 excavations.* CIPICS/South American Quaternary Documentation Program (Printed in Germany, 1984).

Okladnikov, A.P., and G.A. Pospelova
1982 Ulalinka, the oldest Paleolithic site in Siberia. *Current Anthropology* 23: 710-712.

Owen, R.
1984 The Americas: the case against Ice-Age human population. In *The origins of modern humans: A world survey of the fossil evidence,* edited by F.H. Smith and F. Spencer, pp. 517-563. Alan R. Liss, N.Y.

Roe, D.A.
1981 *The Lower and Middle Paleolithic periods in Britain.* Routledge and Kegan Paul, London.

Schmidt, M.
1905 *Indianerstudien in Zentralbrasilien.* Dietrich Reimer, Berlin.

Schnurrenberger, D.W., and A.L. Bryan
1985 A contribution to the naturefact vs. artifact controversy. In *Essays in honor of Don E. Crabtree,* edited by J.C. Woods, M.G. Pavesic and M.G. Plew. University of New Mexico Press, Albuquerque.

Schweger, C.E., J.V. Matthews, Jr., D.M. Hopkins, and S.B. Young
1982 Paleoecology of Beringia — a synthesis. In *Paleoecology of Beringia,* edited by D.M. Hopkins, J.V. Matthews, Jr., C.E. Schweger and S.B. Young, pp. 425-444. Academic Press, New York.

Smith, W.G.
1894 *Man the primeval savage.* Edward Stanford, London.

Steen-McIntyre, V., R. Fryxell, and H.E. Malde
1981 Geologic evidence for age of deposits at Hueyatlaco archaeological site, Valsequillo, Mexico. *Quaternary Research* 16: 1-17.

Stile, T.E.
1982 Perishable artifacts from the Meadowcroft Rockshelter, Washington County, southwestern Pennsylvania. In *Meadowcroft, collected papers on the archaeology of Meadowcroft Rockshelter and the Cross Creek Drainage,* edited by R.C. Carlisle and J.M. Adovasio, pp. 130-141. University of Pittsburgh, Pittsburgh.

White, J.P., and J.F. O'Connell
1982 *A prehistory of Australia, New Guinea and Sahul.* Academic Press, New York.

Yi, S., and G.A. Clark
1983 Observations on the Lower Paleolithic of northeast Asia. *Current Anthropology* 24:181-202.

An Alternative Paradigm for the Study of Early Man in the New World

JOHN ALSOSZATAI-PETHEO
Department of Anthropology
Central Washington University
Ellensburg, Washington 98926
U.S.A.

Abstract

An alternative paradigm is suggested in order to break the conceptual impasse regarding evaluation of the available evidence for people in the Americas before 12,000 B.P. It is suggested that a major reason for the current impasse is a set of implicit assumptions that are a fundamental part of the European/North American world view which emphasizes progress and change. An alternative systemic paradigm is suggested which is based on the world view of modern hunter-gatherers who perceive that they are an integral part of a stable environmental system. With such a paradigm cultural change is necessary only if the environment changes drastically. Therefore, if early Americans entered with a simple material culture of stone and perishable bone artifacts, cultural change would not be archaeologically visible until significant climatic change occurred at the end of the Pleistocene.

INTRODUCTION

Among the various concerns which have contributed to the characteristic flavor of American archaeology, none seem to occupy such a unique niche, be so elusive, or be endowed with such a colorful and distinctive cast of characters and events as the search for the first Americans. The current state of affairs might be loosely termed a "debate," were it not for the fact that its beginnings can be minimally traced to the turn of this century, or even earlier, before the essential character, methodology, indeed the very foundations of American archaeology were laid. Further militating against the use of the term "debate," in this instance, is the extreme polarization of views and the lack of communication between proponents on either side.

Anyone who has become acquainted with the nature of Early Man studies in America, been exposed to the associated literature, and attended the numerous professional meetings where the question of the origins and antiquity of man in the New World have been explored, can attest to the accuracy of this portrayal. Indeed, this picture has become so commonplace that to most professionals it seems inevitable. American archaeology seems to be operating true to form when, aside from the continued discussion of the archaeological evidence for the post-12,000 year old Paleoindian occupation of the continent, the dominant school only briefly deviates from their course to raise their objections and ultimately to discredit the evidence presented by yet one more pre-12,000 year old site.

Over the years, the often-checkered history of such early sites has failed to crystallize into a sensible, acceptable pattern despite occasional attempts by various authors to provide a meaningful sythesis of the evidence presented by such early sites (Krieger 1962, 1964; Willey 1966; Bryan 1965, 1969, 1978; Comas 1972; Wormington 1962; Lorenzo 1967; Irwin-Williams 1969a; MacNeish 1971, 1976; Carter 1980). Yet the evidence remains unheeded, growing yearly as new sites are discovered, adding to the utter sense of bewilderment one gets when confronted with dozens of well-documented, radiocarbon-dated pre-12,000 year old sites in the face of staunch denials by the conservative bastions of American archaeology. This situation may well have led numerous authors to affirm man's presence in the New World long before the acceptable maximum of 12,000 years ago, while cautiously echoing the caveat of archaeological orthodoxy that no "conclusively acceptable evidence" for pre-12,000 year old occupation of America has yet been found! Such contradictory statements are the heights of irrationality and demand some sensible settlement or explanation.

COMPARATIVE HISTORICAL BACKGROUND

The roots of this controversy are deeply buried both in the history of Early Man studies in America, and in the gradually-emerging realization of the underlying assumptions of the archaeologists who have dealt with the unravelling of this complex yet fascinating story. All too often the accounting of the history of such studies overlooks the fundamental cognitive issues which have played as significant a role in the advancement of the research area as has the field data. As long as we limit ourselves to addressing the availability, nature, and validity of the data, we are only addressing half of the problem. The current impasse in Early Man studies in the New World will continue without much change until we begin to evaluate the more fundamental cognitive differences which validate the two widely-divergent interpretations of the available evidence.

In examining these issues, it appeared to me that we were considering the fundamental differences which originate from the interactions of two opposing paradigms. Although a paradigm, as described by Kuhn (1970), is a highly complex concept, for the purposes of this discussion I will consider a paradigm as a network of assumptions, orientations, expectations, and concomitant methodology which acts as a heuristic-cognitive framework aimed at the recovery, analysis, and interpretation of field data. Unfortunately, paradigmatic assumptions are so deeply-rooted that they are often almost impossible to approach directly. For this reason I have chosen to begin with an historical accounting of Early Man studies in America in comparison with the contemporaneous development of Early Man studies in the Old World, including European examples of the reality of cognitive processes and the impact of such processes upon the acceptance and correct interpretation of field data.

In a 1965 article, Wilmsen presented a comprehensive history of Early Man studies in America from its earliest beginnings in 1520 to the time of the publication of his article. He divided this vast expanse of time into six periods characterized by distinctive changing conceptions of the origins of American Indians. The first two periods centered on early speculations and casual inquiries with little rigor yielding few firm facts. By 1859, however, the scientific revolution of Europe was beginning to be felt in America. Wilmsen's introduction to the period from 1859 to 1965 is particularly enlightening. According to Wilmsen:

> From 1859 to 1890, the third period, the concepts of geologic sequence, biological evolution, and cultural succession (all developed in Europe) spurred the search in this country for Paleolithic men of antiquity comparable to those being found in Europe. It was a time when scanty evidence weighed heavily, and this evidence was readily accepted by the untrained men who were devoting their spare time to the problem. As should be expected, this overzealousness produced a reaction, and during the fourth period, 1890-1925, the very thought of local Pleistocene ancestors for modern American Indians was virtually taboo. But this point of view was eventually discarded. The late 1920's are well-known among anthropologists for numerous finds of human remains in indisputable association with extinct fauna. During the fifth period, 1925-1950, the real foundations for Early Man studies on this continent were laid and the first steps towards understanding the earlier prehistory of the Americas were taken. With the development of radiocarbon dating, these steps were given a firmer chronological footing, and the last

period brings us to the present, when the contributions of many natural and life sciences are broadening the scope of Early Man studies (Wilmsen 1965:172).

Unfortunately, there are major aspects and undercurrents of each of these periods which this account omits, and which we must consider here if we are to understand and place Early Man studies in their proper perspective.

As Wilmsen (1965), and Willey and Sabloff (1974) have noted, the original excitement in late-1800s American archaeology about the possibility of finding Paleolithic materials and pre-*Homo sapiens* fossil forms in the New World was a direct reflection of the intellectual ferment and dawning successes of the fields of geology, biology, and paleontology in Europe. As such, American archaeologists shared not only in the excitement and anticipation of additional discoveries, but in a real sense, mirrored many of the biases, failings, and misconceptions of their European counterparts across the Atlantic.

The latter half of the 1800s through the onset of the First World War was a period of intense nationalism in Europe. This period was characterized not only by competition among nations in the economic realm; but large land tracts of the non-industrialized world were actively portioned off and colonized in the name of patriotism, national superiority, and pride. Intellectually, this period also produced the first halting steps in the direction of establishing the field of anthropology utilizing a model of unilineal evolution. These early anthropologists reflected the basically ethnocentric, nationalistic view of their times by proposing an evolutionary scheme of cultural development which placed their respective societies at the pinnacle of human intellectual and cultural development. Similar biases appeared in the undertow of the interpretations of the first fossil hominids found during this period.

Beginning with the finds of the first recognized primitive-looking remains from the Neander valley in 1856, the scientific community was primed for the search for human ancestors by the publication of *The Origin of Species* in 1859 and the subsequent discoveries of the first modern men of Paleolithic age at Les Eyzies, France, in 1868, and the Neanderthal remains from Spy, Belgium in 1886 and La Chapelle-aux-Saints, France, in 1908. Each new discovery was greeted as a national triumph in the highly competitive atmosphere which reigned in the intellectual community of Europe. In this climate of thought, and ignoring lack of evidence for any cultural or biological continuity, popularized interpretations during this period seem to have engaged in a peculiar form of mental pole-vaulting which justified the acceptance of specific finds as evidence for the venerable antiquity and preeminence of the people and nation in whose lands these remains had been found.

These views are both ethnocentric and teleological. They proceed from an intense desire to justify beliefs in nationalistic superiority, and are thus clearly ethnocentric. Conversely, they over-accentuate the significance of field data with the intent of legitimizing individual cultural biases. Together they form a cognitive pattern which pays lip service to the data while perverting the information thus gained to fulfill cultural preconceptions.

Two specific examples of this peculiar trend in early European treatment of Early Man data will serve to illustrate this process. The first example shows how this cognitive set resulted in the ready-acceptance of spurious data, while the second illustrates how valid field data were almost lost forever.

The first of these cases is the well-known and celebrated Piltdown Hoax, perpetrated primarily upon the British scientific establishment, who until 1912 had found themselves in the embarrassing position of lacking a properly authenticated, ancient hominid standard-bearer of the British claim to venerable ancestry. Over the decades following its "discovery," an inordinate amount of literature was generated about the peculiar finds recovered from the gravel pit on an old farm in Sussex, England. The initial near-jubilant reports gradually turned to bafflement as the significance of the traits exhibited by the remains began to be compared with what was already known and was being discovered about human ancestry. Eventually, suspicions about the fraudulent nature of the finds were confirmed through direct testing of the finds, creating a furor of speculation about the identity and intent of the perpetrator which has not yet fully died down. Our focus here, however, does not concern any of these events, but rather it concerns the ease and willingness with which the scientific establishment of the day accepted at face value the validity of the finds. Despite the skill with which the hoax was perpetrated, and despite the relative infancy of the scientific establishment's experience in the study of fossil hominids, it is clear to see that such finds would have met with considerably greater reserve and caution, perhaps leading to an early uncovering of the hoax, if it had not been for the fact that the British scientific establishment of the day had been properly "primed" and ready to accept and believe in the antiquity of man on their own soil!

The second case concerns the discovery of Pithecanthropus, or Java man in 1891. Unlike the French, the Germans, and the Belgians who had found hominid ancestors on their own soil, the Dutch contribution to the story of human evolution came from a distant colony in Java. Beginning with the first cabled reports of his finds, to the arrival, exhibition, and presentation of his evidence and conclusions, Eugene Dubois' contribution was met with open skepticism and opposition:

> Only six weeks after Dubois reached Holland in 1895, he presented Pithecanthropus to the Third International Congress of Zoology at Leiden. Almost at once, a great quarrel broke out over where to place this Java "ape-human" in the scheme of evolution. Opinion seemed to harden along national lines: most German scientists believed that Pithecanthropus was an ape that had humanlike characteristics; most English ones thought it was a human that had apelike attributes; and American experts tended to consider it a transitional form more along the lines Dubois had suggested... In spite of all Dubois' efforts, the attacks on Pithecanthropus continued. Dubois took them personally. Deeply hurt by the refusal of other scientists to accept his interpretation of the bones, he withdrew the remains of Pithecanthropus from the public realm, hid them under the floor of his dining room, and became almost a recluse (Campbell 1982:259).

Dubois remained in isolation, refusing to allow scientists to study the remains of Pithecanthropus until 1923, when he finally allowed Ales Hrdlička to examine them.

Both of these cases have a clear message about the impact of teleological ethnocentrism on the interpretation of scientific data. In the first case, scientists yielded to their eagerness to find the highly-sought data, however unlikely it may have been. In the second case, this same cognitive set caused divisiveness, skepticism, and disbelief in the form of persistent attacks on the validity of the evidence based on purely teleological ethnocentric and nationalistic pre-conceptions. A similar argument can also be made about the effects of such views on the part of the British scientific community toward the first fossil remains of *Australopithecus* found by Raymond Dart in South Africa a dozen years after the discovery and acceptance of the Piltdown remains.

The corresponding period of American Early Man studies was destined not to produce any comparably spectacular discoveries of Early Man materials, thus precluding archaeologists on this continent from participating in the competition of the European states for a position of preeminence in Early Man studies. Furthermore, it was obvious to American archaeologists that any early hominids found in the New World would automatically be excluded from consideration as ancestors of the dominant European population because early Americans were formerly East Asians. This situation ultimately led to a degree of pessimism about the antiquity of man in America.

During this time, America was still in its period of expansion and growth in which the concept of a western frontier and the classification of American Indians by the lay public as mere "savages" was far from being an academic question. This form of popular ethnocentrism led to the opposite conclusion about the antiquity of the American Indian. The pessimism about the likelihood of finding evidence of Early Man, together with the popular ethnocentric attitude about the original occupants of the continent, gradually jelled into an opposition against any attempts to find or present evidence which might contradict this view. For the next three or four decades, American archaeologists would labor under the view of man's relative recency in the New World, while the mere mention of the possibility of greater antiquity was tantamount to professional suicide. Given this orientation, it is not surprising that when the evidence of the antiquity of man in America was finally reported from Folsom, Clovis, and other High Plains sites, it was rejected out of hand by established authorities despite the clear nature of the evidence at multiple locations, uncovered by different researchers, and seen and attested to by a large variety of professional visitor/observers.

Reason dictates that given convincing evidence, clearly demonstrated, under well-controlled excavations, all major objections should have been quickly answered; and research should have proceeded in this new direction. This was not the case, however. Clearly, evidence (however good or controlled) was not sufficient to convince the conservatives of the field. From our historic vantage point, the reason for the rejection of the evidence is obvious. The mind set of conservatives of the day left no room for acceptance and incorporation of the implicit meaning of the field data. It is my contention that in many respects we face a similar situation today.

It is a common truism that those who ignore history are doomed to repeat it. Along these lines a good case can be made for the fact that the role played by W.H. Holmes and Ales Hrdlička (cf. Wilmsen 1965:178-180) in the denial and refutation of evidence for early sites during the first three decades of this century is being repeated by the continued objections of the prime proponents of the 12,000-year school in American archaeology today. Willey and Sabloff, while acknowledging the "negative" impact of men such as Holmes and Hrdlička on the advancement of Early Man studies in America, point out the valuable service they performed to archaeological method of their day by bringing "some degree of rigor and an established mode of validation into archaeological fieldwork and interpretation" (Willey and Sabloff 1974:58). Undeniably their impact on this formative period of American archaeology had its positive side effects. However, this interpretation tends to ignore some of the latent motivations of such men, which become apparent in the face of their continued opposition to the acceptance of valid evidence form the High Plains presented by competent professional colleagues, years after the recovery and presentation of evidence from Late Pleistocene sites had made their objections untenable.

As late as 1942, Herbert Spinden "insisted on a late date for Folsom. He thought that 4000 years was more than sufficient for all of American prehistory, even though he considered Mesoamerican writing, architecture, and astronomy as purely independent developments" (Wilmsen 1965:184). In the same year, Hrdlička "called the Folsom-Yuma finds Neolithic, isolated, and superficial, and said that the Pleistocene animals associated with them were 'the Achilles heel' of American archaeology because they were not long extinct" (Wilmsen 1965:183).

THE CURRENT IMPASSE IN EARLY MAN STUDIES

Today, American archaeologists concerned with the Early Man problem face a similar challenge. Although the crude teleological ethnocentrism of the early part of this century which restricted acceptance of the evidence presented by Early Man sites has ultimately yielded to acceptance of the evidence from the High Plains Paleoindian sites, a similar set of pre-conceived notions have replaced them.

Once again, the conservative leaders of American archaeology are faced with numerous, well-documented sites which pre-date the acceptable early limits set by archaeological orthodoxy. Although not in every case, many of these sites have been excavated with painstaking care, often by highly experienced, trained, and respected professional archaeologists. Unlike the case of the Folsom and Clovis sites, where the association with extinct fauna was the only possible proof of antiquity, these early sites now come with evidence of solid radiometric dates attesting to their great age. Often, many of these sites have been visited by numerous competent archaeologists, and in some instances (such as Valsequillo, Mexico (Irwin-Williams 1969b) and Calico, California), complete segments of matrix containing artifacts have been preserved for widespread professional examination.

In response to this evidence, the reactions of conservative archaeologists seem almost identical to the reactions we saw in the late 1920s and 1930s. In most cases such early evidence has been largely ignored. Occasionally, when directly confronted with the evidence, the reaction has been one of skepticism. Individual sites have been questioned on the basis of the rigor of their excavation methods, their stratigraphic accuracy, the validity of the radiometric dates obtained, the possibility of intrusive materials, the impact of natural agencies shaping purported artifacts, the professional competence of the excavators, and the lack of diagnostic cultural-temporal markers (cf. Dincauze 1984; Owen 1984; Stanford 1982; West 1982).

The history of archaeology provides ample evidence to support the position that such problems must be seriously addressed. In other words, an appropriate level of professional reserve and careful evaluation is always advisable. The evidence suggests that such errors in the past, however, have typically been the product of occasional sloppy work by individual researchers or randomly occurring factors relating to the unique problems encountered at specific sites. On the other hand, never have such factors been so consistently associated with *all* sites of a certain age (in this case, pre-12,000-year old sites) which have been found by a variety of archaeologists, over two whole continents, under widely different conditions, over several decades. This consistent correlation between the purported ultimate lack of validity of sites and their pre-12,000-year dates has never been addressed by the archaeological establishment.

I reject the universal failure of radiometric and other dating techniques when applied to American archaeological sites which pre-date the 12,000-year mark — while functioning adequately in dating non-archaeological materials pre-dating 12,000 years, or early archaeological sites from the Old World. I also reject the assumption that archaeologists who report evidence which predates this age boundary are universally biased, or in error in their interpretations or methodology. I also reject the implication that such early sites are universally disturbed, skewed, or contaminated. Finally, I reject the logic of denying the antiquity, evidence, or validity of early sites simply because they have largely defied past attempts to fit them into well-defined horizons and traditions; or because they lack accepted diagnostic tool types such as projectile points. Without a doubt, the problem lies not with the evidence, but with the biased hearing which such evidence has received by the proponents of the 12,000-year school of American archaeology. I consider all these facts undeniable. However, pointing them out is merely the first half of the task. By itself, without the benefit of offering an alternative viewpoint, such charges take on a petty air; and for the most part tend to be destructive and polarizing rather than being constructive.

THE NEED FOR A NEW PARADIGM

It is my firm belief that we have far overstated the degree of control exercised by Early Man on his environment. We tend to think of man's first appearance on the North American continent as a momentous event in the natural history of the Americas where man, purposeful and unimpeded, overwhelmed the established natural order of two continents covering fifteen million square miles, encompassing thousands of different habitats and ecological communities; and became supreme master of all that his eyes surveyed in a matter of a few hundred years. This view can only be compared to the conquest of the American West, an image which in no small measure may have contributed to our erroneous assessment of early prehistoric events on this continent. The Martin-Haynes hypothesis (Martin 1967, 1973; Haynes 1966, 1967, 1969) is but a single reflection of a widespread phenomenon in which we have subtly imputed into the archaeological record and into our accounting of the culture history of the New World our own teleological world view and value systems.

Although, in the short term, the world view of a culture may often clash with its social and environmental realities, cultural systems generally tend to be unified systems in which various elements ultimately intermesh with considerable internal consistency and agreement. World view is no exception to this rule. Our modern Western world view is largely congruous with our ecological position as food producers who, having harnessed huge energy sources not normally available to biological organisms in more natural settings, have embarked several centuries ago in an exponential growth pattern which ecologists refer to as a "J-type" growth curve (Odum 1975:123-128). This has produced an orientation in our cognitive patterns in which growth and environmental control is normal and expectable, innovation and invention are the norm, and a peculiar optimism which emphasizes the availability of choices and free will and de-emphasizes the long-range impact of obstacles in the path of "progress" seems to us the only possible and logical pattern of life. I submit that much of our maladies in the study, understanding, and interpretation of past cultural realities and events stems from our subconscious attempts to impute this cognitive pattern and "practical" orientation upon the actual evidence presented by archaeological sites. In this view, what could be more natural and self-evident than the proposal that a small band of newcomers arriving on the scene of two continents devoid of human competitors, should engage in a headlong race to occupy and subdue this vast natural expanse, filled with untold "promise" and "opportunity." Perhaps one of the most obvious demonstrations of the tremendous degree of entrenchment of this cultural world view in our society can be gleaned from the pervasively positive connotations associated with the words which we use to describe such an orientation or process. Words such as "growth," "progress," "complexity," "sophistication," "expansion," "control," and others, all carry a heavy burden of positive connotations which underscore our subconscious cultural orientations and preference for our world view.

I would like to propose, however, that there is a very real, alternative interpretation which not only *could* be applied, but which in the face of large quantities of supporting evidence from a wide variety of sources, is ultimately more reasonable and elegant. Occam's razor or the principle of parsimony is my guide in offering the following alternative to the current paradigm of the 12,000-year school of American Early Man studies.

I have chosen the title "An alternative paradigm for the study of Early Man in the New World" in direct, yet independent, recognition of a process perhaps best stated by Binford and Sabloff when they said:

> Archaeological knowledge of the past is totally dependent upon the meanings which archaeologists give to observations on the archaeological record. Thus archaeologically justified views of the past are dependent upon paradigmatic views regarding the significance of archaeological observations (Binford and Sabloff 1982:149).

In this light, Binford and Sabloff (1982) argue that archaeologists in both the Old and New Worlds have projected their own, unique views of cultures into the past, and into their particular evaluations and telling of the story of past cultural events. Under such a system of analysis and study, one's focus is so channeled that expected results are all but a foregone conclusion. As Binford and Sabloff note, "Most archaeological reasoning has been a classic example of inductive argument from archaeological observation; no wonder the past never argued back! Archaeological interpretations have been inductively argued, and hence experience (the archaeological record) is *simply the vehicle for inference*" (Binford and Sabloff 1982:149) (emphasis added).

In the case of Early Man studies in the New World, it is my view that the archaeological record from pre-12,000-year old sites *has* "argued back" now for many decades, albeit unsuccessfully, because the operating paradigm of New World archaeologists was incompatible with the acceptance of such evidence.

In this regard, it is interesting to note that Binford and Sabloff characterize this peculiar American paradigm as one which inevitably sees "the cultural past as a series of growths, followed by declines or collapses" (1982:145). Methodologically, they characterize the New World's paradigmatic approach as one in which:

> the basic unit of observation is the artifact, in a framework of attributes. Types may be recognizable in many data classes. Every class of items does not yield types, for some may be judged so generalized in their distribution as to be 'nondiagnostic,' and as such are most often ignored... In other words, traits which are frequent, not too generalized, and easily recovered from different places are given priority as the defining characteristics of cultures (Binford and Sabloff 1982:146).

This may be fine for the study of the classic Paleoindian sites, where projectile points — for example — have served such a culture-diagnostic function. However, a professional community with such an orientation would likely find themselves at a total loss in the acceptance, integration, and interpretation of earlier sites which lacked such typological markers. Once again, I feel that this is precisely what has occurred in the handling of pre-Paleoindian sites in America. The now-historic failure of the notion and the term "Pre-Projectile Point Horizon" (Krieger 1962, 1964) can be seen as the failure of a "negative definition" which stemmed from this perspective, but which lacked an alternative paradigm which might accommodate the incoming field data into a meaningful cognitive framework.

Binford and Sabloff's alternative to this situation is in direct agreement with my central suggestion for New World Early Man studies. In their view:

> Archaeological literature of the past two decades is replete with arguments which point out that if a major goal of the discipline is to explain culture change, then the traditional ways of looking at the past — the normative paradigms — have not been very productive. A new way of looking to the world, the systems paradigm, has been proposed as a potentially more productive means of reaching this explanatory goal (Binford and Sabloff 1982:139).

AN ALTERNATIVE SYSTEMIC PARADIGM

Although my focus of concern is the presentation of an alternative paradigm which is more consistent with biological, ecological, ethnographic and culture-adaptive processes than our current one, the essential nature of this paradigm is systemic, and as such, lends itself to extensive, detailed field testing by archaeologists. Let us now examine some of the fundamental aspects and assumptions of this alternative paradigm, which for lack of a better name I shall call the 'Steady-State Ecological Equilibrium' paradigm.

Physical anthropologists have long held that population densities prior to the domestication of plants and animals were low and stable. Supporting this claim, they point to the relative rarity of fossil hominid skeletal remains, and to the low population densities which can feasibly be maintained through a hunting and gathering way of life (e.g., Bennett 1979:352). Such ecological balance characterized by stable numbers represents a steady-state equilibrium which is common in natural ecosystems, and represents the latter stages of a "sigmoid-type" growth curve which ultimately levels off at or near the carrying capacity of an ecosystem (Odum 1975:124). The attainment and maintenance of such a balance is directly based on the operation of a variety of feedback mechanisms which permit the maintenance (with some temporary variations) of a given population density, while inhibiting any rapid or disproportionately large growth on the part of any single population in a well-balanced ecosystem.

What are the realities of human beings who traditionally operate under such a steady-state ecological balance? Our answers can probably best be obtained from ethnographic data.

The preponderance of some conscious, overt, or some indirect, unplanned form of birth control practiced by various ethnographic hunting and gathering groups is a matter of record. The infanticide and geronticide of the Eskimos, (Dunn 1968; Birdsell 1968) together with the spacing of births and impact of extended periods of infant nursing practices are all individual examples or instances of cultural-ecological mechanisms which have been responsible for the maintenance of stable numbers promoting a steady-state balance between ethnographic hunter-gatherers and their individual ecological realities.

One particularly enlightening example which allows us to contrast the world view of our modern Western societies with the value system of hunter-gatherers in a balanced, steady-state equilibrium with their environment is afforded by an encounter between Marco Bicchieri and his Hadza informants. During fieldwork among the Hadza, an East African hunting and gathering society, Bicchieri was struck by the relatively low success ratio of hunters in hitting and bringing down game with their arrows. Concerned for their welfare, and eager to share with his informants the benefits of Western achievements and sophistication, Bicchieri arranged to hold a demonstration of archery practice in which he managed to hit the target with repeated accuracy. Having thus demonstrated some of the finer points and refinements of his skill, he was shocked by the response of his informants. Instead of grateful admiration and approval, or even requests for instruction, his informants acted embarrassed, and professed pity for him. Bicchieri was stunned, and inquired as to the reasons for their reaction. Their answer, though simple, holds the key to our understanding of the world view of a people in such a state of environmental equilibrium. Bicchieri's informants told him that "they were sorry for him because he must come from a very poor society"! According to their view, they lived amidst nature's bounty, and they could well-afford to hit their target only once of each ten tries. However, Bicchieri's realities were obviously different, since in their estimation, he must come from such a poor cultural condition where missing the target must be tantamount to disaster (Marco Bicchieri, personal communication 1984)!

Still other ethnographic support for the Steady-State Ecological Equilibrium paradigm still from the work of Gould et al. (1971), who conducted statistical and microscopic studies of ethnographic Australian aboriginal stone tools and their uses to provide analogues for archaeological inventories and interpretations. Although much of their discussion centers on specific tool types and cultural features of Australian aboriginal ethnography and archaeology, they make very clear that their findings can, and should be applied more broadly to the study of archaeological assemblages elsewhere. Among the points which they raise, based on extensive analysis, is that there may well be a tendency by archaeologists to over-classify archaeological materials beyond the practical and perceived limits which governed their use and manufacture in their original contexts (1971:154). From their descriptions, a rather interesting pattern of manufacture and use of stone tools appears which emphasizes local, momentary needs; the casual use of available materials; the lack of emphasis on technological sophistication; the regular discarding of tools after a specific job had been completed; and an attitude which de-emphasizes symmetry, refinement, and systematic continuity in tool types, but instead focuses on the most convenient means of accomplishing the job at hand (Gould et al. 1971). Such specific examples and suggestions along with an understanding of the workings of a steady-state equilibrium between hunter-gatherers and their individual ecological realities must form the model used in the interpretation of the evidence of Early Man's past cultural activites, rather than the model which takes its cue from the growth orientation of modern Western society.

The steady-state equilibrium model does not require the acquisition of new field data beyond what is already known, nor does it require dismissal of the evidence presented by either pre- or post-12,000-year old sites in the New World; but rather it requires the reassessment and testing of existing evidence using this new perspective. Under such a model, it is simple to see that man's arrival caused no major shifts in the ecological balance of the continent. Individual bands of hunter-gatherers adapted singly to successive local environmental realities. This clearly is not a pattern which would yield broadly uniform archaeological horizons covering vast geographical areas over multitudes of locally distinctive ecosystems. Likewise it is unlikely that such adaptations would have produced widespred, culture-diagnostic artifacts. However, they *would* have produced a diagnostic culture-adaptive relationship with individual environments, but with widespread variations congruous with local conditions over long periods of time. In other words, the key to the identification and understanding of regional and temporal cultural modalities among early sites lies not with diagnostic tool types, but with diagnostic culture-adaptive techniques revealed by the study of whole sites in their original environmental contexts. Under this model, man's impact on any single ecosystem would be no greater than the impact exercised by other, naturally occurring foragers and predators. Indeed, instead of dominating and shaping his environment, man would have become incorporated into the local ecosystems upon which he depended. With low population densities over long periods of time, there would have been no need to produce cultural and technological systems of increased efficiency and sophistication. In other words, man's survival depended on the stability of the environment and the maintenance of the balance which had been struck once the human component of individual ecosystems had begun to approach the carrying capacity of the local ecosystem. Thus it was in man's interest to participate in, rather than to control and shape, his environment.

Only long-term, drastic environmental changes could upset this balance, requiring a re-assessment of survival strategies and the adoption of new means of subsistence. In the case of North American archaeology, perhaps no instance illustrates this latter process better than the rise of big-game hunting on the High Plains during Paleoindian times, and the adoption of seasonality and scheduling in the more arid Basin-and-Range province to the west. Thus, from this perspective, the beginnings of the Paleoindian tradition and the "Desert Culture" traditions of North America would represent adaptive responses to changing environmental conditions at the end of the Pleistocene, rather than massive, rapid population movements from the Old World into the New World.

Wherever archaeological remains of this earlier period of occupation have been found, they have been generalized; that is relatively unsophisticated, localized manifestations of flexible adaptive strategies which did not specialize in their subsistence methods or resources, but rather emphasized a wide variety of plant and animal resources, and a "casual" approach to the production and use of technological implements. Specific cases in point supporting this view are so common and obvious that they have been overlooked precisely for their lack of uniqueness.

CONCLUSIONS

As a precautionary note, I would like to make clear that I do not consider the definition of this alternative paradigm complete as presented here, nor do I feel that it should be viewed as a panacea to be misused in widening the chasm which already separates the opposing camps of American Early Man archaeology. I see this paradigm as a beginning, or a conceptual direction to be explored through careful field research and testing. The Steady-State Ecological Equilibrium paradigm still leaves many questions unanswered. These questions will not be resolved from either a purely inductive or deductive line of research; but from careful systematic evaluation of the archaeological record, culture-adaptive systems, human and non-human ecological systems; and insightful assessment of the significance of such information which is not afraid to approach one's own cognitive systems from the most pragmatic viewpoint possible.

As I have indicated at the beginning, the debate over the antiquity and nature of the evidence for the initial peopling of the Americas has been a long and divisive one, while this presentation has failed to produce any new evidence in support of either side of the controversy. However, it is my hope that this discussion has pointed out some of the underlying factors for the lack of communication between the two opposing sides. The reason why evidence, however controlled, exacting, and convincing, has failed to bring resolution to this "debate," is to be found not among the methods, character, or reliability of the evidence, but in the underlying assumptions and cognitive models of each group. After all, as Kuhn (1970) has pointed out, and the case of debate among Catastrophist and Uniformitarian geologists of the previous two centuries demonstrates, real progress in the correct interpretation of the evidence of nature often depends more on our changing mental sets and cognitive models rather than on the accumulation of additional mountains of supporting evidence.

From this perspective, I would like to urge those of us who are actively seeking a solution to this impasse to reevaluate the evidence for early human occupation of the New World from the perspective of a Steady-State Equilibrium paradigm, rather than the current paradigm derived from our own cultural world view. Not only is our current paradigm inconsistent with the realities of past and present hunting and gathering societies; but in the final analysis, the adoption and use of the current model in the interpretation and study of the evidence presented by Early Man sites in America is the final culprit in the inability of American archaeology to incorporate the evidence already uncovered for early human entry into the New World. As long as we accept the implicit premises of the 12,000-year school, or fail to develop an alternative paradigm which could explain all the evidence, both pre-and post-12,000 for man's presence on this continent, we will forever be doomed to failure, forever reliving the mistakes of our past.

REFERENCES CITED

Bennett, K.A.
1979 *Fundamentals of biological anthropology.* Wm. C. Brown, Dubuque, Iowa.

Binford, L.R., and J.A. Sabloff
1982 Paradigms, systematics and archaeology. *Journal of Anthropological Research* 38: 137-153.

Birdsell, J.B.
1968 Some predictions for the Pleistocene based on equilibrium systems among Recent hunter-gatherers. In *Man the hunter,* edited by R.B. Lee and I. De Vore, pp. 22-240. Aldine, Chicago.

Bryan, A.L.
1965 Paleo-American prehistory. *Idaho State University Museum, Occasional Papers* 16. Pocatello, Idaho.

1969 Early man in America and the Late Pleistocene chronology of Western Canada and Alaska. *Current Anthropology* 10: 339-365.

Bryan, A.L. (editor)
1978 Early man in America from a circum-Pacific perspective. *Department of Anthropology, University of Alberta, Occasional Papers* No. 1. Edmonton.

Campbell, B.G.
1982 *Humankind emerging* (third ed.). Little, Brown and Co., Boston.

Carter, G.F.
1980 *Earlier than you think.* Texas A&M University Press, College Station.

Comas, J.
1972 Where did the first Americans come from? *The UNESCO Courier,* August-September: 46-49.

Dincauze, D.F.
1984 An archaeological evaluation of the case for pre-Clovis occupations. *Advances in World Archaeology* 3: 275-323.

Dunn, F.L.
1968 Epidemiological factors: health and disease in hunter-gatherers. In *Man the hunter,* edited by R.B. Lee and I. De Vore, pp. 221-228. Aldine, Chicago.

Gould, R.A., D.A. Koster, and A.H.L. Sontz
1971 The lithic assemblage of the Western Desert Aborigines of Australia. *American Antiquity* 36(2): 149-169.

Haynes, C.V.
1966 Elephant-hunting in North America. *Scientific American* 214(6): 104-112.

1967 Carbon-14 dates and Early Man in the New World. In *Pleistocene extinctions: the search for a cause,* edited by P.S. Martin and H.E. Wright, Jr., pp. 267-286. Yale University Press, New Haven.

1969 The earliest Americans. *Science* 166:709-716.

Irwin-Williams, C.
1969a The problem of the origins of human culture in the New World. Paper presented at the VIII INQUA Congress, Paris.

1969b Comments on the associations of archaeological materials and extinct fauna in the Valsequillo region, Puebla, Mexico. *American Antiquity* 34: 82-83.

Krieger, A.D.
1962 The earliest cultures in the Western United States. *American Antiquity* 28: 138-143.

1964 Early man in the New World. In *Prehistoric man in the New World,* edited by J.D. Jennings and E. Norbeck, pp. 28-81. University of Chicago Press, Chicago.

Kuhn, T.S.
1970 *The structure of scientific revolutions.* (second ed.). International Encyclopedia of Unified Science (2)2. University of Chicago Press, Chicago.

Lorenzo, J.L.
1967 *La etapa litica en Mexico.* Instituto Nacional de Antropologia e Historia, Mexico.

MacNeish, R.S.
1971 Early Man in the Andes. *Scientifc American* 224(4): 36-46.

1976 Early man in the New World. *American Scientist* 64: 316-327.

Martin, P.S.
1967 Prehistoric overkill. In *Pleistocene extinctions: the search for a cause,* edited by P.S. Martin and H.E. Wright Jr., pp. 75-120. Yale University Press, New Haven.

1973 The discovery of America. *Science* 179: 969-974.

Odum, E.P.
1975 *Ecology; the link between the natural and the social sciences.* (second ed.). Holt, Rinehart and Winston, New York.

Owen, R.C.
1984 The Americas: the case against an ice-age human population. In *The origins of modern humans: A world survey of the fossil evidence,* pp. 517-563. Alan R. Liss, New York.

Stanford, D.
1982 A critical review of archaeological evidence relating to the antiquity of human occupation of the New World. In Plains Indian Studies, a Collection of Essays in Honor of John C. Ewers and Waldo R. Wedel, edited by D.H. Ubelaker and H.J. Viola, pp. 202-218. *Smithsonian Contributions to Anthropology* 30. Washington D.C.

West, F.H.
1982 The antiquity of man in America. In *Late Quaternary environments of the United States* (Vol. 1: *The Pleistocene),* edited by S.C. Porter, pp. 364-382. University of Minnesota Press, Minneapolis.

Willey, G.R.
1966 *An introduction to American archaeology* (Vol. 1). W.H. Freeman and Company, San Francisco.

Willey, G.R., and J.A. Sabloff
1974 *A history of American archaeology.* W.H. Freeman and Company, San Francisco.

Wilmsen, E.N.
1965 An outline of Early Man studies in the United States. *American Antiquity* 31: 172-192.

Wormington, H.M.
1962 A survey of early American prehistory. *American Scientist* 50(1): 230-242.

Pleistocene Archaeology in Old Crow Basin: A Critical Reappraisal[1]

RICHARD E. MORLAN
Archaeological Survey of Canada
National Museum of Man
Ottawa, Ontario K1A 0M8
CANADA

Abstract

Some of the bones and antlers found among the rich paleontological deposits of Old Crow Basin in northern Yukon Territory have been interpreted as artifacts made prior to permineralization of the fossils. Previous reports have attempted to show that artificial alterations are exhibited by fossils recovered from an Early Wisconsinan floodplain dated to approximately 80,000 years B.P. In this paper, all such specimens, from Disconformity A and deeper deposits, are reexamined with alternate interpretations presented. The alternate interpretations do not prove that humans were not present in Early Wisconsinan time, but they show that such ancient presence of people cannot be demonstrated on the basis of evidence gathered thus far.

The fact that alternate interpretations can be devised for all specimens from Disconformity A does not obviate the identification of redeposited artifacts found on the modern banks and bars of the Old Crow River. It does mean, however, that none of the definite artifacts has a known stratigraphic context in the Old Crow Pleistocene, although several artifacts have now been directly dated by radiocarbon techniques. The directly dated specimens show that people were present in eastern Beringia by at least 25,000-30,000 years ago.
ago.

INTRODUCTION

The Old Crow River valley of northern Yukon Territory has been intensively studied during the past 16 years because of its rich stratigraphic, paleontological, and archaeological deposits. Downcutting by Porcupine River, to which Old Crow River is tributary, began in Late Pleistocene time and resulted in the exposure of dozens of 30-40 m sections along both rivers (Hughes 1972). The sections reveal complex sedimentary sequences that enclose a wide variety of plant, invertebrate, and vertebrate fossils affording an opportunity to reconstruct a paleoenvironmental history of Old Crow Basin spanning more than 150,000 years. Dissection of these sediments began shortly before 12,000 years ago in Old Crow Basin and reached its

1. The body of this paper was typeset in April 1983 and reflects my thinking at that time.

modern level within a few millennia. During this rapid episode of downcutting, Old Crow River exhumed millions of Pleistocene fossils, washed away or destroyed most of the plant and invertebrate remains, and concentrated many of the vertebrates in its new channels, point bars, and floodplain sediments. As a result, the modern banks and bars of Old Crow River yield remarkable mixtures of fossils representing different ages. Most of them represent the Rancholabrean fauna, but older forms include, for example, the Irvingtonian *Soergelia* (a primitive muskox) and possibly the Blancan *Planisorex dixonensis* (a shrew; Harington 1977; Kurtén and Anderson 1980).

In addition to comprising one of the world's richest paleontological concentrations, the vertebrate remains in Old Crow Basin have posed both challenge and promise for archaeological studies. Hundreds of bones, antlers and tusks exhibit alterations suggestive of cultural practices contemporaneous with the Rancholabrean forms. Since Old Crow Basin was occupied by glacial meltwater during the last major advance of the Wisconsinan (Hughes 1972; Hughes et al. 1981), most such specimens are believed to pre-date that advance. A radiocarbon date to be reported in this paper places the beginning of glacial meltwater inundation around 25,000 years ago in Bluefish Basin and, by extension, in Old Crow Basin, implying that any pre-lake archaeological remains must be of comparable or greater age.

Several reports have provided descriptions of some of the fossils that are interpreted as artifacts (Irving and Harington 1973; Bonnichsen 1979; Morlan 1980, with other references), but interpretations of them have been hampered by the complex history of fluvial redeposition that has destroyed the associations of the fossils with one another and with their primary sedimentary contexts. Archaeological interpretations of redeposited fossils must be based entirely on patterns of alteration exhibited by one or more specimens, and such interpretations are at risk from the many natural processes of bone alteration that can resemble the results of cultural practices. The concept of taphonomy has been borrowed from paleontology in the belief that consideration of all the alterations exhibited by a fossil would enhance the isolation and recognition of those modifications that owe their origin to human activity (Morlan 1980). The larger goal of this approach is the explication of a set of discrete, universal alteration patterns diagnostic of human activity on osseous remains. Realization of that goal would permit archaeologists to recognize the former presence of humans in areas where redeposited fossils are available but primary archaeological sites cannot be found (Morlan 1984).

Continued field observations have offered new insights into the processes and agents that alter fresh and fossil bone, and ongoing laboratory work continues to challenge some of the artifact identifications published previously. Furthermore, critical appraisals of some of the interpretations of Old Crow fossils have begun to appear in the literature, and these invite discussion. The major problems pertain to (1) identification of artifacts among the fossils, (2) stratigraphic context of the fossils, (3) dating of the fossils, and (4) the limitations that accrue to the interpretation of redeposited fossils. For some specimens previously identified as artifacts, alternate interpretations will be suggested in this paper. The stratigraphic framework is under study with detailed reports in preparation. New dates on artifacts are now available, and more dating is planned.

This discussion pertains primarily to specimens housed in Ottawa at the National Museum of Man and the National Museum of Natural Sciences; brief reference is made to a few other specimens that have been described in the literature. To facilitate cross-references with the National Museums collections as well as other publications on Old Crow Basin, localities and specimens are identified by Borden number (e.g., MlVl-2) with alternate designations provided in Table 1; the localities are shown in Figure 1. Only those fossils that have been interpreted as artifacts or "probable artifacts" (Morlan 1980; Bonnichsen 1979; Jopling et al. 1981) are considered here.

The predominant focus on bone from Old Crow valley reflects the nature of recovered samples and should not be construed as a suggestion that a "Bone Age" (e.g., Estabrook 1982) be seriously considered as a technological entity in prehistory. In fact, a number of stone artifacts have been recovered from Old Crow valley but are of unknown age without primary context (Morlan 1980:Chapter 8). Much larger lithic samples, rarely associated with bone remains, have been found in the uplands surrounding Old Crow Basin, and among them may be stone artifacts contemporaneous with the Old Crow valley fossils, if only we knew how to recognize their associations (Irving and Cinq-Mars 1974; Cinq-Mars 1978).

I have continued to seek explanations for the alterations seen on the Old Crow fossils without assuming that people had to be involved. The most important conclusion reached in this paper is that at this time we can neither prove nor disprove the presence of human beings in eastern Beringia in Early Wisconsinan or earlier time. There remains, however, strong evidence of mid-Wisconsinan human occupation in Old Crow Basin. This conclusion differs from earlier statements (e.g., Morlan 1980, 1981, 1983; Morlan and Cinq-Mars

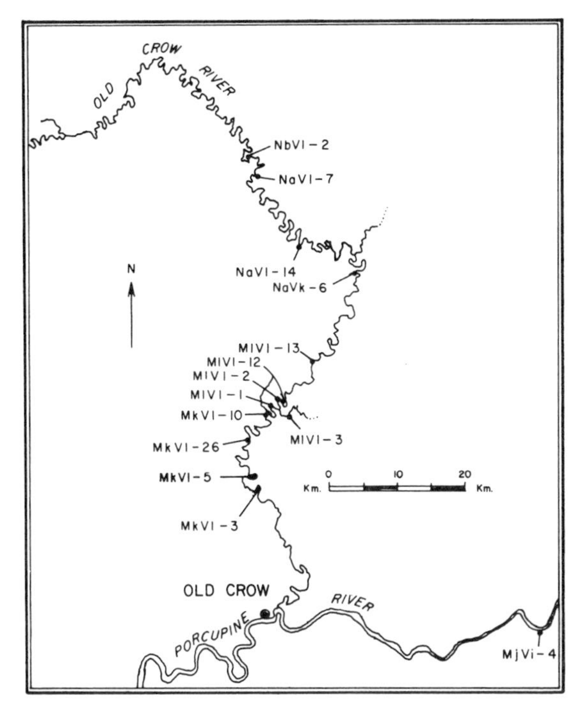

Figure 1. Old Crow River, northern Yukon Territory, showing localities discussed in the text.

1982), but it is not necessarily a retraction of those statements. I have definitely changed my mind about some of my earlier interpretations, but in most cases I am simply trying to enlarge our conceptual framework and to stimulate further observations and discussions.

Table 1. Localities discussed in this paper, showing Borden numbers used by the Archaeological Survey of Canada, National Museum of Man, and alternate designations assigned by various researchers (CRH, C.R. Harington; HH, O.L. Hughes).

Borden Nos.	*Alternate Nos.*
MjVi-4	Cadzow Bluff
MkVl-3	CRH 3
MkVl-5	CRH 4
MkVl-10	CRH 12
MkVl-26	CRH 74
MlVl-1	CRH 14N
MlVl-2	CRH 15
MlVl-3	CRH 70
MlVl-12	CRH 69
MlVl-13	HH 69-21
NaVk-6	CRH 20
NaVl-7	CRH 87
NaVl-14	CRH 94
NbVl-2	CRH 29

PROBLEMS OF ARTIFACT IDENTIFICATION

Elsewhere (Morlan 1984) I have presented an overview of four categories of bone alteration that can be attributed in some instances to artificial causes: cutting, fracturing, flaking, and polishing. Each of them can also be caused by natural processes, the results of which could be confused with artificially induced alterations. Obviously we need a much more comprehensive body of experimental and field data that would provide clear causal links between natural and cultural processes and the specific patterns of alteration they produce on bones.

For example, Thorson and Guthrie (1984) have recently studied river icing and the annual spring breakup of northern rivers with respect to bone alteration. They have argued convincingly that bones can be fractured, flaked, polished and striated by the complex forces and circumstances surrounding fluvial entrainment, transportation, and deposition, and they suggest that such modifications of bone in northern rivers may be the rule rather than the exception. The precise effects of river icing and breakup on bones are difficult to observe directly because of the high incidence of loss from experimentally introduced samples. Observations of altered bedrock and stationary boulder surfaces are of little relevance, in my opinion, since the forces responsible for such modifications utterly destroy both fresh and fossil bones. Thorson and Guthrie (1984) attempted to simulate some of the effects of breakup by dragging frozen ice blocks containing bones behind a truck on a variety of road surfaces. If we assume that these experiments accurately simulate the effects of river ice on bones, the observed alterations are impressive for the hazards they might pose to recognition of artificial alterations among redeposited fossils. However, some critical variables probably were not simulated adequately (e.g., texture and hardness of the substrate, buoyancy of the ice block), and it is noteworthy that many of the experimental bones are more profoundly altered than those recovered from natural environments. Certainly these experiments have not shown that all the altered fossils from Old Crow Basin can be attributed to river icing and breakup, and similar studies should be carried out in a variety of northern streams to enlarge our understanding of the taphonomic influence of river ice on bone.

In the following discussion, I will mention briefly some of the cultural and natural processes thought to produce fracturing, flaking, polishing, and incising on bones along with those attributes that are diagnostic of each process on the resulting fossil or sub-fossil specimen. Many of the processes are time-transgressive with rates of change varying in relation to one or more environmental parameters that may also vary through time. The discussion will highlight those processes that need further investigation and will form the basis for reevaluating fossils from Old Crow Basin.

Fracturing

One dimension of variation in fracture patterns pertains to the condition of the bone at the time of fracture. Relevant attributes are arrayed in Table 2 for limb bones in three conditions: fresh, dry, and fossilized. These conditions represent qualitatively defined stages in the gradual decomposition of the bone which, of course, can be completely destroyed at any point along the continuum from fresh to fossil. The rate of decomposition is dependent upon many environmental variables including circumstances of death, temperature and moisture of air and soil, geochemical and hydrodynamic factors, composition of carrion fauna, and activities of predators and scavengers (including people).

In northern environments, special circumstances of death presumably can enable long-term preservation of bones in a quasi-fresh state as when a carcass is preserved intact with soft parts frozen and gradually mummified. Even defleshed bones lying on the surface in the high Arctic islands may retain considerable structural integrity and be clearly identifiable after many centuries (A.J. Sutcliffe personal communication 1982). Such bones always exhibit split lines, usually exhibit some exfoliation, and often are colonized by lichens and algae; they may be dry, having lost moisture, or fossilized, having undergone mineral loss or replacement. In tropical environments, bones on the surface survive for only a few years, during which many decomposition stages can be recognized (Behrensmeyer 1978).

In most environments, burial is a prerequisite to preservation in a sub-fossil or fossil condition. The time and mode of burial and subsequent sub-surface environmental conditions strongly influence the nature of the potential fossils. Excellent preservation can result from immediate burial in anaerobic sediments as when an animal drowns in a lake or stream. Immediate burial did not ensure bone preservation at Pompeii, however, because of geochemical processes (Maiuri 1958). Erosion and redeposition, especially in fluvial environments, can be quite destructive of bones, depending upon the state of preservation that has been achieved.

Obviously the taphonomic history of bones can be highly variable, and it is necessary to conduct

Table 2. Attributes of limb bone fractures related to bone condition at the time of fracture (modified after Bonnichsen 1979:Table 3; Morlan 1980:Table 3.2; Stanford et al. 1981; personal observations).

Attribute	*Fresh bone*	*Dry bone*	*Fossilized bone*
Negative impact scars (loading points)	Present or absent	Present or absent	Absent
Texture of fracture surface	Smooth	Smooth or rough	Rough ("pebbly")
Angle of fracture with outer surface of bone	Acute, obtuse or right	Acute, obtuse or right	Right
Termination of fracture at epiphyses	At or prior to epiphyses	May cross-cut epiphyses	May cross-cut epiphyses
Color of fracture	Same as outer surface	Little or no contrast with outer surface	Often contrasts sharply with outer surface
Outline form of fracture	Straight diagonal curved, spiral; generally smooth	All outlines seen in fresh and fossil bones, often perturbed by split lines	Usually straight, transverse, or longitudinal; can be curved, rarely spiral, often perturbed by split lines

new studies of taphonomic variables for every major area of paleontological and archaeological research. Old Crow Basin has produced an enormous number of well-preserved Pleistocene vertebrate fossils that owe their preservation to a complex geologic history. Mass spectrographic analysis by Bonnichsen (1979:44-51) and neutron activation analysis by Farquhar and his associates at the University of Toronto (Farquhar et al. 1978; Badone 1980) have shown that the Old Crow fossils are enriched by a variety of minerals and can be regarded as permineralized. The preservation has not been ensured simply by freezing as suggested by some writers (e.g., Haynes 1971; Guthrie 1980). Permineralization has profoundly altered the fracture properties of the bones as shown in the right-hand column of Table 2, and it is relatively easy to recognize post-permineralization fractures on these fossils.

It is somewhat more difficult to separate fresh bone fractures from dry bone fractures, especially on small fragments. Given a suitable microenvironment, a limb bone may survive for many years on the surface in the northern Yukon today. Red and yellow painted horse bones that were placed on the banks of Old Crow River in 1977 were still intact when seen in 1981 although some of them had developed open split lines. Those without open split lines might exhibit many of the fracture attributes of fresh bones. We cannot determine how long the Pleistocene fossils in Old Crow Basin may have lain on the surface prior to burial and permineralization, but some of the fossils were weathered prior to burial, as shown by exfoliation and split lines, and others were not so weathered. We cannot know in most cases whether fractures that produced fragments with the attributes of fresh bones were made before or after desiccation of the specimens, and this fact limits some of our interpretations regarding the processes or agents that may have fractured the bones.

There are many processes that cause fractures in fresh or live limb bones, and some of them result in distinctive attributes. Accidental fractures during life exhibit all the fresh bone fracture attributes shown in Table 2 except negative impact scars. Carnivore-induced fractures may exhibit negative impact scars, but the scars are only slightly larger than the diameter of the tooth contact area and are often associated with identifiable tooth marks on the bone surface; there are also regularities in the positioning of carnivore damage on most limb bones, and these differ from the patterns normally produced by people (see Binford 1981 for a detailed discussion). The use of a hammerstone to break a fresh bone can result in very large negative impact scars as well as distinctive mid-shaft fracture patterns; evidence of periosteum removal by scraping may be visible on the bone surface. River ice probably can fracture fresh bones in a variety of ways (Thorson and Guthrie 1984), but diagnostic attributes that could aid in the recognition of ice-induced fractures have not been isolated. Animal trampling is reputed to fracture large bones, but Haynes (1983:109-111) reports that modern bison appear to induce "insufficient force to cause fracturing when the element is whole and fresh....However, weathered and degreased limb bones, being much more brittle than fresh elements, often fracture when kicked or stepped on by large animals."

In earlier reports, I concluded that proboscidean limb bones are so large and robust that they cannot be broken by carnivores, and I assembled evidence suggesting that elephants rarely fracture their limbs during life. Limited observations suggest that trampling, even by other elephants, is an unlikely explanation for proboscidean bone fracture (Stanford 1982), and we do not yet have direct evidence concerning the effects of river ice on proboscidean bones. On the other hand, it has been shown that people can systematically fracture elephant limb bones through repeated impacts with large hammerstones (Stanford et al. 1981), and the resulting fragments are indistinguishable from the hundreds of green-fractured mammoth bone fragments recovered from Old Crow Basin.

Until recently, however, I have overlooked the interpretive problems that might arise from the fracture of dry bones. Even on small fragments it is often possible to determine whether a bone was broken before or after fossilization, but small fragments do not always exhibit the split line perturbations that are diagnostic of dry bone fracture. Most of the "green-fractured" fossils with known stratigraphic contexts in Old Crow Basin (Table 3) are not large enough to support a confident decision as to their precise condition at the moment of fracture except to say that they were not yet permineralized. Weathered or dried bones are much more brittle (less elastic) than fresh bones, and if the bones were fractured when dry, such processes as trampling, river ice movement, mass wasting of sediment, and even carnivore gnawing might have been responsible for many of the "green-fractured" proboscidean bones recovered thus far from Disconformity A and other stratigraphic contexts along the Old Crow valley (see below).

Flaking

Bone cores and flakes have long been known from various Paleolithic sites in Europe as shown by recent reports on Czechoslovakian collections excavated during the past half century (Valoch

1980, 1982). Experimental studies have shown that many kinds of animal bones can be flaked, and the knapping of bone to shape tools may be a widespread but often overlooked prehistoric technique. Because of their size and very thick cortical walls, proboscidean limb bones are especially well suited to such procedures, and the resulting flakes are sufficiently sharp and durable to be used as butchering implements (Stanford et al. 1981).

There are other processes that detach flakes from bone, and the results can be difficult to distinguish from artificially knapped specimens. For example, the rebound of a bone surface struck with a hammerstone can cause the detachment of a flake that leaves a negative flake scar on the bone, but features indicative of directed force, such as ribs and hackle marks, are normally absent on such scars. Bones tumbled in a stream can be flaked by impacts against rocks, but these will usually be rounded to a noticeable degree. River ice simulation studies suggest that bones can be fractured and flaked by stream action in ways that resemble some kinds of artificial knapping (Thorson and Guthrie 1984). Carnivores can remove flakes from bones but usually leave recognizable tooth marks near the proximal ends of large flakes.

Many bone cores from Old Crow Basin were produced by flaking a platform area across the bone wall and then detaching flakes from the outer surface of the bone by means of impacts on the platform. Such cores must have been rotated 180 degrees about their long axes while maintaining critical angles between platform and core face during flake detachment. Furthermore, some such cores exhibit the kinds of technological problem solving familiar to flint knappers (e.g., removal of hinge fractures, strengthening of the platform by grinding), and it seems more reasonable to interpret them as artifacts than to postulate that natural processes such as river breakup could regularly entail such complex manipulations on an appropriate scale. However, all the possible examples of bone flaking from known stratigraphic contexts in Old Crow valley are open to alternate interpretations that will be outlined below (Table 3).

Table 3. Summary of 30 previously proposed artifacts and probable artifacts given alternate interpretations in this paper (see Morlan 1980:Table 9.1, Plates 6.1-6.17, 7.1-7.8). Unit 2a/2b = Disconformity A.

Unit	*Locality and Cat. No.*	*Plate No.*	*Alternate Interpretations*
"Green-fractured" proboscidean bones			
2a/2b	MlVl-2:5-1, 29-2, 39-1, 42-1, 43-20, 61-1, 79-2, 83-1, 85-1, 102-1, 105-9, 105-11	6.1-6.2	Bones fractured when dry by various natural and/or cultural agencies
2a/2b	MlVl-13:24.3		
Bone and ivory cores and flakes			
2a	MlVl-2:144-38	6.7	Flake detached by river ice, carnivore, or human
2a/2b	MlVl-2:15	6.4	Fractured and flaked by carnivore or human
2a	MlVl-2:131-1	6.8	Flakes detached by thermal stresses or by human
Polished bone and ivory			
2a	MlVl-2:142-37 + 162-8	6.17	Polished by river ice, sub-surface movement, or by human
2a/2b	MkVl-10:28-1	7.7	same as above
2a	MlVl-2:144-21		same as above
Incised bones			
2a/2b	MlVl-2:25-3		Incision is a cut mark but is too small to evaluate
2a/2b	MlVl-2:27-1, 61-13	6.13-6.14	"Incisions" are vascular grooves
2a/2b	MlVl-13:11, 20.1, 21, 26	7.2-7.5	Incisions indistinguishable from cut marks but might have been produced by angular sand grains in frozen sediment
2a/2b	MlVl-2:4-9, 17-7, 75-9	6.9-6.10	Scratches and scrapes could have been made by river ice
2a	MlVl-2:132-8	6.15	"Incision" may be a polished facet produced by movement against an adjacent bone in sediment

Polishing

Less is known about the polishing of bone than about any other category of bone alteration. Ethnographic documentation of bone polishing is very scarce, and very few field observations provide adequate links between pattern and process. Bones become polished during use as hide scrapers but may also be deliberately polished either for the sake of appearance or as a means of enhancing the durability of bone surfaces. Bone fragments licked or partially digested by carnivores may be highly polished, and river ice can produce polished facets on bones that are dragged against bottom sediments (Thorson and Guthrie 1984). There may also be sub-surface mechanisms that polish bone fragments, such as those collected from ice wedge casts (Morlan 1984), but such processes have not been specifically identified. In view of our general ignorance of bone polishing processes, we should be especially cautious of artifact identifications based solely on polishing, and I shall suggest alternate interpretations for all of the polished bone and ivory pieces from known stratigraphic contexts in Old Crow valley.

Incising

The study of surface incisions on bone has been pursued by numerous researchers (Walker and Long 1977; Bunn 1981; Potts and Shipman 1981; Shipman 1981; Binford 1981:44-51). Both the morphology and the placement of incisions on bones provide useful attributes for distinguishing among tooth marks, rootlet etching, vascular grooves, trowel marks, curatorial damage, and stone tool butchering marks. Previously I have described scratches, scrapes, and cuts as separate categories of incision (Morlan 1980), and it is now clear that some kinds of scratches and scrapes can be produced by river ice (Thorson and Guthrie 1984). We do not yet know whether discrete cuts can be made by angular fragments of silica in turbated burial environments. Several distinctly cut bones from known stratigraphic contexts in Old Crow Basin will soon receive further study with electron microscopic techniques to reveal more detailed attributes that might aid interpretation. Several other specimens have been incorrectly identified as artifacts, and their interpretations are revised below (Table 3).

STRATIGRAPHIC CONTEXTS

The great majority of bone, antler, and ivory artifacts from northern Yukon have been found redeposited on the modern banks and bars of Old Crow River, but a few purported artifacts, representing each of the categories discussed above, have been found in stratigraphic contexts for which chronological limits can be given. Nine units were used to summarize the gross stratigraphy of a composite section for Old Crow Basin (Morlan and Matthews 1978; Morlan 1979, 1980), but recent work on exposures farther upstream along Old Crow valley has shown that such detailed numbering cannot be extended to all parts of the basin.

A more general stratigraphic framework can be devised for unconsolidated sediments along Old Crow River, comprising four gross units numbered from bottom to top: (1) a lacustrine silty clay at the base, largely concealed by the river; (2) approximately 18-20 m of sand, silt, and clay representing an interlacustrine sequence; (3) approximately 5 m of glaciolacustrine clay; and (4) silt and peat of variable thickness, usually about 1-3 m, that forms the modern surface. No precise age can be given for Unit 1 or the lower part of Unit 2 (2a). The upper part of Unit 2 (2b) dates to early and middle Wisconsinan time (the Happy and Boutellier Intervals of Hopkins 1982). Unit 3 represents the inundation of Old Crow Basin by classical Wisconsinan glacial meltwater beginning about 25,000 years ago and represents the Duvanny Yar Interval (Hopkins 1982; Hughes 1972; Hughes et al. 1981). The silt and peat of Unit 4 accumulated during the Holocene, which also saw the entrenchment of Old Crow River through the Pleistocene sediments to its present level ca. 40 m below the floor of the late Wisconsinan lake.

Wherever Unit 2 is subdivided, the contact between Units 2a and 2b is marked by a disconformity represented by an erosional contact at some exposures or a gleysol at others. It is not always possible to recognize such a disconformity, especially at exposures in the upstream part of the valley, but along the lower course of Old Crow River a disconformity has been found approximately 5-7 m below the classical Wisconsinan lake at every exposure examined thus far. Correlation of these separate occurrences is greatly aided at five exposures by the presence of Old Crow tephra 30-100 cm below the disconformity which is informally known as Disconformity A.

Five Old Crow Basin exposures have produced fossils interpreted as artifacts or probable artifacts in one or more of the units outlined above. Their gross stratigraphy is outlined in Figure 2, and the specimens in question are summarized in Table 3.

MkVl-10

MkVl-10 exposes an erosional contact between Units 2a and 2b, and this contact can be traced throughout the length of the section as well as along the walls of gullies that dissect it. Unit 2a is largely occupied by the dipping beds of point bars built by an ancient stream system apparently similar in size to the modern river. Near the top of Unit 2a these foreset beds give way to flat-lying topsets that in turn culminate in the erosional contact we have correlated with Disconformity A (see below). We have not been able to find Old Crow tephra at this locality, presumably because it was not preserved as a recognizable layer in these alluvial sediments; possibly the tephra was removed during the construction of the point bar deposits exposed in Unit 2a. Analogous foreset and

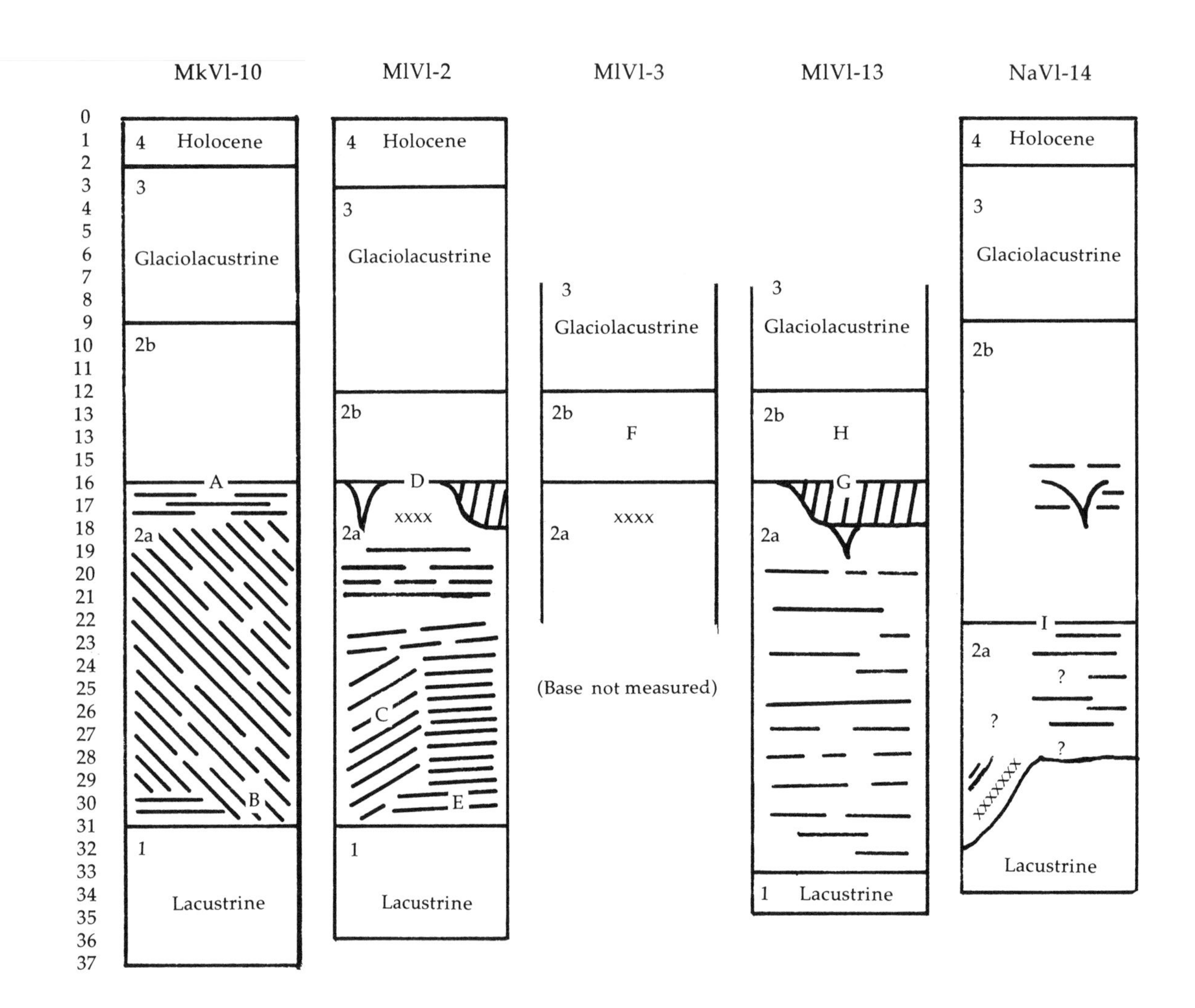

Figure 2. Schematic profiles of five Pleistocene sections in Old Crow Basin. See text for discussion of capital letters. Scale in meters. The river flows from right to left. For MlVl-3 and MlVl-13, where the tops of the sections were not measured, the Unit 2b/3 contact was set level with that at MlVl-2. =tephra; ///=gleysol; V=ice wedge cast; other lines indicate bedding planes.

topset beds can be seen today along Old Crow River where terraces and point bars are exposed in cross-section by progressive migration of meanders. Just as a bone deposited on a modern point bar and another deposited on the nearby floodplain surface are of the same age but at elevations differing by as much as 15 m, so too in the past did alluvial deposition like that seen at MkVl-10 result in burial of like-age specimens at differing depths below the contemporary surface.

We have recovered only three specimens of possible archaeological interest (in addition to surface finds) at MkVl-10, and all were found on Disconformity A (Figure 2A). A proboscidean limb bone fragment (MkVl-10:33) was broken prior to permineralization, but the specimen is too small to reveal whether the fracture occurred before or after desiccation of the bone. On a tusk fragment (MkVl-10:28-1), the "complexity of the polished facet, the occurrence of scraping prior to exfoliation, and the suggestion that the specimen originated as a struck flake prompt me to classify the alterations as probably artificial" (Morlan 1980:225, Plate 7.7); I now believe that this specimen could have been polished and scraped either by river ice or by sub-surface sediment deformation. The double-faceted polished end of a large mammal long bone fragment (MkVl-10:16) might have been either artificially or naturally polished (Morlan 1980:225, Plate 7.6), and river ice could easily produce such faceting (see Thorson and Guthrie 1984).

Several specimens from a stratigraphic context near the base of the bluff (Figure 2B) have been described as artifacts (Jopling et al. 1981), but both the age and the origin of these pieces remain open to alternate interpretations. Three of the four illustrated pieces (Jopling et al. 1981: Figures 12-15) are polished or striated in ways that have been reproduced in river ice simulation experiments (Thorson and Guthrie 1984). The fourth specimen is a fragment of proboscidean bone that appears to have been broken when fresh in that no split line perturbations are apparent along its margins; this is the strongest candidate for artifact status, according to the interpretive framework employed here. A small flake of proboscidean bone is also mentioned but is not described (Jopling et al. 1981:29).

Jopling and his colleagues ascribe the dipping beds at this locality to a pingo that formed after the lake represented by Unit 1 drained and its bed was invaded by permafrost: the artifacts "are inferred to have come from a human campsite, placed near a pingo for both the shelter and the commanding outlook it provided. We may conjecture that the campsite was on the side of the pingo for reasons of drainage, and that the colluviation occurred, at least in part, as a result of a thin vegetative cover having been worn away by foot traffic" (Jopling et al. 1981:30). My colleagues and I have studied this same locality and have concluded that the dipping beds expose the cross-sections of former point bars with the artifacts in question actually having been excavated from an ancient river channel. The dipping beds can be traced upward from the channel bottom to flat-lying topset beds that culminate at the erosional contact we have named Disconformity A. This interpretation supports an age estimate in the Early Wisconsinan time range rather than the late Illinoian suggested by Jopling et al. (1981:18-23).

MlVl-2

MlVl-2 has been extensively discussed (Morlan 1980:Chapter 6), and this summary will reiterate only the archaeologically important specimens. It was here that Disconformity A was first recognized, and this was also the first Old Crow Basin locality to reveal Old Crow tephra, first sampled by O.L. Hughes in Bluefish Basin to the south. Disconformity A is represented by an erosional contact formed by cross-bedded silt and sand resting on blue-gray silty clay at most stations along the exposure, but one small area of Station 2 preserves a gleysol that formed at the surface of the floodplain (N Rutter, personal communication, 1979; C. Tarnocai, personal communication, 1981; Bombin 1980). Portions of Unit 2a are clearly composed of flat-lying fluvial sediments, but some stations may reveal dipping foreset beds like those at MkVl-10 but exposed in longitudinal rather than transverse section so that dip is toward the river (and the excavator). Therefore it is possible that fossils in Unit 2a (Figure 2C) are similar in age to those on Disconformity A despite their vertical separation by 10-12m of sediment (but see below).

Three specimens — a bone flake, a polished ivory piece, and a cut arctic hare radius — have been described as artifacts, and an ivory core(?) and a polished bone have been described as probable artifacts from Unit 2a (Morlan 1980: Plates 6.7, 6.8, 6.15, 6.17). The bone flake (MlVl-2:144-38) is missing its critical proximal end where carnivore tooth marks might have been preserved. The arctic hare radius (MlVl-2:132-8) may not be cut at all but might have acquired its polished, striated concavity through subsurface movement against an adjacent bone. The beautifully polished ivory "flesher" (MlVl-2:142-37+162-8; Morlan 1980: Plate 6.17; Morlan and Cinq-Mars 1982) is the best candidate for artifact status but might have been polished by river ice or a sub-surface process. Such a process might also account for the polished bone previously described as a "probable artifact" (MlVl-2:144-21). I simply do not understand the flaking on a tusk fragment (MlVl-2:131-1), although Semenov (1964:Figure 74-7) has supposedly reconstructed the technique; perhaps freezing and thawing of tusk ivory could also produce this sort of spalling, since it appears similar to the "pot-lids" thought to represent thermal spalling of chert.

Disconformity A (Figure 2D) has produced 13 "green-fractured" proboscidean long bone fragments, two of which were polished after fracture (Morlan 1980: Plates 6.1, 6.2); none of them is large enough to reveal whether the fractures occurred before or after desiccation. Other purported artifacts include three "cut" bones (Morlan 1980: Plates 6.13, 6.14) of which only one (MlVl-2:25-3) appears definitely to have been cut. Improved microscopy and additional reference specimens have shown that the other two "cut" bones (MlVl-2:27-1, 61-13) actually exhibit vascular grooves rather than cuts.

"Probable artifacts" from Disconformity A include a fractured and flaked *Bison* sp. humerus fragment, two scratched bones and one scraped bone (Morlan 1980: Plates 6.4, 6.9, 6.10). The bison humerus (MlVl-2:15) could have been broken and chipped by carnivores, and the scratched and scraped bones (MlVL-2:4-9, 17-7, 75-9) might have been altered by river ice dragging them against bottom sediment.

Jopling and his colleagues mention that seven of 64 specimens from near the base of MlVl-2 (Figure 2E) belong to their "class 2" which is defined as "fragments and whole bones that one might expect to find among camp refuse" (Jopling et al. 1981:24, 29). Without more detailed descriptions we can only note that fragments and whole bones are also found in non-archaeological contexts. My sample, of comparable size, from these deep deposits at MlVl-2 contains no evidence of human activity (see Morlan 1980:Chapter 6, "Basal").

MlVl-3

MlVl-3 is located directly across the Old Crow valley from MlVl-2 within the present course of a tributary called Johnson Creek. A fresh-fractured *Bison* sp. radius was found in Unit 2b, approximately 3.2 m above Disconformity A (Figure 2F; Morlan 1979:138, Figures 6-7). Correlation of the disconformity at this section is aided by the presence of Old Crow tephra only 30 cm below it. Although it is not out of the question that the bison bone was broken by carnivores, its massive size and micro-relief features indicative of dynamic fracture suggest that it was broken by man. The enclosing matrix of organic silt is suggestive of a thaw-lake deposit and yielded a date of >37,000 B.P. (GSC-2792).

MlVl-13

MlVl-13 reveals a simple sequence of flat-lying sediments in Unit 2a, but Disconformity A is only locally recognizable as an erosional contact and is represented at most stations by a complex series of layers among which significant gleysol formation can be recognized (Morlan 1980:Figure 7.1; C. Tarnocai personal communication 1981). Possible artifacts from the gleysol (Figure 2G) include a green-fractured proboscidean long bone fragment, a split and cut large mammal rib, and a butchered (cut) *Bison* sp. innominate (Morlan 1980:Plates 7.4, 7.5). Two additional specimens from the contact between the gleysol and overlying cross-bedded silts and sands include a green-fractured and scraped large mammal long bone fragment and a butchered (cut) *Bison* sp. rib (Morlan 1980:Plates 7.2, 7.3). A green-fractured *Mammuthus* sp. humerus fragment (MlVl-13:47) was recovered from the gleysol in 1980.

The mammoth humerus fragment (MlVl-13:47) is large enough that split line perturbations should have been visible if the bone was broken after desiccation; it was probably broken when fresh and therefore qualifies hypothetically as an artificially broken bone. The cuts and scrapes on the other bones (MlVl-13:11, 20.1, 21, 26) are indistinguishable from those made by stone tools during butchering and defleshing of an animal carcass. These four specimens comprise the most formidable barrier to a global dismissal of our supposed Early Wisconsinan archaeological record from Disconformity A and its equivalents (see Figure 3).[2]

A probable artifact (MlVl-13:10) was found above Disconformity A within Unit 2b, 4.1 m above the surface of the gleysol, at the contact between cross-bedded and massive silts (Figure 2H). It is a proboscidean long bone fragment that occurred as an isolated find and appears to have been shaped by flaking, cutting, and polishing (Morlan 1980:Plate 7.1); alternatively, at least some of the alterations might have been produced by river ice dragging the bone against bottom sediment.

NaVl-14

During the 1981 field season, investigations at NaVl-14 revealed complex and as yet unresolved stratigraphic problems as illustrated at Figure 2I. An erosional contact, similar in appearance to Disconformity A elsewhere, is situated much lower in the section. Flat-lying sediments comprise Unit 2a at most stations, but at the downstream end of the section the unit is apparently occupied by a filled channel at the bottom of which is a thick

2. While this paper was in press, the two cut bones shown in Figure 3 were sent to Dr. Pat Shipman, Johns Hopkins University, for examination under the scanning electron microscope. The marks were examined with reference to a collection of more than 1000 documented marks on bones, and the provenience of the specimens was not made known until after the marks had been identified. The surface of the large mammal long bone fragment (Figure 3, lower) is damaged and difficult to evaluate, but Dr. Shipman positively identified the mark on the Bison rib (Figure 3, upper) as a tool mark. These studies will be reported in more detail elsewhere.

Figure 3. Two incised bones from MlVl-13, Disconformity A. The bison rib (top; MlVl-13: 21) exhibits four incisions of which the enlarged one is 4.1 mm long. The large mammal long bone fragment (bottom; MlVl-13: 20.1) is extensively incised, and the most prominent incision in the enlargement is 9.3 mm long.

and continuous lens of tephra. The channel fill was too deeply covered by partially frozen slump to permit observation of its contents, and the tephra differs from Old Crow tephra in its shard habit (J. Westgate personal communication 1982; it has not yet been dated or characterized chemically). It would appear that the channel removed most of Unit 2a and that the upper portion of the unit, spanning only a meter or less elsewhere, is telescoped into a thickness of nearly 10 m in the channel fill. The disconformity at the top of the channel fill produced a proboscidean bone core (Morlan 1983) as well as several other green-fractured proboscidean bones. However, the bone core does not have a preserved platform and might represent the kind of flaking that can result from the movement of river ice. Such flaking might occur if a bone protruded from frozen sediment below the seasonal high water level or if the bone protruded from an ice block that tumbled against the bank or bottom of the stream. The green-fractured proboscidean bone fragments are generally too small to evaluate with respect to their condition at the time of fracture except to note that they were not yet permineralized.

CHRONOLOGY

In addition to aiding correlation of the sections, Old Crow tephra also provides chronological evidence. Continued analysis has pushed back its maximum age to less than or equal to 120,000 years ago on the basis of fission track dating, and a radiocarbon date of >56,000 years ago from one of several Alaskan exposures of this tephra provides a minimum limit on its age (Westgate, et al. 1983; Naeser et al. 1982). Since the tephra occurs just below Disconformity A, the fission track date provides a maximum age for the latter. A minimum age is given by a radiocarbon date of 41,100±1650 B.P. (GSC-2574) on *Salix* sp. wood from an autochthonous peat 0.6-1.0 m below the Unit 3 clays and 3.0-3.2 m above Disconformity A at MlVl-2. These dates bracket Disconformity A between 41,000 and 120,000 years ago while the upper levels of Unit 2b have produced radiocarbon samples as young as 31,300±640 B.P. (GSC-1191) in Old Crow Basin. Such dates can be readily reconciled with paleoenvironmental evidence that will be reported elsewhere.

Uranium-Thorium Dates

Since Disconformity A is clearly beyond the range of radiocarbon dating, I asked James L. Bischoff, U.S. Geological Survey, if he could attempt to date our fossils by means of the Uranium-Thorium series. He kindly agreed to examine a few specimens on a trial basis and soon found that the uranium content of Old Crow fossils is very low. Dates have been calculated for ten fossils from three localities in Old Crow Basin (Table 4), but only two of these contained enough protactinium to check for concordance with the thorium dates.

Of five bones dated from MlVl-2, four were recovered from Disconformity A and provided dates ranging from 49,000 to 118,000 years ago. The isotope ratios in these samples provided no criteria to reject the dates, and there was no evidence of contamination. All four of the dates fall within the 41,000 to 120,000 year time range projected from radiocarbon and fission track dates above and below the disconformity, respectively. Despite their variation, it is possible that all these dates are correct since the bones were found on an erosional contact where they could represent a lag concentrate.

Obviously the dates lack precision, but they are certainly of the right order of magnitude. In a recent synthesis of the paleoecology of eastern Beringia, Hopkins (1982) states that ice-wedge growth on Disconformity A occurred during the Happy interval either near the end of isotope stage 5 or during stage 4. The thawing of the wedges probably occurred early in the subsequent stage 3 warming, i.e., during the Boutellier interval. The stage 4/5 boundary is dated at about 75,000 years, and stage 3 began 60-65,000 years ago (Andrews and Barry 1978:Figure 2).

All four of the dated bones were fractured prior to permineralization. A *Bison* innominate (MlVl-2:103-3) exhibits extensive gnawing and was probably broken by a carnivore. A *Bison* humerus fragment (MlVl-2:15) has previously been described as a probable artifact on the basis of its fracture patterns (Morlan 1979:Figures 4-5; 1980:Plate 6.4), but it might have been broken by a carnivore. Two proboscidean long bone fragments (MlVl-2:102-1, 105-11) were interpreted as "green-fractured" and therefore artificially fractured (Morlan 1980:Plates 6.2a, 6.2b), but they are too small to exhibit the split line perturbations that would indicate dry-bone fracture possibly by a natural agency.

A bone from Unit 2a at MlVl-2 was the first one dated in this series and seemed at the time to provide an unacceptable result. I had argued that the enclosing sediments represented point bar deposits dipping toward the river and that the sediments might be comparable in age to Disconformity A despite their 12 m vertical separation (Morlan 1980:Chapter 6). During subsequent field seasons we have been unable to confirm (or deny) that interpretation, and our data could permit us to take the date of 163,000 years ago at face value. However, the high content of Th-232 in this bone represents a kind of contamination that could render the resulting age calculation too old. The bone sample (MlVl-2:147) is part of a nearly intact mammoth femur that was associated with a mammoth bone flake and a highly polished tusk fragment previously interpreted as artifacts (Morlan 1980:Plates 6.7, 6.17); alternate interpretations have been suggested above.

The two samples from MlVl-13 provided opportunities to check for concordance between ages based on Th-230 and Pa-231. Finite concordance between these two decay systems was found in the case of MlVl-13:37 but not in the case of MlVl-13:47. We may place highest confidence in the former, but the latter is questionable with the Thorium date more likely correct. MlVl-13:37 is a *Bison* sp. femur shaft fragment from Disconformity A, and its dates of 77,000 and 72,000 years ago fall within the bracketing dates discussed above. MlVl-13:47 is a mammoth humerus fragment from the gleysol preserved on the Disconformity A floodplain, and its thorium date of 106,000 years ago falls within the interval discussed above for Disconformity A at MlVl-2.

Neither of these specimens has been previously described in the literature. The bison femur could have been broken by a carnivore although its surface does not exhibit tooth marks. The mammoth femur is a hypothetical artifact in that it does not exhibit split line perturbations despite its large size and was therefore broken when fresh, presumably by man. The gleysol in which it was found also yielded the bison bones on which probable butchering marks have been observed.

Of three dates from NaVl-14, two have overlapping one-sigma limits with one another and with MlVl-2:147. The third date from NaVl-14 is significantly older and might represent a specimen redeposited from more ancient units. The reference to "Contact A" in Table 4 signifies general

Table 4. Chronometric dates on selected fossils from northern Yukon Territory (see Morlan 1980: Table 9.4 for other radiocarbon dates). All errors=1 sigma; radiocarbon dates calculated with Libby half-life.

Specimen	*Description*	*Age x 10^3 yrs.*	*Position, comments*
Uranium-Thorium (J.L. Bischoff, U.S. Geological Survey)			
MlVl-2:15	*Bison* sp. bone	49+2-2	Disconformity A, good agreement with Unit 2b C-14 dates
MlVl-2:102-1	Proboscidean bone	118+8-7	Disconformity A, overlaps date on subjacent Old Crow tephra
MlVl-2:103-3	*Bison* sp. bone	66+4-3	Disconformity A, good agreement with other dates
MlVl-2:105-11	Proboscidean bone	81+8-7	Disconformity A, good agreement with other dates
MlVl-2:147	Proboscidean bone	163+38-28	Unit 2a, possible Th-232 contamination
MlVl-13:37	*Bison* sp. femur	77±3	Disconformity A, good agreement and concordance (see below)
MlVl-13:47	*Mammuthus* humerus	106±5	Disconformity A, good agreement but discordant (see below)
NaVl-14:2	Proboscidean bone	290+55-35	"Contact A," no criteria to reject
NaVl-14:3	Proboscidean bone	165+12-10	"Contact A," no criteria to reject
NaVl-14:4	Proboscidean bone	183+15-12	"Contact A," no criteria to reject
Uranium-Protactinium (J.L. Bischoff, U.S. Geological Survey)			
MlVl-13:37	*Bison* sp. femur	72+15-12	Disconformity A, good agreement and concordance (see above)
MlVl-13:47	*Mammuthus* humerus	142+62-27	Disconformity A, no agreement and discordant (see above)
Radiocarbon/apatite/proportional counting (Geochron)			
MlVl-1:1	*Rangifer* tibia	27+3-2	GX-1640; from modern river bank; flesher
MlVl-1:2	Proboscidean bone	25.75+1.8-1.5	GX-1568; from modern river bank; transverse core
MlVl-1:3	Proboscidean radius	29.1+3-2	GX-1567; from modern river bank; longitudinal core
Radiocarbon/collagen/proportional counting (Isotopes)			
MkVl-3:9	Proboscidean bone	29.3±1.2	I-11050; from modern gravel bar; fractured when fresh
Radiocarbon/collagen/accelerator (Chalk River)			
MjVi-4:37	Proboscidean tusk	25.17±.63	NMC-1232; 50 cm below Unit 3 clay in Bluefish Basin
MkVl-5:13	Proboscidean bone	30.49±.55	NMC-1235; from modern gravel bar; fractured when fresh
MkVl-26:1	Proboscidean bone	41.46+5.56-3.29	NMC-1219; from modern river bank; transverse core
MlVl-1:143	Proboscidean bone	25.97±.56	NMC-1234; from *Anodonta* phase deposits; longitudinal core
NaVl-7:1	Proboscidean bone	13.335±.39	NMC-1218; from modern gravel bar; longitudinal core
NbVl-2:6	*Rangifer* antler	24.8±.65	NMC-1233; from *Anodonta* phase deposits; polished pestle

similarities to Disconformity A as described elsewhere, but no correlation is intended. "Contact A" is an erosional contact on which vertebrate fossils are concentrated, but no ice wedge casts were seen, and an underlying tephra differs markedly from Old Crow tephra at least with respect to shard habit (J.A. Westgate personal communication 1982). Perhaps fission track dating of this tephra will aid in assessing the Uranium-Thorium dates from NaVl-14. At the present time, there exist no criteria demanding rejection of them.

Figure 4. Proboscidean limb bone core, fractured when fresh, and flaked transversely at the upper right edge of the right-hand view (MkVl-26: 1). A radiocarbon date (41,460+5670-3290; NMC-1219), obtained by counting carbon atoms from the collagen fraction in the Chalk River accelerator, is thought to be spuriously old due to an organic contaminant in the sample.

Radiocarbon Dates

Nine artificially altered specimens have been directly dated by various radiocarbon techniques (Table 4). The three dates based on apatite were obtained a decade ago (Irving and Harington 1973) but were regarded with suspicion because of hazards related to the sub-surface preservation of the apatite component of bone (Hassan et al. 1977; Morlan 1980:261). Furthermore, these three dates, falling between 25,000 and 29,000 years ago, seemed unreasonably young when compared with radiocarbon dates based on organic materials other than bone in Old Crow Basin. The latter group of dates seemed to imply that the northern Yukon basins were already flooded with glacial meltwater by 30,000 years ago, rendering it unlikely that vertebrate fossils, whether artificially modified or not, could be expected to date to the 25,000 to 29,000 year-old period. The date of 25,170±630 B.P. on tusk only 50 cm below the glaciolacustrine clay at MjVi-4 in Bluefish Basin shows that these apatite dates could be correct, but nevertheless doubts concerning the reliability of apatite-based radiocarbon dates must urge caution in adopting them.

The only collagen-based proportional counting date (29,300±1200 B.P.) on an artificially modified bone (a fresh-fractured proboscidean long bone; Morlan 1980:Plates 4.1-4.2) suggests that the specimen was derived from Unit 2b sediments.

The Chalk River accelerator has provided dates on five artifacts from redeposited contexts in Old Crow Basin (Andrews et al. 1980; Brown et al. 1983). The oldest is on a transversely flaked proboscidean bone core (Figure 4; MkVl-26:1; Morlan 1980:Plate 4.16) from the surface of the modern river bank, dated to 41,460+5670-3290 B.P., implying that it may have originated from sediments within Unit 2b; the error on this date is an order of magnitude greater than that of any other date in the series, and evidence of organic contamination was discovered during excessive hydrolysis of this bone (R.M. Brown personal communication, 1982).

Three of these dates fall between 24,000 and 30,000 years ago. MkVl-5:13 is an enormous fresh-fractured proboscidean limb bone fragment (Morlan 1980:Plate 4.6) dated to 30,490±550 B.P. The other two specimens were recovered from *Anodonta* phase deposits (see below) and provided dates of 25,970±560 B.P. on a longitudinally flaked proboscidean bone core with a polished platform remnant (MlVl-1:143; Morlan 1980:Plate 4.8) and 24,800±650 B.P. on a polished caribou antler "pestle" (NbVl-2:6; Harington et al. 1975:48). The one-sigma limits of these dates overlap that of the tusk fragment from MjVi-4, suggesting that these artifacts were eroded from sediments near the top of Unit 2b.

A longitudinally flaked proboscidean bone core (NaVl-7:1; Morlan 1980:Plate 4.11) from a modern river bar was dated to 13,335±390 B.P. This date is consistent with the unusually light-colored stain on the core and implies that bone flaking techniques were still in use in final Wisconsinan time as well as earlier. Furthermore the date is older than the presumed time of final drainage and downcutting of the classical Wisconsinan lake (ca. 12,000 B.P.), suggesting that the lake may have undergone late Wisconsinan fluctuations well below its maximum highstand and permitted colonization of newly exposed areas of its floor near the margins of Old Crow Basin. Perhaps the hiatus imposed by the lake will prove to be of shorter duration than expected.

OTHER STRATIGRAPHIC CONTEXTS

The foregoing discussions suggest that alternate interpretations should be considered for every proposed artifact from Disconformity A and Unit 2a. The best candidates for interpretation as artifacts are four bones from MlVl-13 that appear to have been cut with stone tools. It has been evident for several years that these stratigraphic contexts are older than most of the redeposited fossils that have been directly dated by means of radiocarbon. Sixteen collagen-based radiocarbon dates older than 13,000 years ago, analyzed by proportional counting, have been obtained on bones from Holocene terraces, gravel bars and the modern banks of the Old Crow River (Morlan 1980:Table 9.4). These range between 22,600 B.P. and >42,000 B.P. with 12 of them between 25,000 and 37,000 years old. Only one of these (I-11050, Table 4) was obtained on a bone thought to have been artificially modified (a fresh-fractured proboscidean long bone, MkVl-3:9).

Elsewhere I have argued that these dates reveal "the expected picture of a cessation of large mammal life in the Old Crow basin for a period of at least 10,000 years, and perhaps as much as 16,000 years, during which the glacial meltwater of the Laurentide advance continued to maintain a large lake in the basin" (Morlan 1980:267). I suggested that the lake formed approximately 28,000 years ago and that a few of the dated bones indicated that some animals may have fallen through the ice that would have covered the lake at least during the winter. This line of argument was developed because of an apparent peak in the distribution of dates around 28,000 years ago and because of our failure to find non-osseous organic materials younger than 30,000 years ago in the Old Crow Basin exposures. However, a new date on proboscidean tusk only 50 cm below the lake clays in Bluefish Basin has been obtained from the Chalk River accelerator (Table 4, MjVi-4:37): 25,170±630 B.P. This date shows that the Bluefish and Old Crow basins were not inundated prior to 25,000 years ago and we need attribute only one available date (22,600±600 B.P.; I-3573) to an animal that might have fallen through the ice.

The new date also provides support for an earlier suggestion put forward to explain the apparent lack of organic materials other than bone younger than 30,000 B.P.: "that the vertebrate remains recovered from redeposited contexts in the Old Crow valley preserve a record of radiocarbon ingestion which has not been preserved in the form of plant fossils of the same age" (Morlan 1980:265). In other words, animals were still eating non-woody plants after trees and woody shrubs had been eliminated from the region by climatic changes. Hence all of the radiocarbon dates in Table 4 fall within a reasonable period of time according to the sequence of events now envisioned for the northern Yukon basins.

These dates also imply that the upper part of Unit 2b is probably the source of many redeposited vertebrate fossils and therefore many of the artifacts found on the modern banks and bars. Since 1978, we have searched intensively for vertebrate remains in Unit 2b and have found very few, generally small, isolated fossils. The deposition of Unit 2b seems not to have entailed fluvial processes that would concentrate vertebrate fossils. We have searched for concentrations of vertebrate remains in these sediments on the presumption that they would represent either animal carcasses or archaeological sites. Among the isolated finds from Unit 2b are two specimens that may have been artificially altered (see MlVl-3 and MlVl-13:10, above).

One other stratigraphic context is important for providing minimum ages on its fossil contents. Old Crow River meanders across a valley floor that varies from 1.5 to 3 km in width, and along its winding course it has constructed terraces of several elevations as well as extensive sand and fine gravel bars. Cross-sections of the terraces reveal that some of them rest on Unit 1 clay while others reveal Holocene alluvium to levels well below the surface of the river. Apparently the Unit 1 clay extends toward mid-valley as buried benches, implying that dissection of the river was temporarily arrested by the relatively resistant clay. The articulated, intact valves of a large mollusk, *Anodonta beringiana*, can be found in growth position in alluvium overlying the benches of clay, but in terrace sequences away from the clay benches the remains of *Anodonta* are usually fragmentary. Radiocarbon dates on the shells consistently fall between 10,000 and 11,000 years ago, and it is possible that this mollusk became extinct in the Old Crow drainage after that time. Hence the intact mollusk valves mark a delimited phase in the redeposition history of vertebrate fossils, and objects found in deposits of this "*Anodonta* phase" are probably no less than 10,000 years old.

A number of artifacts have been found in the *Anodonta* phase deposits at four localities (Table 5). They represent all the categories discussed above with reference to Disconformity A, but these specimens are not so susceptible to the kinds of alternate interpretations I have described. For example, the green-fractured proboscidean bone fragments are large enough to show that the fracture fronts were not perturbed by split lines, and two of the cores exhibit platform remnants retouched across the bone wall prior to 180 degree rotation for the detachment of flakes from the outer face of the bone. One core that generally lacks a platform preserves a small remnant on which a polished facet may represent an attempt to strengthen the platform area (MlVl-1:143; Morlan 1980:304). Likewise the five sub-parallel cuts on an innominate fragment are much too long and deep to have been created by the movement of angular quartz grains (Morlan 1980:339), and the polishing and cutting on the two antler specimens are not only complex but are interpretable in terms of artifact functions (Morlan 1980:335). In addition, there are two stone artifacts from these deposits.

Clearly the *Anodonta* phase preserves a small, albeit redeposited, archaeological record with a minimum age of 10,000 B.P. Many of the fossils and some of the artifacts in these deposits were undoubtedly derived from sediments below the classical Wisconsinan lake clay of Unit 3 and are therefore more than 25,000 years old; our first demonstration of this mode of redeposition has come from direct dates on the antler pestle from NbVl-2 and the bone core from MlVl-1 (Table 4, NbVl-2:6, MlVl-1:143).

Only imprecise geochronological limits can be placed on deposits of artifacts and fossils found in other Holocene alluvial contexts, and evidence of chronology must be derived from the identification of extinct forms, from inferences based on the extent and nature of permineralization and staining, and through the direct dating of specimens by means of techniques such as radiocarbon analysis.

Table 5. Artifacts recovered from deposits of the *Anondonta* phase, Old Crow valley, northern Yukon Territory. See Morlan (1980) for table, page and plate references.

Cat. No.	*Description*	*Table*	*Plate*
MlVl-1			
124	Green-fractured proboscidean bone	B1	
142	Transversely flaked bone core	B8-9, 17	4.15
143	Longitudinal core with polished platform remnant (dated to 25,970±560 B.P.)	B2-3, 17	4.8
144	Green-fractured proboscidean bone, scraped	B1, 16	
MlVl-12			
2	Longitudinal core, platform lost to recent fracture	B2-3	
4	Longitudinal core, retouched platform	B6-7	4.12
6	Green-fractured proboscidean bone	B1	
8	Green-fractured proboscidean bone	B1	
9	Green-fractured proboscidean bone	B1	
10	Green-fractured proboscidean bone	B1	
11	Green-fractured proboscidean bone	B1	4.4
15	Green-fractured proboscidean bone	B1	
16	Longitudinal core, platform lost to ancient fracture	B2-3	
36	Green-fractured proboscidean bone	B1	
37	Longitudinal core, retouched platform	B6-7	
NaVk-6			
1	Obsidian biface	8.1	8.1
NbVl-2			
5	Chert flake	8.1	8.1
6	Antler pestle (dated to 24,390±500 B.P.)	p.335	
12	Cut innominate fragment	B19	4.29
15	Antler billet	p.335	4.27

DISCUSSION

In a previous report, I had concluded that approximately 1% of the fossils from the middle of Unit 2a and from Disconformity A at MlVl-2 had been altered in ways that could be interpreted as evidence for human occupation, but that none of those from the base of Unit 2a exhibited such alterations (Morlan 1980:Table 6.22). I attempted to show that a comprehensive taphonomic analysis of fossil assemblages could reveal evidence of human activity at what we might call "trace levels." During the past two field seasons, excavations at other localities in Old Crow Basin have reinforced the pattern observed at MlVl-2 but have begun to cast doubt on the previous interpretation of an artificial origin of some of the alterations. We now have several samples of fossils from the so-called basal deposits just above the Unit 1/2a contact, which dates sometime before the Early Wisconsinan Stadial. All of these bones are devoid of the kinds of alterations that might indicate human activity. On the other hand, additional samples from Disconformity A consistently exhibit the kinds of alterations that have been discussed at length in this paper. Hence the impression is stronger than ever that new processes or agents were introduced into the taphonomic histories of fossil bones in Early Wisconsinan time.

Previously I have argued that the arrival of people would account adequately for the alterations that first appear at the time of Disconformity A, and one purpose of this paper has been the consideration of alternate explanations. If people were not present, the altered bones must be explained by other taphonomic variables. Most of the alternate explanations proposed in this paper involve ice or cryodeformation in some way. It may be noteworthy that the alterations in question appear in association with evidence of a large meandering river flanked by floodplains on which cold climate subaerial weathering produced cryoturbation and ice wedges. This is not to say that meandering streams and cold climate weathering first occur in Old Crow Basin at the time of Disconformity A. However, the bluffs we have examined apparently record active aggradation below Disconformity A, reduced aggradation and floodplain weathering and erosion to form Disconformity A, and relatively slow overbank sedimentation above the disconformity (as well as subsequent glaciolacustrine inundation). Hence both edaphic and climatic changes may have influenced the taphonomic histories of vertebrate remains, and the distinctive bone alterations that might suggest artificial activity are associated with a weathered floodplain and relatively fine-grained overbank sediments. I cannot relate the bone alterations to specific processes in these sedimentary environments, but we should devise experiments and field studies that would reveal the potential for cold climate weathering to alter bones that are entering the fossil record. Likewise we need more studies of the role of river ice in the transport and alteration of bones.

It remains possible that human groups reached eastern Beringia in Early Wisconsinan time, but we have not demonstrated that they did so. The earliest date on an indubitable proboscidean bone core is 41,460 B.P., but this assay is flawed by evidence of organic contamination and by an inordinately large error. Seven other dates, including three based on apatite, indicate that artifacts were made in Old Crow Basin between 24,000 and 30,000 years ago. Thereafter the basin was flooded by glacial meltwater, but human occupation may have continued in adjacent uplands. By at least 15,500 years ago, people utilized the Bluefish Caves, about 65 km south of Old Crow Basin (Cinq-Mars 1979, 1982; Morlan and Cinq-Mars 1982), and one proboscidean bone core from the Old Crow valley has been dated to the period when the Bluefish Caves are known to have been occupied by man (Table 4, NaVl-7:1).

Some writers would apparently ignore or suspend judgement on these directly dated artifacts on the grounds that they were not found in primary archaeological sites (e.g., Griffin 1979; Müller-Beck 1981). In my opinion this is tantamount to ignoring an animal bone merely because it was isolated from the rest of the carcass. Just as a redeposited but directly dated animal bone can indicate the former presence of a species, so can a dated but redeposited artifact tell us when people were present in the area and can even provide an important clue to the technology in use at a given time. As Hopkins (1982) has recently noted, proof of the mere presence of people is of great significance with respect to paleoecology even if other evidence of anthropological importance cannot be obtained.

Furthermore, it is important to distinguish among several kinds and degrees of redeposition. In a few of our excavated samples, especially several concentrations of mammoth bones at the base of Unit 2a, redeposition is probably minimal. More often, the fossils may have been moved a considerable distance but are nonetheless enclosed in sediments of essentially the same age as the fossils; most samples from Disconformity A are in this category. The most famous Old Crow specimens, such as the caribou tibia flesher, are totally divorced from their original sedimentary environments and differ markedly in age from the fluvial contexts in which they were found. These variable degrees of redeposition have significantly complicated the task of reconstructing taphonomic histories, because all pre-burial (perthotaxic) and many primary burial (taphic) alterations have been overprinted or even erased by fluvial redeposition (anataxic) alterations (Morlan 1980:Figure 3.1). In addition, the loss of primary associations among the fossils restricts the analysis to individual bone alterations and renders suspect most considerations of assemblage composition.

Hence our identification of redeposited artifacts must be cautious and give due regard to alternate interpretations, and the purpose of this paper has been to indicate where some of the pitfalls may lie. The cold climate of northern Yukon may entail natural processes that can alter bones in misleading ways, and it also seals the fossils in perennially frozen sediment that has defied our efforts to discover primary archaeological sites of mid-Wisconsinan or older age. Large bone fragments, including the clearly artificial bone cores and flakes, move rapidly downslope as soon as they are brought to the face of a bluff by progressive erosion; the primary archaeological sites from which some of them must have been derived are literally torn apart by gravity and their contents introduced to the modern river. And so, every year we have continued to find more artifacts on the banks and bars of Old Crow River. These artifacts must have originated in archaeological sites of some kind or another: kill sites, habitation sites, perhaps even ceremonial sites. Primary archaeological sites must have existed in Old Crow Basin since at least mid-Wisconsinan time; otherwise artifacts directly dated to more than 25,000 years ago would not be found on the modern banks and bars of the river.

There are approximately 60 Pleistocene exposures along Old Crow River and its tributaries where sediments of relevant age and origin can be examined. We have devoted five field seasons to detailed studies of a dozen of these as well as intensive investigations in Bluefish, Bell and Bonnet Plume Basins (Hughes et al. 1981). Most of our resources have necessarily been devoted to the development of a more detailed stratigraphic framework, to the recovery of samples for dating and paleoenvironmental analysis, and to studies of modern geological and biological phenomena that provide the basis for reconstructing the past by means of uniformitarian principles. Such studies have included careful attention to the processes and agents that alter bones, and similar research is being undertaken in many other areas of the world by numerous investigators. I am confident that we will soon know whether it is feasible to define reliable, universal criteria for the recognition of human alteration of bone in redeposited contexts and to use such criteria in areas where primary archaeological sites cannot be found (see Morlan 1984).

Simultaneously we should continue the search for primary materials at all of the fifty-odd localities along Old Crow River that have not been carefully examined in recent years. To find a primary site in the perennially frozen sediments of Old Crow Basin soon after it is exposed by erosion and before it tumbles down the 20 meter incline to the modern river stands as one of the great challenges of North American archaeology.

ACKNOWLEDGEMENTS

A paper bearing the title of this one was presented to the symposium, "Taphonomic Analysis and Interpretation in North American Pleistocene Archaeology," organized by E.J. Dixon for the 9th Annual Meeting of the Alaska Anthropological Association, Fairbanks, April 1982. The collected papers from the symposium are to be edited by Dixon for *Quaternary Research*. The title of my contribution has been revised as shown under Morlan (1984).

This paper has resulted from extensive reading, discussion with colleagues, and many hours of field and laboratory work supported by the National Museum of Man, Ottawa. I sincerely thank my colleagues in the Yukon Refugium Project for permission to borrow freely from our communal data base: C.R. Harington, O.L. Hughes, J.V. Matthews, Jr., N.W. Rutter, and C.E. Schweger. I am grateful to them and to many other colleagues for discussions of bone alteration: L.R. Binford, R. Bonnichsen, A.L. Bryan, E.J. Dixon, P. Egginton, R.D. Guthrie, G.A. Haynes, D.M. Hopkins, D.L. Johnson, P. Shipman, D. Stanford, A.J. Sutcliffe, R.M. Thorson, and J.V. Wright. Special thanks go to Jacques Cinq-Mars with whom every detail of this paper has been extensively discussed. J.L. Bischoff and C. Miller undertook the time-consuming processing of the Uranium-Thorium samples, and Bischoff devoted a week of his 1981 summer to field observations in Old Crow Basin. The opportunity to date several Old Crow fossils with the Chalk River accelerator was certainly most timely, and I thank R.M. Brown for his patience in answering my questions and processing the samples. Obviously I am solely responsible for any errors of interpretation that remain despite the efforts of these advisers and friends. This is Contribution No. 69 of the Yukon Refugium Project.

REFERENCES CITED

Andrews, H.R., G.C. Ball, R.M. Brown, W.G. Davies, Y. Imahori, and J.C.D. Milton.
1980 Progress in radiocarbon dating with the Chalk River MP tandem accelerator. *Radiocarbon* 22:822-829.

Andrews, J.T., and R.G. Barry
1978 Glacial inception and disintegration during the last glaciation. *Annual Review of the Earth and Planetary Sciences* 6:205-228.

Badone, E.
1980 Neutron activation analysis of fossil bone from Old Crow, northern Yukon Territory. MA thesis, University of Toronto.

Behrensmeyer, A.K.
1978 Taphonomic and ecologic information from bone weathering. *Paleobiology* 4:150-162.

Binford, L.R.
1981 *Bones: ancient men and modern myths.* Academic Press, New York.

Bombin, M.
1980 Early and mid-Wisconsinan paleosols in the Old Crow Basin (Yukon Territory, Canada). *Abstracts of the Sixth American Quaternary Association Conference.* pp. 37-39. Orono.

Bonnichsen, R.
1979 Pleistocene bone technology in the Beringian refugium. *National Museum of Man, Archaeological Survey of Canada, Mercury Series, Paper* No. 89. Ottawa.

Brown, R.M., H.R. Andrews, G.C. Ball, N. Burn, W.G. Davies, Y. Imahori, J.C.D. Milton, and W. Workman
1983 Recent Carbon-14 measurements with the Chalk River Tandem Accelerator. *Radiocarbon* 25:701-710.

Bunn, H.T.
1981 Archaeological evidence for meat-eating by Plio-Pleistocene hominids from Koobi Fora and Olduvai Gorge. *Nature* 219:574-577.

Cinq-Mars, J.
1978 Northern Yukon Research Programme: survey and excavation. *Abstracts of the Fifth American Quaternary Association Conference*, pp. 160-161. Edmonton.

1979 Bluefish Cave I: a Late Pleistocene eastern Beringian cave deposit in the northern Yukon. *Canadian Journal of Archaeology* 3:1-32.

1982 Les grottes du Poisson-Bleu. *Geos* 11:19-21.

Estabrook, B.
1982 Bone age man. *Equinox* 1(2):84-96.

Farquhar, R.M., N. Bregman, E. Badone, and B. Beebe
1978 Element concentrations in fossil bones using neutron activation analysis. Paper presented to the 1978 symposium on Archaeometry and Archaeological Prospecting, Bonn, Germany.

Griffin, J.B.
1979 The origin and dispersal of American Indians in North America. In *The first Americans: origins, affinities, and adaptations,* edited by W.S. Laughlin and A.B. Harper, pp. 43-55. Gustav Fischer, New York.

Guthrie, R.D.
1980 The first Americans?: the elusive arctic bone culture (review of Bonnichsen 1979). *Quarterly Review of Archaeology* 1:2.

Harington, C.R.
1977 *Pleistocene mammals of the Yukon Territory*. Unpublished Ph.D. dissertation, University of Alberta, Edmonton.

Harington, C.R., R. Bonnichsen, and R.E. Morlan
1975 Bones say man lived in Yukon 27,000 years ago. *Canadian Geographical Journal* 91:42-48.

Hassan, A.A., J.D. Termine, and C.V. Haynes
1977 Mineralogical studies on bone apatite and their implications for radiocarbon dating. *Radiocarbon* 19:364-374.

Haynes, C.V.
1971 Time, environment, and early man. *Arctic Anthropology* 8:3-14.

Haynes, G.A.
1983 Frequencies of spiral and green-bone fractures on ungulate limb bones in modern surface assemblages. *American Antiquity* 48:102-114.

Hopkins, D.M.
1982 Aspects of the paleogeography of Beringia during the late Pleistocene. In *Paleoecology of Beringia,* edited by D.M. Hopkins, J.V. Matthews, Jr., C.E. Schweger, and S.B. Young, pp. 3-28. Academic Press, New York.

Hughes, O.L.
1972 Surficial geology of northern Yukon Territory and northwestern District of Mackenzie, Northwest Territories. *Geological Survey of Canada Paper* 69-36.

Hughes, O.L., C.R. Harington, J.A. Janssens, J.V. Matthews, Jr., R.E. Morlan, N.W. Rutter, and C.E. Schweger
1981 Upper Pleistocene stratigraphy, paleoecology, and archaeology of the northern Yukon interior, eastern Beringia, 1. Bonnet Plume Basin. *Arctic* 34:329-365.

Irving, W.N., and J. Cinq-Mars
1974 A tentative archaeological sequence for Old Crow Flats, Yukon Territory. *Arctic Anthropology* 11 (supplement):65-81.

Irving, W.N., and C.R. Harington
1973 Upper Pleistocene radiocarbon-dated artefacts from the northern Yukon. *Science* 179:335-340.

Jopling, A.V., W.N. Irving, and B.F. Beebe
1981 Stratigraphic, sedimentological and faunal evidence for the occurrence of Pre-Sangamonian artefacts in northern Yukon. *Arctic* 34:3-33.

Kurtén, B., and E. Anderson
1980 *Pleistocene mammals of North America.* Columbia University Press, New York.

Maiuri, A.
1958 Pompeii. *Scientific American* 198:68-78.

Morlan, R.E.
1979 A stratigraphic framework for Pleistocene artifacts from Old Crow River, northern Yukon Territory. In *Pre-Llano cultures in the Americas: paradoxes and possibilities,* edited by R.L. Humphrey and D. Stanford, pp. 125-145. Anthropological Society of Washington, Washington, D.C.

1980 Taphonomy and archaeology in the Upper Pleistocene of northern Yukon Territory: a glimpse of the peopling of the New World. *National Museum of Man, Archaeological Survey of Canada, Mercury Series, Paper* No. 94, Ottawa.

1981 Big bones and tiny stones: early evidence from the northern Yukon territory. *Union Internacional de Ciencias Prehistôricas y Protohistoricas, X Congreso, Comisiôn XII:* pp. 1-26. Mexico, D.F.

1983 Pre-Clovis occupation north of the ice sheets. In *Early Man in the New World,* edited by R. Shutler, Jr., pp. 47-63. Sage Publications, Beverly Hills.

1984 Toward the definition of criteria for the recognition of artificial bone alterations. *Quaternary Research* 22:160-171.

Morlan, R.E., and J. Cinq-Mars
1982 Ancient Beringians: human occupations in the Late Pleistocene of Alaska and the Yukon Territory. In *Paleoecology of Beringia,* edited by D.M. Hopkins, J.V. Matthews, Jr., C.E. Schweger, and S.B. Young., pp. 353-381. Academic Press, New York.

Morlan, R.E., and J.V. Matthews, Jr.
1978 New dates for early man. *Geos* Winter 1978:2-5.

Müller-Beck, H.
1981 Review of Morlan (1980). *Science* 213:1241-1242.

Naeser, N.D., J.A. Westgate, O.L. Hughes, and T.L. Péwé
1982 Fission-track ages of late Cenozoic distal tephra beds in the Yukon Territory and Alaska. *Canadian Journal of Earth Sciences* 19:2167-2178.

Potts, R., and P. Shipman
1981 Cutmarks made by stone tools on bones from Olduvai gorge, Tanzania. *Nature* 219:577-580.

Semenov, S.A.
1964 *Prehistoric technology: an experimental study of the oldest tools and artifacts from traces of manufacture and wear.* Barnes and Noble, New York.

Shipman, P.
1981 *Life history of a fossil: an introduction to taphonomy and paleoecology.* Harvard University Press, Cambridge.

Stanford, D.
1982 Experimental taphonomic studies—Ginsberg. Paper presented to the 9th Annual Meeting of the Alaska Anthropological Association, Fairbanks.

Stanford, D., R. Bonnichsen and R.E. Morlan
1981 The Ginsberg experiment: modern and prehistoric evidence of a bone flaking technology. *Science* 212:438-440.

Thorson, R.M., and R.D. Guthrie
1984 River ice as a taphonomic agent: an alternative hypothesis for bone "artifacts." *Quaternary Research* 22:172-188.

Valoch, K.
1980 Knochenartefakte aus dem Micoquien (Schicht 7C) in der Kulna-Höhle im Mährischen Karst. *Acta Musei Moraviae* 65:7-18.

1982 Die Beingeräte von Predmostí in Mähren (Tschechoslowakei). *Anthropologie* 20 :57-69.

Walker, P.L., and J.C. Long
1977 An experimental study of the morphological characteristics of tool marks. *American Antiquity* 42: 605-616. 16.

Westgate, J.A., T.D. Hamilton, and M. Gorton
1983 Old Crow tephra: a new Late Pleistocene stratigraphic marker across Alaska and the Yukon Territory. *Quaternary Research* 19:38-54.

Indications of Pre-Sangamon Humans near Old Crow, Yukon, Canada

W.N. IRVING
Department of Anthropology
University of Toronto
Toronto, Ontario M5S 1A1
CANADA

A.V. JOPLING
Department of Geography
University of Toronto
Toronto, Ontario M5S 1A1
CANADA

B.F. BEEBE
Department of Anthropology
University of Toronto
Toronto, Ontario M5S 1A1
CANADA

Abstract

Excavations at Old Crow Locality 12 have yielded flaked, polished and cut bones of mammoths and other large mammals from an erosional surface interpreted from its stratigraphic position and faunal associations (particularly *Dicrostonyx henseli*, which is thought to have become extinct at the end of the Illinoian) to be pre-Sangamon in age. These deposits underlie 15 m of Sangamonian sediment, near the top of which is the Old Crow Tephra, tentatively dated to 80,000 B.P. The bone and ivory artifacts for the most part show little evidence of transport, although all have been recovered from alluvial or colluvial sediments. The paucity of abrasion or broken specimens supports the hypothesis that the breaking was done by man rather than by natural impact forces.

INTRODUCTION

Yukon Territory, in extreme northwestern Canada, has produced abundant artifactual, radiometric and stratigraphic evidence for the presence of humans before the time of the Late Wisconsinan glacial advance which climaxed at about 18,000 years ago. In this paper we summarize our investigations in the Old Crow Basin, where Recent downcutting has exposed fine-grained, fossiliferous sediments that represent a large part of Upper Pleistocene time. We refer to our 4-unit depositional sequence, and in particular to the occurrence of artifacts near the top of our Unit 1. We will summarize our reasons for thinking that Unit 1 correlates with the Illinoian glaciation.

Discussion is taken from our joint paper of 1981 (Jopling et al.), on data presented by Morlan (1979; this volume) and Westgate (1982), who summarize and interpret radiometric data bearing on chronology, and on our field work in 1981 and 1982 (e.g., Julig et al. 1983). Our treatment of bone artifacts is discussed in Irving and Harington (1973); Bonnichsen (1979); and Jopling et al. (1981); and is the subject of continuing study.

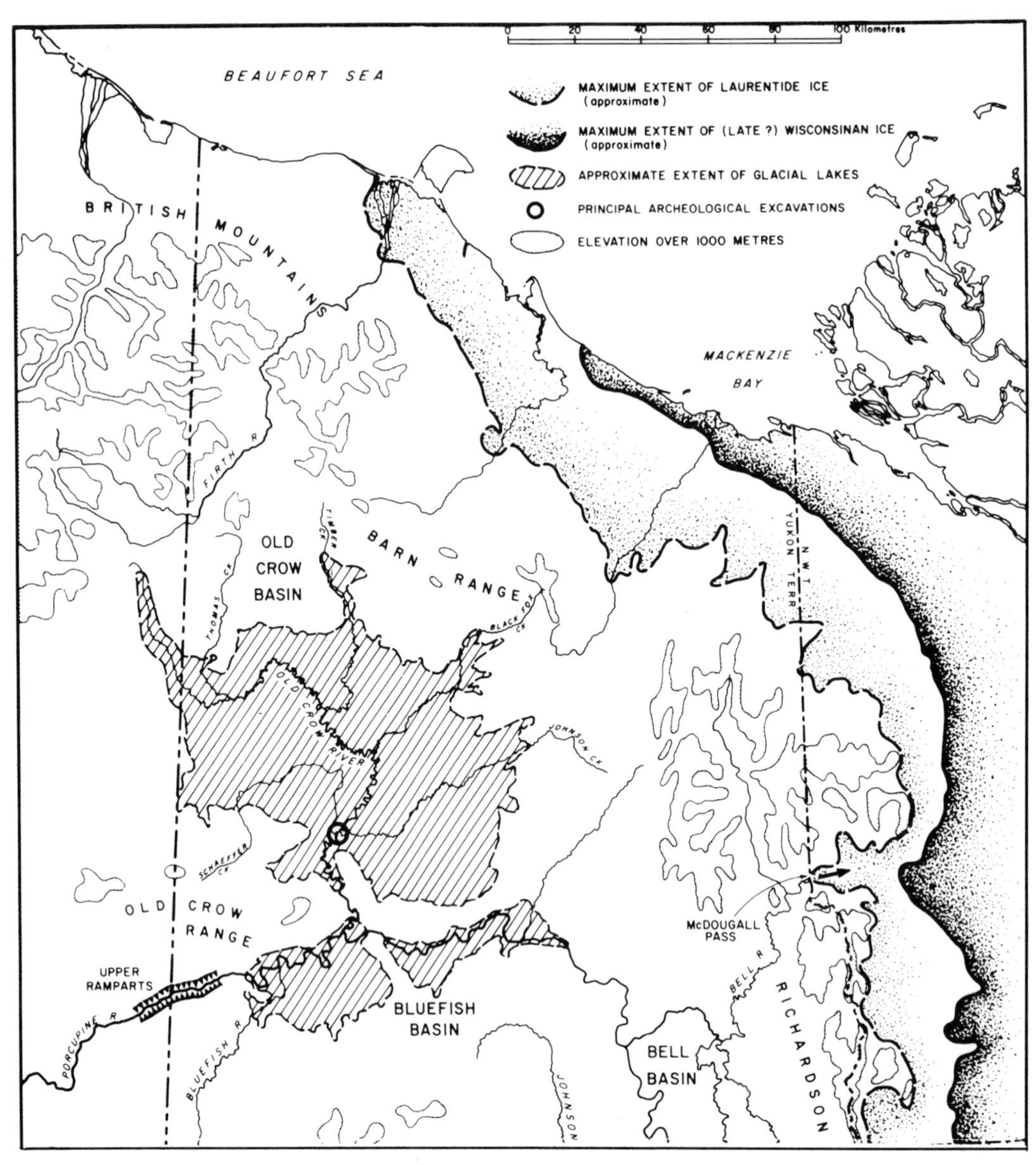

Figure 1. Location map of study area showing Laurentide ice margins and extent of proglacial lakes in Old Crow and Bluefish Basins (from Jopling et al. 1981).

REGIONAL SETTING AND STRATIGRAPHY

The Old Crow Basin is noteworthy both because it has not been touched directly by Pleistocene glacial ice and because it is filled with fine-grained sediments of Quaternary age which are consolidated by permanently frozen conditions. Both of these distinctions have contributed to the preservation of perishable materials dating from Pleistocene times.

The Late Quaternary depositional history of the Basin appears to have been one of aggradation by the ancestral Old Crow River, with intervals of stabilization and possibly shallow down-cutting, until around 10-12,000 years ago, when erosion through 35 m of sediments, down to the present river level, occurred during a very few thousand years. Early in the Holocene, aggradation may have resumed briefly, if so, this was followed soon by renewed downcutting to the early Holocene base level at which the river is now stabilized. This synopsis of late Quaternary sedimentation may be too simple; however it is consistent with all of the presently known facts.

Downcutting during the last 10,000-12,000 years has exposed some 35 m of sediment along the middle and lower course of the Old Crow River. It also has left behind a lag concentrate on dozens of river beaches that includes tens of thousands of vertebrate fossils, almost all of Upper Pleistocene age. Our study, since 1980, has focussed both on the thousands of fossils from Recent deposits, and on the stratigraphy of the deposits from which they came. These deposits all appear to be of Upper Pleistocene or Recent age. The very few specimens that are either older or younger than Upper Pleistocene are of interest in themselves but they bear but indirectly, if at all, on our present interpretation of the chronology of Upper Pleistocene depositional units that we have identified.

We have grouped our field observations of the stratigraphy of the 35 m high bluff exposures along the lower course of the Old Crow River, our principal study area, into four sequential sedimentary units which can be diagrammed thus:

Unit 4: Alluvial, subaerial, small pond
Unit 3: Lake
Unit 2: Alluvial, subaerial, small pond
Unit 1b: Transitional
Unit 1a: Lake

This grouping was based initially on field observations of changes in sedimentation pattern; it has proven useful, but of course it can be supplemented or superseded.

The alternation of deposition with downcutting by the Old Crow River during the Upper Pleistocene appears to have been caused primarily by changes in the Porcupine River, into which the Old Crow drains. Survey data and regional geomorphic studies still are incomplete, but as yet we have seen no evidence of major tectonism having affected the Basin itself in Upper Pleistocene times. Minor, local displacements, however, have been abundant; these are discussed with examples in Jopling et al. (1981). The present base-line control of drainage within the Old Crow Basin is the Palaeozoic bedrock in the Canyon area some 40 km by the river below our study area.

All presently available data point to factors outside the Old Crow Basin as having been responsible for the alternation of depositional environment from deep lakes to shallow, slow rivers, with episodes of stabilization at different levels, and occasional episodes of downcutting. This general pattern was first recognized by Hughes (1969), who proposed a regional synthesis and a correlation of events with the sequence of continental (Laurentide) glaciation east of the Rocky Mountains. Our interpretation of the Old Crow Basin sedimentary sequence fits well with the synthesis proposed by Hughes, if allowances are made for changes required by faunal and chronometric evidence recovered since his initial study.

The regional synthesis proposed by Hughes invokes continental glaciation as a primary cause of changes in drainage of the Porcupine River, which in turn resulted in two episodes during which meltwater proglacial lakes filled the Old Crow Basin and several adjacent basins, to at least 360 m asl (the lower course of the Old Crow River, in our study area, now runs at 268 m asl). These proglacial lakes are represented in the Old Crow Basin by thick deposits of blue clay (our Units 1 and 3), separated by 15-18 m of fine grained alluvium. Two or more sets of high strand lines are visible around the margins of the Old Crow Basin, and elsewhere as well. The correlatives of our Units 1 and 2 are not readily recognized outside the Old Crow Basin, but may nevertheless be present.

INTERPRETATION OF THE STRATIGRAPHIC COLUMN AT LOCALITY 12

The geological history of Locality 12, interpreted as of our 1979 investigations has been outlined by Jopling et al. (1981). Here, we amplify our discussion with the results of work in 1981 and 1982,

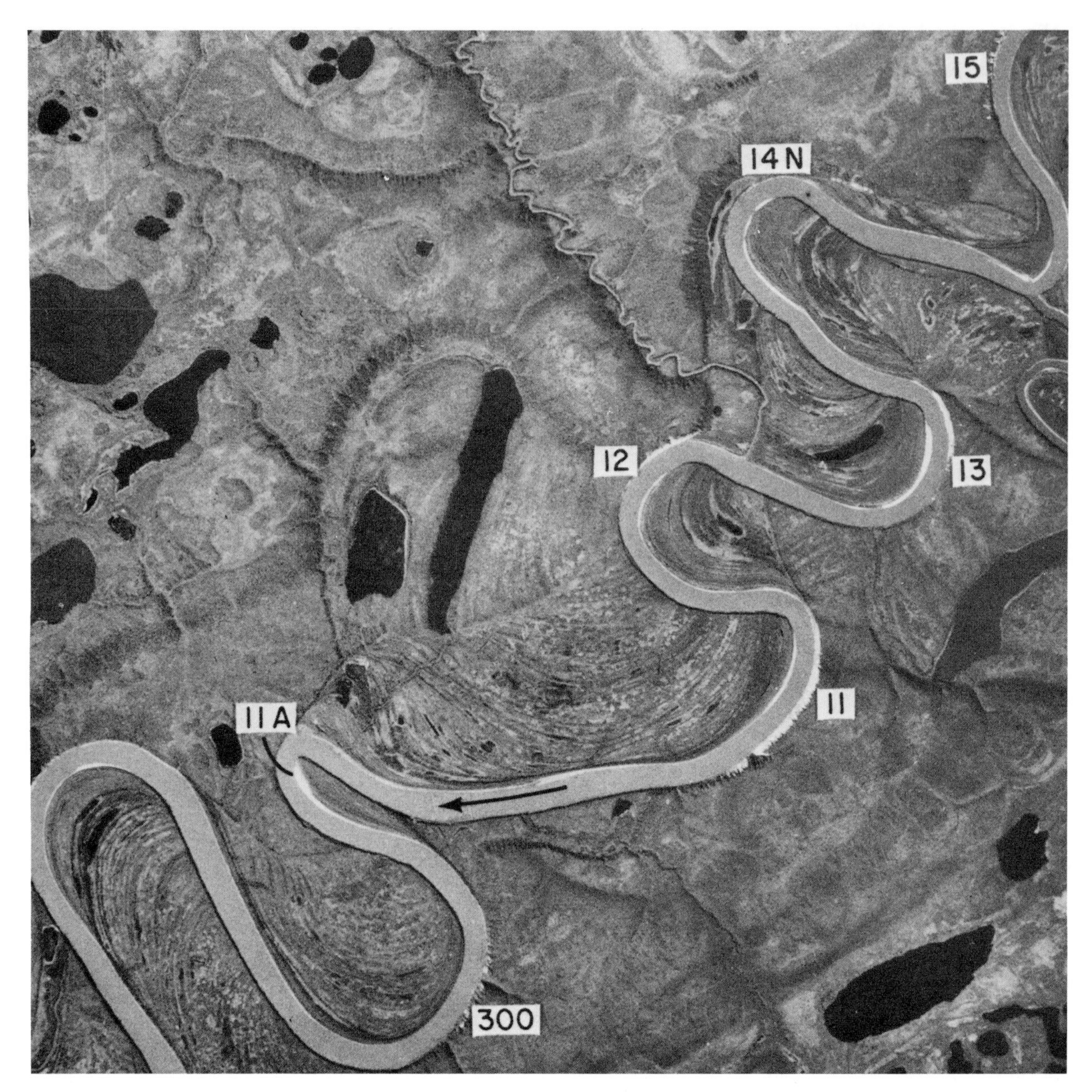

Figure 2. Airphoto enlargement of study area. Note erosional scarps defining sides of the Old Crow Valley (photography by E.M.R., Ottawa).

which confirm our previous findings, and indicate to us that our Unit 1b is stratigraphically beneath all of Unit 2, and therefore older than that deposit. The relevant stratigraphic sections are reproduced in Figure 2 and Figure 3.

According to our interpretation, the lowermost beds excavated in controlled fashion, Unit 1b, belong to the reworked Lower Lake deposits (field terminology used by us and by members of the Bering Refugium Project). Unit 1b was deposited during and immediately after the waning stages of Glacial Lake Old Crow, a large proglacial lake that inundated the Old Crow Basin. We correlate the deepwater phase of the lake (Unit 1a) with an Illinoian glacial advance. The lake might be older, if there is a significant unconformity in the deposits we designate Unit 1b. We see no evidence for such an unconformity

The fine clastics of Unit 1a are predominantly dark blue-grey clays with interbedded silts and occasional sands and fine gravels. Organic material, mostly twigs and plant detritus, is present in minor

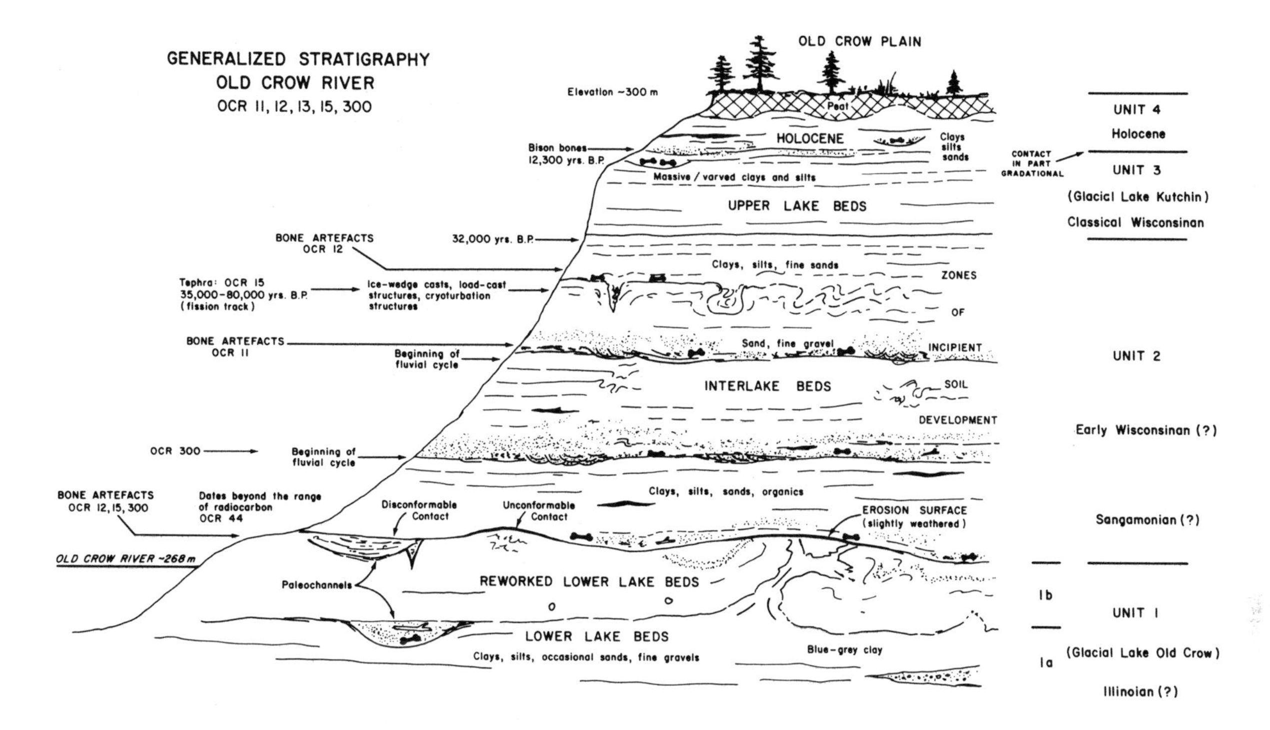

Figure 3. Generalized stratigraphy and suggested correlations in the study area and beyond (from Jopling et al. 1981).

quantities in many parts of the unit. Gravel and sand beaches on mountain sides around the basin rise to some 70 m higher than the top of the unit; this suggests the depth of Glacial Lake Old Crow, but a definite correlation of the beaches with the deepwater deposits of Unit 1a has not been made.

Because of its position near the base of most exposures in our area, Unit 1a is not easily studied. However, "jet drilling" with a light, gasoline driven water pump to a depth of about 14 m has revealed no detectable stratigraphic break, and we think that the massive clays of Unit 1a at Localities 11 and 12 have a thickness of more than 17 m. Near the margins of the Basin, for example, at Locality 300, facies change may account for the presence of coarser materials — silt and sand; however, the effects of local deformation and of alluviation in a potentially more dynamic environment than that of Locality 12 remain to be evaluated.

The deposition and modification of Unit 1b at Localities 12, 13, 15 and 300 was correlative with a terminal phase of Glacial Lake Old Crow, when the lake was drained and erosion products were transported in from the margins of the basin. It appears to have followed Unit 1a without stratigraphic interruption, but the complexity of this rather thin (3 to 5 m) deposit may mask significant stratigraphic events. For the present, we attribute all of the complexity of the deposit to processes related to the disappearance of the former lake, the presence of easily deformable clays, sands and silts, and the re-entry of permafrost. In view of the labile character of the sediments, and the evident presence of mammoth (see faunal list), we must wonder about the effect of the latter upon the former.

In the closing stages of deposition of Unit 1b, the old lake floor was traversed by shallow streams carrying fine to medium sand, silt and clay, together with organic detritus. The channel sands were fairly "clean." Because of the cohesive nature of the lake sediments (mainly clays and silts) and the gentle stream gradients, these streams were only shallowly incised into the underlying material. We surmise that much of the sand was derived from nearby beach lines of Glacial Lake Old Crow as the water receded. Without detailed provenance studies, however, it is difficult to assess the relative contributions from local beach sources and upstream sources.

These sand-bearing streams imported some of the animal bones found near the top of Unit 1b, because on some of these specimens there is unmistakeable rounding, sometimes accompanied by weathering. Such bones are of allochthonous origin; however, we think it unlikely that they have been transported more than a few tens of meters, because the modern Old Crow River moves experimental bones but very short distances, even during times of maximum flood (David Parama, personal communication). Other bones in the same deposit are fresh and angular, which indicates an in-site or autochthonous origin. Local concentrations of both allochthonous and autochthonous bones are found at Locality 12 in deposits resulting from soft-sediment flowage into minor basins in the irregular, emergent landscape.

These small, periodically water-filled depressions served as "sinks" that collected colluvial detritus eroded from the high points squeezed up by soft-sediment flowage or ground-ice growth. The sinks also collected detritus from fluvial activity, perhaps as over-bank deposits which alternated with those of colluvial origin. Concentrations of bones are sometimes embedded in a ferruginized matrix of silt, clay and sand with plant detritus, suggesting accumulation in wet, near-stagnant conditions with appreciable iron concentration in the water. Such conditions can be seen at certain sites along the modern Old Crow River.

These deposits yielding relatively large numbers of fragmented vertebrate fossils, and "artifacts" attributed to humans, are best documented at Locality 12. Here they have been excavated from deposits related to a complex erosion surface that extends along 200 m of the bluff exposure (Figures 4, 5, 6).

Soft-sediment flowage and/or other minor deformational movements continued on into the basal part of Unit 2, the beginning of which is represented at Locality 12 by the onset of aggradation. During the transition from Unit 1b to Unit 2, lag deposits incorporating fossil bone (including artifacts) and small cobbles — the latter most unusual in this area — formed, we infer, as a result of stream action on previously deposited sediments. These basal lag deposits, in the transition between Unit 1b and Unit 2, were found at Locality 12 to truncate, with minor unconformity, a very small channel deposit and an ice-wedge cast, which we think are of terminal Illinoian age. A single unweathered flake of proboscidean bone, considered by us to be of human manufacture, came from the bottom of the

Figure 4. Excavations at Locality 12 in 1982, in Unit 1b. Continuity of stratified deposits on both sides of the gully was confirmed. Most bones and artifacts were recovered earlier from a diamicton to the left of the gully (cf. Figure 5). Excavations to the right of the gully are described in Julig et al. (1983). Most of the relief of buried surfaces is due to deformation: stake to the left of gully is 68 m from upstream datum.

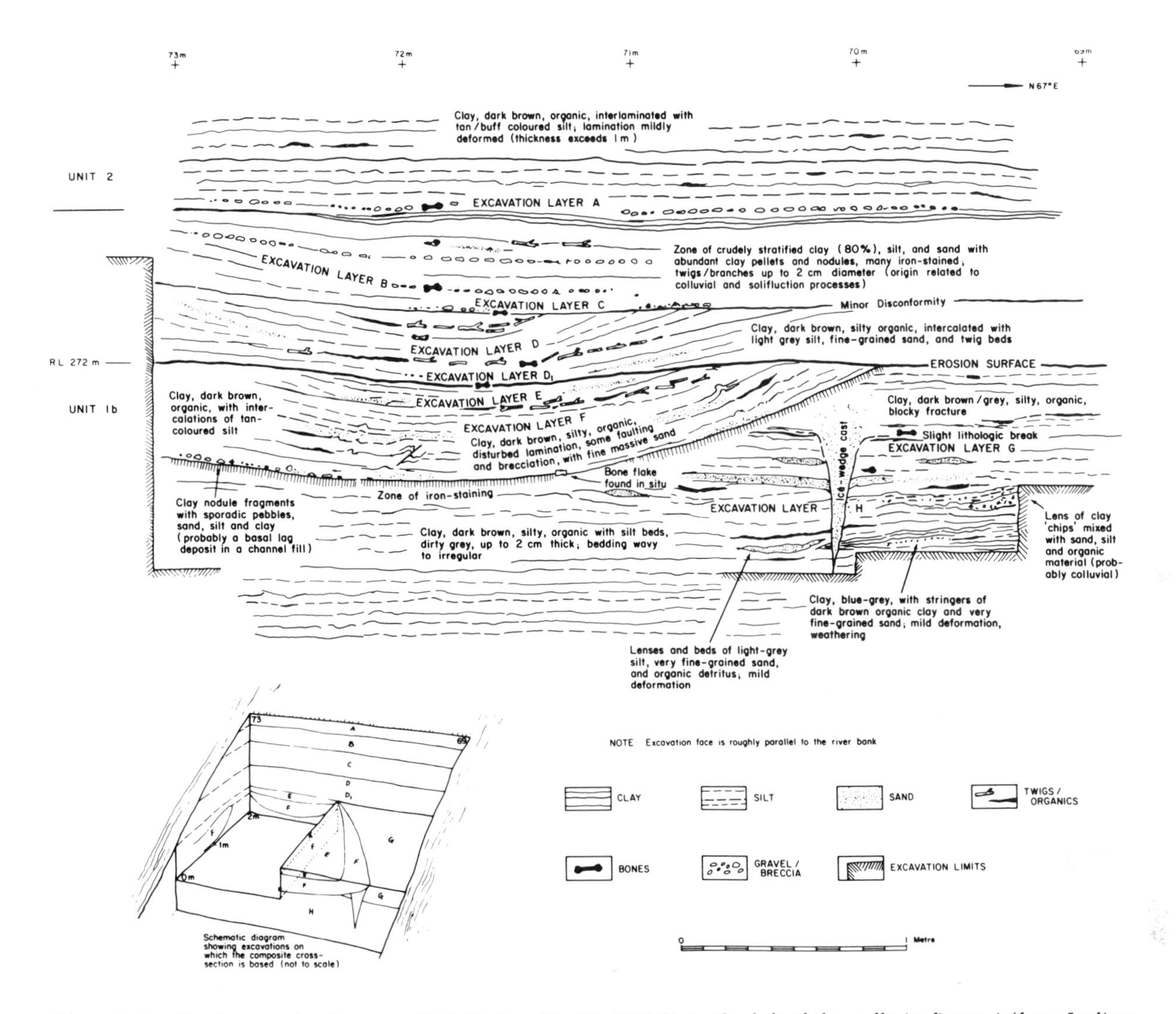

Figure 5. Details of excavation in upper Unit 1b, Locality 12, 1977-79, to the left of the gully in figure 4 (from Jopling et al. 1981).

channel deposit. The flake fits well in the assemblage of bones and artifacts found in the overlying layers — discussed later — and is mentioned here as suggestive of the brief interval of time represented by the minor unconformity near which most of the excavated fossils and artifacts were found.

Several fossil frost wedges and injection structures were found in the lower part of Unit 2, but the main part of this sequence at Locality 12 appears to be devoid of such structures. The absence of frost features may be consistent with the warmer climate of an interglacial period, but it is hardly compelling evidence for such an interpretation.

Frost wedges and thermokarst features occur farther up in the stratigraphic section near an horizon designated by Morlan (this volume) as Disconformity B. This diastem is at an elevation of 290-291 m asl, where it has been studied at Locality 12 — along some 40 m of horizontal exposure directly above the archaeological excavations in Unit 1b and lower Unit 2.

Three or four m below Disconformity B in this same exposure is Disconformity A, at 287 m asl. This minor depositional interruption in places subtends a zone of weathering. On this weathering zone, fragments of vertebrate fossils, mostly quite small, are distributed in apparently random fashion; some of these exhibit polished facets, flaking, and other attributes of human workmanship; they will not be discussed further at this time.

The dating of Disconformity A is discussed effectively by Morlan (this volume) and we have little to add, apart from the additional artifacts just mentioned. We agree that the best estimate that can be given now for its age is in the range of 60,000-80,000 years. This raises the possibility that Disconformity B represents a cold period, perhaps that of an early Wisconsinan glaciation.

Disconformity A exhibits a certain amount of relief, and at Locality 12, between our base-line reference of 45 m, directly above the archaeological excavations, and a point near the downstream end of the bluff at about 190 m on the base line, it drops fairly gradually from 287 m to about 275 m asl — the stratigraphic picture at the downstream end of the bluff is not entirely clear. What is clear, however, is that Disconformity A does not intersect any part of Unit 1, and at the site of archaeological excavations in Unit 1b and lower Unit 2 at Locality 12, Disconformity A in the upper part of Unit 2 is separated from the bottom of Unit 2 by approximately 15 m of alluvial sediment.

In the area between baseline 40 and 80 m, where most of our work has been concentrated and where the accompanying stratigraphic section was taken, in the 15 m of Unit 2 between the top of Unit 1b and Disconformity A there are five or six fluvial sedimentation units. These correspond to different spatial positions of the ancestral Old Crow River. These units, which can be delimited on the basis of strike and dip observations, and on lithological and textural grounds, were deposited during a long interval of time which we suggest correlates with the Sangamon. The long, sloping cross beds traceable along the bluff in Unit 2 at locality 12 represents transverse and oblique sections of point bar deposits laid down by a meandering stream system. At nearby Locality 11, the dimensions of a nicely preserved river cross section in Unit 2 indicate that the principal drainage during Unit 2 was a river approximately the same size as the modern Old Crow.

Figure 6. Excavations at OCR Locality 12, in fossiliferous diamicton — lag, the source of most artifacts recovered in 1982 and previously.

The attribution of that part of Unit 2 below Diastema A to the Sangamon is the most plausible one now apparent, in the light of chronological considerations discussed below. However, it is not the only one possible, as we will point out.

The style of fluvial sedimentation at Locality 12 is uniform over the vertical interval bounded by the base of Unit 2 and Disconformity 1b (272.5 and 291 m asl, respectively). Sedimentation units typically show a fining-upward sequence beginning with very fine-grained sand and grading upwards into interbedded clays and silts, with occasional beds of twigs and other plant detritus. Vertebrate fossils are rare in this zone at Locality 12, but are fairly abundant in the coarse sands of Locality 11 and at Localities 15 and 300. The first sedimentation unit of the Sangamon sequence at Locality 12 is four m thick; the overlying units are thinner. The unusual thickness of the lowest fluvial sedimentation unit may be due to local conditions, for example, downwarping or channel cutting by lateral stream migration.

The relatively fine clastics comprising the stratigraphic column up to the level of Disconformity B connote a low to moderate energy environment for the meander system that prevailed at Locality 12. Representative velocities may have been several tens of cm per second. However, the presence of fine gravel beds in the talweg zone of channel cross-sections at nearby Locality 11 suggest somewhat higher velocities than this in what may be presumed to have been the main channel of basin drainage during Unit 2 times. But with fine gravel the coarsest of the observed deposits, and this fraction being relatively mobile as part of a bed load, we think that the overall evidence favors no more than a moderately dynamic regime for the Sangamon stream.

The sedimentation units mapped in Unit 2 at Locality 12 are separated by minor depositional discordances, only, and not by weathered contacts. The contact designated Disconformity A may be an exception, but even in this case, in which some apparently erosional relief was developed and traces of weathering are seen, there was no interruption of the prevailing Unit 2 pattern of sedimentation. These observations suggest a depositional episode of some stability, and a fairly uniform rate of floodplain (basin-wide) aggradation. Quite possibly the base level of the ancestral Old Crow River was controlled by factors downstream from the present bedrock control at the "Canyon"; however, this is beyond the scope of the present discussion.

The later sedimentary history of the Old Crow Basin appears to be very complex and warrants separate study; apart from the exposure of our sections by recent downcutting, however, the recent geomorphic history of the Basin does not bear on our present discussion. Therefore we will not describe the sediments of Units 3 and 4 in detail, because they are summarized by Morlan (this volume).

ARCHAEOLOGICAL EXCAVATIONS AND STRATIGRAPHY

At OCR Locality 12 we have excavated systematically in Units 1 and 2 during several field seasons, and have recovered artifacts in alluvial deposits that are integral parts of these units, in each season. Excavations in the same stratigraphic units at Localities 11, 15 and 300 also have produced artifacts. This is significant because it shows that the occurrence of artifacts in these stratigraphic positions at Locality 12 is neither unique nor exceptional. However, we have concentrated much of our attention on Locality 12 because the deposits there have been productive and amenable to study; we think, also that they represent the Old Crow Basin stratigraphy, as we now understand it, if one makes allowances for facies changes and special local conditions.

Vertebrate fossils may have been found *in situ* at OCR Locality 12 by Harington as early as 1966, when he designated the locality. In 1967 and again in 1970 Irving's crews recovered fragmented bone from Units 1 and 2 of this exposure, but these were not regarded as being significant (it is worthwhile to record that in 1967, Lazarus Charlie of Old Crow thought that the bone fragments he found near the upstream gully *were* archaeologically significant; Irving and the other crew members did not, so excavation was discontinued). It was not until 1977 that it became apparent that bone artifacts as well as midden-like bone fragments could be excavated from the stratigraphic layer we now call Unit 1b. In 1977, also, we found fragmented bone and bone artifacts at Locality 12 in the upper part of Unit 2, associated with what we now call Diastema A. Our excavations and observations at Diastema A and in Unit 1b were continued in 1978, 1979, 1981, 1982 and 1983.

Our excavations at and near Diastema A have been exploratory, because in 1977 our attention was drawn to the more productive and stratigraphically more interesting deposits in Unit 1b. At Locality 12 we have carried on limited, reconnaissance excavations at Diastema A, which have produced artifacts and fragmented bone which would not appear out of place in Unit 1b, according to our present analytical criteria. It is interesting however, that analysis of bone fragments and artifacts as sedimentary particles shows that those from Diastema A were deposited in a sedimentary environment somewhat different from that which prevailed during the deposition of Unit 1b (Julig et al. 1983). Elsewhere, for example at OCR 11 and 300, our excavations have produced many hundreds of vertebrate fossils and several dozens of bone artifacts, coming from alluvium in Unit 2. It is possible that some of these correlate with Diastema A.

Horizontal and vertical positions have been recorded in relation to a datum point set at the intersection of the modern floodplain with the erosional scarp, at the upstream end of the bluff exposure, about 5 m above the low water mark. The datum point has been described and located by conventional instrumental survey. The "baseline" from this datum is in fact an arc of variable radius, which follows the curve of the river bank. A few durable but impermanent, reference points have survived until 1982.

Our excavations in Unit 1b have extended from 0+38 m to 0+220 m, from the datum point. Only in the 0+38 m-80 m portion of the baseline have concentrations of cobble-size vertebrate fossils been found, and only in the basin-shaped deposits between 0+66 and 0+80 have concentrations of both large and small bone fragments, suggesting an absence of sorting by water, been found. The deposits under discussion all appear to be related to a minor erosion surface, which has an irregular profile along most of the exposed length of Unit 1b. The relief of the surface does not exceed five m, and for a distance of 100 m it is nearly flat and nearly horizontal. In the 66-80 m section defined above, the erosion surface appears to have been deformed to create a small synclinorium in which a little over a meter of predominantly fine sediments accumulated; these contain bone fragments and artifacts, and quantities of plant detritus. Between 66 and 38 m, the erosion surface continues, but different factors have affected it. The surface itself is contorted between 60 and 65 m, and the lag deposit on it is thin — 10 cm thick. The lag deposit in this section includes large bone fragments, only, and small cobbles — the only occurrence of particles of this size and density at Locality 12.

The archaeological deposit in the section between 66 and 80 m has been interpreted as colluvium resting on a slope of structural origin caused by soft-sediment deformation in the presence of permafrost (Jopling et al. 1981). Traces of bedding due to alluviation are evident as stringers of sand and layers of plant detritus. In general, the numerous vertebrate fossils, all of which are fragmentary, occur with the sands, silts and poorly sorted layers taken to be colluvial, rather than in the lighter vegetation mats or the thin layers of silty sand. This deposit apparently has undergone subsequent deformation by tilting on an axis that approximately parallels the baseline. The top of the deposit grades upward into the layered sands and silts of Unit 2, containing progressively less vegetal debris and smaller bone fragments, until both disappear. Beneath the deposit the transition to the contorted sands of Unit Ib is sharp. A very small channel fill was found to be archaeologically sterile, except for a flake of proboscidean bone, which shows the channel fill to be close in time to the overlying colluvium in which bone fragments are numerous.

Beyond the 80 m point, the colluvium, which has become thinner, grades laterally into a thin series of sands resting disconformably on a more or less continuous surface. This surface is archaeologically sterile, so far as is known, from 80 to 220 m.

From about 65 to 55 m the colluvial deposit pinches out and is displaced by a lag resting on an irregular surface. This has been described in some detail (Julig et al. 1983). The lag comprises coarser particles than are found commonly in other parts of Locality 12. These consist of fragments of bone — proboscidean and horse, with a few others — and an assortment of small cobbles. Most of the cobbles are well-rounded; they are lithologically diverse. The bone fragments also are diverse, in as much as some are more weathered than others. However, in terms of flatness, sphericity and rounding they are quite different from the cobbles. In places they form part of a fabric, the orientation of which is consistent with deposition by downslope movement under conditions of permafrost.

ARTIFACTS

It is our intention here to provide only enough descriptive information to document the occurrence of artifacts among the specimens excavated from Unit 1b at Locality 12. Fuller discussion of the artifactual nature of excavated proboscidean and horse bone fragments is given in Jopling et al. (1981) and Julig et al. (1983).

It is important to note that among the many thousands of large vertebrate fossils found along the Old Crow River, it is possible to distinguish evidence for human alteration on a fairly large, but unspecifiable number of them. The reasoning is summarized as follows: the recognition of green-bone fracture, which in general occurs only under very limited conditions, raises the possibility of human activity, but it does not by itself prove anything. Because of their rarity in nature,

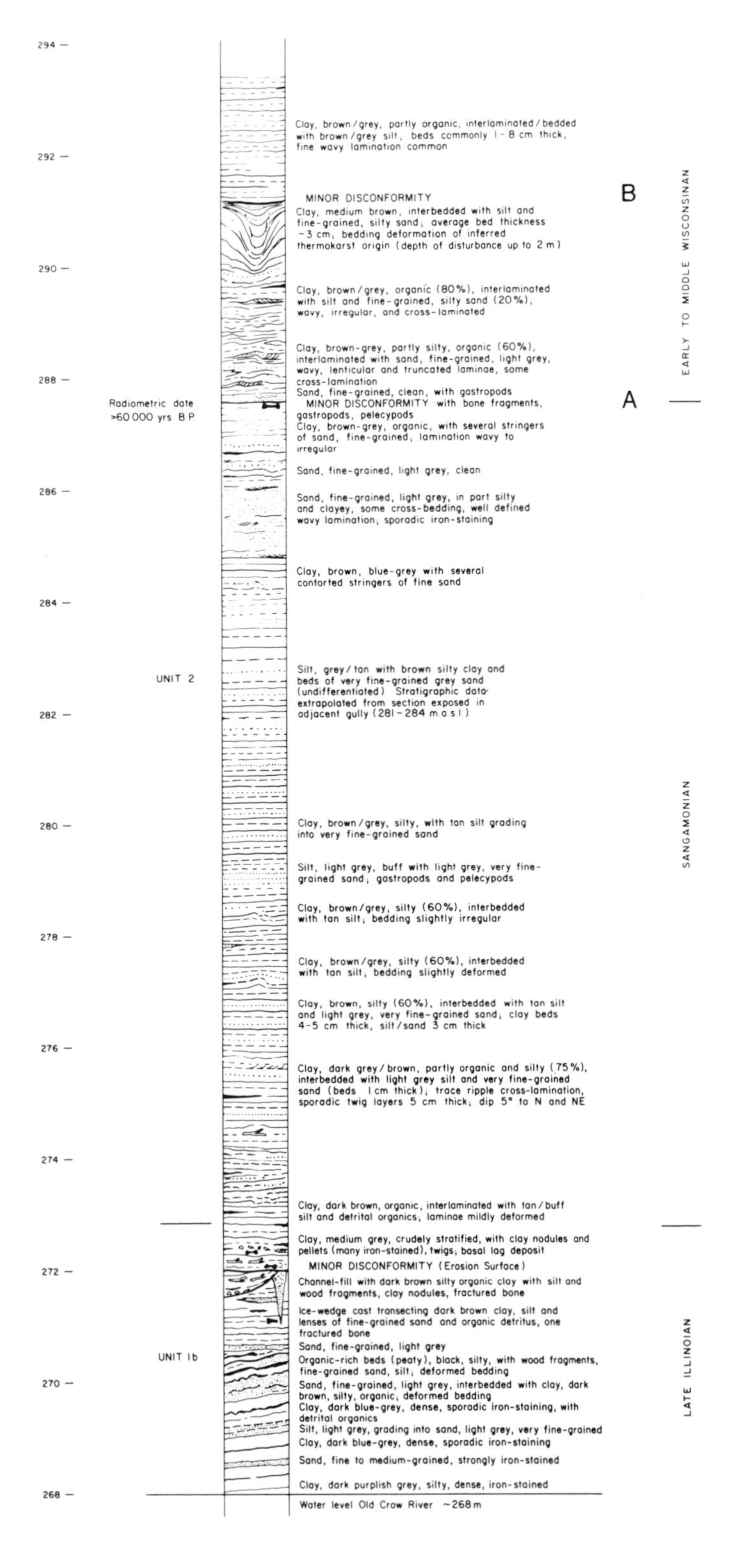

Figure 7. Stratigraphic section at OCR Locality 12, showing relations between Units 1b and 2 (272.8 m asl), and minor disconformities near the top of Unit 2. Section taken between 68 and 80 m downstream from datum.

Figure 8. Ivory flake excavated from Unit 1b.

green-bone fractures, when they occur in large numbers without marks of gnawing by carnivores, are strongly indicative of human involvement, and should not be dismissed or ignored. However, they still are not conclusive evidence for the presence of humans, at least according to our present understanding of this phenomenon.

We think that we can specify two kinds of definite evidence for human alteration that can be recognized on individual specimens. One is the fracture, while fresh, of proboscidean long bones, which we believe can only be done by a human with a hammerstone. The other is evidence of systematic fabrication — an apparently simple judgement, which, however, has often proven troublesome in discussion. The diagnosis of "systematic fabrication": must be made in the light of experience with tools made of stone, and a minor amount of experimentation with bone (e.g., Stanford et al. 1981). Both controlled percussion and controlled abrasion are recognized as diagnostics of human alteration of Old Crow bone specimens.

We have excavated specimens showing each kind of definite evidence for human alteration, from Unit 1b and Diastema A at Locality 12. Examples of these are shown in Figures 6, 7, 8, 9, 10 and 11.

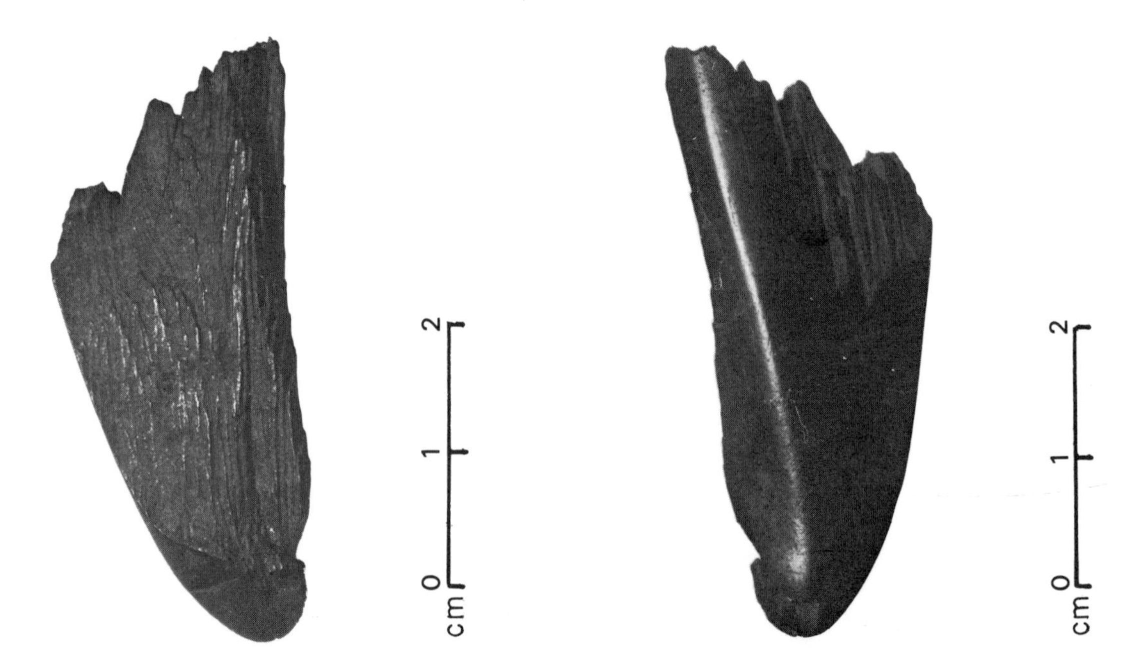

Figures 9 and 10. Carved ivory point fragment, interpreted as a pry-bar for use in disarticulating large mammal skeletons.

CHRONOLOGY

Hughes (1972) suggested that the two lacustrine units in the Old Crow Basin which we designate Units 1 and 3 represent the meltwater of Laurentide glacial ice, which advanced westward to the Richardson Mountains, impounding the drainage of the ancestral Old Crow and Porcupine rivers which formerly passed through the mountains and into the Mackenzie Valley. This hypothesis has not been challenged or superseded. Although it may require modification in the future, we accept it as the best hypothesis now available to account for the presence of these two clay-rich, deepwater deposits in the Old Crow Basin exposures. This hypothesis also accounts for the series of sandy, gravelly beaches that are readily seen on mountainsides around the basin, at elevations up to about 340 m asl.

Most workers in the area agree with Hughes' suggestion that the uppermost of the two glacio-lacustrine deposits — our Unit 3 — is to be correlated with the Late Wisconsinan glacial advance. Hughes' reconnaissance in the Richardson Mountains suggests a date of about 18,000 years B.P. for the maximum extent of ice at this time, a figure that is consistent with other estimates for the age of the Late Wisconsinan glaciation in other parts of North America. This attribution of age, which is entirely plausible from the point of view of reconnaissance geology and terrain analysis, receives support from radiocarbon dates: bison bones from Unit 4 at OCR Locality 11 are a little more than 12,000 years old (Harington 1977), and several dates on plant and vertebrate fossils from the upper part of Unit 2 range upwards from 30,000 B.P. (Morlan, this volume). The clays of Unit 3 are locally banded (varved?) and appear to be of glacial origin; the Late Wisconsinan Laurentide Maximum seems to be the only likely glacial source. This correlation is fundamental to our chronological reasoning.

Our research and publication prior to 1980 assumed that the glacio-lacustrine unit below Unit 3 — our Unit 1 — probably would turn out to be of Early Wisconsinan age, perhaps 50-70,000 years old. Despite suggestions from colleagues that this might not be true, we persisted in this assumption during the excavation of artifacts from the lower part of Unit 2 and the upper part of Unit 1a at Locality 12 in the 1977, 1978, and 1979 field seasons. The first indication to us that these deposits might be significantly older came with the recognition by Beebe of numbers of teeth of the extinct lemming species *Dicrostonyx henseli* taken from artifact-bearing deposits near the transition between Units 1 and 2. This species is thought to have become extinct before the onset of the Wisconsinan glaciation; its occurrence in a stratigraphic position far below the Late Wisconsinan Unit 3, in close proximity to deposits of an older lake, thought also to have been proglacial, led us to the conclusion that the older glaciation was a part of the Illinoian, rather than the Early Wisconsinan. At the same time, the occurrence of horse (*Equus* sp.) and giant beaver (*Castoroides* sp.) in the upper part of Unit 1a makes an age greater than Illinoian unlikely.

Recently, Westgate (1982) has summarized data on the "Old Crow Tephra," a volcanic ash found in many parts of Alaska and Yukon Territory. In the Old Crow Basin it has been identified at Locality 15, two km from locality 12; it occurs there just below the minor disconformity near the top of Unit 2 called Disconformity A (vide supra, and Morlan, this volume). Westgate concludes, on the basis of extensive studies, that the Old Crow Tephra dates to about 80,000 years ago. More detailed studies report the estimated age of the Old Crow Tephra as falling between 60,000 and 120,000 (Westgate et al. 1981).

These pieces of evidence, together with the review presented by Morlan (this volume) seem to us to support an Early Wisconsinan age for the upper part of Unit 2 and an early Sangamon or terminal Illinoian age for the transition between Unit 1b and Unit 2.

Our treatment of the chronology of these deposits cannot be definitive until more radiometric and biostratigraphic data bearing on the age of Unit 1a have been secured. Thus, it must remain uncertain just how long a period is represented by our Unit 1b. The "short scenario" presented above seems to us the most likely one, but a much longer interval seems within the realm of possibility. In any event, all of the evidence available to us shows Unit 1b to be older than all of Unit 2, the upper part of which is most plausibly correlated with an Early Wisconsinan stage. Thus, our attribution of the bottom of Unit 2 and the top of Unit 1b to about the time of the Illinoian-Sangamon transition seems reasonable. It must be borne in mind, however, that we have no local evidence either for the duration of the Illinoian or for the duration of episodes within our Unit 1.

DISCUSSION AND CONCLUSIONS

In our view, the Old Crow River exposures, represented by Locality 12, where we have focussed our description in this paper, provide unambiguous stratigraphic evidence for the occurrence of artifacts in Unit 1b. We see no reason not to believe that all of the materials that we have excavated from Unit 1b are stratigraphically below, and older than, all of the events associated with the deposition of Unit 2.

The foregoing statement leaves two major questions unanswered: what is the nature of evidence for human activity — "where are the artifacts?" and "how old is Unit 1b?" We and our colleagues have addressed these questions in several papers, and we are continuing our research. We can summarize the conclusions drawn from our collective work at this time, and hope that this will be useful.

The existence of a previously undescribed industry with associated debitage and incidental products, in which bone was purposely modified by percussion, is shown by fragments of mammoth bone broken systematically while they were still fresh (e.g., Irving and Harington 1973; Bonnichsen 1979; Stanford et al. 1981). Observation and experiment show that no agency other than an adept man with a stone hammer can be invoked to account for the fracture of fresh proboscidean bones. Furthermore, there are cases of systematic, patterned fracture that resulted in the production of recognizable implements (cores, flakes, and scrapers) (e.g. Bonnichsen 1979; Jopling et al. 1981). These kinds of modification, which exemplify the industry, are best seen in specimens picked from many hundreds of thousands of vertebrate fossils found along the Old Crow River. However, it should not be inferred from this that these examples could have been produced by other natural processes, which might occasionally throw out a "core" or "scraper." If natural processes had produced these results, then the results should be found wherever bone fragments are found. This however, does not happen. Extensive and intensive testing of assertions of this kind is in progress by Haynes (e.g. 1983), and D'Andrea and Gotthardt (1983), and we await their further findings with interest. We think that it is most unlikely that natural forces, including carnivores and other animals, frost activity, and the action of river ice during spring break-up, can account for the fracture of fresh proboscidean bones, or for the systematic modification of these and the bones of other species. A detailed exposition of this point of view is given in the papers mentioned earlier; much further work is in progress.

The question, "how old is Unit 1b?" is intriguing and cannot be answered definitively at this time. Our best estimate places it during the transition from Illinoian to Sangamon time, in round numbers about 150,000 years ago. The reason for saying this is that we are confident that Unit 1b is much older than the Old Crow tephra, which overlies it in Unit 2 and is dated to 80,000 B.P. (90,000-120,000). All of our lithostratigraphic and biostratigraphic evidence is consistent with a Sangamon age for lower Unit 2, and a pre-Sangamon age for Unit 1. The uncertainty arises from the fact that we have no direct indication as to which part of the Illinoian — a very long interval — we are dealing with.

The significance of these conclusions is profound. Man, still and always a tropical animal in physiology, arrived in Beringia — with his family — 150,000 or more years ago. This means that man, tropic-adapted, was meeting and overcoming challenges posed by harsh northern environments, with intelligence and technology, earlier than previously envisioned.

The further implications of these findings for the prehistory of the Western Hemisphere are equally significant, for they show that Man of some sort had gotten to Beringia at a time when *Homo erectus* was the prevailing form in most parts of the World. Indeed, it is beginning to appear more and more likely that Man moved into Beringia from Asia during the great exhange of fauna between the two hemispheres that took place during the Illinoian glaciation, a suggestion made to us by M. Yoshizaki (personal communication, 1977).

REFERENCES CITED

Bonnichsen, R.
1979 Pleistocene bone technology in the Beringian refugium. *Archaeological Survey of Canada, National Museum of Man, Mercury Series Paper* No. 89. Ottawa.

D'Andrea, A.C., and R. Gotthardt
1983 Wolf kills of horses: a study of taphonomy and bone alteration. Paper presented to the 16th Annual Meeting of the Canadian Archaeological Association, Halifax, April.

Harington, C.R.
1977 *Pleistocene mammals of the Yukon Territory*. Unpublished Ph.D. dissertation. Department of Zoology, Univeristy of Alberta, Edmonton.

Haynes, G.
1983 Frequencies of spiral and green-bone fractures on ungulate limb bones in modern surface assemblages. *American Antiquity* 48: 102-114.

Hughes, O. L.
1969 Pleistocene stratigraphy, Porcupine and Old Crow rivers, Yukon Territory. *Geological Survey of Canada, Paper* 69-1. Ottawa.

1972 Surficial geology of northern Yukon Territory and Northwest District of Mackenzie, Northwest Territories. *Geological Survey of Canada Paper* 69-34. Ottawa

Irving, W. N., and C. R. Harington
1973 Upper Pleistocene radiocarbon dated artifacts from the northern Yukon. *Science* 179: 335-340.

Jopling, A. V., W. N. Irving, and B. F. Beebe
1981 Stratigraphic, sedimentological and faunal evidence for the occurrence of pre-Sangamonian artifacts in northern Yukon. *Arctic* 34: 3-33.

Julig, P., A. Jopling, B. Beebe, J. Alcock, C. D'Andrea, and W. Irving
1983 Excavation report on an *in situ* bone assemblage from Locality 12, Old Crow River, Northern Yukon. *Northern Yukon Research Programme Contribution* No. 52. Toronto.

Morlan, R.E.
1979 A stratigraphic framework for Pleistocene artifacts from Old Crow River, northern Yukon Territory. In *Pre-Llano cultures in the Americas: paradoxes and possibilities*, edited by R.L. Humphrey and D. Stanford, pp. 125-145. Anthropological Society of Washington, Washington D.C.

Stanford, D., R. Bonnichsen, and R. Morlan
1981 The Ginsberg experiment: modern and prehistoric evidence of a bone flaking technology. *Science* 212: 438-440.

Westgate, J. A.
1982 Discovery of a large-magnitude, late Pleistocene volcanic eruption in Alaska. *Science* 218: 789-790.

Westgate, J. A., J. V. Matthews, Jr., and T. D. Hamilton
1981 Old Crow tephra: a new Late Pleistocene stratigraphic marker across Alaska and the Yukon Territory. *94th Annual Meeting, Geological Society of America, Abstracts with Programs*, p. 579. Cincinnati.

The Mission Ridge Site and the Texas Street Question

BRIAN REEVES
Department of Archaeology
University of Calgary
Calgary, Alberta T2N 1N4
CANADA

JOHN M.D. POHL
Department of Anthropology
University of California,
Los Angeles, California 90024
U.S.A.

JASON W. SMITH
New Orleans, Louisiana
U.S.A.

Abstract

Archaeological studies in 1977 in the Mission Valley of San Diego included the brief excavation of a surface site on a mid-Pleistocene terrace. Fractured quartzite cobbles, including the "bipolar cores" recognized by George Carter at the Texas Street site, also in Mission Valley, as well as percussed flakes, and other unifacial cobble/core tools were recovered. Fractured quartzite and "bipolar cores" can be found on high Pleistocene mesas, where they apparently result from ongoing thermal stress and spalling of quartzite cobbles. At Mission Ridge, the bipolar objects were fresh and unrolled, and significantly lacked associated spalls, suggesting that the bipolars had been brought in by man. People also manufactured the percussed tools and flakes found in the site. The materials are technologically unrelated to the Holocene San Dieguito or La Jollan complexes, implying that they are of Pleistocene age. At Texas Street, the same assemblage occurs in alluvial fans of Wisconsinan age overlying the San Diego River alluvial fill, which is dated to the last major interglacial (Sangamon).

INTRODUCTION

Were actual artifacts uncovered at Texas Street, and is the site really Last Interglacial in age? The Texas Street question has been the focus of much acrimonious and often bitter debate for more than two decades (Greenman 1957; Haury 1959; Johnson and Miller 1958; Krieger 1958, 1962; Oakley 1959; Patterson 1977; Roosa and Peckman 1954; Simpson 1954; Taylor and Payen 1979; Wormington 1957). Because of the weight of critical "evidence" presented by established archaeologists, the senior author, like most other archaeologists, accepted the positions of the skeptics uncritically, dismissing the sites and the objects as natural phenomena.

While visiting San Diego in 1976 the senior author had the opportunity to view some of George Carter's and Herb Minshall's (a local avocational archaeologist) collections from Texas Street (Carter 1952, 1954, 1957) and nearby Buchanan Canyon (Minshall 1976) in Mission Valley (Figure 1). Among the fractured quartzite cobbles were many objects that appeared to Reeves and R.S. MacNeish (who examined them at the same time) to be culturally-produced, modified and utilized quartzite cobble artifacts.

In 1977, while on sabbatical leave in San Diego, Reeves decided to examine the Texas Street question independently, to study the processes involved in creating the fractured quartzites, the landforms of Mission Valley in which Texas Street is located, and the question of their age. Reeves' objective was to assess the nature and age of the Texas Street materials uncritically; not an easy task under the best of working conditions, which a highly urbanized environment is not.

What remains today of the Texas Street type site (Carter 1952, 1954, 1957) is an alluvial fan overlying a Sangamon alluvial terrace fill. The front of the fan/terrace was destroyed by borrow pitting in the early 1950s. These terrace fans were the characteristic Late Pleistocene landform in the valley. All of the 20 face sections Reeves examined contained fractured quartzites in secondary depositional contexts. The materials were derived from the mesa tops or high terraces above. The latter are Early to Middle Pleistocene marine terraces, characterized by unfractured and some fractured cobbles. Having identified the source, Reeves then directed his field studies to finding an undisturbed area on one of the high river terraces. In Mission Valley, practically all valley fills and terraces of Sangamon-Wisconsinan age have been destroyed by borrow pitting and urban development, precluding any excavations or intensive study other than observation of exposures, which can be a tricky business, what with dogs, interstate and arterial highways and irate developers.

One primary site, Mission Ridge (Figures 1,2,3), was found and hurriedly excavated in September 1977, immediately before its loss to a condominium development. A local archaeologist had already officially cleared the site, and the developer — a laundry corporation based in Miami Beach — was under no further legal obligation. Ten days work was allowed only after we applied certain media and political pressures. The results of this field excavation and subsequent field studies in 1981 will be discussed here.

GEOLOGICAL SETTING

The Mission Ridge site at Mission San Diego was located on a 60 m terrace of Middle Pleistocene Age on the south side of the Mission Valley, 10 km east of the valley mouth (Figures 1-4), where the valley is 2 km wide and entrenched to a depth of 120 m. Above lies the Linda Vista Mesa, which consists of a series of Early Pleistocene marine abrasion platforms, paleontologically dating as early as 1.2 million years (Kennedy 1973). It is from these surfaces that the San Diego River cut its valley following the Mission Valley Fault line. Bedrock units of Eocene and Pliocene age outcrop along slopes. One particularly important member is the Stadium Conglomerate (Kennedy 1975), the primary source of the quartzite cobbles.

A series of mid-Pleistocene river terraces lies within the valley. The Mission Ridge Terrace was 60-70 m high. This surface is very poorly preserved along Mission Valley. It may correlate to the west around San Diego Bay with a marine platform which has an amino acid racemization date of ca. 500,000 years ago (Demeré 1981). Below the mid-Pleistocene terrace is a well-preserved Sangamon or Late Quaternary (Kennedy 1975) river terrace standing at ca. 40 m above the Holocene flood plain. This terrace consists of a basal Sangamon sands and gravel fill which may correlate with the marine Bay Point formation dated 120,000 years ago (Kern 1977). Wisconsinan slope wash and fan deposits containing buried soils overlie the basal fill.

"Artifacts" (fractured quartzite and oxidized clasts) occur both in the back ends of the remnant fans, as at Texas Street, and as "isolates" or "floors" in the very heavily indurated fan sands and soils overlying the Sangamon fill on the terrace fronts. Near the mouth of Mission Valley, the Brown Site, excavated in 1977 by James Moriarity and Herb Minshall (Minshall 1981), is the only accessible terrace of this age left in Mission Valley. Within the fluvial deposits, a well-defined quartzite cobble/core/flake industry was found stratigraphically below a complete sequence from Spanish

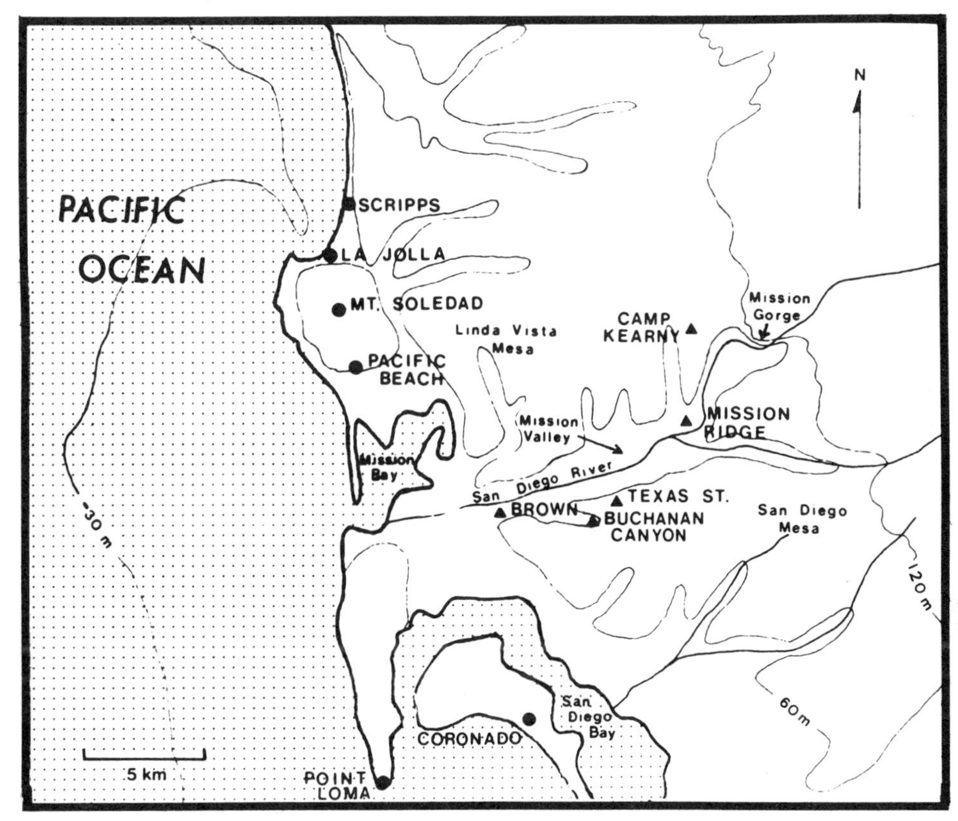

Figure 1. Index map of San Diego area showing Mission Valley sites and collecting locales.

Figure 2. Mission Valley view north of Mission San Diego, Mission Ridge Site on high terrace.

Figure 3. Mission Ridge, site view during excavation and terrain modification for condominium development.

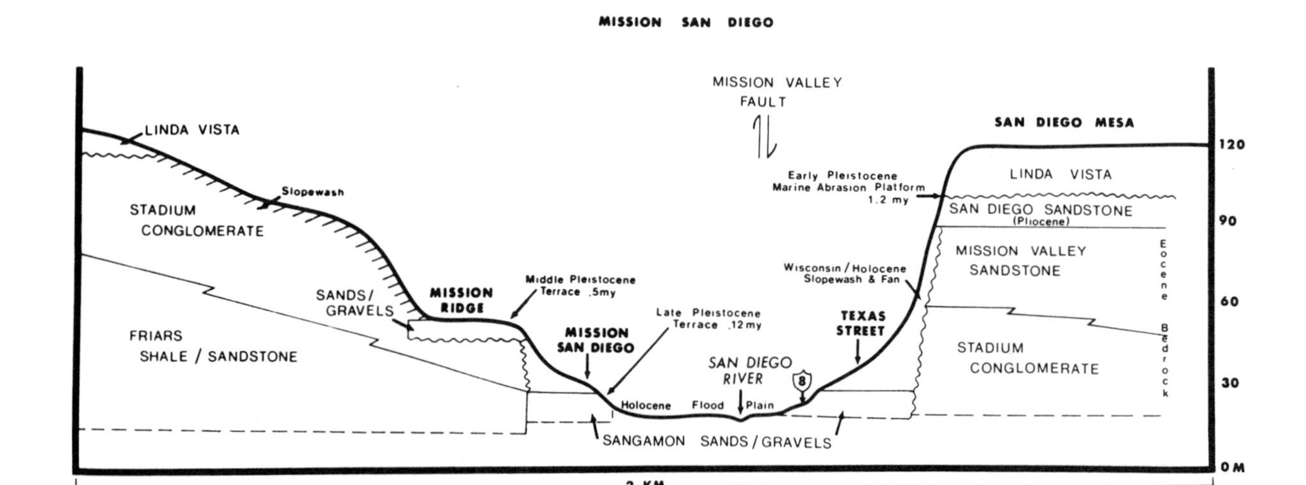

Figure 4. Mission Valley geological cross section at Mission San Diego.

through La Jollan to San Dieguito. Although no radiocarbon dates were obtained on the early deposits, geochronologically the early quartzite core and flake industry must date to early mid-Wisconsinan times (Minshall 1981). As the Brown site had also previously been officially cleared, further excavations at this significant site were not permitted.

MISSION RIDGE EXCAVATIONS

The Mission Ridge site, once part of the Mission San Diego lands, had at the time of discovery in 1977 already been transformed into choice condominium sites along its southern edge (Figure 2). The only excavatable remnant area lay 100 m back from the edge. A roadcut through the terrace sediments exposed percussed quartzite flakes and oxidized igneous rocks in association with what appeared to be a rockfilled "hearth."

The excavations were carried out in September 1977 (Figure 3). A 3x8 m area was excavated utilizing a 2 m horizontal grid. Excavations were all undertaken by volunteers — students, avocational archaeologists, and professionals. Excavations were by hand, utilizing hammers and chisels as the soil was cemented. Rigourous vertical and horizontal controls were maintained and detailed recording and mapping carried out. Excavations were taken to the "B" horizon of the soil ca. 40 cm below surface. All backdirt was screened.

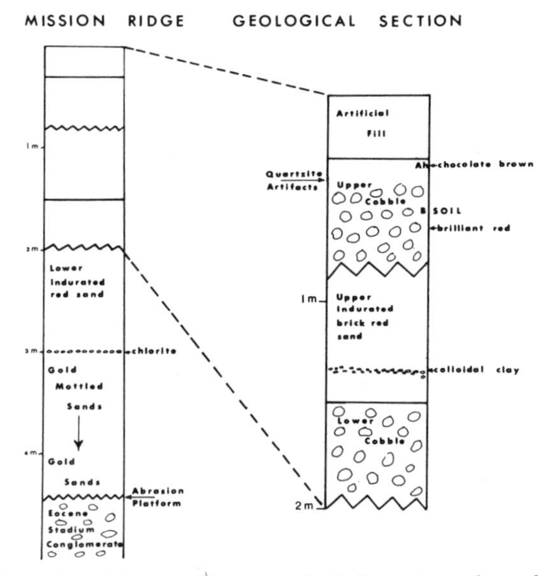

Figure 5. Mission Ridge geological section through mid-Pleistocene terrace.

MISSION RIDGE STRATIGRAPHY (FIGURE 5)

The terrace fill consisted of 4.5 m of fine sands containing two coarse gravel members. These sediments were heavily iron-stained and cemented by water percolation of the iron oxides, which increases towards the surface to form the typical indurated brilliant red soils of the higher terraces and surfaces of the San Diego area. A layer of artificial fill overlies the surface soil. This fill was composed of Pliocene bedrock dumped onto the site from basement excavations. Four stratigraphic units were distinguished in the cutbank adjacent to the site.

Surface Soil and Upper Cobble Layer

The surface soil, 40-50 km thick, consisted of a 10-30 cm thick chocolate brown Ah horizon underlain by a brilliant, heavily indurated, red B horizon, 10-30 cm thick. The soil consisted of weathered gravels and cobbles in a sandy matrix. This unit was a lag alluvial deposit incorporating earlier Pleistocene gravels and cobbles. Many small cobbles were heavily rolled, indicating transport at one time or another in a high velocity fluvial environment. While of fluvial origin, this unit represented a lag surface in which coarser fractions had concentrated through time from the effects of wind, water, chaparral fires and changing climates over the past 500,000 years. The fractured quartzites occurred in the Ah horizon. An erosional surface separated the upper cobble deposit from the underlying sands.

Underlying Sands and Cobbles

The upper sands, 60 cm thick, were brick red in color and heavily indurated. Gravel stringers and caliche zones occurred in the sands, and a 10 cm thick chocolate brown heavy clay layer containing salt crystals. A 10-100 cm thick fill of cobbles lay between the upper and lower sands. The cobbles represented a high velocity channel fill. Its contact with the underlying sands was irregular, indicating erosion of the earliest sands during deposition of the fill. The lower sands, 2.5 m thick, consisted of a 1 m thick upper reddish indurated sand, grading through a gleyed zone into gold colored sands. An illite layer, suggestive of deposition in a brackish water environment, occured at ca. 3 m below surface.

Figure 6. Rolled quartzite cobbles and fragments from Mission Ridge excavations.

Figure 7. Bipolars.

Interpretation

The stratigraphic column outlined above suggests the following sequence of events over the last 500,000 years: 1. Formation of an erosional bedrock surface on the Stadium Conglomerate during mid-Pleistocene times; 2. river deposition of some 2.5 m of sands in a brackish water enviromment; 3. channel scouring by the San Diego River and deposition of 1 m of fine alluvial sands; 4. deposition of a gravel cap on the sands, formation of a lag concentrate on the surface, and formation of a soil.

FRACTURED QUARTZITES

Two hundred quartzite pieces were collected from the excavations. These included rolled (Figure 6) and unrolled, faceted, fractured and unfractured quartzite cobbles. For analysis they were first divided into rolled and unrolled categories. The unrolled category included 96 fractured pieces, which included any quartzite fragment with a sharp edge, whether or not it exhibited evidence of multiple stages of fracturing and spalling. The megaquartzites range in color from white to yellow, grey, pink, dark red, and black. Most were derived from the Stadium Conglomerate. They range in size from elliptical pebbles (3-5 cm long) to large cobbles (20 cm in length).

The fractured cobbles are of various sizes and shapes and are difficult to categorize. They range from relatively regularly shaped ellipsoidal spalled forms with spalls longitudinal to the axis (Figures 7, 15), termed "bipolar blade cores" by Carter to "rectangular" forms (Figure 8) fractured along the "grain" of the rock. These fragments may have spalled before or after transverse fracture. Thin "flakes" or shatter are infrequent (Figure 2). Five flakes (Figure 10) with striking platforms, bulbs and other features demonstrating production by percussion were recovered. For convenience, we will categorize the objects into the following groups.

"Bipolar"/Spalled Quartzites (N=37) (Figure 7)

The bipolar specimens range in length from 4 to 10 cm, are generally ellipsoidal, cylindrical or hemispherical in form, and are characterized by multiple spalled fractured surfaces. Cobble cortex may be present. The spalls tend to be "bidirectional," having spalled from opposite ends on the ellipsoidal specimens. Multidirectional spalls occur on the more spherical or rectangular forms. Remnant cortexes on some consist of a weathered or oxidized rim.

The fractured surface edges are usually sharp and relatively regular in form; the ends are often "pointed." Thin (ca. 1-2 cm thick) "blade like spalls" have often popped off at mid point on the specimens (Figures 7, 15), sometimes parallel to internal bedding planes. It is important to note that the spalls were not found at the site. Evidence indicative of human manufacture in the form of "human" prepared platforms, negative bulbs and other features are absent on the "bipolar" objects. The edges and ends are often battered and step-flaked, a pattern which may represent cultural use.

Quartzite Fragments (N=26) (Figure 8)

This group consists of thick rectangular to irregular forms, 3-13 cm in length, often with a partial cortical surface. They have fractured at one or more points transversely to the major cobble axis. Some are bipolar forms which have subsequently broken (three bipolar specimens could be reconstructed), while others appear to be more spherical or tabular cobbles broken transversely into fragments, sometimes with subsequent longitudinal spalling. Edges are sharp, tending towards obtuse angles.

Split Pebbles/Cobbles (N=8)

Eight specimens consist of roughly rectangular cobbles and pebbles, (4-6 cm in length) which have fractured transversely or obliquely; cortex covers most of the external surface.

"Tabular" Pieces and "Flakes" (N=13) Figure 9)

Two wedge-shaped flakes were recovered (one complete, 9 cm long, 12 mm wide). They are characterized by a thick triangular cross-section and a cortical ventral surface.

Eleven "tabular" specimens, ranging from 3-14 cm in length, were found. Others, rectangular to irregular in form, are fragmented bipolar objects or cobble fragments. Little or no cortex remains, edges are steep (ca. 80°-90°) and edge nibbling is common. common.

Figure 8. Quartzite cobble fragments.

Figure 9. Tabular pieces and flake.

Flakes with Bulbs (N=6) (Figure 10)

Six specimens (35-65 mm long, 44-55 cm wide) had defined biconvex natural fracture planes used as percussion platforms. Bulbs of percussion, erraillures (N=3), impact scars (N=2), and fissures were present; all indicative of percussive removal from quartzite cobbles. Two specimens had cortical dorsal surfaces. Three are step-flaked along the lateral edge.

Figure 10. Percussed flakes.

Tabular "Core"

A 4.5 cm thick, cortical-backed, roughly triangular-shaped quartzite fragment was recovered. It had been split along two bedding planes, the edges of which then appear to have been used as platforms for intentional removal of flakes bidirectionally.

"Horseshoe" Chopper (Figure 11)

A thick quartzite fragment (5x5x5 cm) was recovered. It was characterized by a flat slightly convex ventral surface, a vertical "back" and steep-spalled dorsal surface/face. Edge retouch is present along the "working" edge. This specimen is very similar in size, form and modification to a specimen illustrated by Berger (1980:Figures 4, 5) from the Woolley Mammoth site on Santa Rosa Island.

Figure 11. Horseshoe Chopper, frontal view showing use retouch along working edge.

Edge Battering and Flaking

The edges of a number of specimens exclusive of the percussed flakes are battered, dulled or step-flaked.

Edge nibbling ranges from a relatively regular patterning along an edge 5 cm in length to irregular nibbling of a straight to slightly convex edge. Nibbling is generally unifacial, occurring on the acute angle facet. Some bifacial edging (N=3) has occurred. One elongate specimen (Figures 12, 13) with bifacial nibbling and a constricted "stem" area is very similar in size, form and modification to a specimen illustrated by Berger (1980:Figures 4, 5) from the Woolley Mammoth site.

Concave edges or double notches (N=4) of the kind produced in slope movement through rotation and compression are very rare in comparision to the number of unaltered sharp thin edges in the sample.

Porphyritic Cobbles

Porphyry cobbles were common in the cobble/soil in Mission Ridge. They constitute 90% of the Stadium Conglomerate. The cobbles ranged from highly oxidized weathered and fractured specimens with sharp edges, to unweathered "fresh" specimens. Similar specimens from Linda Vista Mesa and Texas Street had been identified as "fire-cracked rock" by Carter.

One large porphyry cobble appeared to have been "struck" from above, causing the central area to disintegrate and the rock to fracture into two pieces; clearly post-depositional fracture.

Six porphyritic cobbles exhibited "wear" facets or surfaces. These consist of a generally smoothed and slightly polished flat surface, characterized by sets of striations. In contrast, other cobble surfaces are convex, pitted, weathered or irregular. One triangular fragment of broken porphyritic rock had a sharp, step-flaked edge.

Spatial Patterning

Plotting of the fractured quartzite did not show any readily apparent clustering. The excavated floors (10 cm levels) were generally homogenous. While one circular cluster of cracked porphyritic stone was logged, it would be presumptuous to designate this as a hearth as it could be an artifact of the excavators.

The fractured quartzites were carefully examined to determine if spatially dispersed pieces from the same rock could be recognized and fitted together. The attempt was successful in one case. Three pieces of a quartzite fragment, one of which was 100 cm distant from the other two, which were 25 cm apart, were fitted together (Figure 14). Two other pieces from the same rock occur within 20-30 cm.

ORIGIN OF THE FRACTURED QUARTZITES

The fractured quartzite and oxidized porphyries at Mission Ridge are similar to those found on/in other Mid/Late Pleistocene surface and fan deposits throughout the valley. The widespread occurrence of this phenomenon suggested to Reeves in spring 1977 that "oxidized" porphyries had a natural origin. Reeves was also unable to explain the fine spalling characteristics of the so-called "bipolar cores."

The majority of Mission Valley's secondary deposits had their origin on the Linda Vista Mesa, and this was the logical place to look. Fortunately, the abandoned Second World War military Camp Kearny, 3 km north of Mission Ridge, provides a suitable field study locale.

The area contains three erosional surfaces. The Linda Vista Formation is preserved here only as a series of beach ridges containing coarse cobble storm beaches. Below these are two erosional surfaces. One developed on the Stadium Conglomerate is exposed in erosional slopes. A lower surface is developed on the Friars Formation, a shale sandstone.

Surface inspection revealed differential weathering patterns associated with quartzite or porphyritic cobbles. On the Linda Vista Mesa, bipolar quartzites and fractured cobbles are common, as are their spalls and flakes. These tend to be heavily oxidized, cracked, spalled and split, presenting the appearance of fire-cracked rocks.

In contrast, on later Stadium and Friars surfaces, and along gulley sides, bipolars were rare and the porphyries relatively "fresh." Quartzites, eroding out of the gulleys developed in the Stadium Conglomerate, were characterized by large exploding or "spalling" boulders. Here we were able to reconstruct original cobbles by replacing the spalls on their parents, which might lie up to 3 m away. Large "bipolar-like" spalled forms are developing on this surface. In contrast, on the Friars the quartzite cobbles are not "spalled" or exploded. Rather, scattered percussed flakes, deliberately formed cobble choppers and other artifacts occur.

In summary, in our opinion the "bipolar cores" (Figure 15) at Mission Ridge can best be explained at this time as a natural phenomenon — the end product of the continuous spalling over tens of millennia of suitably shaped quartzite cobbles exposed by erosion from the Linda Vista Formation or Stadium Conglomerate outcrops. It is an exfoliation process — the earlier stages of which are observable on erosional slopes where spalls are being flung off variously shaped quartzite cobbles. Most likely, it is the result of physical-chemical stresses built up between the weathering cortex and the unweathered matrix, resulting in a spall eventually being thrown off. Stresses along other fracture planes build up and assist in eventual reduction of the quartzite cobble. Salt crystal formation and chaparral fires have played important if not key roles in spall formation.

The origin of the "bipolar cores" is a complex issue. Both Carter and Minshall consider some to be culturally produced. Some may well be the result of human reduction, because Minshall and others have been able to replicate identical bipolar forms. Probably the category includes both natural

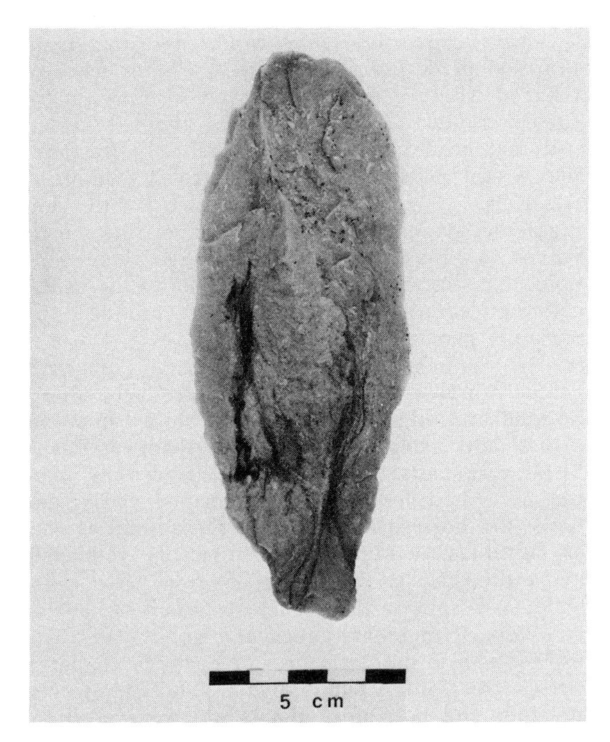

Figure 12. Wedge shaped flake with bifacial edge retouch. Dorsal view.

Figure 13. Wedge-shaped flake. Ventral view.

Figure 14. Reconstructed exploded quartzite fragment. Pieces 25 cm-100 cm apart found during excavations.

Figure 15. Bipolar and spall, when found spall was still attached. Fracture subsequently occurred along initial stress plane.

and man-made forms, and it can be exceedingly difficult to distinguish them.

A similar phenomenon occurs with the porphyries. Through oxidation of the surfaces, stresses are set up and eventual spalling and splitting of the cobbles occurs. Seismicity may also play a role in the Mission Valley formational process, prestressing cobbles imbedded in cemented bedrock.

Spalling and splitting of quartzites is probably a relatively common and widespread natural phenomenon where suitable environmental conditions and source rock exists. Since 1977, Reeves has observed spalled cobbles in a number of locales, including those broken up by historically recent fires in the Rocky Mountains. Reeves has found the spalls on terrain in the Rockies and High Arctic in definitely non-cultural environments (e.g., steep-sided slopes of Middle or Early Pleistocene age and isolated mountain tops). Time is clearly a major factor in their formation. In contrast to those in Mission Valley, the bipolars in these isolated locales tend to vary in size and be irregular, reflecting natural variations between the constituent cobble conglomerates.

HUMAN INTERACTION

Deliberately formed unifacial cobble tools, percussed flakes and cores with negative flake scars do, however, occur on the Linda Vista surface, thereby suggesting their use by man. At some period in the past people used the sites as quarry sources, both for quartzite cobbles and the useful naturally fractured sharp-edged pieces.

The bulk of the fractured quartzites recovered from Mission Ridge were, in our opinion, naturally broken but collected elsewhere and brought to the site by man for use use primarily as ready-made expediency tools. The bipolar objects did not spall in place at Mission Ridge, as no spalls were found. These spalls occur on the Linda Vista Mesa. If they had spalled naturally at the site, the spalls would be present. If the objects had been introduced by fluvial activity, we would not expect sharp edges, which are characteristic. Spalled, heavily-abraded bipolar quartzites were present (Figure 6); however, they are easily distinguishable from unabraded objects (Figure 7).

Suggestive evidence of human intervention, we think, is the situation where quartzite fragments from the same cobble were recovered locally dispersed (Figure 14). We find it difficult to perceive of geological or biological processes other than human behavior which would result in lateral transport more than 100 cm in a non-depositional environment within a highly cemented soil.

More conclusively, we would note that fresh, unrolled percussed flakes with eraillures were recovered. There is no natural mechanism to account for the production of these flakes, given their depositional environment. To be naturally fractured, one would require a very steep stream gradient or waterfall. (High velocity streams with low gradients, even though coarse channel fills are involved, will not produce flakes). Also, we would note that according to François Bordes (personal communication), eraillures are not produced on naturally percussed materials.

In summary, the Mission Ridge quartzite cobble complex includes naturally produced sharp-pointed and edged bipolar cores, blocky quartzite pieces and irregular-shaped sharp-edged flakes. These fragments were not only utilized by man, but also modified into more formed flakes and tools (the horseshoe chopper, for example) as well as culturally manufactured, unifacially retouched and utilized flakes.

The porphyries are easily explained as naturally oxidized fragmented rock and not the result of use by man in domestic fires or burnt in chaparral fires. This interpretation does not, however, preclude the possibility that some were used for hot rock roasting, steaming or stone boiling.

DATING

Mission Ridge is not directly datable. The original surface is most probably mid-Pleistocene in age, around 500,000 years old. The quartzite assemblage occurred in the upper 30 cm along with a piece of Blue Willow, a .22 shell and a Coca-Cola cap.

The quartzite cobble assemblage is not technologically relatable to either the San Dieguito or the La Jollan cultural complexes of Holocene age. The latter have quite different technologies, and San Dieguito does not utilize quartzite to any extent. San Dieguito, representing post-Pleistocene hunters, is characterized by an evolved bifacial technology utilizing felsite; the La Jollans were littoral peoples utilizing a basic split cobble/spall and flake technology.

Although it cannot be demonstrated at this time, by the process of elimination the fractured quartzite industry evidently was Late Pleistocene in age. It evidently predates the Brown site, which on the basis of geological and pedological evidence, dates to early mid-Wisconsinan times (Minshall 1981). The early Brown site assemblage is a quartzite cobble core and flake industry containing well-formed tools and specific tool types, including choppers and chopping tools. Redeposited bipolar cores occur in gravel channel fill at the base of the fill in the Brown site.

Although functional differences exist between Brown and Mission Ridge, the Mission Ridge assemblage is, we think, technologically the precursor of the Brown assemblage, dating (as George Carter suggested in 1957) to Sangamon or earlier times. This fractured quartzite industry (essentially the same as the Texas Street industry) occurs in alluvial fans (as at Texas Street) overlying Sangamon river sands and gravels.

CONCLUSION

The fractured quartzite complex, as first claimed by Carter, is part of a Middle to Late Pleistocene quartzite cobble core/unifacial flake tradition of Pacific coastal-adapted peoples. In the New World, this long-standing coastal adaptation, utilizing simple cobble and flake tools, extended into Holocene times. In contrast, cultures represented by tools and technologies adapted to terrestrial hunting of large game existed in the interior, where early unifacial blade and flake industries (e.g., Calico, Valsequillo) later evolved into Late Pleistocene bifacial industries (China Lake, San Dieguito I, Manix Lake).

In closing, we would note that more than three decades ago George Carter first proposed the Texas Street Industry and its probable age. Had Carter's claims been taken seriously enough by professional archaeologists to undertake detailed field studies instead of simply dismissing them, we would have had a major body of data on Late Pleistocene North American coastal settlement in the tectonically emergent San Diego area. Unfortunately, this research opportunity is now essentially precluded by subsequent intensive urban development.

REFERENCES CITED

Berger, R.
1980 Early man on Santa Rosa Island. In *The California Islands*, pp. 73-78. Santa Barbara Musuem.

Carter, G.F.
1952 Interglacial artifacts from the San Diego area. *Southwestern Journal of Anthropology* 8:444-456.

1954 An interglacial site at San Diego, California. *The Masterkey* 28(5):164-175.

1957 *Pleistocene man at San Diego.* John Hopkins Press, Baltimore.

Demeré, T.A.
1981 Review of age and stratigraphy of Marine Pliocene and Pleistocene deposits in and around the San Diego metropolitan area. San Diego Museum of Natural History, San Diego. Manuscript in possession of author.

Greenman, E.F.
1957 An American Eolithic? *American Antiquity* 22:298.

Haury, E.W.
1959 Review of *Pleistocene man at San Diego. American Journal of Archaeology* 63:116-117.

Johnson, F., and J.P. Miller
1958 Review of *Pleistocene man at San Diego. American Antiquity* 24:206-210.

Kennedy, G.L.
1973 Early Pleistocene invertebrate faunule from Linda Vista Formation, San Diego, California. *San Diego Society of Natural History, Memoir* 21:72.

1975 Geology of the Del Mar, La Jolla and Point Loma quadrangles, San Diego metropolitan area, San Diego County, California. *California Division of Mines and Geology, Bulletin* 200A.

Kern, J.P.
1977 Origin and history of Upper Pleistocene marine terraces, San Diego, California. *Geological Society of America, Bulletin* 88:1553-1566.

Krieger, A.D.
1958 Review of *Pleistocene man at San Diego. American Anthropologist* 60:974-979.

1962 The earliest cultures of the western United States. *American Antiquity* 28:138-143.

Minshall, H.L.
1976 *The broken stones: the case for early man in southern California.* Copley Press, San Diego.

1981 The geomorphology and antiquity of the Charles H. Brown archaeological site at San Diego, California. *Pacific Coast Archaeological Quaterly* 12:39-57.

Oakley, K.P.
1959 Review of *Pleistocene man at San Diego. Man* 59:183.

Patterson, L.W.
1977 Comments on Texas Street lithic artifacts. *Anthropological Journal of Canada* 15:15-25.

Roosa, W.B., and S.L. Peckman
1954 Notes on Third Interglacial artifacts. *American Antiquity* 19:280-281.

Simpson, R.D.
1954 A friendly critic visits Texas Street. *The Masterkey* 28(5):174-176.

Taylor, R.F., and L.A. Payen
1979 The role of archaeometry in American archaeology: approaches to the evaluation of the antiquity of *Homo sapiens* in California. In *Advances in archaeological method and theory* 2:239-283. Academic Press, New York.

Wormington, H.M.
1957 Ancient man in North America. *Denver Museum of Natural History, Popular Series* No. 4.

Geoarchaeology at China Lake, California

EMMA LOU DAVIS
Great Basin Foundation
1236 Concord Street
San Diego, California 92106
U.S.A.

Abstract

The Late Pleistocene lake basins of the Great Basin must have provided some of the most equable paleoenvironmental conditions for occupation by early Americans. Several early sites have been reported in the literature, but only those dating after about 11,000 B.P. have been generally accepted by skeptical archaeologists. Suggestions are presented as to how to search for early sites in this desert region where most sites lay exposed by erosion.

Intensive foot and low altitude aerial surveys can yield a great deal of useful information. An example is given of the Basalt Ridge mammoth site found by intensive survey in the Pleistocene China Lake Basin in the Mojave Desert. Two flakes were found associated with a mammoth tooth, which produced a uranium series date of 42,350±3300 B.P.

INTRODUCTION

Solutions to the complex problems of how and when early people came to the Americas have and will depend on two contending models which have led to extreme polemic positions.

1. No humans entered the hemisphere until late in the Wisconsinan glacial/pluvial, because all claims for earlier entry have never been proven. Since there ARE no early remains to be found, any search is useless and the intellectual veractiy of advocates of really early man is suspect (e.g., McGuire 1982).

2. People were in the Americas for hundreds of thousands of years (e.g., Steen-McIntyre et al. 1981). Since the site of Hueyatlaco has a uranium series date of 245,000±40,000 years, very ancient Native Americans are at least an intriguing possibility, if not a probability.

Since there were people with a well-developed stone technology in Soviet Central Asia 250,000 years ago (R.S. Davis et al. 1980:130), and even earlier in North China (Aigner 1978), why not also the New World? In my opinion, the real reason this idea seems so outrageous to American archaeologists is because we have been looking in the wrong places with the wrong methods. Lake valleys are the most promising places and geoarchaeology provides the most heuristic methodology (Butzer 1975). Geoarchaeologists must postulate connections between fossil people and highly selective fossil landforms of appropriate ages. Geoarchaeology also requires use of aerial photography, remote sensing and machine trenching, among other techniques.

THE FOSSIL ANTHROPOLOGY OF A LAKE VALLEY

Traces of ancient people will not be found in recent landforms. Instead, it is necessary to reconstruct ancient landscapes; find their residual landforms; and look there for bogs, paleosols, and/or calcic horizons. An ideal place to commence this work is a tectonically active Pleistocene lake basin such as the Panamint Valley (Davis 1970; E.L. Davis et al. 1980) and China Lake Valley (Davis 1978) in the Mojave Desert.

Fossil bogs on old shorelines are the most likely paleoenvironments to contain kills, butchering remnants, and faunal remains. Fossil soils that represent weathering intervals are likely to contain camp debris of stone and bone. Bog-traces will be found in buried and/or exhumed fossil shorelines, which are occasionally identifiable in aerial photographs and can then be revealed by bulldozer trenching. They can also be exposed by erosion (e.g., the 3,500,000 year old African shores found by Johanson (1981:block diagram opp. p. 128)).

China Lake now has a uranium series date of 42,350 years on mammoth tooth enamel in direct contextual association with two sophisticated finishing flakes (Davis et al. 1981). Also in the Mojave desert, the Calico site (lowest levels with

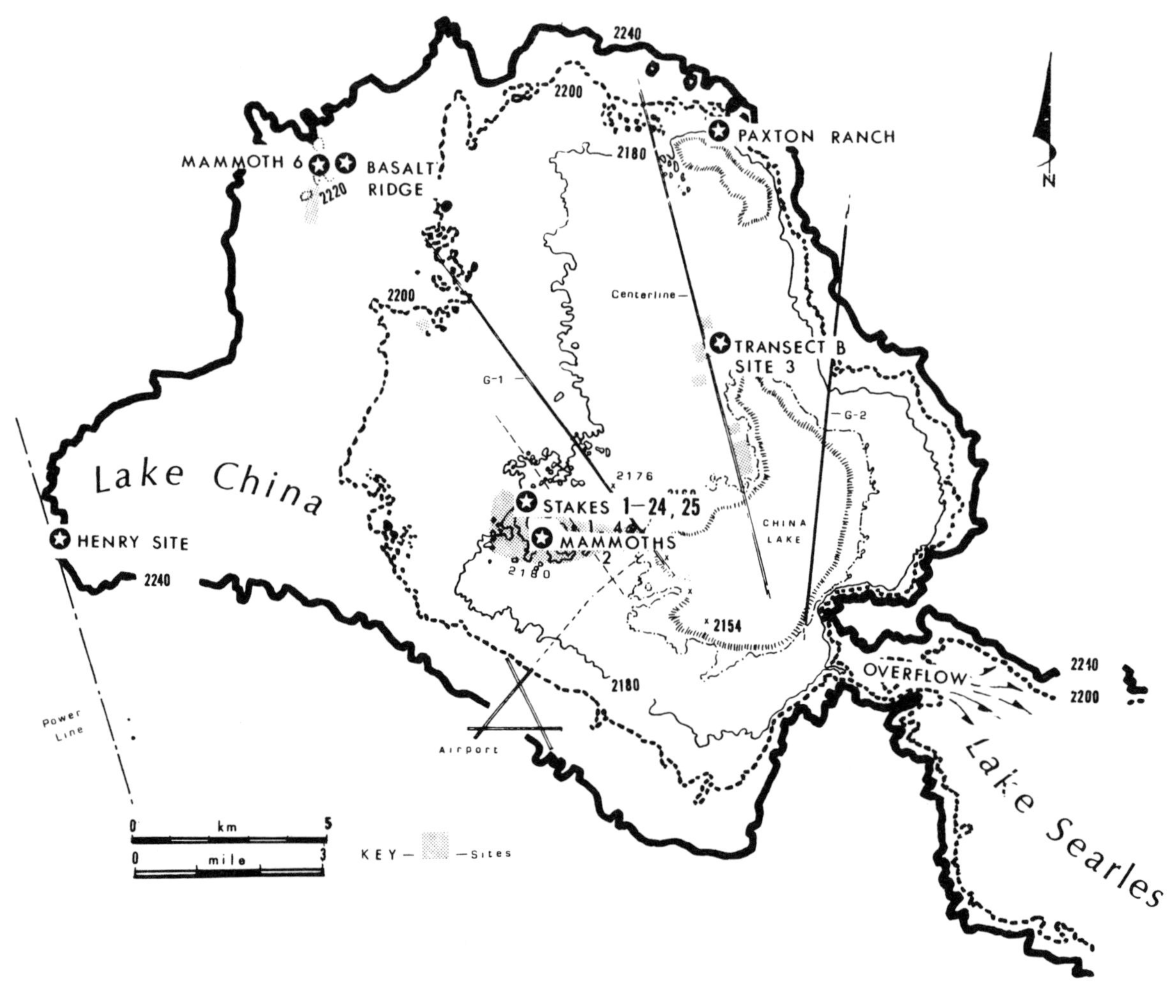

Figure 1. China Lake elevations and sites.

convincing artifacts) has a uranium series date of about 200,000 years (Simpson et al., this volume).

But the long-held 13,000 year deadline for human arrival in the New World still bobs around like a cork on the surface of archaeological thinking, distracting attention from the true problem: archaeological concepts and training in the western United States are inadequate to cope with the problem of really early man. We are run through graduate programs that make us into cultural anthropologists who know how to dig isolated holes chosen at random, and to classify statistically manipulated little pieces of humanly altered stone that come out of them; but we do not practice geoarchaeology — prehistory as an earth science — that visualizes geological/climatic metamorphoses with enormous time depth. In this vast theater of earth and time, people have moved, adapted and lived. But we archaeologists in America do not find their "sign" because we do not work systematically.

One of the best places to look for early sites is in the open "exposed archaeology" of the western Pleistocene lakes country (Davis 1978). Aerial photographs and bulldozer stratigraphic trenches are the keys to discoveries when they are combined with regional rather than restricted hole-in-the-ground thinking.

If you want to find 50,000 year old sites you must reconstruct 50,000 year old landscapes and sample them. Archaeologists should be asking questions like, What forces shaped or destroyed the ancient landforms?; What resources attracted people to a particular landform?; Where are the old marshes buried?; What fossil landforms (that emerge occasionally through miles of alluvial deposits) are of suitable ages to contain sites and artifacts of 10,000, 50,000; or even 200,000 years? Low level aerial photographs will reveal these landforms, and a bulldozer can probe them stratigraphically.

We have found little because we have been doing it all wrong, unsystematically, unscientifically. The Bureau of Land Management recently spent several million dollars to inventory surface sites and pot-drops in the California deserts. Not one of the contractors with whom I worked as a consultant between 1978 and 1981 ever made productive use of aerial records available through Eros Data Center, U.S.G.S., Department of Agriculture, and other sources. Not one contractor put in exploratory trenches around dry lakes because no one was thinking like a geoarchaeologist.

When a trenching program is undertaken in Pleistocene lake valleys, we shall begin finding Pleistocene people. These folks were People of the Marshes — the food baskets. We need only read *Lucy* (Johanson 1981) to see how richly this geoarchaeological understanding paid off in the Afar Triangle of Africa.

Skeptics will ask "Why bother? You will find nothing." Not true. The skeptics do not cruise the fans, shores, scarps, — and residual landforms of the desert as some of us do. The Desert Rats have long since become geomorphologists-without-portfolio, able to spot ancient exposures from afar and to find here and there upon their surfaces rare, familiar, weathered, and encrusted pieces of stone that once were altered by human hands. Many pieces are physically/chemically weathered to the point of being almost (but not quite) unrecognizable. As Don Tuohy has said, "they're obliterated."

Having scrambled from a lakeshore over miles of boulders to stand at last on a residual or exhumed feature, you occasionally see, with excitement, one of the strange stones at your feet. A solitary person looks (and knows he/she is looking) over unbelievable years at an ancestral relic.

By establishing new geoarchaeological systematics, by using aircraft, airborne camera, and bulldozer as tools, such accidental discoveries can become regularized episodes in studies of the ancient people of the New World. Accidental finds can become a coherent body of knowledge put together by teams of specialists.

If we would work in our desert and think as contemporary Africanists do in theirs (Johanson 1981), we would make surprising discoveries. I know for one thing that we would dissolve the Clovis blockage. The evidence should be massive. Further, the greater than 42,000 year date on a mammoth tooth with flakes at China Lake offers a good chance that discoveries should go much further back into the Quaternary.

In this report a model is presented of the theory and method essential for discovering the still buried archaeological deposits of a Pleistocene dry-lake valley in the California desert. China Lake is used for modelling. Since 1969 China Lake has been tentatively explored (Davis 1978; E.L. Davis et al. 1980, 1981). Geoarchaeology is seen as a theoretical and methodological base for continuing and expanding this work.

Desert prehistory cannot rest on "tool-types" alone but must be placed in a dynamic geological/climatic framework (a geochronology of what the lake was doing). Aerial camera, coring rig, and bulldozer trenching are required tools for locating marsh and alluvial deposits that contain or support the lithic and faunal record of geoarchaeology. People used the marshy bogs as food sources and traps for large animals. The same people lived and left their camp debris on the nearest dry ridge with a good overview of the country.

The archaeologist must use airborne cameras to understand the anatomical history of the chosen landscape. Next, a bulldozer must be used to make long stratigraphic cuts at right angles to the high ridges. Hand excavations can then be stripped sideways from the trench walls.

ELEVATED LANDFORMS

When high or emergent landforms are being reduced by erosion, archaeological materials such as animal bones, stone tools, and flakes are exposed in linear or arcuate distributions that outline temporary stratigraphic margins or contours. Low level aerial photos taken from tethered balloons

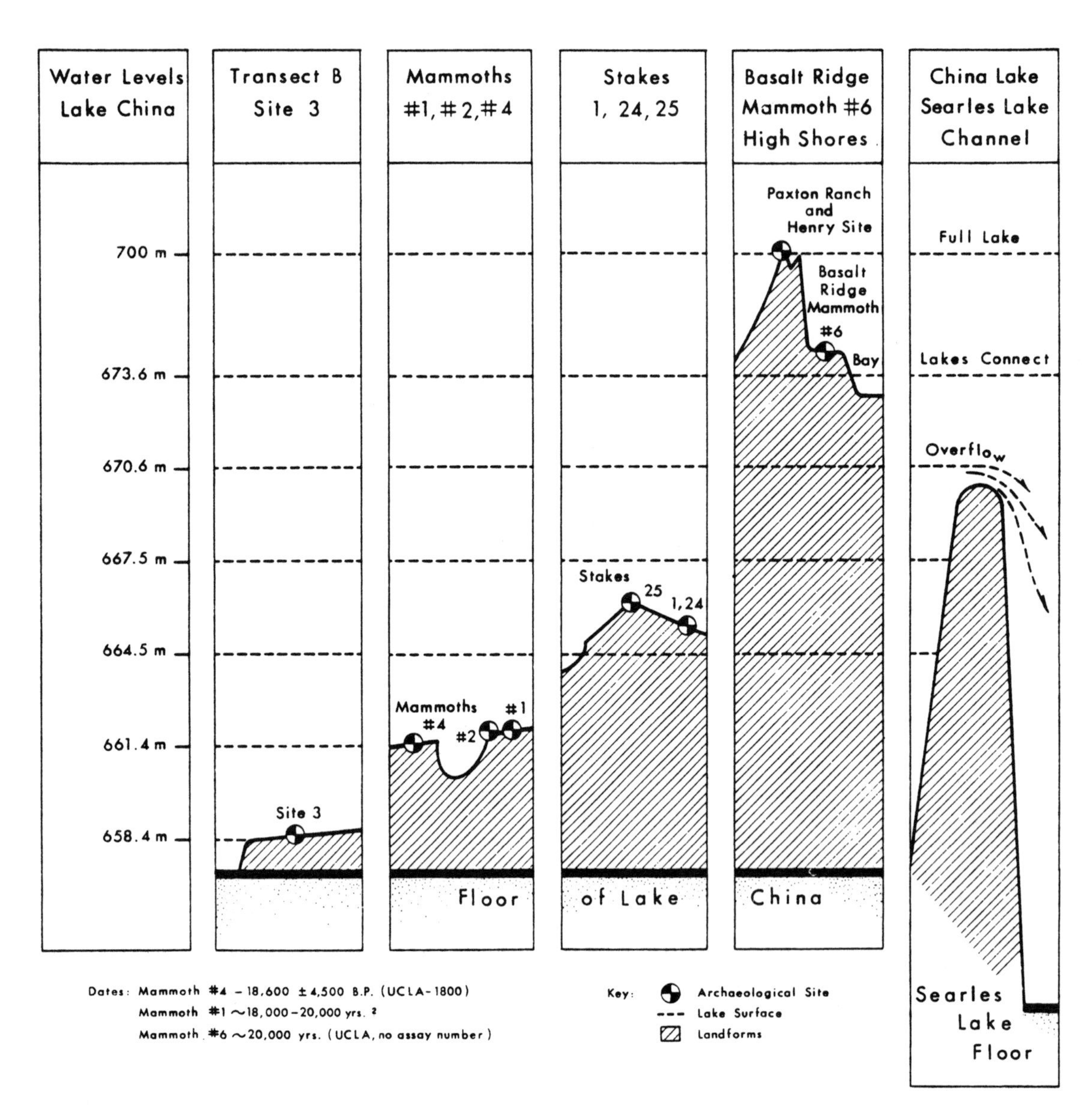

Figure 2. Temporal correlations of lake levels, dates and sites.

reveal shorelines, faults, uplift blocks, and dissected drainages which may lie hidden under several layers of aeolian sand. Within the shoreline gravels there are buried bones, artifacts, and calcic layers containing calcareous plates, root casts, and probably most of the bones and stone materials. There appears to be a close relationship in the surface distribution of the calcareous forms and archaeological material. However, these configurations change. The whole landscape, like all this desert, is a study in relative metamorphoses due to tectonics, deposition by wind and water, and erosion by these same elements. We are looking at a dynamic prehistoric scene in which landforms are constantly changing. This is geoarchaeology, where artifacts can only be understood in relation to a climatic case history of earth anatomy that is constantly growing, shifting, and being destroyed.

SHORELINES

By studying satellite and other aerial photographs, strategic places around lake perimeters can be selected for exploratory trenching with a bulldozer, as at Panamint (Davis 1970). The aerials and straticuts again create a probability background for customary hand excavations that strip from known to unknown away from trench wall exposures. You know what you are working with before you start.

The proposed geoarchaeological methodology is especially applicable at the Basalt Ridge. The Basalt Ridge embayment and lower slopes should be trenched in this manner. Shoreline trenches, as at Panamint (Davis 1970), should be cut at right angles across the shore in order to reveal the interfingering of lacustrine and alluvial deposits.

Table 1. The Rancholabrean faunule of China Lake.

In connection with surface mapping (Davis 1975), over 800 faunal specimens have been analysed by David Fortsch, including the following taxa:

Mammuthus sp.	
Hemiauchenia sp.	
Camelops sp.	—large grazers
Bison cf. *antiquus*	
Equus sp.	
Odocoileus (?) sp.	—a browser
Canis cf. *latrans*	
Canis dirus	—carnivores
Phalacrocorax sp.	
Anas sp.	
Aythya sp.	
Branta sp.	—birds (chiefly waterfowl and waders)
Oxyura sp.	
Aquila sp.	
Grus sp.	
Order Chelonia (unid. turtle)	

Species frequencies were as follows:

1) *Hemiauchenia* (?) sp. .... 2 fragments
2) *Camelops* sp. .... 375 fragments
3) *Bison* cf. *antiquus* .... 187 fragments
4) *Equus* sp. .... 126 fragments
5) *Mammuthus* cf. *columbi* .... 42 fragments

Implications of this faunule and its large herbivore frequencies are extensively discussed in Davis (1978:18-21).

THE BASALT RIDGE

The most important series of localities at China Lake is associated with the Basalt Ridge ("La Brea Tarpits with people") because of its wealth of butchered animal remains associated with lithics. A uranium series date on one association, a dated paleosol, and two dates on Late Pleistocene tufas place lake episodes in time.

Tools and animal bones are distributed downslope from datum (2234 feet AMSL) into swamp deposits at 2215 feet. A Late Pleistocene basalt flow formed a dam across the northwest corner of the Pleistocene Lake China catchment basin. Because of this dam, water from the Owens Valley drainage had to enter the basin from the south, go past the High Bench Area, and into Lake China, which filled (numerous times) from south to north (Figures 1, 2; Davis 1978:90-149).

The U.S. Navy did not permit trenching of Basalt Ridge in order to confirm the stratigraphy. The exposed stratigraphy consists of upper dune sand draped over a basalt core. An argillaceous paleosol is exposed about halfway downslope over alluvial gravels and a lower calcic paleosol (dated 10,800±310 B.P. [GX-3446]). Beneath that is a sequence of blocky marsh deposits that contain disarticulated remains of extinct animals, including the 42,000 year old mammoth, camel and horse, in addition to coyote, vole, shorebirds, and fish.

Cultural Associations

Lithics appear to represent a long sequence from a Core Tool Tradition on upper exposures through Lake Mojave/Clovis at intermediate levels, to an undefined post-Clovis transitional culture at shoreline level. Marls and clays in the lakebed contain redeposited materials.

Dates and Materials

1. 10,800±310 B.P. (GX-3446) on calcareous earth, lower calcic horizon, site 1.
2. 12,200±120 B.P. (UCLA-1911 B) on upper layer of tufa deposit (Tufa B) developed on basalt near datum stake. This date may be 500-1000 years too old (Rainer Berger, personal communication, 1975).
3. 13,300±150 B.P. (UCLA-1911 A) on lowest layer of tufa (Tufa A) deposit near datum stake. Tufa A was growing on the basalt during the penultimate stage of Lake China.
4. 42,350±3300 B.P. A uranium series date (Davis et al. 1981) on enamel from a mammoth tooth excavated with two finishing flakes. This material was embedded in Unit D.

Recommendations

The "Rock Pile site" should be mapped and excavated. A trench should be made from the summit of the ridge into lake deposits of the embayment. The exposed profile should be logged by a geologist and used as a stratigraphic reference in excavating the bone and artifact beds at Site 1 (Davis 1978:89-96: Figures 63C-F). Uranium series and radiocarbon dates should be obtained on recovered animal remains to establish the geochronology and sequences of the lacustrine depositional events.

CONCLUSIONS

Aerial photography (obtained mainly from special, low-altitude flights), bulldozer straticuts (for soil/bog profiles), and, primarily, geoarchaeological thinking are required in order to exploit the scientific potential of China Lake basin and to answer, among others, the following questions:

How can we best construct a geochronology of the main China Lake basin? (Remember, what the lake was doing tells us where the people were active.) How far back in the stratigraphic record can human/extinct animal associations be traced?

Any scientific investigations that seriously address Lake China's human prehistory must use the holistic effort of a scientific team. This should include a geomorphologist/soils expert, a paleontologist, a paleobotanist (including an expert in plant phytoliths), an expert in local tectonics, a paleomalacologist (gastropod population changes are indices of climatic change), and an individual experienced in regional ethnology and folklore.

In summary, China Lake basin and mountains are a treasure house of ancient American history. The extensive bonebeds at the Basalt Ridge embayment are particularly important. They contain the answers to the cultural evolution from pre-Clovis to Clovis and post-Clovis.

REFERENCES CITED

Aigner, J.S.
1978 Important archaeological remains from North China. In *Early Paleolithic in south and east Asia,* edited by F. Ikawa-Smith, pp. 163-232. Mouton, The Hague.

Butzer, K.W.
1975 The ecological approach to archaeology: Are we really trying? *American Antiquity* 40:106-111.

Davis, E.L.
1970 Archaeology of the north basin of Panamint Valley, Inyo County, California. *The Nevada State Museum Anthropological Papers* 15:83-142. Carson City.

1975 The "exposed archaeology" of China Lake, California. *American Antiquity* 40:1-26.

Davis, E.L. (editor)
1978 *The ancient Californians.* Natural History Museum of Los Angeles County, Science Series 29, Los Angeles.

Davis, E.L., K.H. Brown, and J.B. Nichols
1980 *Evaluation of early human activities and remains in the California Desert.* Great Basin Foundation, San Diego.

Davis, E.L., G. Jefferson, and C. McKinney
1981 Man-made flakes with a dated mammoth tooth at China Lake, California. *Anthropological Journal of Canada* 19: 2-7.

Davis, R.S., V.A. Ranov, and A.E. Dodonov
1980 Early man in Soviet central Asia. *Scientific American* 243:130-137.

Johanson, D.C.
1981 *Lucy, the beginnings of humankind.* Simon and Schuster, New York.

McGuire, K.R.
1982 A reply to Gruhn and Bryan's comments on "Cave sites, faunal analyses, and big-game hunters of the Great Basin: A caution." *Quaternary Research* 18:240-242.

Steen-McIntyre, V., R. Fryxell, and H.E. Malde
1981 Geologic evidence for age of deposits at Hueyatlaco archaeological site, Valsequillo, Mexico. *Quaternary Research* 16:1-17.

Lithic Technology of the Calico Mountains Site, Southern California

RUTH D. SIMPSON
San Bernardino County Museum
Redlands, California 92373
U.S.A.

LELAND W. PATTERSON
418 Wycliffe
Houston, Texas 77079
U.S.A.

CLAY A. SINGER
San Bernardino County Museum
Redlands, California 92373
U.S.A.

Abstract

A new, more objective approach has been inaugurated to present the evidence for the presence of people at the Calico Mountains site, which has recently received a uranium-series date of 200,000 years. Objections to the acceptance of the flaked objects as artifacts, carefully excavated at Calico, are countered by ongoing functional analyses of flake edge wear and comparative flake scar analyses based on replication experiments. Preliminary quantitative analyses of a significant sample of all excavated broken clasts (including all items kept, and reexamination of items stockpiled by excavation levels at the time of excavation), has revealed more than 3000 percussion flakes with clearly visible bulbs, 300 probable unifacial tools, several cores with striking platforms, and several simple bifacial tools. It is argued that this simple unifacial industry is related to the Early Paleolithic of East Asia.

INTRODUCTION

The Calico Mountains site in southern California is perhaps the best known example of the proposed evidence of very early man in the New World. Work was started on this site in 1964 and is continuing (Simpson 1979, 1980). The site was recorded by Simpson in 1958. She brought Louis S.B. Leakey to the area in 1963. He selected the excavation locations; and, with Simpson, directed the excavation until his death in 1972. Since then Simpson has been in charge of the Calico Project. The site is situated on Bureau of Land Management land. The project is sponsored by the San Bernardino County Museum and proceeds under permit from the Federal Government.

The Calico Mountains site is located in the Mojave Desert of southeastern California at 34° 56′ 54″ north latitude, 116° 45′ 37″ west longitude. It is situated at the west end of the Pleistocene Lake Manix basin, near the town of Barstow and just east of the Calico Mountains (Figure 1).

During late Miocene times, lake sediments and volcanic materials accumulated in the region north of the Transverse Range. These deposits have been designated as the Barstow formation. Ultimately, the Barstow formation was uplifted, inverted, and distorted to form the Calico Mountains. The Calico Mountains lie north of the Transverse Range which forms the southern boundary of the Great Basin. The Mojave River has its origin in the Transverse Range and flows northeastward. During Pleistocene pluvial periods, this river filled Lake Manix and overflowed to fill Lake Mojave and Death Valley. Pleistocene alluvial fans, like dry-land deltas, built eastward from the Calico Mountains into the Lake Manix basin. A major fan built at the mouth of Mule Canyon as the result of mud flows has been designated as the Yermo formation (Figure 2). Movement along a fault at the east end of the Calico Mountains isolated the fan and terminated fan building.

Although there has been subsequent earth distortion and erosion of the Yermo fan surface, 10 to 30 m of the fan remain. The remaining

Figure 1. Calico site location.

Figure 2. View of alluvial fan.

Figure 3. View of Calico site.

units of the fan are stratified and evince no secondary movement. The stratified fan deposits contain volcanic and siliceous rocks ranging in size from pebbles to large boulders. The siliceous materials were suited to production of lithic tools by Early Man as well as by late prehistoric Indians.

The Calico Mountains site, also called the Calico Early Man site, is situated on the Yermo alluvial fan (Figure 3). Clements (1979) and Bischoff et al. (1981) believe that the fan formed quite rapidly. Ultimately, earth movement separated the fan from the source of the lithic materials in the Calico Mountains. Since then, the fan has been subjected to gradual erosion.

There have been fillings of Lake Manix since the fan formed. Erosional prehistoric beaches have been cut into the fan surfaces. The organic tufa on one beach has been dated at approximately 20,000 years B.P. by the radiocarbon method at the UCLA laboratory (UCLA-121). Artifacts recovered there are of somewhat more sophisticated varieties than the very early man-made specimens recovered from deep within the fan. There has been concern in accepting Calico as an archaeological site because of uncertainties in determining its age and questions as to whether or not the lithic specimens are man-made. A uranium-thorium date of 200,000 B.P. was obtained in 1980 for the deeper levels of this closed, stratified site. Details of dating by the uranium series and soil-geomorphic methods are given in separate papers (Shlemon and Bischoff 1981; Bischoff et al. 1981).

There are psychological barriers to acknowledging very early archaeological sites such as Calico in the New World. It is accepted dogma in the literature that people did not arrive in the New World before the *Homo sapiens sapiens* stage of development. Acceptance of a site dating from 200,000 B.P. would change present concepts of the times and numbers of migrations of Early Man from Asia into North America.

This paper presents evidence that the Calico site contains man-made lithic artifacts with the technological level of the concurrent upper portions of the Lower Pleistocene period in Asia. In the analysis of the Calico collection, emphasis is placed on manufacturing patterns, morphology and consistently recurring attributes that are not likely to have been reproduced often by natural forces. Also, the probability of natural forces simulating an entire early man tool kit is even more remote. This paper presents some objective criteria used in determining the man-made nature of the Calico lithic specimens, thus minimizing the use of subjective opinions by individual investigators. It should be noted that, in publications discussing how nature fractures rock simulating man-made patterns, many statements are not based on physical evidence, and others are not applicable to the type of paleoenvironment or rock types present at the Calico site in the Pleistocene.

EXCAVATION METHODOLOGY

The early artifacts are being recovered from the Yermo formation alluvial fan which affords a closed, stratified context. Erosion has removed more of the alluvial deposit in the vicinity of Master Pit I than at Master Pit II. Hence, in Master Pit I, the significant artifact-bearing deposit commences approximately one m below the present surface, and four m below the present surface in Master Pit II. Surface specimens are not included in the analysis of the excavated assemblage, and few significant specimens have been recorded in the upper levels. The horizontal layout of the major excavations is shown in Figure 4. In Master Pits I and II, the artifact-bearing deposit extends to the base of the fan, nine to ten m below the surface. The total area of Master Pits I and II is 69.7 m^2. Excavation in Master Pit III is now in progress but has attained a depth of only 2 m.

Excavation began in the northwest corner of Master Pit I (P-19) and was extended in five-foot increments forming two trenches in the P and 19 series of units, i.e., along the west and north walls of the pit. As soil profiles were defined, five-foot units were opened (P-20, Q-20, etc.). However, a witness column consisting of three units was retained intact for use as needed in future tests and analyses. A witness column of a single unit was retained in Master Pit II.

All excavation of the main pits and related test pits and trenches has been accomplished with small hand-tools. Nothing larger than hammers and quarter-inch chisels has been used in the main pits and trenches; nothing larger than hand-mattocks in the outlying pits. Soil removed from all the excavation is screened through 1/2", 1/4",

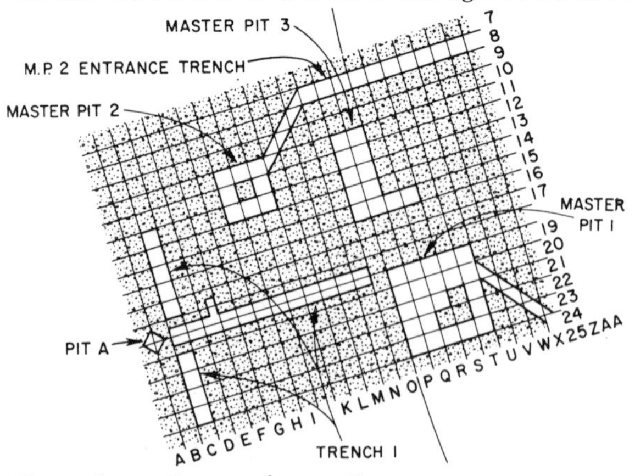

Figure 4. Layout of excavation.

and 1/8″ screens. Excavation was and is in three-inch levels unless soil changes dictate a lesser increment. Specimens recognized in the pits are triangulated from the northwest and northeast corners of the individual unit and depth is measured from the datum for the unit. All specimens, including debitage, are counted, measured, plotted if relevant, recorded and stockpiled. Permanent field numbers were assigned at the field laboratory until 1972. Since then, cataloguing has been done at the San Bernardino County Museum. All potential archaeological specimens are shipped to the County Museum for the in-depth analyses which are basic to consideration of morphology, attributes, distribution and wear-pattern studies.

The vertical distributions of artifacts from sections H-13 and I-13 in Master Pit II are shown in Figure 5. One of the key points in demonstrating human activity at the Calico site is documentation of the concentrations of lithic artifacts, including debitage, recovered from the limited areas within the Master Pits. This is a non-random distribution that would not be expected if the occurrence were of naturally fractured rock.

Excavations have been made in a careful, controlled manner. Hence, ongoing stratigraphic and horizontal data are available. There is no problem in determining the context of early specimens found in the excavations at this site.

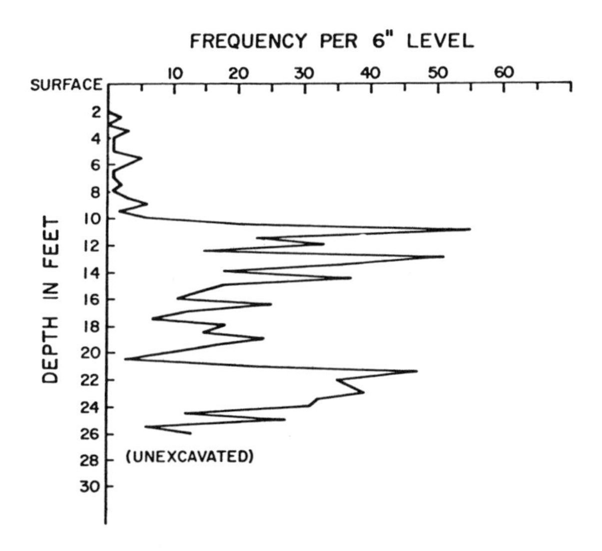

Figure 5. Vertical distribution of tools and other worked pieces from Units H-13 and I-13 (total frequency — 798 pieces).

LITHIC TECHNOLOGY

Introduction

At the present time, the central issue concerning the Calico site is whether or not there are man-made lithic specimens in the fan deposit. While there has been ever-increasing support for the evaluation of specimens as artifacts by outside investigators, led by the late Louis S.B. Leakey, others continue to regard the Calico lithic collection as the product of naturally fractured rock. This paper addresses the opportunity to demonstrate that the Calico site does yield man-made lithic artifacts. Examples pictured in this paper are lithic specimens having definitive morphological characteristics and technological attributes which support the proposition of human craftmanship. Preliminary analyses have included many marginal specimens, which are easy for critics to use selectively when questioning the man-made nature of the assemblage. However, frequent critical statements have been made to the effect that an overall sample of all broken stone should have been used, and that, instead, there has been a selection of objects that appear to be man-made, resulting in a biased sample. This criticism is based on the assumption that, if natural forces fracture enough rock, objects similar to man-made tools and flakes will be produced. As noted, this assumption has not been verified by physical evidence for the geological context of the Calico site.

All rocks were examined in the field. Those regarded as non-artifactual were stockpiled by excavation levels and are available for study. The stockpiled non-diagnostic lithic materials from the excavations contain many pieces of flakeable siliceous materials, but relatively few pieces that are well-fractured. This fact demonstrates that this alluvial fan was not functioning as a giant "gravel crusher," as many people speculate. Non-diagnostic fractured rock specimens in the museum collection will be included in final quantitative analyses. It should be noted that lithic reduction using hard percussion produces significant quantities of non-diagnostic fractured stone as well as diagnostic specimens.

Selection of specimens that appear to be man-made is the only practical strategy for initial study where there are large quantities of miscellaneous rock (Patterson 1980). It is then left for the investigator to demonstrate by detailed study that collected specimens are actually man-made. This can be done by showing consistent patterns of technological attributes and morphological characteristics that are typical of lithic manufacture by man; patterns which random natural forces would not be likely to reproduce very often (Patterson

1983). Primary lithic reduction produces large quantities of miscellaneous debris which has little value for demonstrating the man-made nature of the knapping process. Recognition of diagnostic tool types and technical flakes then becomes the accepted practice observed in the analysis of any lithic collection, and is the study method used with the Calico material.

Flaking experiments using hard percussion on siliceous materials from the Calico site area have produced much shattered non-diagnostic debitage in addition to diagnostic percussion-made flakes. This experimental evidence is one demonstration that the presence of a significant quantity of non-diagnostic fractured stone in the Calico lithic collection should not detract from the acceptance of other Calico specimens which exhibit characteristics diagnostic of human flaking.

This paper discusses the range of types of the most diagnostic lithic materials from the Calico site. A program is underway to make a complete quantitative analysis of the Calico lithic assemblage. This study will be the subject of a future detailed report, which should remove much of the concern regarding use of a biased sample.

Man-Made Flakes

In the development of an alluvial fan there are limited opportunities for natural forces to do percussion flaking (Carter 1980:Chapter 4), while very early human lithic manufacturing would mainly be by percussion flaking. Natural fractures of stone in mud flows, such as at Calico, would expectably be mainly by pressure, developed from the weight and movement of materials, which does not usually produce concentrations of flakes having distinct ventral surface attributes characteristic of percussion flaking. Natural earth movement in a fan would not usually produce flakes with prominent ventral face ripple lines or force bulbs with bulb scars which are typical fracture attributes from percussion flaking, as shown for the Calico specimen in Figure 6. Mud flows inhibit the physical conditions that would be necessary for stone fracturing by percussion. The high viscosity medium of the mud does not permit high velocity impact situations between rocks to occur.

It can be readily demonstrated by experimental flintknapping that little controlled flaking can be done with striking platform angles greater than 90°, with this angle defined as the angle between the striking platform and adjacent core face or dorsal flake surface (Patterson 1981). All of the percussion flake specimens with evidence of platforms discussed here for Calico have striking platform angles equal to or less than 90°. When there is a large concentration of flakes with acute angle striking platforms, such as at Calico, this attribute is an indication of consistent patterning caused by man-made flaking. However, the analysis of striking platform geometry in relation to man-made flaking activity is widely misunderstood (Patterson 1981). The tabulation of flakes with both acute and obtuse striking platform angles has little meaning, since unlike attributes are being compared. Controlled flaking by man and simulations of man-made flaking by nature both require acute striking platform angles. Flakes labelled as having obtuse angle striking platforms usually do not have correct identifications of residual striking platforms.

Crushing of residual striking platforms on flakes is another indication of percussive flaking. A significant percentage of the flakes so far examined in the Calico collection display crushing of residual striking platform edges. Crushed striking platforms from percussion flaking characteristically have many small scars with step fracture terminations.

The lack of cortex on residual striking platforms of flakes can indicate purposeful striking platform preparation in lithic manufacturing, and multiple faceted platforms may indicate extensive striking platform preparation. A high percentage of Calico flakes with platforms have no remaining cortex on striking platforms and many have multiple faceted platforms. Some of the percussion flake specimens also display evidence of striking platform edge preparation, in the form of small trim flake scars. Quantitative data on these attributes will be produced by the new study being done.

Attributes on the dorsal faces of flakes can indicate manufacturing by man. Remnant multiple facets on dorsal faces indicate prior flake removals from a core, a feature which is more indicative of human flaking than of natural fractures. Random natural forces are not likely to remove many flakes in series from a single core. Flakes produced by natural forces would also be expected to have more cortex remaining on dorsal surfaces. In the Calico assemblage, many flakes have multifaceted dorsal faces, and a high percentage have no remaining cortex on dorsal surfaces.

While there may not be a real prismatic blade industry at Calico, more than 400 specimens have been classified as prismatic flakes and blades (Figures 7, 8) are more likely to be produced by controlled flaking with a prepared core (Figure 9) than by random natural forces. It is common to produce many prismatic flakes and low percentages of prismatic blades when doing experimental flintknapping not related to purposeful production of prismatic blades (Patterson and Sollberger 1978: 110).

Many of the flakes from the Calico collection have sharp edges, which demonstrates that they

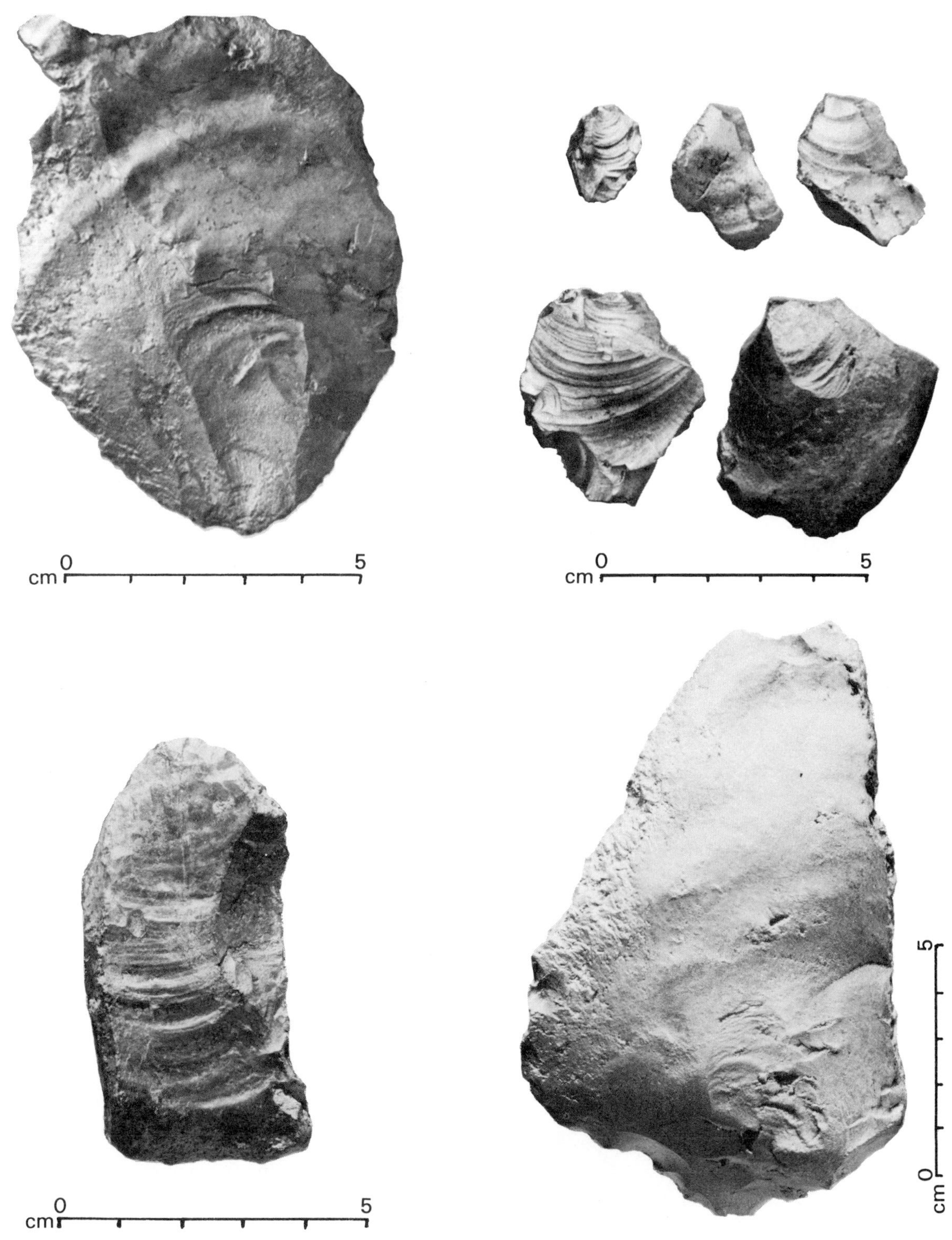

Figure 6 a, b, c, d. Percussion flakes ventral face attributes.

Figure 7. Prismatic flakes and blades.

Figure 8. Prismatic blade.

Figure 9. End and side scraper.

have been subjected to minimal rolling action in this alluvial fan, and that they were produced at this location.

Unifacial Tools

Natural forces in a mud flow would be expected to give mainly bidirectional random damage to flake edges. It would be difficult for nature to produce many specimens resembling man-made unifacial tools, with completely unidirectional edge retouch done in a uniform, directed manner. The Calico site has yielded many completely unifacial stone tools with uniform edge retouch. These include end scrapers (Figure 10), side scrapers (Figure 11), and gravers (Figure 12). Some gravers have bifacial retouch on points, which can be expected in even unifacial flake tool industries. Calico specimens identified as gravers have uniformly retouched points, and are not just flakes with fortuitously pointed shapes. Many unifacial tools in the Calico collection are selectively flaked only on dorsal surfaces, which is the usual pattern for man-made stone tools, where the flat ventral surface is used as the striking platform for force application.

It would be extremely difficult for nature to reproduce the completely unifacial scraper/graver

Figure 10 a, b. End scraper.

Figure 11. Side scraper.

Figure 12. Graver.

Figure 13 a, b. Scraper-graver.

Figure 14 a, b. Nosed-tool.

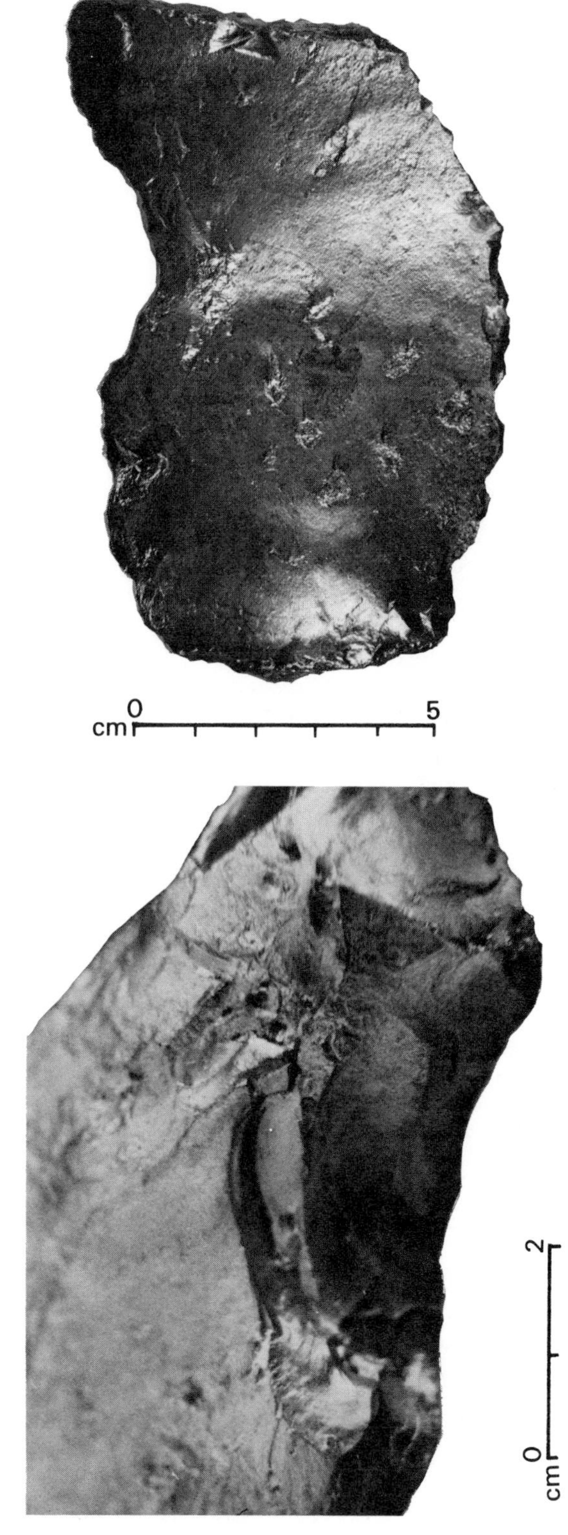

Figure 15 a, b, c. Notched tools.

as shown for the Calico specimen in Figure 13. Fine, parallel unifacial retouch on a nosed tool is shown in Figure 14. The edge retouch pattern of the scraper shown in Figure 11 has been replicated on the same type of material by use of a small quartzite hammerstone as a percussor.

There are also concave scrapers, and notched tools (Figure 15) with completely unifacial retouch done in a uniform manner. Some specimens with unifacial serrated or denticulated edges are also present (Figure 16). A unifacially trimmed cutting tool is shown in Figure 17.

Many of the Calico lithic specimens exhibit selective patterned flaking of some edges, with little random damage on other edges. Patterned flaking selectively restricted to specific edges is an indication of controlled flaking by man.

It would be difficult for nature to produce many completely unifacially retouched objects, especially with uniform retouch on long edge intervals. It follows that it would be even more difficult for nature to produce multiple examples of the several unifacial tool categories comprising a tool kit such as is found at the Calico site.

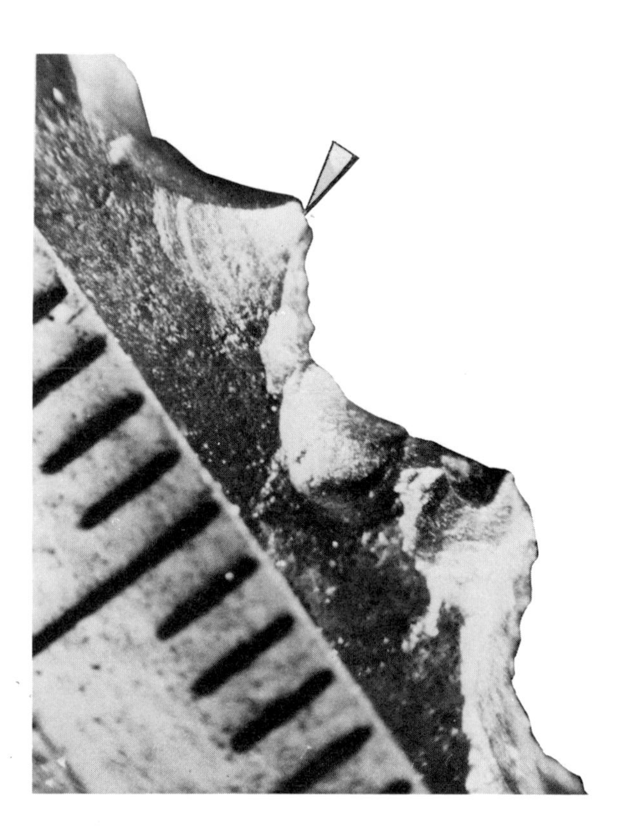

Figure 16 a, b. Denticulated edged tools.

Figure 17. Unifacially trimmed cutting tool.

Bifacial Tools

While natural forces can produce objects with random bifacial flake scars, and amorphous-shaped bifacial objects, nature would not be expected to produce many bifacial objects with patterned flaking and definite shapes. The Calico lithic industry collection contains several bifacially flaked specimens with definable shapes.

Bifacial tool categories include chopping tools (Figure 18), and picks (Figure 19). Large, relatively crude bifacial tool types such as these have been in use for a very long time. Bifacial artifacts are a relatively small category at the Calico site, compared to unifacial artifacts. The overall lithic manufacturing industry here is primarily based on percussion-made flakes rather than core tools.

Lower Paleolithic type bifacial tools of large, crude forms would be expected to have 10 to 50 flake scars, according to Ronen (1975:41). Four Calico bifacial specimens have been examined, and each have 22 to 37 flake scars. Flakes removed during bifacial thinning by man will tend to have patterned flake size distributions (Patterson and Sollberger (1978:111). This pattern can be demonstrated experimentally for the distribution of flake scar sizes on product bifaces, also. Three Calico bifacial specimens have this exponential curve of flake scar sizes, skewed toward smaller size scars.

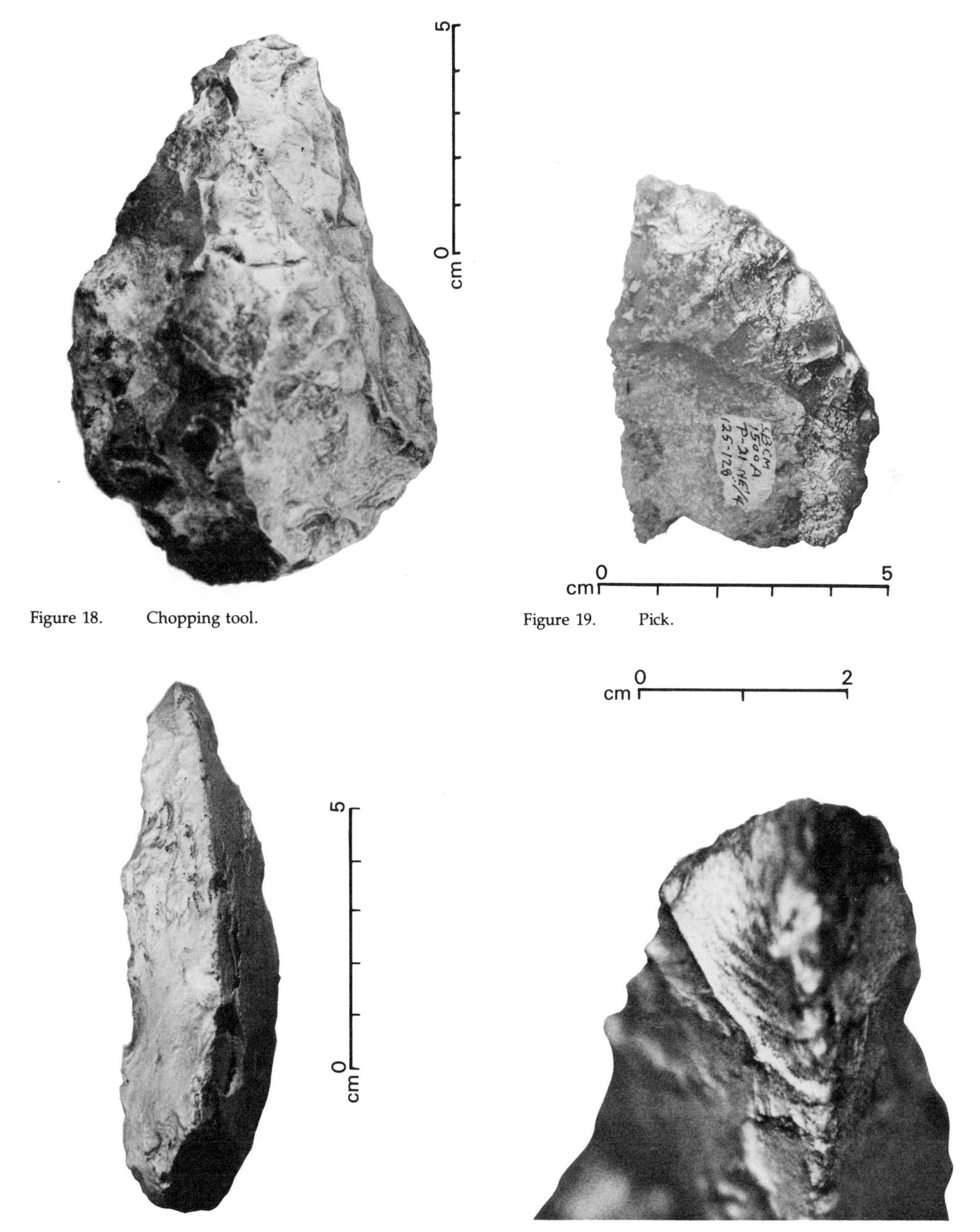

Figure 18. Chopping tool.

Figure 19. Pick.

Figure 20 a, b. Reamer.

There are also smaller tools in the Calico collection with purposeful bifacial retouch. These include rotational tools, such as reamers (Figure 20).

Prepared Cores

In lithic industries with any degree of sophistication, the knapper would be expected to have produced residual cores which have a patterned nature instead of simply having random flake scars. The Calico collection contains cores with definite striking platforms, as in Figure 21. Such cores display a series of patterned flake removals by definite percussion force application that would not be expected of naturally fractured stones. Stones subjected to much natural fracturing will have rounded edges and amorphous shapes. Flake scars on Calico cores are strongly negative conchoidal forms typical of percussion flaking. Individual force-application points are apparent on clearly defined striking platforms.

In the Calico excavated lithic materials, there is little evidence of flakes with corresponding cores of siliceous materials that are products of natural fracturing of stone. If many of the flakes from this site were produced by natural causes, there should have been evidence of related cores.

Figure 21. Core with striking platform.

Spatial Considerations

When man manufactures stone tools, the resultant by-product debitage is concentrated in lithic workshop areas. A site where stone tools have been manufactured would be expected to contain such workshop concentrations or clusters. There have been more than 200 significantly large clusters identified at Calico containing percussion flakes and miscellaneous debitage. They range from .5 m to 2 m or more in diameter. There is an average of about 75 items per cluster. Natural forces would be expected to produce broken stone randomly scattered. Naturally broken materials are present in and around the Calico site, but not in concentrations as are most of the tools and debitage. Aside from individual clusters of percussion-formed debitage within each test pit, the total concentrations of diagnostic lithic materials in the relatively small area of each test pit is significant indication of non-random human activity.

Selective Use of Raw Material

Calico artifacts were fashioned from siliceous materials, including chert, chalcedony, jasper, etc. None of these materials have natural cleavage planes which could simulate directional flaking. It has been noted for some time (Singer 1979:55) that Calico lithic artifacts were made preferentially from these siliceous materials. This selectivity would be expected of man's work, but not for random fracture of stone by natural forces.

Functional Edge Damage

There are on-going studies by Singer (1979) of functional edge wear on tools at Calico, but the present paper does not permit detailed discussion of this complex subject. Use-wear patterns typical of cutting, scraping and rotational functions are being recorded as microscopic examination progresses. These evidences of utilization can be reproduced experimentally (Tringham et al. 1974).

Another example of functional edge damage that would be difficult for nature to simulate is edge wear from hard chopping action. Natural edge damage usually consists of shallow, random bifacial fracture scars and does not usually feature a significant amount of step-fracture terminations. In contrast, chopping edge damage results from force being directed into the mass of the tool edge.

This process creates numerous short step-fracture terminations and a wide variation in flake-scar sizes. This edge damage pattern can be demonstrated experimentally (Patterson 1982). The Calico collection has some large flake choppers that exhibit this type of edge damage selectively on single edges, as shown in Figure 22. Selective damage on a single edge with a specific wear pattern is another attribute which would not be expected to occur often from randomly applied natural forces.

DETAILED ANALYSIS OF THE CALICO COLLECTION

Work is now underway to prepare a report that will present a complete qualitative and quantitative analysis of a significantly large sample of the Calico lithic collection. As the analytical work is still in a preliminary stage, the analyses to be discussed must be considered as tentative in nature. As an example of one analytical category, the occurrence of concentrations of true percussion flakes are taken as an important indication of the man-made nature of the Calico lithic collection. Percussion flakes have clearly visible bulbs of force on their ventral faces. Almost 4000 true percussion flakes, representing approximately 18% of the total fractured rock specimens, have so far been identified from a 10 test square sample of Master Pits I and II.

Lithic flake samples are being analyzed to obtain both metric data and data on a number of key qualitative attributes. While the presence of bulbs of force is the key attribute being used to identify diagnostic percussion-type flakes, other attributes being noted include: ripple lines on ventral faces, crushing of striking platforms, number of major dorsal face facets, amounts of remaining cortex on striking platforms and dorsal faces, indications of striking platform edge trimming, and the presence of bulb scars (eraillures). All flakes above 15 mm^2 are being included in tabulations. One sample from test squares P-19 and Q-20 in Master Pit I and J-13 in Master Pit II includes 3310 flakes, of which 364 (11%) are diagnostic percussion flakes. A total of 48 diagnostic flakes (13% of the percussion flakes) have intact residual striking platforms. These striking platforms are flat and evidently prepared because they lack cortex. Employment of cores with striking platforms of this type can be indicative of a man-made manufacturing pattern. Striking platform angles on these diagnostic flakes are less than 90°.

Figure 22 a, b. Flake chopper.

Analyses have so far identified over 300 prime specimens of possible unifacial tool types in the Calico collection, and over ten times this number of specimens that can be given at least marginal consideration as representing man-made tools. Analytical criteria being used for stone tools are as discussed in this paper. It is now apparent from even preliminary analysis that the Calico lithic collection is not confined to a few specimens of possible diagnostic value, but rather includes hundreds of prime candidates for being man-made flake tools, and thousands of diagnostic percussion flakes.

CONCLUSIONS

Work continues at Calico, although currently with restricted funding. There are other potential very early man sites in the New World, but Calico remains the one that has been most thoroughly investigated, and which also is now dated. A complete early man stone tool kit has been presented here for Calico. This stone tool assemblage is similar to the technological level of Lower Paleolithic collections from China (Jia 1975, 1980; Institute of Vertebrate Paleontology 1980; Aigner 1978) such as Zhoukoudian and Tingtsun, and Central Asia (Davis et al. 1980).

Data presented here and in future detailed reports should move arguments concerning the man-made nature of the Calico collection to a more objective level. Critics must now consider the demonstrated nonrandom, consistent patterns of attributes on Calico lithic specimens. Natural forces would not be likely to consistently reproduce multiple specimens of each of the Calico stone tool types, and concentrations of well-made percussion flakes. The probability of natural forces simulating an entire early man lithic tool kit (Figure 23) is even more remote. The data base for very early man in the New World is growing rapidly, and can no longer simply be ignored, because it does not fit current models of prehistory in the New World.

With the present data gaps that exist in our knowledge of the prehistory of man in the New World, any current proposed 'final' solutions to the early origins, migrations, and cultures of Pleistocene man in the New World are premature. At the present state of knowledge in early man research, there is a need for flexibility in thinking to assure unbiased peer reviews.

ACKNOWLEDGEMENTS

Appreciation is expressed to Daniel J. Griffin of San Bernardino, California for producing the photographic illustrations for this paper.

REFERENCES CITED

Aigner, J.S.
1978 Important archaeological remains from North China. In *Early Paleolithic in south and east Asia,* edited by F. Ikawa-Smith, pp. 163-232. Mouton, The Hague.

Bischoff, J.L., R.J. Shlemon, T.L. Ku, R.D. Simpson, R.J. Rosenbauer, and F.E. Budinger
1981 Uranium-series and soil-geomorphic dating of the Calico archaeological site, California. *Geology* 9:576-582.

Carter, G.F.
1980 *Earlier than you think.* Texas A&M University Press, College Station.

Clements, T.
1979 The geology of the Yermo fan. In Pleistocene man at Calico, edited by W.C. Schuiling. *San Bernardino County Museum Association Quarterly* 26(4):21-29.

Davis, R.S., V.A. Ranov, and A.E. Dodonov
1980 Early man in Soviet central Asia. *Scientific American* 243(6):130-137.

Institute of Vertebrate Paleontology
1980 *Atlas of primitive man*. Science Press, Beijing. Distributed by Van Nostrand and Reinhold, New York.

Jia (Chia), Lan-po
1975 *The cave home of early man*. Foreign Languages Press, Beijing.

1980 *Early man in China*. Foreign Languages Press, Beijing.

Patterson, L.W.
1980 Comments on a statistical analysis of lithics from Calico. *Journal of Field Archaeology* 7:374-377.

1981 The analysis of striking platform geometry. *Flintknappers' Exchange* 4(2):18-20.

1982 Experimental edge damage on flint chopping tools. *Bulletin of the Texas Archaeological Society* 53:175-183.

1983 Criteria for determining the attributes of man-made lithics. *Journal of Field Archaeology* 10:297-307.

Patterson, L.W., and J.B. Sollberger
1978 Replication and classification of small size lithic debitage. *Plains Anthropologist* 23(80):103-112.

Ronen, A.
1975 *Introducing prehistory*. Cassell, London.

Shlemon, R.J., and J.B. Sollberger
1981 Soil-geomorphic and uranium-series dating of the Calico site, San Bernardino County, California. *X Congreso, Union Internacional de Ciencias Prehistoricas y Protohistoricas, Comisión* XII:41-42. Mexico, D.F.

Simpson, R.D.
1979 The Calico Mountains archaeological project. In Pleistocene Man at Calico, edited by W.C. Schuiling. *San Bernardino County Museum Association Quarterly* 26(4):9-19.

1980 The Calico Mountains site: Pleistocene archaeology in the Mojave Desert, California. In, *Early Native Americans*, edited by D.L. Browman, pp. 7-20. Mouton, The Hague.

Singer, C.A.
1979 A preliminary report on the analysis of Calico lithics. In Pleistocene man at Calico, edited by W.C. Schuiling. *San Bernardino County Museum Association Quarterly* 26(4):55-63.

Tringham, R., G. Cooper, G. Odell, B. Voytek, and A. Whitman
1974 Experimentation in the formation of edge damage: a new approach to lithic analysis. *Journal of Field Archaeology* 1:171-196.

Preliminary Report on Archaeological and Paleoenvironmental Studies in the Area of El Cedral, San Luis Potosí, Mexico 1977-1980

JOSÉ LUIS LORENZO and
LORENA MIRAMBELL
Instituto Nacional de Antropología e Historia
Departamento de Prehistoria
Moneda 16
Mexico 1, D.F.
MEXICO

Abstract

Geological, archaeological, and paleontological research at El Cedral, S.L.P., northern Mexico, revealed evidence of a complex series of lacustrine, marsh, and artesian spring deposits spanning the late Pleistocene and early Holocene periods. Evidence of human utilization of the Rancho La Amapola site was found together with bones of extinct animal species in undisturbed stratified deposits on horizons radiocarbon-dated at 33,300 B.P., 31,850 B.P., 21,960 ± 540 B.P. and older than 15,000 B.P.

INTRODUCTION

The research in the El Cedral area was part of a major project intended to investigate Pleistocene and early Holocene events in basins with interior drainage and featuring lakes or lake sediments. The perspective was as much archaeological as paleontological, with the objective to obtain a record of paleoenvironmental changes and to examine evidence for the presence of early man in the region.

The archaeological/paleontological site of Rancho La Amapola is located on the outskirts of the town of El Cedral in the northern part of the state of San Luis Potosí, Mexico (Figure 1). The site is situated only a few kilometers north of the Tropic of Cancer, at 23° 49′ north latitude and 100° 43′ west longtitude, at an elevation of 1700 m above sea level. The Rancho La Amapola site is situated in a small structural depression with almost level floor, once a lake but now a semi-desert. The area of El Cedral now features *matorral* vegetation of xerophytic type, with cacti, shrubs, and small-leaved trees. Common plants include mesquite (*Prosopis juliflora*), nopal (*Opuntia*), *Yucca, Acacia, Berberis, Cassia,* and various grasses. There are also a few scattered nogales (*Juglans*), and junipers (*Juniperus*). The climate is warm, with a prolonged dry season and short wet season.

The area of El Cedral has long been known for Pleistocene fossils. In the past there were many artesian springs, all of them now dry; but evidently until the end of the nineteenth century sufficient flow existed to form a fairly extensive lake, surrounded by thick vegetation totally distinct from the regional vegetation of today. Now there are only a few junipers; and the springs are only a memory, as at present one can obtain water from only two or three of the springs by means of a well and pump. In the past the springs in the area of El Cedral attracted many animals, and bones of animals which died at the waterholes were incorporated into the sediments. At Rancho La Amapola, many bones now lie scattered on the surface, removed by historic excavations of deep wells as the water table retreated.

With the visible abundance of remains of Pleistocene fauna, the site at Rancho La Amapola appeared to be particularly promising for investigation. Paleontological excavations were therefore initiated at this site, with archaeologists present to watch for cultural evidence, and geologists, soil scientists, and botanists collecting data for paleoenvironmental interpretations. The participants in the field work included José Luis Lorenzo, Lorena Mirambell, Azucena Angulo, and Jésus Narez as archaeologists; Rodolfo Casamiquela, Ticul Alvarez Solorzano and Oscar Polaco Ramos as paleontologists; Manuel Reyes Cortés as geologist and surveyor; Antonio Flores Díaz as soil scientist; and Fernando Sánchez Martínez and Juan Gonzales Solís as botanists.

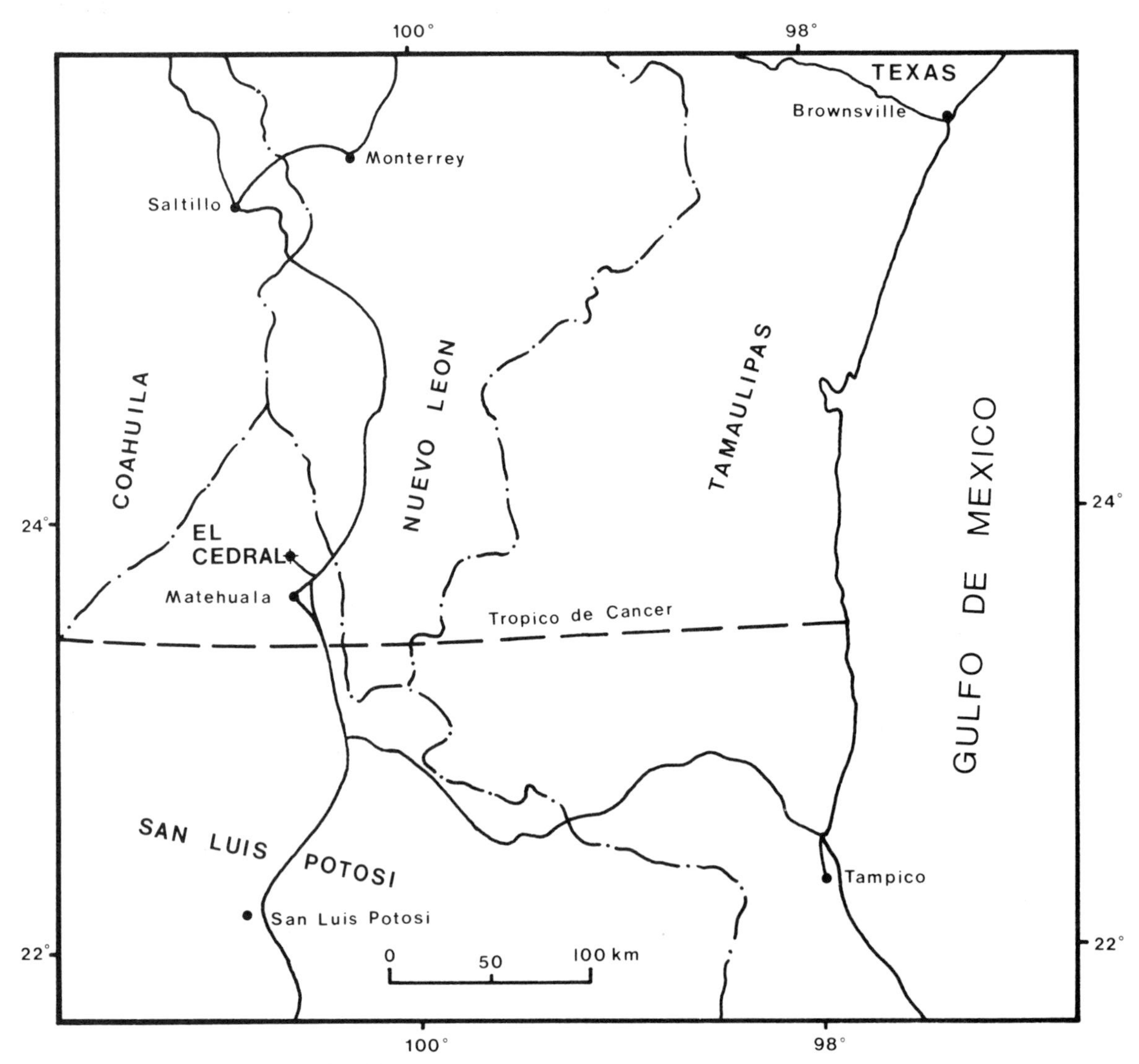

Figure 1. Location of El Cedral within Mexico.

DESCRIPTION OF RESEARCH

There were four seasons of excavation from 1977 to 1980 at the now-dry spring on the Rancho La Amapola. At the same time as the excavations, extensive geomorphological survey work was carried out within the old lake basin in order to determine the paleohydrologic system of the region.

Scientific excavations at the spring site on the Rancho La Amapola began with the removal of a thick layer of dark earth and debris which consisted of backfill and slumped material from historic excavations in the spring in search of the retreating water table. This layer of earth and debris contained many fragments of bone and wood, which were salvaged by screening. After the central part of the spring area had been cleaned of this disturbed material, a grid system was established and careful excavation began into undisturbed sediments. Altogether, in the four seasons of field work, an extensive area was excavated to a maximum depth of 4.5 m. Bones, stones, and other features exposed *in situ* were mapped and photographed; detailed profiles of the stratigraphy were recorded. Sediment samples for palynological and soils analyses were collected; and samples of organic materials for radiocarbon dating were obtained. Fragments of bone and wood were collected for identification, and lithic materials analyzed. Study of the materials recovered is continuing, but a preliminary report of the results to date is possible at this time.

GEOMORPHOLOGY

Geomorphological study of the ancient lake basin in the area of El Cedral produced evidence of a very ancient stream system which once extended through the area in a general west to east direction. This system discharged into the Matehuala depression, a karst zone. Changes in the ancient stream system were related to climatic changes in the region, with cycles of cutting and filling due to fluctuations in effective precipitation. A severe erosional cycle resulted in the formation of large alluvial fans which blocked this drainage system in its lower reaches, and the channels filled.

Close study of the small basin in which the site at Rancho La Amapola is located produced evidence for a chain of small lakes and two or three terraces, evidently old strand lines at different lake levels. It appears that an old stream system which once drained the area was converted into a lake or series of lakes when the drainage was impeded by alluvial fan expansion downstream. The ancient stream system has been covered by calcareous sediments deposited by a fluctuating or ephemeral lake, with dry intervals marked by erosional disconformities or paleosol horizons.

STRATIGRAPHY

The stratigraphy exposed in excavations at the Rancho La Amapola site is the result of complex relationships between lake deposition in relatively wet phases and spring discharge in drier periods. Fifteen strata have been recognized on the basis of variation in color, texture, and chemical precipitates, and organic material in the stratum (Figure 2). A tentative model of the sequence of events has been constructed. It is hypothesized that strong flow from the spring resulted in erosion of its banks and redeposition of clayey and calcareous sediments by overflow. At an early stage in the cycle, a forest grew up in the immediate vicinity of the spring. Preliminary identification of the wood preserved indicates that most of the trees were juniper. As the water flow increased and a shallow lake formed, this vegetation was flooded and drowned. Numerous tree trunks, branches, and even stumps traceable to the old land surface were buried in the clayey lake sediments; and samples of this organic material indicate a radiocarbon age over 40,000 years (I-10434) for this event. Subsequently a severe drought set in, and the lake dried. The sediments indicate a subsequent cycle of deposition by spring or ephemeral lake, with several intervals of complete desiccation and erosion. While the spring continued to flow, animals were attracted to the waterhole, and a great number of bones were incorporated into the deposits at the edge of the spring. Bone fragments were recovered from three stratigraphic horizons. A series of 14 radiocarbon dates on organic materials from the complex sequence of lake sediments and spring deposits indicates that the recorded cycle of periodic overflow and periodic drought extended from more than 40,000 years ago until about 2000 years ago.

FAUNAL REMAINS

Many of the bones recovered from the site at Rancho La Amapola were salvaged from the disturbed material removed from the spring area at the beginning of excavations, and thus were not found *in situ*. However, more than 50% of the bones were recovered from undisturbed sediments, with some bones found in anatomical position (Figure 3). The range of classes is broad: represented are large and small mammals, various

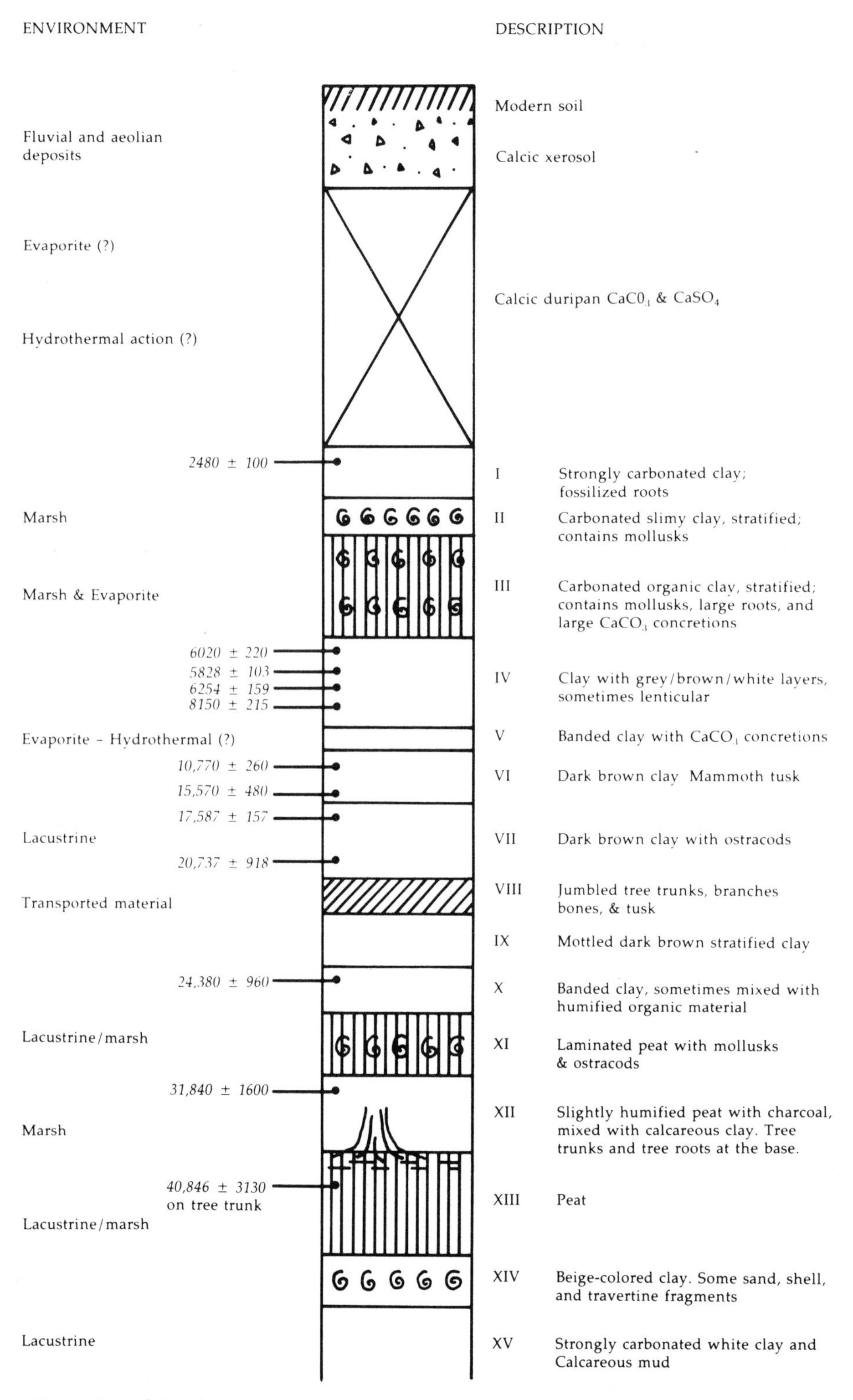

Figure 2. Composite stratigraphic column.

Figure 3. Exposed proboscidean bones.

birds, and reptiles. Remains of ostracods and mollusks were also recovered from the sediments. When all remains have been identified, a good data base will be available for paleoenvironmental interpretations. Most notable are the Pleistocene megafauna — glyptodont and mylodont, dire wolf (*Canis dirus*), short-faced bear (*Arctodus*), lion (*Felis atrox*), mastodont and mammoth, two species of horse (*Equus*), and camelids (*Camelops*).

EVIDENCE FOR EARLY MAN

One would expect that man would also be attracted to the spring or waterhole in order to obtain drinking water, to bathe, and to take advantage of any opportunity to kill game. However, it is unlikely that a campsite would have been situated in the immediate vicinity of the waterhole, a dangerous location due to the movement of numerous large herbivores and carnivores. One would expect, then, that artifacts or other material evidence of human activity in the immediate area of the spring would be rare.

Thus only a few artifacts were found *in situ* in the sediments excavated. One is a fragment of a horse tibia broken and used as a point (Figure 4), found *in situ* in a stratigraphic zone dated at 21,960±540 B.P. (I-10436). Another is a discoidal scraper of chalcedony, measuring 49 x 45 x 16.5 mm, made on a primary flake by complete edge trimming on the dorsal surface by direct percussion (Figure 5). This specimen was found *in situ* in a stratum dated 33,300+2700-1800 B.P. (GX-7684). Found *in situ* in strata dating older than 15,000 B.P. was a limestone nucleus with evidence of use as a hammerstone (measuring 66 x 60 x 44 mm) (Figure 6), and a chert blade (measuring 67 x 27 x 21 mm). The nearest source of chert is several kilometers distant, and there is no evidence of water transport on the specimen. A stratum dated between 6000 and 8000 B.P. yielded a pebble nucleus; and a chert nucleus was recovered below a zone of gypsum dated 2480±180 B.P. (GX-6637).

A hearth exposed *in situ* (Figure 7) in the sediments has yielded a radiocarbon date of 31,850 ±1600 B.P. (I-10438). This undisturbed feature consisted of a circle of proboscidean tarsal bones surrounding a zone of charcoal about 30 cm in diameter and 2 cm thick. Another feature consisted of a pit approximately 90 cm in diameter and 85 cm deep excavated into the deposits; the stratum of its origin can be dated between 6000 and 8000 years ago.

Figure 4. Utilized horse tibia.

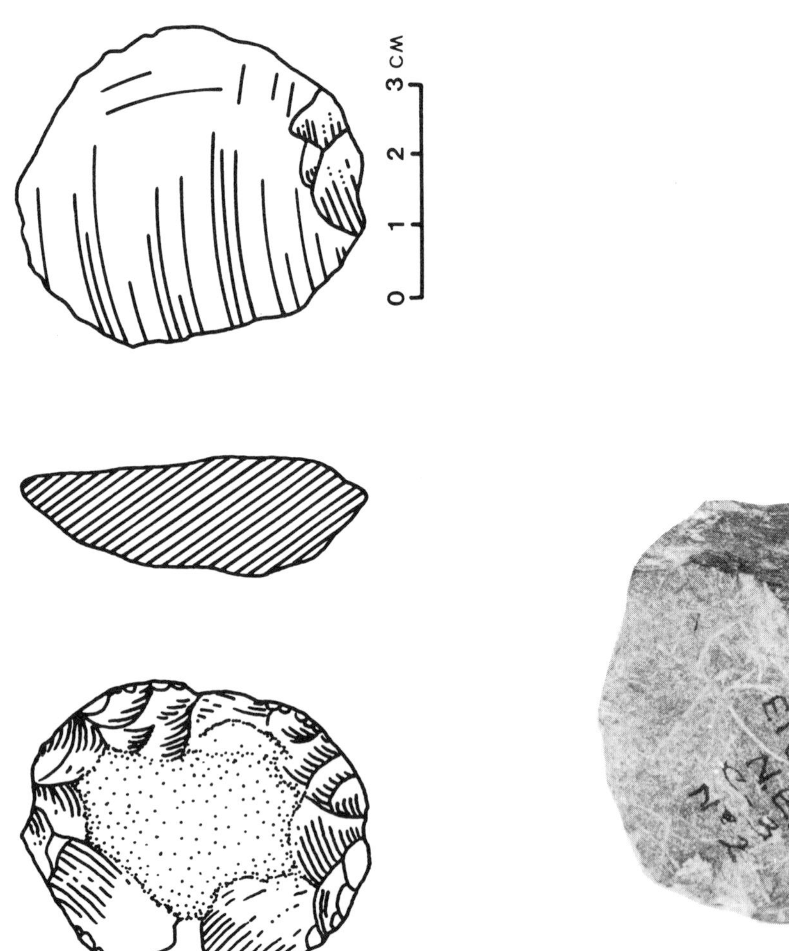

Figure 5. Discoidal scraper.

Figure 6. Limestone nucleus used as a hammerstone.

Figure 7. Hearth encircled by proboscidean tarsal bones.

In addition to these finds *in situ*, nine bone fragments recovered from the zone of disturbed material appear to show evidence of human modification. Three identifiable bone fragments are horse. All specimens are now undergoing careful study in order to determine the nature of the modification.

CONCLUSION

Regional geomorphological studies combined with excavations at a spring locality in the area of El Cedral have produced evidence which will permit reconstruction of paleoenvironmental changes in late Pleistocene and early Holocene times in this small structural depression situated on the border between the tropical and temperate zones. Below deep lake sediments, researchers have found traces of a very ancient stream system, indicating a climatic regime with much more regular precipitiation than at present. A severe erosional cycle produced large alluvial fans which impeded this drainage, and the stream channels were filled and buried. Subsequently, after 40,000 years ago, a cycle of lake deposition in moist periods and artesian spring flow in drier periods was initiated, reflected in the complex sequence of sediments exposed in the excavations at a spring on the Rancho La Amapola. In the drier phases, bones of animals utilizing the waterhole accumulated in the spring deposits. Pleistocene megafauna, including mammoth and mastodont, horse and camel, glyptodont and giant ground sloth, are well represented in the paleontological collection

The excavations at Rancho La Amapola found definite evidence of early man *in situ* in undisturbed stratified deposits. A discoidal scraper of chalcedony was found *in situ* in a stratum dated at 33,300+2700-1800 B.P. A fireplace consisting of a lens of charcoal surrounded by a circle of proboscidean tarsal bones yielded a radiocarbon date of 31,850±600 B.P. The modified fragment of horse tibia was found *in situ* in strata dating older than 15,000 B.P.

The excavations at Rancho La Amapola, then, have produced definite evidence for the presence of man in north-central Mexico by 33,000 years ago. It is hoped that further research in the El Cedral area will provide evidence of early settlement and subsistence patterns in the context of paleoenvironmental changes in the region in late Pleistocene and early Holocene times.

ACKNOWLEDGEMENTS

This article is a condensation of preliminary reports on the field seasons at El Cedral in 1977, 1978, 1979, and 1980, and a paper presented at the X Congress of the IUPPS in Mexico, 1981; edited and translated by Ruth Gruhn.

Apuntes Sobre el Medio Ambiente Pleistocénico y el Hombre Prehistórico en Colombia

GONZALO CORREAL URREGO
Instituto de Ciencias Naturales, Universidad Nacional de Colombia
Bogotá
COLOMBIA

Abstract

Archaeological and paleoenvironmental research in Colombia in the past decade has provided a record of human occupation and climatic change extending over the past 13,000 years. At present the best information is available from the Sabana de Bogotá, a high basin in the central highlands, where three major archaeological sites have produced evidence of man in stratigraphic horizons which may be correlated with palynological zones reflecting a sequence of major changes in climate and vegetation since the late Pleistocene.

The paleoenvironmental sequence on the Sabana de Bogotá has been established by the research of Thomas van der Hammen and his associates. At the close of the Pleistocene the climate was relatively cold and dry, and open vegetation of páramo type covered the floor of the basin. Between 13,000 and 11,000 B.P., in the Guantiva interstadial, the basin floor was invaded by Andean forest. The first definite evidence of man on the Sabana de Bogotá dates to this period. From 11,000 to 10,000 years ago, in the Abra stadial, the climate was relatively colder and drier, and the forest retreated, with the formation of a sub-páramo environment. After 10,000 years ago the climate became warmer, and an Andean forest was established. The Hypsithermal phase commenced about 8500 years ago, and there was a severe dry period around 5000 years ago. The archaeological record shows that the early human inhabitants of the Sabana de Bogotá were greatly affected by this sequence of climatic changes.

The earliest definitely dated evidence of man in Colombia has been recovered from rockshelters in a small valley called El Abra, in the northern sector of the Sabana de Bogotá. Stratigraphic zone C3 of Rockshelter no. 2 at El Abra yielded a number of small tools made by unifacial edge retouch on cores and flakes of chert, associated with charcoal which produced a radiocarbon date of 12,460±160 B.P. Unifacial artifacts of the same tradition, now known as the Abriense industry, were found in three concentrations with bone fragments and rocks in stratigraphic zone 3A at the open site of Tibitó, in a small basin not far from El Abra. Identifiable bone from these three activity areas, which were adjacent to a very large sandstone boulder, indicated the procurement of horse (*Equus (Amerhippus)*), mastodon (*Haplomastodon*), and deer (*Odocoileus virginianus*). Many of the bones were split, and some were charred. A radiocarbon date of 11,740±110 B.P. was obtained from a sample of bone from zone 3A at Tibitó. There is a later occupation of the site, with a continued association of the Abriense industry with extinct horse and mastodon, in the overlying zone 3, a stratigraphic unit which can be correlated with the regional palynological sequence and referred to the cool Abra stadial, dated between 11,000 and 10,000 B.P.

Rockshelters at Tequendama, at the southwest edge of the Sabana de Bogotá, contain well-stratified deposits with a record of human occupation extending back to the Abra stadial. Occupation I, the earliest occupation zone, with a radiocarbon date of 10,920±260 B.P., has yielded small core tools and flake tools of the Abriense industry. The tools can be interpreted as scrapers, cutting tools, and perforators. Distinctive artifacts found on this horizon include single examples of a fragmentary bifacial projectile point, a leaf-shaped biface, and a keeled scraper. There were also some lamellar blades. Bones associated with Occupation I at Tequendama were predominantly those of deer (*Odocoileus* and *Mazama*), plus an assortment of small mammals. Also recovered were five human phalanges which were split longitudinally and calcined. In Occupation II, dated between 9500 and 8500 B.P., deer and small mammals continued to be the game animals; rodents increased in frequency relative to deer. Remains of terrestrial snail shells were also abundant in the occupation refuse. The Abriense lithic tradition continued, with changes in frequency of artifact types; spokeshaves were common. In addition, numerous bone artifacts were associated with Occupation II at Tequendama.

A generalized hunting and collecting economy is indicated for the early occupants of the rockshelters on the Sabana de Bogotá. Between 8500 and 5000 B.P., this general way of life continued, with shifts in frequency of particular species hunted, and changes in frequency of specific artifact types within the Abriense tradition. Pounding and grinding implements may indicate increasing emphasis on plant foods. There is evidence of a gradual increase in the human population of the Sabana de Bogotá over the millennia, until the severe drought around 5000 B.P. reduced or scattered the population in the area. At this time, in the northern lowlands of Colombia, hunting/collecting populations concentrated on the coast or near marshes and lagoons, and shellmounds accumulated. In this context on the Caribbean coast pottery appeared by 5000 B.P. Ceramics appeared by 2300 B.P. on the Sabana de Bogotá, associated with the establishment of sedentary agriculture.

INTRODUCCION

Las investigaciones arqueológicas relacionadas con la etapa que antecedío en varios milenios a las poblaciones aborígenes portadoras de la agricultura intensiva, organizadas en aldeas y enriquecidas por una compleja gama de manifestaciones culturales, fueron incrementadas en Colombia hace dos década, cuando el hallazgo de sitios precerámicos estratificados permitió hacer la lectura de acontecimientos pretéritos guardados en las páginas borrosas de las capas de la tierra pisadas por el hombre más antiguo en Colombia. Se empieza a descorrer el telón que permite escudriñar el panorama que tuvo ocurrencia hacia finales del Pleistoceno, hace algo más de doce mil años; pero esta visión inicial, sólo nos permite identificar acontecimientos fragmentarios; en unos sitios, los restos humanos nos cuentan algo sobre las características físicas del hombre prehistórico o nos ilustran acerca de sus prácticas funerarias; en otros, la presencia de abundantes artefactos de piedra nos facilitan la reconstrucción de sus actividades como cazadores y recolectores, en algunas áreas los restos de animales y las partículas de polen fósil de plantas permiten establecer la relación medio ambiente-hombre.

Solamente una paciente y prolongada investigación en el futuro, permitirá la reconstrucción de los capítulos de la prehistoria de Colombia que aún quedan por descubrir. Por ello, sólo queremos, confiando en la generosidad de nuestros lectores, presentar estas notas resumidas sobre el precerámico en Colombia.

ANTECEDENTES Y ESTADO ACTUAL DE LAS INVESTIGACIONES SOBRE EL PRECERAMICO

En Colombia, ya desde 1951 el doctor Luis Duque Gómez (1955:100) planteaba la posibilidad de la existencia de un substrato prechibcha en la Sabana de Bogotá, basado en el hallazgo, durante investigaciones arqueológicas adelantadas en el municipio de Mosquera, de un utillaje de piedra asociado a fragmentos cerámicos de aspectos atípicos en la morfología conocida para estos elementos en el complejo cultural muisca.

El análisis de conjunto del condiciones ambientales ocurridas durante el Pleistoceno en nuestro territorio, permitía postular que estas fueron favorables para el tránsito humano hacia Suramérica, a través del Istmo de Panamá; entre los múltiples factores que podrían facilitar este ingreso menciona Reichel-Dolmatoff (1965a) la existencia de'un puente de tierra formado por el descenso del nivel marino. A estas hipótesis y planteamientos se sumaron los hallazgos superficiales de instrumentos de piedra que recordaban las tecnicas planteamentos del estadio ''pre-puntas de proyectil'' en áreas de la Costa Atlántica (Canal del Dique), en las regiones de San Nicolás en el Bajo Sinú, Also río Baudó, río Jurubidó, río Chorí y Bahía de Utría en la Costa Pacífica y también en las bocas del Carare en el área de Magdalena Medio; estos artefactos, de pedernal en unos casos, de flint y piedras cuarzosas en otros, mostraban indudables señales de manufactura humana. En otros sitios, como en El Espinal (Tolima), Manizales (Caldas), Restrepo (Valle) y en la Costa Atlántica en Mahates, Santa Marta y Betanci, fueron halladas puntas de proyectil líticas, elementos éstos, que nos dan testimonio de la presencia de cazadores muy primitivos en esas áreas (Reichel-Dolmatoff 1965a:46-50).

Las hipótesis anteriormente formuladas acerca de la existencia de grupos nómadas cazadores, recolectores y pescadores que antecedieron a los autores de nuestros desarrollos culturales formativos y preclásicos en nuestro altiplano andino, quedaron definitivamente comprobadas

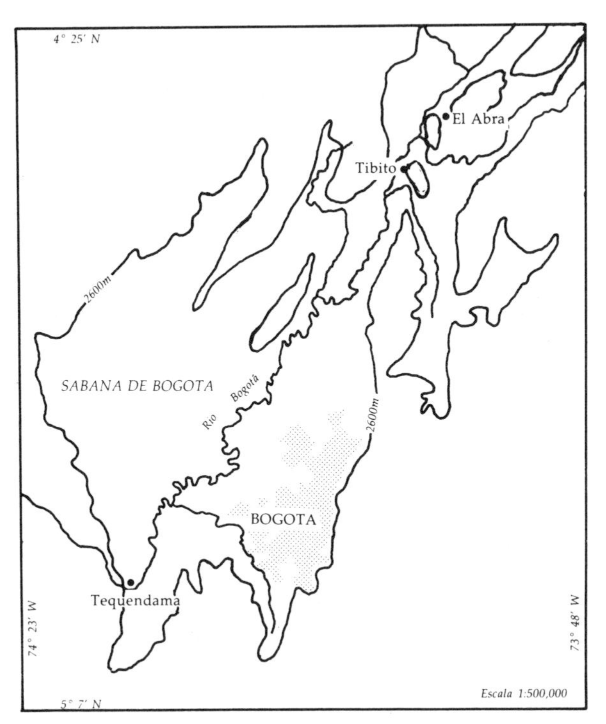

Figura 1. Mapa de Sabana de Bogotá.

con el hallazgo, durante investigaciones adelantadas por el Instituto Colombiano de Antropología, en los abrigos rocosos de El Abra al sureste del municipio de Zipaquirá, en el cual se encontraron los vestigios arqueológicos más antiguos hasta ahora conocidos en Colombia y fechados mediante el procedimiento de radiocarbono en 12.460 años antes del presente. Estas investigaciones fueron continuadas y ampliadas con la colaboración de la Universidad de Indiana durante el año de 1969 en desarrollo del proyecto "Medio ambiente pleistocénico y Hombre Prehistórico en Colombia" (Correal et al. 1970; Correal et al. 1977; Hurt et al. 1972, 1976).

Una serie de exploraciones practicadas en la Cordillera Oriental durante el año de 1976, nos permitió identificar nuevos sitios precerámicos en los municipios de Madrid, Ubaté, Bojacá y Soacha; en este último, con el apoyo de la fundación Neerlandesa para el fomento de estudios tropicales (WOTRO), se practicaron excavaciones arqueológicas sistemáticas en el sitio denominado Tequendama 1, al suroeste de la cabecera municipal siendo posible allí reconocer una secuencia cultural continua, que abarca desde el XI milenio hasta el año 5.000 antes del presente (Correal and Van der Hammen 1977). A estos hallazgos, se añadió un nuevo sitio fuera del altiplano sobre las terrazas aluviales situadas al noroeste del municipio de Neiva, en predios de la hacienda Boulder.

Investigaciones más recientes nos permitieron, en el término comprendido entre 1973 y 1978, localizar una serie de 20 estaciones precerámicas superficiales, en las cuales se registraron huellas inequívocas del paso de los primeros pobladores de nuestro territorio. Estas estaciones, se extienden desde la Península de la Guajira (Serranía de Cosinas) hasta el departamento del Huila. La continuidad de estas investigaciónes, adelantadas con el apoyo financiero de la Fundación de Investigaciones Arqueológicas Nacionales del Banco de la República en colaboración con el Instituto de Ciencias Naturales y Departamento de Antropología de la Universidad Nacional, nos permitió, luego de exploraciones sistemáticas adelantadas en el sítio Sueva 1 al noroeste del corregimiento del mismo nombre (municipio de Junín), obtener una nueva fecha de 10.090 años antes del presente y reconstruír una secuencia continua precerámica. En la misma vertiente del Guavio, en trabajos practicados por el Instituto de Ciencias Naturales en colaboración con el Departamento de Antropología de la Universidad Nacional, fueron localizados vestigios culturales precerámicos al noroeste del municipio de Gachalá, obteniéndose una fecha de 9.360 antes del presente. Posteriormente, un sitio excavado en las proximidades del municipio de Nemocón, en la vereda Piedecuesta situada al norte de la cabecera municipal, nos permitió nuevamente identificar un complejo lítico asociado a abundantes restos de fauna; estas evidencias culturales, en uno de los estratos fechado mediante carbono 14, nos permitieron datar la parte baja de la secuencia en 7.530 años antes del presente. Otro hallazgo relativamente reciente consistió en una punta de proyectil acanalada, cuyo registro se efectuó en Bahía Gloria (Golfo del Darien), durante una exploración arqueológica en aquella área con los auspicios del Ministerio de Educación Nacional, a todos estos hallazgos nos referiremos en forma mas detallada a lo largo de estas notas.

CARACTERISTICAS COMUNES DE LOS SITIOS PRECERAMICOS

Todos los sitios arqueológicos precerámicos localizados en nuestro altiplano andino, en la Sabana de Bogotá y sus alrededores, presentan un común denominador: se encuentran en parajes pintorescos, protegidos de los vientos y adornados por rocas areniscas duras que se proyectan formando abrigos o aleros; están situados cerca de las corrientes fluviales y nichos ecológicos favorables para la obtención de recursos de subsistencia.

Los abrigos rocosos de El Abra, Tequendama, Sueva, Nemocón y Gachalá (Departamento de Cundinamarca), constituyeron el albergue natural de nuestros primeros pobladores; allí, pudieron protegerse de las inclemencias del tiempo, alrededor de los fogones integraron sus actividades, prepararon sus alimentos, elaboraron los instrumentos necesarios para sus excursiones de cacería y recolección, y enterraron con místico respeto a sus muertos.

En el Valle del Magdalena y en los pisos térmicos cálidos, los cazadores recolectores y pescadores precerámicos escogieron, como sitios preferenciales para sus estaciones temporales, lugares próximos a las ciénagas y terrazas aledañas a los ríos, como lo demuestran los hallazgos consistentes en herramientas muy primitivas que se suceden desde la Costa Atlántica hasta el departamento del Huila.

MEDIO AMBIENTE Y CARACTERISTICAS CULTURALES

Los estudios palinológicos practicados en los sitios de El Abra, Tequendama y otros lugares del altiplanicie, demuestran que hacia el final del Pleistoceno durante el décimo tercer milenio antes del presente el clima era frío y relativamente seco haciéndose más húmedo hacia el final de dicha fase; el record palinológico indica una vegetación abierta de tipo páramo con el límite del bosque situado por lo menos 1.300 m más abajo que el actual; es en este momento cuando aparecen las primeras huellas del hombre en la Sabana de Bogotá en la unidad estratigráfica C-2 del abrigo No. 2 del Abra se encontraron 29 lascas de chert que sugieren la presencia del hombre durante esta época en esta área.

El record palinológico de la Unidad C3 del Abra fechada entre 13.000 y 11.000 años antes del presente y que se pueden correlacionar con la zona de ocupación 1 del Tequendama, demuestra que durante esta época los bosques invadieron la hoya de la sabana y las faldas de las montañas circundantes y en la altiplanicie alrededor de los lagos y lagunas existían bosques marginales de aliso. Hacia 12.500 años antes del presente, el clima ha mejorado tanto que el área de la altiplanicie de Bogotá y sus alrededores fueron invadidos por el bosque andino y el límite altitudinal de los bosques se encontraba a unos 400 m por debajo del actual. Este clima más calido de la Sabana de Bogotá combinado con bosques que incluyen el encenillo (*Weinmannia*), robles (*Quercus*) y abundantes arbustos que incluyen *Myrica* y *Symplocos* así como la abundancia de reductos lacustres indica un habitat favorable a los animales de caza; todos estos factores aumentan la probabilidad de la presencia del hombre en el Abra durante esta época que es denominada interestadial de Guantiva. Artefactos obtenidos en la unidad C3 del Abra y fechados en 12.460 ± 160 B.P. nos muestran que los artefactos de esta época fueron elaborados mediante percusión simple y se caracterizan por la preparacion de un lado de trabajo. Las formas mas comunes de chert usadas en su preparación son fragmentos tabulares que se encuentran en afloramientos cercanos al sitio siendo también usados en menor proporción guijarros que pueden obtenerse en las terrazas al pie de los ríos. Entre los artefactos de esta época se encuentra un raspador unifacial de chert negro. El reducido número de artefactos del Abra durante esta época indica tan sólo la presencia de un grupo poco denso de individuos que ocuparon durante un corto lapso de tiempo los abrigos del Abra. Situación similar revela la unidad estratigráfica 5a del sitio Tequendama 1. Ya hacia finales el Pleistoceno las cambiantes condiciones climáticas marcaron nuevos rumbos en el acontecer prehistórico. Hace aproximadamente 11.000 años el clima se tornó nuevamente frío y considerablemente seco; esta condición se prolongo casi un milenio. La vegetación de la Sabana de Bogotá adquirió entonces el carácter del bosque de páramo. La vegetación local contiene Ciperaceae, Hydrocotyle, *Geranium* y Caryophyllacea.

Tal situación ecológica favoreció en la Sabana de Bogotá la permanencia de los primitivos cazadores. Cuando en los bosques y praderas abiertas proliferaron abundantes especies de animales, sólo algunos restos de animales fueron encontrados en los estratos correspondientes del Abra mientras que en los abrigos rocosos del Tequendama abundan restos de venados (*Odocoileus* y *Mazama*) cuyos vestigios constituyen el mayor porcentaje y en menor proporción animales más pequeños: como el ratón silvestre (*Sigmodon*), el curí (*Cavia*), el conejo (*Sylvilagus*), el armadillo (*Dasypus*), el zorro patón (*Tayra*), el perro de monte (*Potos flavus*). Durante esta época conocida como el interestadial del Abra, los pocos artefactos encontrados en los abrigos rocosos de esta denominación son del tipo abriense, predominado la percusión como técnica y elaboración, pero es interesante señalar que la serie de artefactos del sitio Tequendama correspondientes a la zona de ocupación 1 — muestra herramientas de mejor elaboración tecnológica; una punta de proyectil fragmentada, una hoja bifacial de cuarcita y un raspador aquillado muestra finos retoques sobre su superficie; se registran en este sitio también núcleos con preparación previa de plataforma y lascas prismáticas que sugieren un inicial manejo de la técnica del astillado prismático.

Una serie de cuchillos laminares junto con raspadores de diferentes tipos se relaciona con la limpieza y preparación de pieles y tratamiento de las presas de cacería. Sobre el total de los artefactos, más del 50% son cortantes, 30% raspadores y 7% perforadores; raspadores concavos son muy escasos. Este conjunto de evidencias nos demuestra que los abrigos del Tequendama estuvieron habitados durante el estadial del Abra en el lapso cronológico comprendido entre 11.000 y 10.000 años B.P. por grupos de cazadores más o menos especializados que se habían adaptados a los terrenos semi-abiertos de la altiplanicie de Bogotá, Del físico del hombre de esta época sólo podemos saber lo que nos muestra cinco falanges con fracturas longitudinales parcialmente calcinadas. Mientras los abrigos rocosos del Tequendama muestran una habitación densa durante este período, los abrigos rocosos del Abra fueron visitados sólo ocasionalmente, lo cual sugiere que fueron ocupados probablemente por los mismos cazadores que ocuparon el Tequendama o grupos similares contemporáneos.

Artefactos elaborados en basalto encontrados en el sitio Tequendama nos están indicando desplazamientos entre el Valle del Magdalena y áreas aledañas de la Cordillera Central y la altiplanicie oriental.

EVIDENCIAS CULTURALES ASOCIADA A MEGAFAUNA

Investigaciones arqueológicas practicadas durante 1980 por el Instituto de Ciencias Naturales de la Universidad Nacional con el apoyo financiero de la Fundación de Investigaciones Arqueológicas Nacionales del Banco de la República, permitieron en el sitio TIBITO 1 Sabana de Bogotá (municipio de Tocancipá), el hallazgo, por primera vez, en Colombia de artefactos asociados a megafauna, (mastodonte) junto con otros restos de caballo y venado.

La unidad estratigráfica 3A, fechada en 11.740±110 B.P. se corresponde con el interestadial de Guantiva. La vegetación, de acuerdo con los polen-diagramas obtenidos, incluye alisos (*Alnus*), robles (*Quercus*), trompetos (*Bocconia*) y compuestos alternando con praderas (Gramineae).

La presencia de algas como *Botriococcus* nos muestra que el área fue propensa a las inundaciones durante la época en que se depositó esta unidad estratigráfica.

Las concentraciones de restos óseos de *Equus* (A), de mastodonte (*Cuvieronius hyodon* y *Haplomastodon*) y, en menor escala, venado (*Odocoileus virginianus*) junto con artefactos en la periferia de la roca central, nos permiten identificar el sitio como una estación de matanza y ubicar este complejo arqueológico dentro de la étapa paleoindia (lítica superior), aunque no se hallaron puntas de proyectil. La carencia de éstas no excluye su manufactura por los cazadores de este período, como tampoco excluye el uso de materiales perecedores como la madera y el hueso.

Son muy significativas las características de los depósitos 1-2-3 incorporados en esta unidad; es evidente en ellos, la acumulación selectiva de restos de fauna, caballo, mastodonte y venado, junto con artefactos. Este hecho sumado al registro de restos parcialmente calcinados, partículas de carbón, presencia de huesos con fracturas longitudinales probablemente con fines de extracción de la médula, el registro de hueso con incisiones y en un caso mostrando aparentes señales de ruptura ocasionada por arma punzante contudente, y la acumulación de fragmentos de roca arenisca, nos muestra la relación de estas evidencias con actividades de cacería.

Este conjunto de resultados sumados a la ausencia de evidencias similares en las zanjas exploratorias y el corte adjunto (Cuadrícula H), descartan en amplio grado la posibilidad de acarreo. Apoya también estas premisas la concordancia tipológica de algunos artefactos con series ya estudiadas anteriormente en la Sabana de Bogotá.

Si bien estas actividades, pueden representar la simple respuesta a necesidades básicas de subsistencia (tasajeo, limpieza y preparación, descuartizamiento y cocción), no se descarta, que dado el carácter de las acumulaciones en los depósitos, éstos representen contenidos culturalmente más elaborados que se manifiesten en superestructuras de carácter ritual, como lo podría indicar la acumulación selectiva y calcinación en los depósitos 1-2-3.

Entre los pocos artefactos asociados a esta unidad estratigráfica 3A, predominan elementos del tipo Abriense, siendo mayor la frecuencia de fragmentos de núcleo, y lascas con señales de uso, que corresponden básicamente a instrumentos cortantes, los cuales representan el 41% de los artefactos obtenidos en esta unidad. Un cuchillo (raspador) y raspador aquillado nos muestran una tecnología, por lo menos en este último, que la asimila a la zona de ocupación 1 del Tequendama. Los artefactos fueron elaborados usando como materia prima el *chert*.

Las evidencias paleontológicas, nos muestran una densidad más alta de restos de *Haplomastodon* y en menor proporción restos de *Cuvieronius hyodon* y venado, durante la época en que se depositó la unidad estratigráfica 3A.

Durante la época en que se depositó la unidad estratigráfica 3, las evidencias palinológicas nos revelan un descenso de la temperatura; la vegetación que incluye *Calamagrostis*, *Geranium*, *Lofhosoria* y *Valeriana* entre otras especies, nos indica la presencia de un sub-páramo. Esta situación concuerda con estudios adelantados en la Sabana de Bogotá, que nos permiten situar esta unidad estratigráfica frente al estadial del Abra y cronológicamente entre 11.000 y 10.000 años antes del presente.

Aunque durante esta época de ocupación no se registran estructuras de acomodamiento o áreas de fogón o de taller con una localización específica, nuevamente se registran concentraciones de restos óseos y de artefactos en el sector de la periferia de la roca, pero parcialmente al W. Los artefactos de este período son del tipo Abriense, desapareciendo instrumentos con finos retoques; continúan predominando fragmentos de núcleo o instrumentos cortantes y disminuye también la frecuencia de lascas retocadas, hecho que nos puede indicar una decadencia en las técnicas del trabajo de la piedra.

La disminución de las frecuencias de los artefactos y de los restos óseos en la unidad estratigráfica 3 en comparación con la 3A, nos puede indicar una disminución de la actividad de cacería especializada en esta área.

La disminución de restos de *Haplomastodon* puede relacionarse con los cambios ecológicos

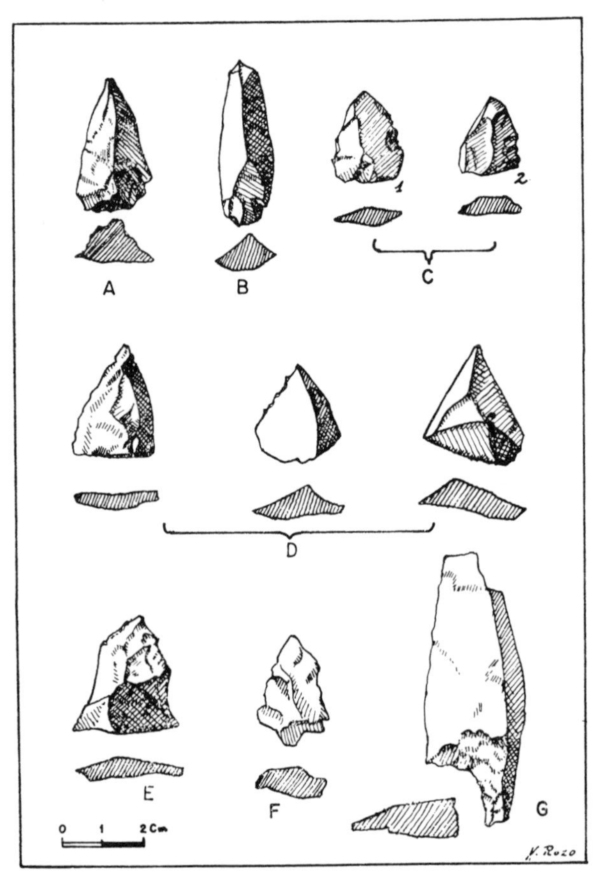

Figura 2. Lascas modificadas de la industria Abriense.

Figura 3. Tibito. Vista de las excavaciones desde el su roeste.

ocurridos durante este período. La carne utilizada en la alimentación como lo denotan los restos óseos, era prevista por proboscidios y, en menor escala, por venados. Similar a lo ocurrido durante la primera ocupación, durante el período en que se depositó la unidad 3, es notoria la mínima densidad de huesos largos. Esta situación no

Figura 4. Tibitó. Depósito 1 con vestos de mastodonte, caballo, venado, 4 artefactos de piedra.

excluye el carácter del sitio como estación de tasajeo y preparación de elementos obtenidos durante actividades de cacería (huesos rotos, y en algunos casos cremación).

Aunque en el estado actual de nuestra investigación se carece de fecha de carbón 14 para la unidad estratigráfica 2, por correlación con otros sitios de la Sabana de Bogotá, sabemos que tanto ésta, como la parte más alta de la secuencia, se sitúan en el Holoceno, como también se sabe, hacia el comienzo de éste hace unos 10.000 años A.P. el clima comenzó a mejorar en forma definitiva (Correal and Van der Hammen 1977:168) quedando la Sabana de Bogotá en la zona del propio bosque andino. No se registran evidencias arqueológicas ni paleontológicas en esta unidad estratigráfica (2) en el sitio Tibitó 1; es interesante señalar que simultáneamente en los sitios de Tequendama y El Abra, para los comienzos de Holoceno se nota una disminución de la población en esos abrigos rocosos, hecho que debe tener relación con el cambio climático ocurrido al finalizar el Pleistoceno; presión ecológica que debió determinar el desplazamiento o la adaptación del hombre a las nuevas circunstancias; posteriormente, en las áreas de Tequendama en la zona de ocupación II (9500 a 8500 B.P.), se puede identificar una población mejor adaptada a las nuevas circunstancias; no se trata ya de cazadores superiores especializados, sino de cazadores de especies menores como el venado y los roedores; para este período también la abundancia de restos de caracoles indica actividades de recolección.

En estos mismos sitios fuera de Tibitó, hacia 8500 años B.P. al principio del hipsitermal, el carácter más denso de los bosques y la reducción de las áreas abiertas pantanosas debió conducir a una nueva disminución de la población de algunas áreas de la Sabana de Bogotá (Tequendama, El Abra etc.).

Una población más densa se registra en el lapso entre 7000 y 6000 años, marcándose una nueva disminución hacia el 5000 B.P. hasta donde hemos podido seguir la secuencia precerámica de nuestra altiplanicie; durante este último lapso cronológico, como ha sido establecido (Van der Hammen 1973), debió influir en esta situación un período de fuerte sequía (ocurrido cerca de 5000 años B.P.) que se pudo reconocer por medio de diagramas de polen, no sólo en la Sabana de Bogotá, sino también en las planicies de clima tropical de los Llanos Orientales.

Las evidencias palinológicas de Tibitó 1 nos muestran que mientras se depositaba la unidad 2, la vegetación incluía los tipos reconocidos en el bosque andino (*Quercus, Dicksonia, Podocarpus,* Cyatheaceae).

La ausencia de registros palinológicos y arqueológicos, nos puede indicar, que el sector de nuestra investigación perdió importancia, durante la época en que se extinguieron especies como el mastodonte y el caballo; por otra parte el carácter mismo anegadizo de estos terrenos, y la falta de abrigos rocosos suficientemente amplios, no garantizaban la presencia estacionaria del hombre adaptado a las nuevas condiciones del sitio Tibitó 1.

Se cierra aquí un pequeño capítulo de nuestra prehistória, con la expectativa de que futuras investigaciones nos permitan ampliar la informaci- ón sobre evidencias culturales asociadas a megafauna durante el Pleistoceno en nuestro territorio colombiano.

EL HOLOCENO

Hace unos 10.000 años antes del presente, al finalizar el Pleistoceno y comenzar nuestro período actual el Holoceno, el clima mejoró considerablemente a consecuencia del ascenso térmico. El límite altitudinal del bosque se trasladó más hacia arriba quedando entonces la sabana de Bogotá en la zona del bosque andino, abundando en las áreas bajas *Alnus*, en los alrededores de las faldas bajas montañosas *Weinmannia* (probablemente abundante en la parte plana), *Ilex, Borreria,* Myrtaceae, Melastomataceae, y en los suelos secos de la altiplanicie abunda *Myrica*. La relativa abundancia de Dodoneae indica disturbios locales y erosión de suelos en los alrededores de los abrigos rocosos probablemente originada en el desmonte causado por el hombre.

La ocurrencia de fenómenos volcanicas nos es revelada por la presencia de acumulaciones de ceniza volcánica asociada a estratos fechados mediante carbono 14 en el período cronológico anteriormente señalado. La baja densidad de

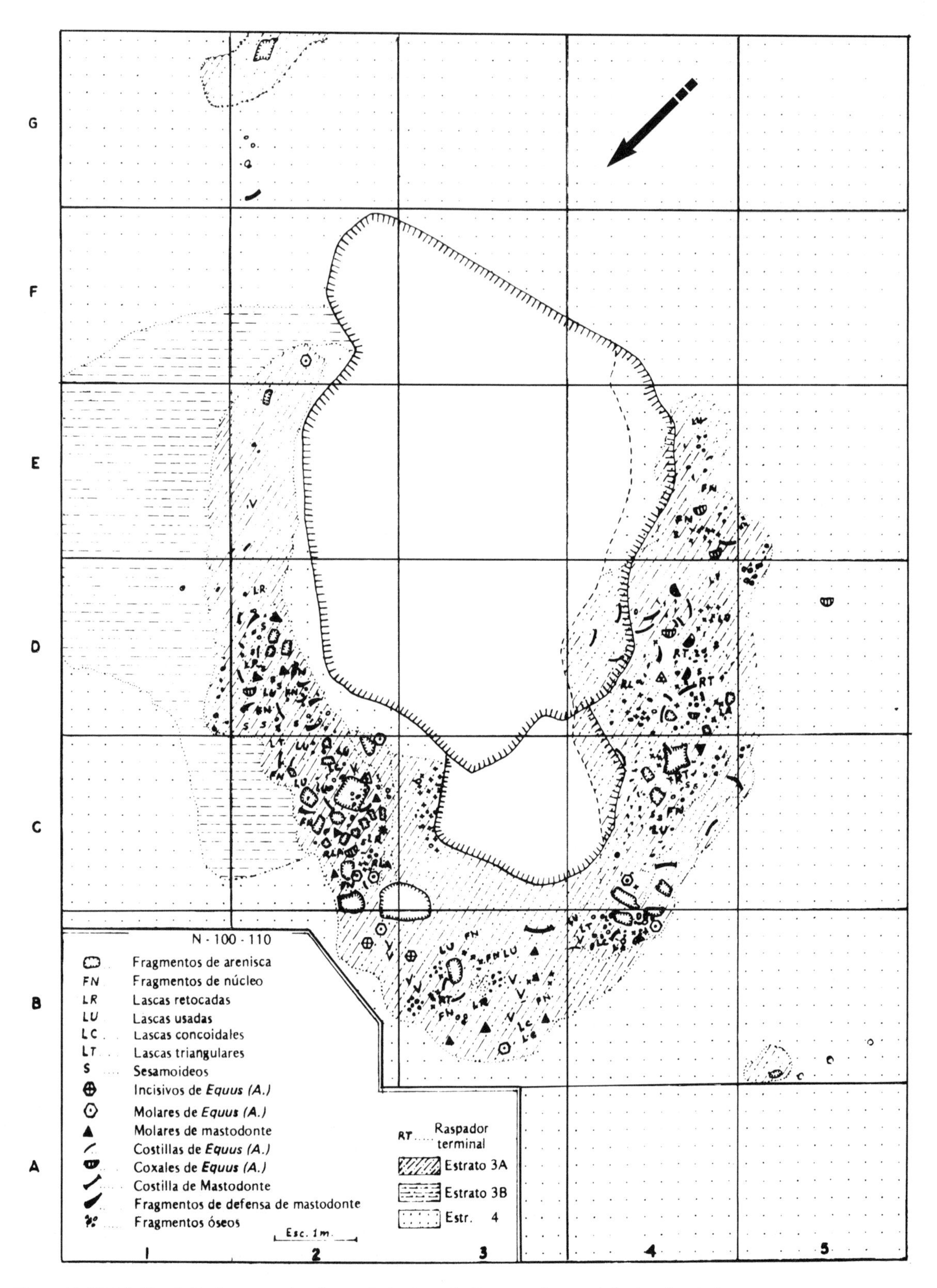

Figura 5. Tibitó. Depósito 1 con restos de mastodonte, caballo, venado, y artefactos de piedra.

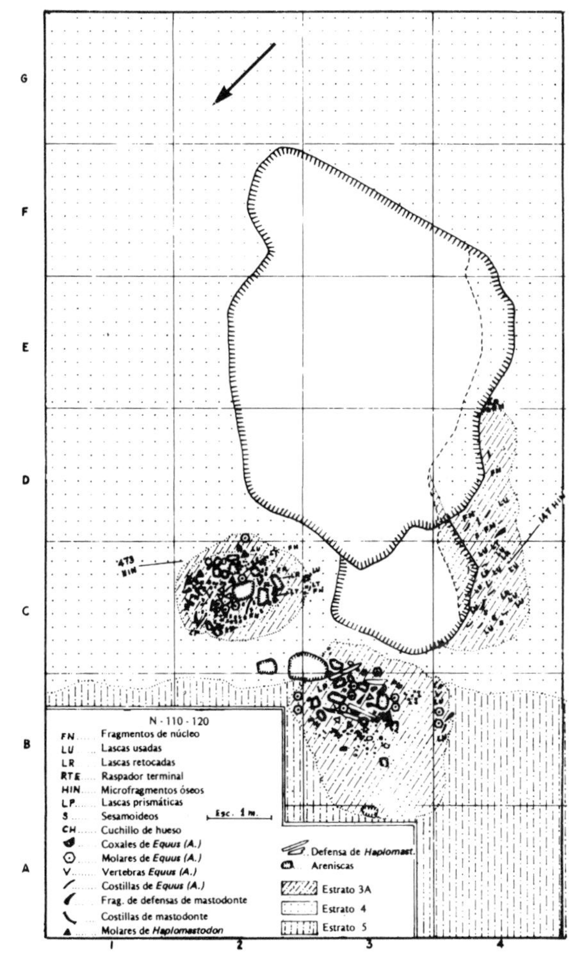

Figura 6. Tibitó. Plano de estrato 3A, nivel 110-120 cm.

artefactos en los abrigos de la Sabana, nos muestra una considerable disminución de la población.

Con posterioridad a estos acontecimientos, las evidencias del Tequendama nos muestran una ocupación por grupos más densos adaptados a nuevas condiciones ambientales; en el período comprendido entre 9500 y 8500 años antes del presente, aunque en la sabana existían áreas pantanosas, los bosques densos de encenillos y robles dominaban las florestas, y su límite altitudinal ascendió entonces por encima de los 3000 m como lo demuestran estudios palinológicos. Bajo los abrigos se pueden apreciar áreas de fogones y una gran concentración de huesos y artefactos alrededor de éstos; las acumulaciones de lascas indican que los artefactos fueron allí elaborados.

Durante este período, se continúa conservando el tipo abriense en los artefactos, pero a través de los registros arqueológicos del Tequendama, ya se manifiesta una decadencia en la elaboración de los artefactos de piedra, hecho que se traduce en la desaparición de los instrumentos con finos retoques y técnica de presión, pero aun subsisten raspadores aquillados y cuchillos laminares; la alta frecuencia de raspadores cóncavos, durante este período, bajo los abrigos rocosos del Tequendama y Nemocón, nos indica un incremento del trabajo de las maderas ya que estos elementos, son útiles adecuados para remover la corteza y aguzar extremos en los implementos de este material.

La industria de huesos alcanzó su apogeo durante este período, a juzgar por la gran cantidad de artefactos de este material, entre los que se distinguen punzones, perforadores, raspadores y cuchillos, elementos prácticamente ausentes en los períodos anteriores. La persistencia de materiales como el basalto, nos indica que todavía se producían desplazamientos o migraciones amplias.

De acuerdo con los resultados de las excavaciones arqueológicas practicadas en el sitio de Tequendama 1, se pueden identificar

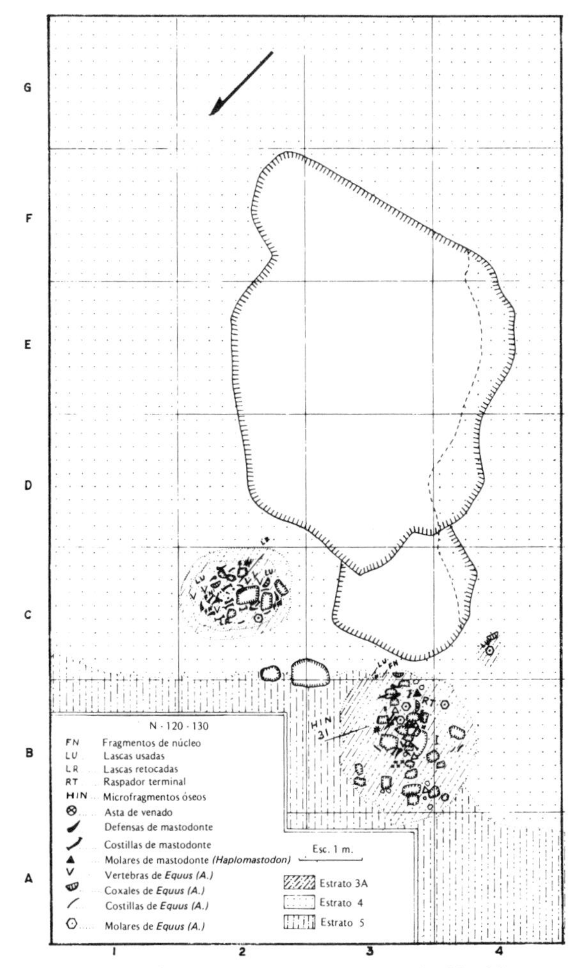

Figura 7. Tibitó. Plano de estrato 3A, nivel 120-130 cm.

modificaciones de la fauna, que se manifiestan en un aumento en la proporción de roedores, como el ratón silvestre, el borugo o tinajo y el guatín, aunque si bien en menor proporción, los restos de mamíferos nos demuestran que cazaban todavía venados. En la unidad estratigráfica 3 de Nemocón, cuya edad se calcula entre 9000 y 8000 años antes del presente, la fauna es muy variada, estando compuesta por venados, conejos, curíes, ratones y armadillos, a los que se añaden la nutria, el saino y el aullador, especies que nos indican la gran movilidad de nuestros cazadores, que debieron desplazarse frecuentemente hacia pisos térmicos cálidos en busca de medios de subsistencia.

Hacia el VII milenio antes del presente, los abrigos rocosos de la Sabana de Bogotá y vertiente del río Guavio muestran una población densa. El calentamiento gradual que comenzó hacia el X milenio antes del presente, probablemente alcanzó su máximo hace unos 7000 a 3000 años, como lo demuestran estudios palinológicos practicados por Van der Hammen.

En las unidades que integran la denominada zona de ocupación 3 en el corte arqueológico Tequendama 1, los roedores continúan siendo la presa más apetecida, aunque los venados formaban todavía parte de la dieta. El mismo fenómeno se registra en El Abra, en donde durante este período los restos de curí son mucho más abundantes que los de venado, registrándose junto con estos también un aumento significativo de lascas y artefactos.

Las herramientas continúan siendo elaboradas mediante percusión simple; en el sitio del Tequendama, pueden distinguirse una serie de artefactos que se adaptaban más a las presas menores y entre los que se pueden diferenciar navajas prismáticas con más frecuencia que en períodos anteriores, siendo notable también el aumento de raspadores cóncavos.

Entre la fauna de este período junto con la fauna antes mencionada pueden diferenciarse restos correspondientes a la nasua, el runcho y abundantes restos de caracoles terrestres, que indican su consumo como complemento de la dieta durante este período. La presencia de golpeadores y de cantos rodados con señales de desgaste por uso, nos está indicando que durante este período los cazadores se adaptaban más a la vida de los bosques y añadían a la cacería actividades de recolección de tubérculos, semillas y raíces.

En los abrigos rocosos de Sueva, el registro palinológico asociado a este período nos permite reconstruír un paisaje en el que alternan bosques de alisos y granizos con áreas de praderas; en el estrato fechado frente a este período, se pudo reconstruir un piso de habitación en el que se pueden diferenciar, hacia la parte interior del abrigo, concentraciones de artefactos y fogones asociados a restos de fauna; muestras de carbón vegetal tomadas en esta zona de ocupación nos permitieron obtener la datación de 6350 años antes del presente. Como rasgo particular se encontró, asociado a este piso de habitación, un depósito relleno de carbón, artefactos líticos y restos de venados.

En el sitio Nemocón 4, encontramos todavía una ocupación densa en la zona de ocupación asociada a la unidad estratigráfica 5 y fechada entre 7530 y 6825 años antes del presente; los polendiagramas nos revelan que los alrededores del abrigo estaban cubiertos por una vegetación abierta de clima frío. Se puede diferenciar en esta unidad una zona de ocupación en la que se encuentran concentraciones de restos óseos de animales alrededor de los fogones y, al igual que en el sitio del Tequendama, el estudio de los artefactos nos muestra un aumento de instrumentos cortantes elaborados sobre lascas (cuchillas triangulares, prismáticas y concoidales). Al clasificar los restos de fauna, se encuentra un aumento en la densidad de los venados, es interesante el hecho de la baja frecuencia de huesos largos de las extremidades; este hecho puede indicarnos que el sitio Nemocón 4, representa una estación de cacería, probablemente un epicentro de las actividades cinegéticas de los grupos densos que se asentaron en los abrigos rocosos de El Abra próximos a este sitio; otro factor favorable para la cacería en esta zona, es la presencia de una amplia terraza que se extiende sobre la parte alta del abrigo Nemocón 4; las praderas, alternando con vegetación arbustiva, debieron construír un nicho favorable para la proliferación de esta especie. Es interesante en este sitio, también, el hecho de la presencia de fracturas múltiples y transversas al eje de los huesos de la fauna, no causadas por golpes o percusión con fines de extracción de la médula; este tipo de fracturas múltiples, relacionada muy probablemente con prácticas de cacería usuales desde los más remotos tiempos del paleolítico europeo y que consistían en ahuyentar las presas hacia los sitios con escarpas o accidentes del terreno, lugares en los cuales eran dominadas con facilidad o podían ser obligadas a despeñarse, siendo posible dominarlas con mayor facilidad luego de las fracturas y traumatismos producidos por su caída.

El sitio Nemocón 4, nos muestra otro hecho de gran interés en la reconstrucción de los comportamientos del hombre prehistórico de la Sabana. Aquí, como en el sitio del Tequendama, hay un aumento de los restos de roedores entre los que pueden identificarse el curí, el conejo, el borugo y los ratones, sumándose a éstos en menor proporción el armadillo y el zorro; pero a estas evidencias paleontológicas se suma el hallazgo de otras especies como el fara, el tigre, el tigrillo, el perro de monte y el papagayo, cuyos restos nos indican que nuestros cazadores de la Sabana de Bogotá en sus incursiones de cacería se desplazaban, franqueando las estribaciones de nuestra cordillera, hasta descender por vías naturales de acceso hacia pisos térmicos diferentes al de la altiplanicie. Durante este período también encontramos bajo los abrigos rocosos de Nemocón, entre los restos de la fauna, evidencias de la cacería del águila y del cangrejo, que formó parte de la dieta, junto con gasterópodos como los caracoles de tierra firme.

Después de estos acontecimientos, los elementos obtenidos en estratos fechados entre 6000 y 5000 años antes del presente, nos revelan una disminución considerable en la población de la altiplanicie de Bogotá y sus alrededores. Entre tanto ocurrían nuevos cambios de clima y medio ambiente; las temperaturas medias anuales llegaron a ese tiempo a su máximo y, alrededor del año 5000 antes del presente, 3000 antes de Cristo, se puede conocer por medio de diagramas de polen un período de fuerte sequía, no sólo en la Sabana de Bogotá sino también en la planicie tropical de los Llanos Orientales, fue entonces cuando en la Costa Atlántica se desarrolló la etapa cultural denominada arcaica, que marca los comienzos cerámicos, y durante la cual se produjeron asentamientos de pueblos marcadamente recolectores a lo largo del litoral y junto a las ciénagas, como lo demuestran grandes acumulaciones de conchas asociadas a la cerámica más antigua conocida en América. Posiblemente esta misma presión ecológica determinó la reducción de la población de la Sabana de Bogotá, o su desplazamiento hacia otras zonas de medios de subsistencia, adaptándose el hombre a nuevas condiciones ambientales, que debieron conducir a formas de producción diferentes. En la zona zoológica 3 de El Abra, situada entre 7000 y 2500 años antes del presente, los huesos del conejillo de Indias son 5 veces más numerosos que los del venado, hecho que puede relacionarse con la domesticación incipiente de esta especie. En la última fase de la unidad estratigráfica 3a de El Abra, hacen su aparición herramientas completa o parcialmente pulidas, hechas de basalto, entre las que se peuden diferenciar cabezas de mazas en forma anular y hachas trapezoidales.

Alrededor del abrigo rocoso de Nemocón, como lo revela el estudio de pólen fósil sobre muestras de la unidad estratigráfica 6, cuya edad se calcula entre 6000 y 2500 años antes del presente, es reconocible una vegetación abierta de clima frío; la subsistencia allí, como lo indican los restos de fauna, continuaba basándose en la cacería de venados y, en menor proporción, de roedores, Este hecho debe estar determinado por las condiciones ecológicas favorables a las que aludimos anteriormente; el curí, el borugo, el ratón y, en menor proporción, el armadillo, continuaba formando parte de la dieta.

Los artefactos de esta área, continúan siendo el tipo abriense; la ausencia de martillos en el estrato correspondiente a este período relaciona como una disminución de la actividad recolectora en esta área, fenómeno que también puede estar determinado por el período de sequía ocurrido hacia el año 3000 A.C. En este mismo sitio de Nemocón, el estudio palinológico nos revela que durante este período en el abrigo y sectores aledaños dominaban bosques de robles, granizos y cucharos, pero también compuestas, helechos y en los pantanos subyacentes, vegetación de tipo paludal. En este sitio de Nemocón el estudio de los restos de fauna correspondientes al período al que nos referimos, nos muestra que el venado y el conejo continuaban siendo parte de la dieta, y los artefactos corresponden al tipo abriense.

En los abrigos rocosos de Peñitas sitio Chía 1 (municipio de Chía, Departamento de Cundinamarca), se obtuvo una fecha correspondiente al primer milenio A.C. para una estación abierta compuesta por artefactos de chert que pueden incluirse en su totalidad a la serie ABRIENSE. Entre los artefactos del sitio Chía 1, las frecuencias más altas corresponden a martillos, golpeadores, raspadores cóncavos e instrumentos de corte; cuchillos laminares y prismáticos. Las evidencias arqueológicas muestran en este sitio, que la subsistencia se basó preferencialmente en la recolección y subsidiariamente la cacería de especies menores como el curí (*Cavia porcellus*) cuyos huesos alcanzan las máximas frecuencias; también fueron identificados en este conjunto de artefactos restos de venado (*Odocoileus* sp.), (Ardila 1984).

SITIOS PRECERAMICOS Y FECHAS MAXIMAS DE CARBONO CATORCE

Sitio	
EL ABRA	12.460 ± 160 B.P.
TIBITO	11.740 ± 110 B.P.
TEQUENDAMA	10.920 ± 260 B.P.
SUEVA	10.090 ± 90 B.P.
GACHALA	9360 ± 45 B.P.
NEMOCON IV	7530 ± 100 B.P.
ESPINAL	3780 ± 95 B.P.
CHIA 1	3120 ± 210 B.P.

CARACTERISTICAS RITUALES DE LOS ENTERRAMIENTOS

Como quedó anotado, del hombre de los primeros tiempos de nuestro precerámico sólo sabemos por el sitio de Tequendama, lo que nos pueden decir cinco falanges con fractura longitudinal y parcialmente calcinadas, que representan los restos más antiguos encontrados hasta ahora en Colombia y que corresponden al Pleistoceno tardío. Este hallazgo inicial nos hacía contemplar la posibilidad de prácticas de incineración asociadas al ritual funerario; en efecto, a este hecho se sumó el hallazgo, en el mismo sitio, en la zona de ocupación 2 fechada entre 9500 y 8500 años antes del presente de una serie de restos óseos calcinados (entierro No. 14) y el registro de molares y huesos aislados en estado de calcinación en varios estratos del Tequendama 1.

En el sitio de Nemocón también tenemos evidencias de la costumbre de incineración de restos humanos asociada al ritual funerario; allí se efectuó el hallazgo de un hueso ilíaco parcialmente calcinado junto con vértebras lumbares y un sacro que también presentan señales de haber sido sometidos al fuego. Aparte de las evidencias arqueológicas registradas bajo los abrigos rocosos, por los estudios etnográficos sabemos hoy de la costumbre de incinerar humanos y de la práctica del endocanibalismo ritual funerario entre diferentes grupos aborígenes; Serries (1960) citando a Oviedo y Valdés, describe esta práctica entre los guayupes y saes que vivían en los llanos del sur de Venezuela y Colombia, y de acuerdo con el mismo autor, entre los cazadores primitivos, sobre todo en el norte de Asia, fue muy frecuente el mezclar la ceniza de la calcinación de restos con bebidas. De acuerdo con Serries: "las raíces espirituales del endocanibalismo, cuyo contenido significa la incorporación de las virtudes y esencia vital del muerto en los seres supervivientes de la comunidad, se encuentra en la época de los cazadores y recolectores". Barandiarán (1967) anota esta costumbre entre grupos indígenas venezolanos como los Sanema-Yanoama, Chirianas y Guaicas del Alto Orinoco y sureste de Venezuela y también entre los Tucanos de otros tiempos.

Personalmente pude comprobar hace algunos años, la práctica de incineración asociada al ritual funerario entre los Cuibas del río Ariporo.

Esta costumbre de incinerar los restos, puede estar asociada a entierros primarios o secundarios; refiriéndose al hallazgo de restos humanos calcinados en el sitio de La Loma (Mosquera). Anota el arqueólogo Duque Gómez (1967) "La costumbre de quemar los restos óseos en entierros de segunda fase, si bien fue común en otros grupos chibchas como los Guanes, indios de la Sierra Nevada de Santa Marta e indios Zapías, parece, sin embargo, que no fue muy frecuente en el área de la Sabana de Bogotá". Los entierros del Tequendama, Sueva y Gachalá también nos ilustran sobre las costumbres funerarias practicadas durante la etapa lítica; durante largos períodos, se practicaron entierros en los abrigos del Tequendama y bajo las rocas cercanas. Los entierros de los adultos fueron practicados colocando el cadáver en posición decúbito lateral (apoyados sobre el costado derecho o izquierdo) o en decúbito dorsal (apoyados sobre la espalda) y con los miembros flexados, los antebrazos y brazos apoyados contra el tórax y las piernas apoyadas contra el abdomen. La inhumación de adultos se practicó en fosas de planta aproximadamente oval o redondeadas y los restos infantiles eran colocados en un pequeño pozo de forma cilíndrica, en posición fetal o en posición de cuchillas.

Frecuentememte se colocó junto al cadáver el ajuar funerario, consistente en artefactos líticos, instrumentos de asta de venado, hueso, fragmentos de ocre (óxido de hierro), hematita especular (hierro), cuarzo, pequeños guijarros de río y comida, representada por presas de cacería como lo denotan restos de venados y roedores asociados a las tumbas.

El ocre cobra especial significado en los enterramientos bajo los abrigos rocosos en el Tequendama y Gachalá; este elemento, desde tiempos del paleolítico europeo se encuentra estrechamente asociado a las prácticas funerarias; al incorporarse al ajuar póstumo de nuestros cazadores paleolíticos, este mineral se colocaba sobre los despojos, molido o en forma de pequeños fragmentos; al desaparecer las partes blandas los pigmentos impregnaban la superficie de los huesos con un tonalidad rojiza; en esta forma después de varios siglos podemos encontrar los cráneos del hombre prehistórico del Tequendama y Gachalá coloreados en su superficie exterior así como también algunos huesos largos.

ESTUDIO ANTROPOMETRICO Y CARACTERISTICAS FISICAS

El estudio de los restos óseos del hombre precerámico en Colombia, nos permite hoy definir sus características físicas generales; tanto los cráneos del Tequendama como los de Gachalá y Sueva, presentan marcado alargamiento en sentido anteroposterior; esta elongación es denominada en antropología física bajo el término de tipo craneal dolicocéfalo. Los resultados de radiocarbono nos permiten determinar la antiguedad de este rasgo morfológico predominante en los restos precerámicos; de acuerdo con las fechas obtenidas para la unidad estratigráfica 3, a la cual se encuentra asociado el entierro No. 1, hallado bajo el abrigo rocoso de Sueva, podemos situarlo en una edad que remonta al X milenio antes del presente.

Los entierros hallados en sitio Guavio 1, municipio de Gachalá, están asociados a un estrato fechado en 9360 años antes del presente y los entierros del Tequendama se pueden situar también con base en resultados de radiocarbono y análisis de fluor entre el VII y V milenio antes del presente.

Todos estos esqueletos presentan, como carácter común, su cráneo marcadamente dolicocéfalo, rasgo morfológico al que se suman fuertes desarrollos mandibulares, manifiestos en la presencia de una rama ascendente ancha y con marcadas superficies de inserción, que denotan fuertes desarrollos musculares. A estos rasgos, se añade la presencia de superficies oclusales, o de masticación, de las piezas dentarias con marcado desgaste (abrasión). Todos estos caracteres, a la luz de las investigaciones adelantadas por Oliver, pueden relacionarse con factores exógenos como un régimen de alimentación duro y una elaboración primitiva dada a los alimentos, en un régimen de subsistencia basado en la cacería y recolección.

Por sus rasgos craneales debemos señalar que el tipo humano hallado bajo los abrigos rocosos de la Sabana de Bogotá y vertientes del Guavio difiere en alto grado del muisca de la cordillera oriental, que en su mayoría es braquicéfalo, esto es con un cráneo más ancho y redondeado, muestra además menor prognatismo y mayor anchura de los arcos cigomáticos (pómulos más salientes).

En términos generales, puede decirse que los caracteres morfológicos del cráneo de los autores de nuestra etapa lítica recuerdan los rasgos ya anotados por Rivet para las series de Lagoa Santa en el Brasil, aunque en estos últimos la altura craneal es mayor, el prognatismo alveolar más acentuado y menor la proyección a nivel de los pómulos. Dando un vistazo general al conjunto de otras series americanas, encontramos que estos caracteres apuntan en la dirección de los tipos craneales descritos en Paltacalo, Mata Molley y Tierra del Fuego por Stewart y Newman (1950).

En las actividades de cacería debieron también influír la morfología de este hombre prehistórico; al estudiar los restos del esqueleto postcraneal encontramos que éstos acusan fuertes desarrollos musculares, principalmente a nivel de la cintura escapular, huesos largos y cintura pélvica.

Si basados en los resultados osteométricos obtenidos en los restos humanos comprendidos entre el X y V milenio antes del presente, hiciésemos una reconstrucción de sus principales características físicas, nos encontraríamos con individuos de talla media (aproximadamente 1,65 m de estatura) contextura atlética, cabeza alargada, nariz estrecha, pómulos algo salientes y rostro alargado.

CARACTERES PATOLOGICOS

El estudio de los restos óseos del hombre del Tequendama, Gachalá, Sueva y Nemocón, nos revela la presencia de lesiones óseas que derivan de procesos patológicos relacionados con enfermedades como la artritis, periostitis y espondilitis. Tenemos ejemplos casi universales de una amplia distribución de estas dolencias, que aquejaron a la humanidad desde el paleolítico.

En los restos óseos del paleolítico en Colombia las lesiones derivadas de procesos artríticos tuvieron particular incidencia sobre las vértebras, superficies articulares de los huesos largos, y también en la articulación témporo-mandibular.

Tanto en piezas dentarias correspondientes a los restos óseos del Tequendama como a los de El Abra, Sueva, Gachalá y Nemocón, se pudo identificar abrasión; sobre esta característica dentaria en nuestros grupos prehispánicos de suramérica existen estudios amplios entre los que pueden mencionarse los de Santiana y a la luz de los cuales se puede establecer la correlación entre un régimen de alimentación duro y este desgaste sobre las superficies de masticación dentarias.

A juzgar por la edad manifiesta en los restos humanos el término de vida debió ser relativamente corto (cerca a los 40 ó 50 años) en nuestro hombre del precerámico y el índice de mortalidad infantil fue muy alto durante todo este período, como lo denota la alta frecuencia de restos de este tipo.

EL GOLFO DEL DARIEN

Una exploración hacia la Bahía Gloria (Golfo del Darién), nos permitió el hallazgo durante 1973 de una punta de proyectil acanalada cuya morfología recuerda los rasgos señalados para ejemplares hallados en el área del Lago La Alajuela (Bird y Cooke 1977). El artefacto de Bahía Gloria fué localizado en inmediaciones de la formación geológica conocida como Cueva de los Murciélagos. Una exploración posterior nos permitió rescatar una punta de proyectil lanceolada; este artefacto, fue encontrado dentro del estrato 3 (arcilloso-amarillento) reposando inmediatamente encima de él (55 cm). El hallazgo de estos elementos sugiere la presencia en el área del Darien Chocoano de por lo menos 2 tradiciones de punta de proyectil. Hasta el momento no ha sido sin embargo, posible la datación de estos elementos, cuya procedencia podemos atribuir a cazadores superiores. En Colombia se tienen noticias de la presencia de puntas de proyectil tipo cola de pescado solamente en el sector referido y a través de un ejemplar que proviene de Manizales (Reichel-Dolmatoff 1965a:46).

La punta de proyectil lanceolada encontrada en la Cueva de los Murciélagos (Golfo del Darien) concuerda en sus rasgos generales con un ejemplar proveniente de la Cordillera Occidental en la localidad de Restrepo, Valle descrito por Reichel-Dolmatoff (1965a:18), y con ejemplares del Departamento de Antioquia los cuales muestran características muy similares (Ardila 1982).

INVESTIGACIONES EN LA COSTA ATLANTICA Y VALLE DEL MAGDALENA

Las condiciones propuestas para el Istmo de Panamá durante el Pleistoceno posibilitaban el tránsito humano hacia nuestro territorio; el hallazgo en Bahía Gloria (Golfo del Darién) de una punta de proyectil acanalada cuya morfología nos recordaba los rasgos tipológicos establecidos para el complejo Lago La Alajuela en Panamá, (Correal 1973), nos sugería contactos o desplazamientos verificados desde el norte. Partiendo de este orden de consideraciones, adelantamos una exploración a lo largo de la Costa Atlántica y Valle del Magdalena, a partir de 1973, con la ayuda financiera de la Fundación de Investigaciones Arqueológicas Nacionales del Banco de la República y en colaboración con el Instituto de Ciencias Naturales de la Universidad Nacional.

De acuerdo con los estudios sobre el Cuaternario de la Costa, de que se dispone en la actualidad, durante el Pleistoceno el mar Caribe cubría las áreas occidentales de los departamentos de Córdoba, Sucre, Bolívar y sector-occidental del departamento de Antioquia hasta el límite natural que se prolongaría aproximadamente desde Ovejas hasta la Serranía de San Jerónimo y dentro de un arco extendido desde este último punto hasta la bahía de Santa Marta. Actuando dentro de este marco de condiciones, la prospección arqueológica se verificó sobre áreas libres de inundación. En efecto, la exploración de la Serranía de San Jerónimo nos permitió el registro de sitios con artefactos líticos que indican la presencia de grupos de cazadores y pescadores en aquella área. Los bosques densos tropicales del Alto Sinú, debieron ser pródigos en recursos de caza menor y la pesca constituyó un complemento apreciable de la dieta. De la baja densidad de elementos líticos se puede inferir un poblamiento estacional y disperso de pequeños grupos; los artefactos del Alto Sinú, muestran una percusión mal controlada y fueron elaborados utilizando el flint, que abunda en afloramientos junto al río. Situación similar se da en sitios localizados en el departamento de Bolívar, (estaciones de Puerta Roja y Villa Mery) en donde también aparecen junto a los artefactos relacionados con la limpieza de los productos de pesca, una serie de raspadores y artefactos de cacería.

Esta tradición, probablemente se desarrolló hacia el Holoceno en sectores adyacentes a la Serranía de San Jacinto, y en áreas costaneras ya libres de inundación. Las tradiciones precerámicas se prolongan hasta la etapa arcaica (5000 B.P.) como se infiere de los elementos registrados en Puerto Hormiga por Reichel-Dolmatoff (1965b). La dispersión de estas industrias debió tener cierta continuidad a lo largo de la costa; más al norte, en medio de la llanura seca esteparia de la Guajira, al S.E. de la Serranía de Cosinas, nuevamente aparecen huellas inequívocas de los primitivos cazadores que antecedieron en varios milenios a los belicosos Cosinas. En esta área, se añade a los útiles de cacería, una serie de golpeadores, que indican actividades de recolección. En la elaboración de la hierramientas se utilizaron rocas locales como rhiolitas, basaltos y dasitas. Los abrigos rocosos de Media Luna en el departamento de Cesar, ofrecen particular interés en futuras investigaciones, dada la presencia en estratos profundos de artefactos de areniscas y cantos rodados con señales de calcinación junto con partículas abundantes de carbón. Pero las evidencias más significativas de la presencia de cazadores, recolectores y pescadores muy primitivos, fueron registradas sobre terrazas aluviales a lo largo del Valle del Magdalena y sectores aledaños, o en áreas próximas a las ciénagas o confluencias de los ríos, lugares propicios para la consecución de recursos de subsistencia. Estas evidencias se prolongan desde Yanacué al sur del departamento de Bolívar, pasando por la cueva de las rocas calizas duras al NE del corregimiento de la Susana y el municipio de Puerto Berrío y la Dorada, hasta el departamento del Huila.

A juzgar por la densidad de las hierramientas y elementos líticos, una permanencia más estable y una cohesión social se dió en los grupos que se asentaron en el Valle del Magdalena Medio. Es interesante también el hecho del registro de una bien definida industria de choppers en sitios junto a los ciénagas de San Silvestre, Chucurí y Puerto Carare.

La densidad de desechos de talla, indica que estos utensilios fueron elaborados "in situ", utilizando como materia prima el "chert" (silice anhidrida), que abunda en forma de cantos rodados en esta área; lascas con huellas evidentes de uso, bien pueden relacionarse con la preparación de productos de pesca y un alto porcentaje de percutores fragmentados indica igualmente actividades de recolección. Es importante el hecho de la presencia entre los utensilios de esta área, de raspadores aquillados. El registro de raspadores, junto con los elementos antes descritos, nos está indicando una subsistencia basada en actividades de pesca, cacería y recolección; esta situación se repite en el sitio de Porto Belo (Antioquia), aunque los choppers están ausentes allí; las evidencias de industrias líticas, se prolongan a lo largo de sitios como el Eden, Bocas de Palagua y Guayaquil en Puerto Boyacá, donde junto con desechos abundantes de talla, aparecen navajas laminares y elementos relacionados con una menor actividad recolectora; esta característica se manifiesta también en los sitios de Porvenir y Pipinta en Caldas, en donde aparecen estaciones con baja densidad de elementos líticos, localizadas sobre las colinas o terrenos semiondulados.

En el departamento del Huila, como lo demuestran los sitios localizados en las proximidades de los municipios de Neiva y Villavieja, los asentamientos humanos fueron practicados por grupos más numerosos. La estación temporaria abierta y el poblamiento disperso, son los patrones característicos de la vivienda en esta área. La estación Hotel al N.W. de Neiva, recuerda los rasgos ya descritos para el sitio de Boulder también localizado en este municipio (Correal 1973), en donde nuevamente aparecen instrumentos como el choppers tan característicos del Magdalena Medio, junto con raspadores de distintos tipos, hecho que podría indicar un remoto parentesco entre las industrias de estas dos regiones. Los artefactos de la Argentina y Pachingo en Villavieja, muestran características tecnológicas afines con los de Neiva y permiten definir la presencia de grupos cazadores, recolectores y, en menor grado, pescadores en esta área; las hierramientas de andesitas, basaltos y rhiolitas locales, reposan sobre las terrazas del Pleistoceno Superior, hecho que podría indicar, basados en estimaciones cronológicas, su pertenencia a tal edad, lo que no se contradice con los hallazgos efectuados en la Sabana de Bogotá. En el orden de ideas anteriormente expuesto, ya se profilan varias direcciones de poblamiento y se pueden establecer correlaciones tipológicas.

LA EPOCA CERAMICA

Hasta aquí nos hemos referido a las evidencias arqueológicas relacionadas con el paleolítico en la Sabana de Bogotá y sus alrededores. Aunque no es nuestro propósito en estas notas presentar acontecimientos posteriores a la etapa lítica, conviene señalar que en los sitios de El Abra, Tequendama y Sueva, en la parte más alta de las secuencias estratigráficas, fueron registradas evidencias cerámicas y elementos relacionados con la agricultura intensiva como metates y manos de moler. Una fecha de carbón 14 asociada al estrato cerámico del Tequendama nos permite situarlo hacia el año 2335 antes del presente. Durante la época cerámica tenemos también evidencias de la domesticación del curí en las áreas de El Abra y Tequendama y del sedentarismo y organización en aldeas en sectores aledaños a los abrigos rocosos. Los tipos cerámicos reconocidos en el sitio del Tequendama recuerdan las características establecidas para la cerámica muisca de la Sabana de Bogotá (Broadbent 1971). El más alto porcentaje de tipos está representado por una cerámica stemuisca ya descrita en otras áreas de la Sabana de Bogotá como la laguna de la Herrera (Broadbent 1971) y Zipaquirá (Cardale de Schrimpff 1981). Elementos líticos pulimentados como el hacha, cabezas de maza y volantes de huso decorados, también son característicos de esta época. Algunos de los rasgos estudiados en la cerámica de Sueva recuerdan ciertas características ya anotadas en el área de Zipaquirá y otros semejan formas comunes en la céramica muisca clásica; así mismo se pueden reconocer rasgos que concuerdan con la zona suroccidental de la Sabana de Bogotá (Facatativá), Mosquera y Funza. En el sitio Tequendama 1 se pudo reconocer, asociado a los estratos cerámicos, un piso de piedra sobre el que aparecen huecos de postes para vivienda.

Un sitio recientemente excavado en el municipio de Zipacón (Cundinamarca) nos permitió determinar una asociación de abundantes restos de microfauna, junto con artefactos abrienses y cerámica entre la cual aparecen tipos anteriormente reconocidos como premuiscas por Cardale de Schrimpff (1981). En este abrigo rocoso se Zipacón, se obtuvo una fecha de carbono catorce correspondiente a 3270 ±30 B.P. (GRN-11125).

REFERENCIAS CITADAS

Ardila Calderón, G.

1982 Las puntas de proyectil de Colombia; dos collecciones de Antioquia. Ms. en posesión del autor.

1984 Chia, un sitio precerámico en la Sabana de Bogotá. Fundación de Investigadores Arqueológicos Nacionales. Banco de la Repúbica, Bogotá.

Barandiarán, D. de

1967 Vida y muerte entre los indios Sanema Yanoama. *Antropológica* 21:1-65. Caracas.

Bird, J., and R. Cooke

1977 Los artefactos más antiguos de Panamá. *Revista Nacional de Cultura* 6. Panamá.

Broadbent, S.

1971 Reconocimientos arqueológicos de la laguna de La Herrera. *Revista Colombiana de Antropologia* 15: 171-213 (1970-1971).

Cardale de Schrimpff, M.

1981 *Las salinas de Zipaquirá su explotación indígena.* Fundación de investigaciones arqueológicas Nacionales, Banco de la República, Bogotá.

Correal, G.

1973 Artefactos líticos en la Hacienda Boulder, Municipio de Palermo, Departamento del Huila. *Revista Colombiana de Antropología* 16:195-222. Bogotá.

Correal, G., and T. Van der Hammen

1977 *Investigaciones arqueológicas en los abrigos rocosos del Tequendama; 11.000 años de prehistória en la Sabana de Bogotá.*Banco Popular, Bogotá

Correal, G., T. Van der Hammen, and W.R. Hurt

1977 La ecología y tecnología de los abrigos rocosos en El Abra. *Revista de la Dirección de Divulgación Cultural* No. 15, Bogotá.

Correal, G., T. Van der Hammen, and J.C. Lerman

1970 Artefactos líticos de abrigos en El Abra, Colombia. *Revista Colombiana de Antropologia* 14:9-46.

Duque Gómez, L.

1955 *Colombia.* Monumentos Historicos y Arqueológicos, Instituto Panamericano de Geografía e Historia, Vol. 1. Editorial Fourmier, Mexico.

1967 Tribus indígenas y sitios arqueológicos. *Historia Extensa de Colombia,* Vol.I(2), Editorial Lerner, Bogotá.

Hurt, W.R., T. Van der Hammen, and G. Correal

1972 Preceramic sequences in the El Abra Rock Shelters, Colombia. *Science* 175: 1106-1108.

1976 The El Abra Rockshelters, Sabana de Bogotá, Colombia, South America. *Indiana University Museum, Occasional Papers* No. 2. Bloomington.

Reichel-Dolmatoff, G.
1965a *Colombia*. Thames and Hudson, London.

1965b Excavaciones arqueológicas en Puerto Hormiga (Departamento de Bolivar). *Antropología* 2. Ediciones de la Universidad de Los Andes, Bogotá.

Serries, O.
1960 El endocanibalismo en América del Sur. *Revista do Museu Paulista* (Nova Serie) 12:125-175. São Paulo.

Stewart, T., and M. Newman
1950 Anthropometry of South American Indian skeletal remains. *Bureau of American Ethnology, Bulletin* 143. *Handbook of South American Indians*, Vol. 6, Smithsonian Institution, Washington.

Van der Hammen, T.
1973 The Quaternary of Colombia: Introduction to a research project and a series of publications. *Palaeogeography, Palaeoclimatology, Palaeoecology* 14: 1-7.

Early Man Projectile Points and Lithic Technology in the Ecuadorian Sierra

WILLIAM J. MAYER-OAKES
Department of Anthropology
Texas Tech University
Lubbock, Texas 79409
U.S.A.

Abstract

Four distinctive projectile point types (*Fell's Cave Stemmed, Paijan, El Inga Broad Stemmed* and *Ayampitin*) identify and characterize the Early Man assemblage of artifacts from the El Inga site. With the exception of *Paijan*, each type includes "fluted" versions in addition to non-fluted ones. The El Inga lithic industry associated with these points is primarily unifacial, distinguished by presence of blades and burins. It is hypothesized that this is a reasonable lithic context for the *in situ* emergence and development of the concept of fluting. The presence of this initial, emerging stage of fluting is supported by the contemporaneity of both fluted and non-fluted examples of the same type of point. It is also supported by the occurrence of fluting in both stemmed and lanceolate point styles. Examination of the detailed attributes of El Inga fluting suggests a four-stage developmental continuum: non-fluted, basal thinned, fluted and pseudo-fluted. The fluted attribute itself is further distinguishable into important varieties. Predominance of early elements of this continuum at El Inga further supports the hypothesis of the indigenous development of fluting in Ecuador. Because stemmed points predominate over lanceolate forms, the Ecuador development is viewed as independent of North American fluting on lanceolate forms. Separate consideration of morphology, technology and function lead to comparative conclusions which produce a theoretical model of seven "ideal" diagnostic Early Man projectile point types for South America, four of which are fluted.

INTRODUCTION

Robert Bell (1960) not only recognized the "fluted" nature of some of the projectile points collected by Allan Graffham at the El Inga site in Ecuador, but Bell should also be credited with taking the first step to follow up on this truly revolutionary series of observations. When I visited Norman, Oklahoma, in the spring of 1959 he not only showed me the El Inga materials in the Graffham collection, he invited my interest and participation in a follow-up project. Joining the faculty at the University of Oklahoma a few months later, as Director of the Stovall Museum, I was in a strong position to act on this generous research invitation.

In January 1960, together, Bell and I initiated Oklahoma's entry into the field of South American Early Man archaeology. Our joint fieldwork in 1960 was followed in 1961 by a joint effort, funded by the National Science Foundation, which saw Bell carrying out meticulous excavations at El Inga and publishing his report in 1965. My part of the joint project was the study of our extensive 1960 surface collection.

Study of the surface collection made in 1960 at the El Inga site in highland Ecuador has proceeded in irregular fashion since then. A detailed descriptive study has been submitted for publication (Mayer-Oakes n.d.). Since 1960, I have paid serious attention to the topics of dating, projectile points, lithic technology, and the general field of New World Early Man studies.

As plans for renewed fieldwork and intensive investigation in Ecuador are now coming to fruition, I wish in this paper to look broadly, but selectively, at important aspects of the evidence for Early Man in Ecuador and its interpretation. The focus of attention is the highlands or "Sierra"; I am interested in the area of most concentrated study, the Ilalo region of the upper Guayllabamba Basin, east of Quito. The substantive basis for my ideas presented in this paper is three-fold: study of the artifact (not debitage) segment of the extensive 1960 El Inga surface collection; study of the surface collections from more than a dozen sites in the Ilalo region; study of the extensive surface collections made at the San José site in 1965, 1967, and 1968; and the excavation data gathered in 1971 from San José.

CHRONOLOGY

Nearly all of the work published to date on Early Man in the Ecuadorian Sierra is primarily culture historical in nature. In my recent literature survey (Mayer-Oakes 1979) and summary (Mayer-Oakes 1982a) the conclusion is drawn that initial problems of chronology have not yet been resolved. While the application and use of typological classification has been fundamental to the work accomplished so far in Ecuador by Bell and others, the application and use of the "principle of chronology" is as yet incomplete, i.e., our knowledge of the temporal dimension is unsatisfactory. A secure chronology, perhaps the most fundamental aspect of culture historical studies, has not yet been accomplished. With this fact in mind it seems obvious that much future work of a culture historical kind must be accomplished before we can go very far with processual or lifeway reconstruction studies. To this end, we are planning (Mayer-Oakes 1982b) to apply dating techniques along a new front.

No seriation studies have as yet been accomplished, very little significant stratigraphy has been recorded, and little systematic effort has been made to accomplish cross-dating. Although radiocarbon dates exist for both the El Inga (Bell 1965) and San José (Mayer-Oakes 1982a) sites, there are problems of internal stratigraphic consistency at El Inga and/or "appropriateness" for the San José dates. Some inconsistencies are evident in the obsidian hydration dating done at the Manitoba laboratory (Mayer-Oakes 1970), but results can be read in general support of the typological dating proposed by Bell (1965) and Mayer-Oakes (1963, 1966). Use of hydration dating as a relative dating mechanism at El Inga (Bell 1977) has provided a possible corrective to inconsistent radiocarbon dating. In general, we still lack complete and satisfactory control of the chronological dimension.

The latest development in establishing a chronology for the Ilalo Early Man complex is the series of hydration dates (obtained in early 1982) from the 1971 San José site excavation. Obsidian flake samples were selected from all the levels of the excavation below the plow-disturbed topmost level. The range of dates is from 9321 to 11,248 B.P. As the 13 dates were found to be in reasonably consistent stratigraphic sequence, we have for the second time a consistent series of dates which can be used to indicate temporal relationships to other known Early Man sites elsewhere in both South and North America.

The 13 dates were produced by Joseph Michels at Mohlab, using atomic absorption spectroscopy to ascertain that all samples were of the same chemical type. An induced hydration rate was determined for this obsidian type and was used in calculating the date derived from hydration rim measurements. Level I was plow-disturbed and not dated. Level II dates are: 9321, 9408 and 9430. Level III dates are: 10,228, 10,570 and 10,918. Level IV dates are: 10,387, 10,387, 10,501, 10,616, 10,825, 10,825 and 11,248. (All dates are in years B.P.)

This series of dates can be compared with the much younger radiocarbon dates run from the same levels at the same excavation (Mayer-Oakes 1982a:Table 15.3). Level I was plow-disturbed and not dated. Level II had four dates ranging from 3430 to 4345 B.P. Level III had nine dates ranging from 3875 to 6410 B.P. Level IV had three dates ranging from 5870 to 6445 B.P. Level V had two dates of 4860 and 5420 B.P.

All these dates may be suspect because they were derived from concentrations of charcoal specks in soil samples. However, except for level V, they are stratigraphically consistent, so we are left with a conundrum. If we want to date San José as "late" we can refer to the radiocarbon dates. If we want to interpret it as "early" we can use the hydration dates. At the moment, I prefer to give more weight to these latest hydration dates. They are more likely to be accurate in the light of the generally good fit of hydration dates with radiocarbon dates on charcoal at other sites. I would also argue that the San José assemblage is essentially a pointless duplicate of the El Inga assemblage which has one 9030±144 B.P. (R-1070/2) radiocarbon date and typologically early projectile points. But, there are obviously important problems of chronology that have yet to be solved.

PROJECTILE POINTS

From San José we have collected only one projectile point, a bifacially worked lanceolate form (Figure 1). I believe that it should be treated as an "isolated find" rather than as part of the assemblage from San José. Studies of the large San José artifact and debitage surface collection (Mayer-Oakes 1970) indicate no bifacial (or any other) projectile points. In fact, there are no bifaces, and only a small amount of bifacial edge retouch is evident.

The El Inga site has become widely known for its large sample of several projectile point forms. Although the El Inga site may have been occupied for a long time (Mayer-Oakes 1963; Bell 1965), most scholars have seized on the distinctive Early Man points and technology and think of the site in terms of its earliest period of occupation (Willey 1971).

The definitive study of surface collections (Mayer-Oakes n.d.) has established named projectile point types based on those presented in the preliminary reports (Mayer-Oakes 1963, 1966). Points which are "fluted" or are otherwise dated as Early Man by typology are the major topic of concern in this paper. A limited number of factors concerning each of four projectile point classes is discussed below.

The most distinctive and probably least variable classes are the two named types of stemmed points, *Fell's Cave Stemmed* and *Paijan.*

Fell's Cave Stemmed (Figure 2)

This point type is well known from a number of Early Man contexts in South America. Often called "fishtail" or "cola de pescado" points in the literature, we prefer the term we have used consistently since the 1966 formal description was published. The most exhaustive comparative statement about this point type has been published by Bird (1969) in his attempt to identify the type at El Inga. In addition, Schobinger (1969) surveyed the South American literature for evidence of this type.

Style and Variation

The most common of *Fell's Cave Stemmed* "ideal forms" represented at El Inga is shown in Figure 2. The point is a "stemmed" point in the normal use of this term in Americanist archaeology. The most important variations are in total length of point (short "reworked" specimens are frequent), and in the precise character of the "shoulders" and "ears." Most consistent stem characteristics appear to be the concave curve of the stem and basal edges. A second (and rare mode) for the form (Figure 3) is distinctive primarily for the lengthening of the stem portion and modification of the stem edge curve to a more nearly "tapered" form.

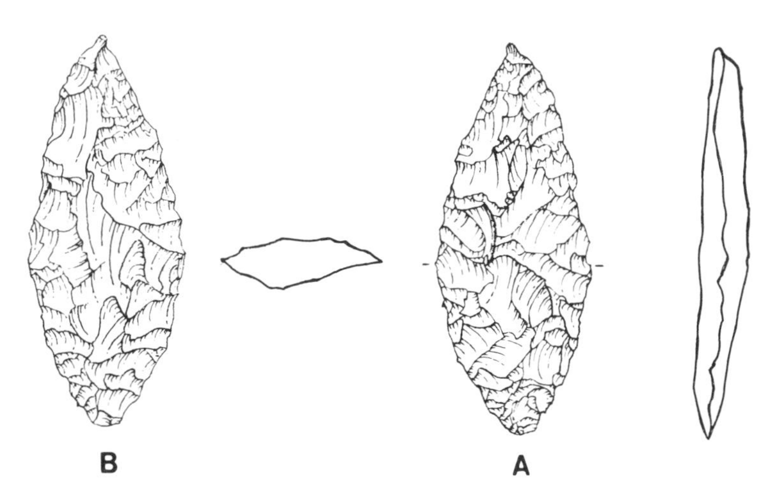

Figure 1. San José point (natural size).

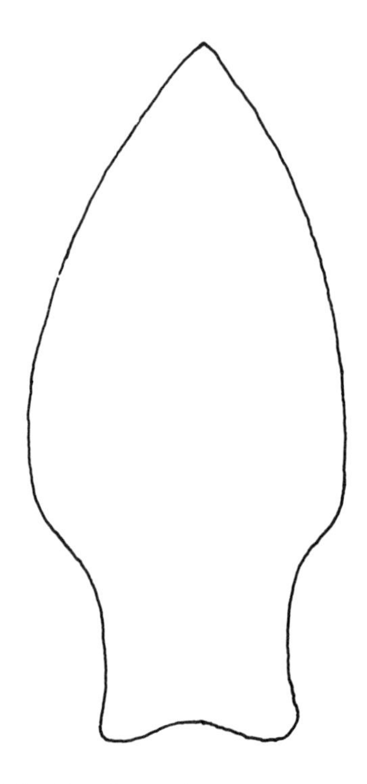

Figure 2. *Fell's Cave Stemmed* point, stereotype.

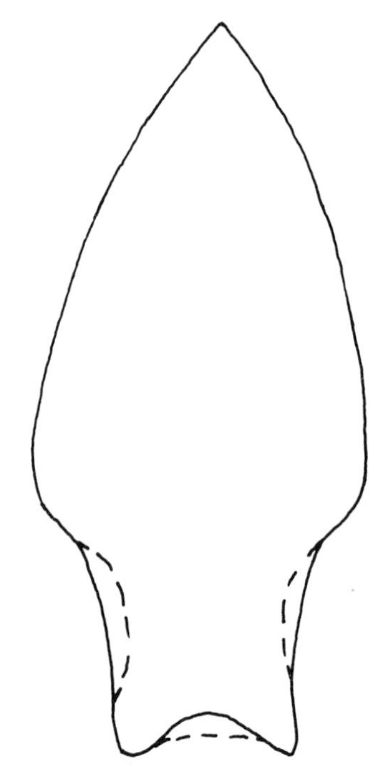

Figure 3. *Fell's Cave Stemmed* point, variant.

Fluting Variations

The presence of several kinds of "fluted" versions of this point in the original surface collections made by Bell and me (Mayer-Oakes and Bell 1960a, 1960b) initiated a renewed interest in Magellanic materials and the first documentation of the fluting characteristics of this previously established type (Mayer-Oakes 1963:120, 1966:Figures 7-10; Bell 1965:Figures 10, 11). The special features of facial chipping recognized in the 1966 point description included "fluting" and "basal thinning" (Mayer-Oakes 1966:Table 3). In the specific individual point descriptions (Mayer-Oakes 1966: 650-652), the evidence cited for "fluting" was the presence of a "channel flake scar," while evidence for "basal thinning" was presence of "short basal thinning flake scars." Less frequently seen evidence for "fluting" is the presence of a prepared platform or "nipple" on the basal edge, and in the "hinged" termination of channel flake scars. Because we have prepared better line drawings and also photographs of these same specimens for our definitive final report (Mayer-Oakes n.d.), it is now possible to document our original description more adequately. (The illustrations in this paper utilize some of the drawings and photos prepared for the final report).

A third variation of facial chipping recognized in the early 1960s was "original flake facet surface" which gave the appearance of fluting. We began to call this feature "pseudo-fluting" in 1969, and formalized it in our draft final report in 1972. Since then we have used this concept in both private and public discussions of the El Inga points. Others are now using this concept (Rovner 1980), but apparently without reference to the context of variation in form of fluting of projectile points which occurs at El Inga.

The El Inga examples of the *Fell's Cave Stemmed* point type are now seen to include unfluted, basally thinned, fluted, and pseudo-fluted varieties. There is, of course, much variation within the categories of basal thinning and fluting. At some point the distinction between these two attribute complexes is arbitrary; but the presence of prepared platforms, patterns of multiple channel flake removal, and channel flake hinge terminations are very similar to the general Clovis fluting technique. The presence at El Inga of pseudo-fluting (utilizing the extant concave facet of a flake or blade "blank"), sometimes in combination on the same specimen with "true" fluting, further complicates a far from simple aspect of New World projectile point technology. The presence of all of these theoretically possible variants of the "fluting" or "grooving" of a point base (and face) on a stemmed point form in South America suggested to me a new theoretical model which could be used to test and refine our current ideas about the diagnostic Early Man technological marker of "fluting."

Associations at El Inga

The *Fell's Cave Stemmed* form is one of three dominant projectile point forms found on the surface of the El Inga site (18% of all points). In the excavation reported by Bell (1965) it was similarly dominant (17% of all points). The vertical distribution at the El Inga excavation suggests that this point type is present at all levels of the site; and it is the major type found in every level except the level from 8" to 12" (Bell 1965:Table 4), where it is co-dominant with the contracting and broad stem types.

Bell reported the five radiocarbon dates derived from El Inga soil samples (1965: 311-12). While there are internal inconsistencies in these dates (see Bell 1977: 68), some direct and indirect associations between the dates and the *Fell's Cave Stemmed* points should be noted. For three of the five dates there are no point associations; for sample R-1070/2 (for which the date is 9030±144 B.P.), there are two associated *Fell's Cave Stemmed* points: Bell's 1965 Table 7 lists a point at the 20"-22" level from which the radiocarbon sample was derived; the caption for Bell's 1965 Figure 11 lists item c, a *Fell's Cave Stemmed* point from a depth of 23.5". Thus, this earliest radiocarbon date from the site is associated directly with a *Fell's Cave Stemmed* point and directly overlies another point of the same type. This context suggests that one example of the point type is as old as the 9030 B.P. date, while another example is an unknown amount older. However, sample R-1070/3 (with a date of 7928±132 B.P.) underlies by at least four inches the only *Fell's Cave Stemmed* type point found in the excavation unit. This evidence suggests that in this part of the site the point type occurs at an indefinite time later than 7928 B.P. In 1977 Bell published the obsidian hydration studies which had led him to discard the three youngest radiocarbon dates. By means of the relative chronology established from a tabular stratigraphic comparison of hydration rim thickness measurements, Bell added a new dimension to the research approaches to the question and problems of dating the El Inga site and materials.

Comparisons

The resemblance of the El Inga examples to the points from Fell's Cave in Chile was first noted by the author in 1959, and was reinforced on our first field trip to El Inga in 1960. In the two 1960 publications which Bell and I co-authored as notices of our finds at El Inga, we stated that the points were related to Fell's Cave specimens. But it was not until the 1963 publication I prepared for *Scientific American* that a detailed and illustrated statement was made of the correlation between Ecuador and the Magellanic region. Subsequently, detailed technical studies with illustrations were made (Bell 1965; Mayer-Oakes 1966) interpreting this style as very similar to the Fell's Cave finds. Bird's (1969) definitive comparative study, based upon a re-examination of all the specimens from southern Chile which he had recovered in the 1930s, supported the typological correlation Bell and I had proposed. Willey (1971) and Schobinger (1973) helped to spread the news and document other examples of this early South American "marker" or diagnostic projectile point. Bird has also investigated the occurence of similar points in Panama (Bird and Cooke 1977, 1978). Nearby Costa Rica has also recently produced an example of this point style (Snarskis 1979:Figures 2F, 3A); as has Peru (Chauchat and Zevallos 1979; Ossa 1976).

Paijan (Figure 4)
(*El Inga Long Stemmed* variant)

The distinctive long stemmed and barbed points found on the surface of El Inga in 1960 were at first considered unique to the site. However, they were soon recognized (Mayer-Oakes 1963, 1966) as being like the much smaller sample of north coast Peruvian points from Pampa de los Fosiles (Lanning 1963). As more information about the *Paijan* style points from Peru became available (Ossa and Moseley 1972; Chauchat 1976, 1978), it became clear that this is also a distinctive (or "marker") Early Man style, known so far only from Peru and Ecuador. There are distinctive differences in form between the two sets of samples, but these specimens now seem best considered as simply regional variations of a single form stereotype.

Paijan point associations in Peru are with an apparently early post-Pleistocene, rather specialized hunting and gathering environmental adaptation (Ossa and Moseley 1972). Radiocarbon dates of 10,500 B.P. and older have been obtained (Ossa and Moseley:13-15). Figure 5 shows a series of points of this style from Peru for visual comparison with the El Inga series.

Bell's excavations at El Inga produced up to 10 examples assignable to this style (Bell 1965: Figure 11n-t; Figure 14f, g, h), while the surface collection (Mayer-Oakes n.d.) produced 22 examples. Although Bell referred to this type as "contracting stemmed," in the sample from the excavation he noted that several of the specimens had parallel rather than contracting (tapered) stem edges. From his illustrations it appears that four of the seven specimens represented by stem portions are in fact characterized by parallel sides.

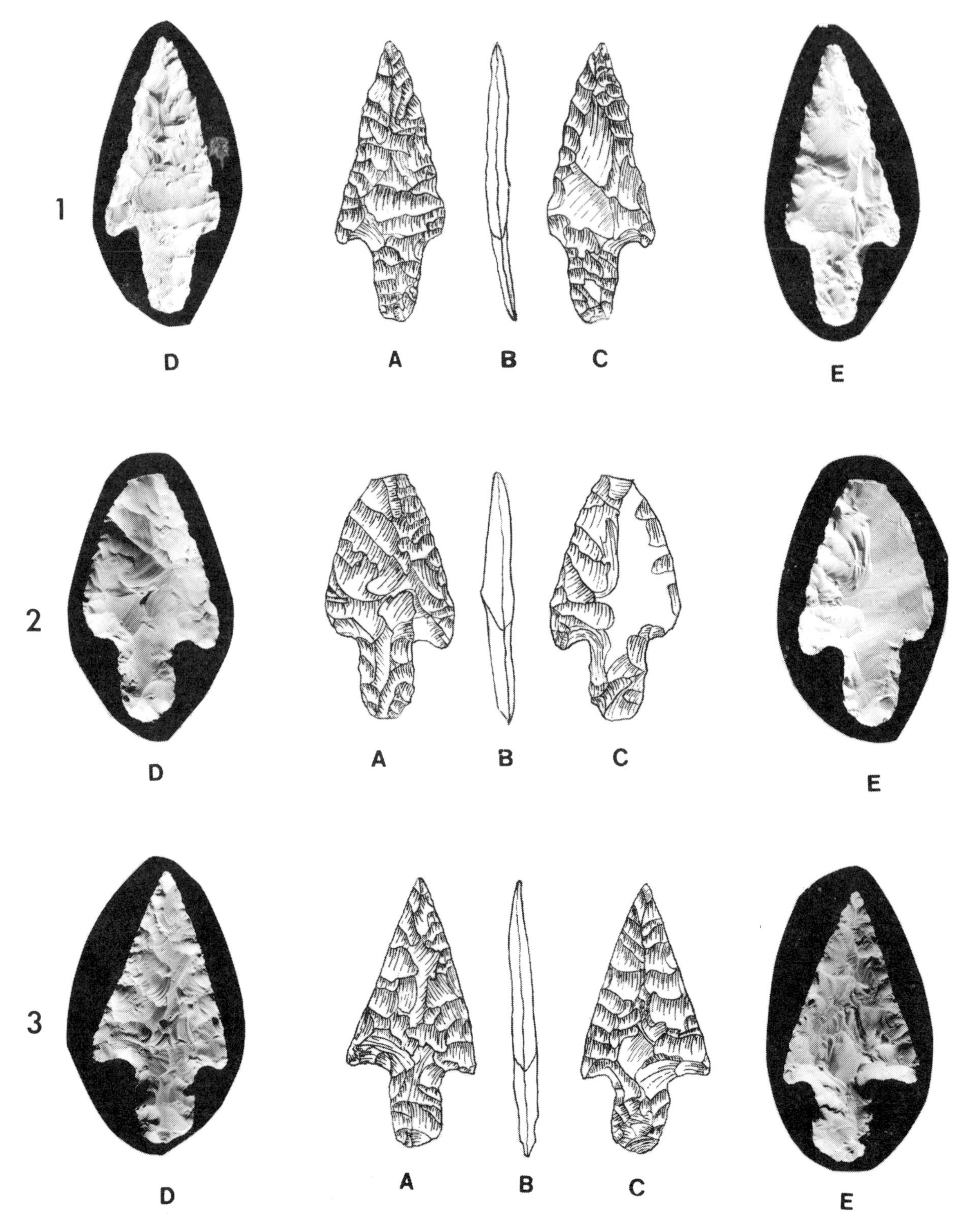

Figure 4. *El Inga Long Stemmed* point, (a *Paijan* type variant) from El Inga surface collections (natural size).

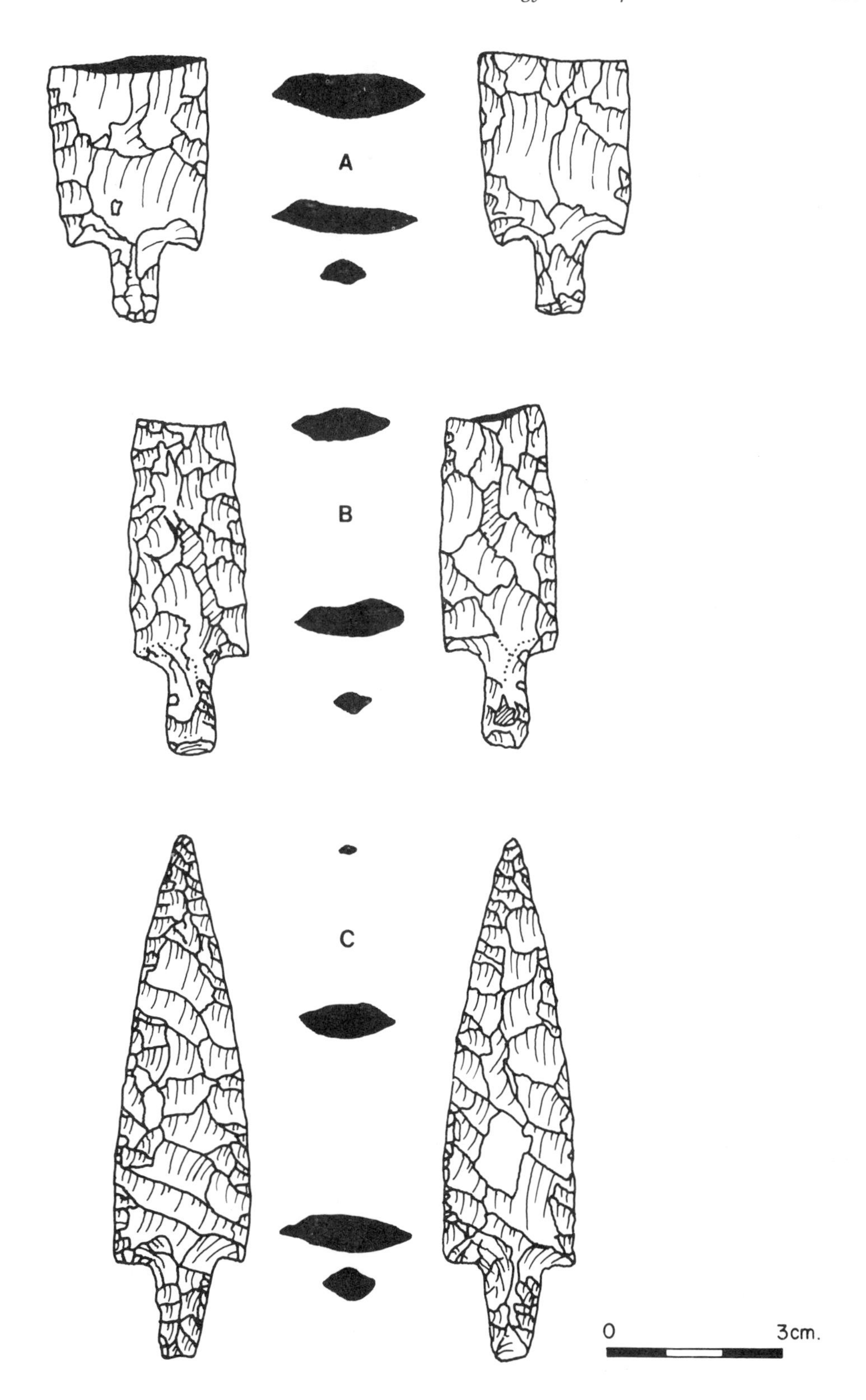

Figure 5. *Paijan* points from La Cumbre, Peru: A) Ossa and Moseley (1972:Figure 9); B) Ossa and Moseley (1972:Figure 12); C) Ossa and Moseley (1972:Figure 10). (90% of natural size).

In the El Inga surface collections, the *Paijan* style is the most common point type (see Table 1). Of a total of 51 stemmed points, 22 are classifiable as *Paijan* (3 are unifaces, 19 are bifaces). The *Paijan* type accounts for 21% of all classified points, and is in fact the most frequent type at El Inga (*Fell's Cave Stemmed* is represented by 19 points or 18%, while *El Inga Shouldered Lanceolate* is represented by 20 points or 19%). See Table 1.

El Inga Broad Stemmed (Figure 6)

Bell's 1960 initial publication on El Inga points showed only one example (Figure 3U) of the type we have termed "Broad Stemmed" (Mayer-Oakes 1966:Figure 12a, b). The survey carried out by Carluci (1963), however, illustrated two of these points in their most distinctive version (Carluci 1963: Figure 3 H, I) as well as three other closely similar forms or preforms (Carluci 1963: Figure 3 E, F, G). The excavation collection reported by Bell (1965:Figure 13a, 3e, i and Figure 16a, g) illustrated five very distinctive examples of this type. In addition, Bell illustrates segments or bases which are probably attempts to achieve the stereotype for (Bell 1965:Figure 13c, f, h, j). Our 1960 surface collection study (Mayer-Oakes n.d.) produced no distinctive complete examples (Figure 6-1, -2) which are probably preforms for this type. In addition, the 1960 surface collection did yield a number of basal fragments which could represent completed versions of this point style (Figure 6-3, -4, -5). Phagan (1970) was not able to study projectile points, but his Plate 22 (Phagan 1970:96) illustrates two good examples of this type. Bonifaz (1978) also illustrates one very carefully made example of this type (lower right corner specimen on page 53). This specimen has been cut for a hydration rim measurement which is listed as 9.5 microns thick (dated by U.S.G.S. at 15,050 B.P.).

We have seen at least six additional good examples of this type in its stereotype form from Ilalo region and the area near Otavalo. Though complete specimens are rare at El Inga, the type

Table 1. Projectile point types, El Inga surface collections.

A. *Stemmed*

type	F	NF	BF	UF	f	%S	%TP
Fell's Cave	X	X	19	0	19	37%	18%
El Inga Long	0	X	19	3	22	43%	21%
El Inga Broad	X	X	6	0	6	12%	6%
El Inga Pointed	0	X	4	0	4	8%	4%
			48	3	51	100%	49%

B. *Lanceolate*

type	F	NF	BF	UF	f	%L	%TP
Ayampitin	0	X	9	3	12	23%	12%
El Inga Asymmetrical	X	X	8	0	8	15%	8%
El Inga Shouldered	X	X	20	0	20	38%	19%
El Inga Fluted	X	0	3	0	3	6%	3%
El Inga Crude	X	X	6	3	9	17%	9%
			46	6	52	99%	51%

KEY

F = fluted
NF = non-fluted
BF = biface
UF = uniface
f = frequency
%S = percent of stemmed class
%L = percent of lanceolate class
%TP = percent of total points

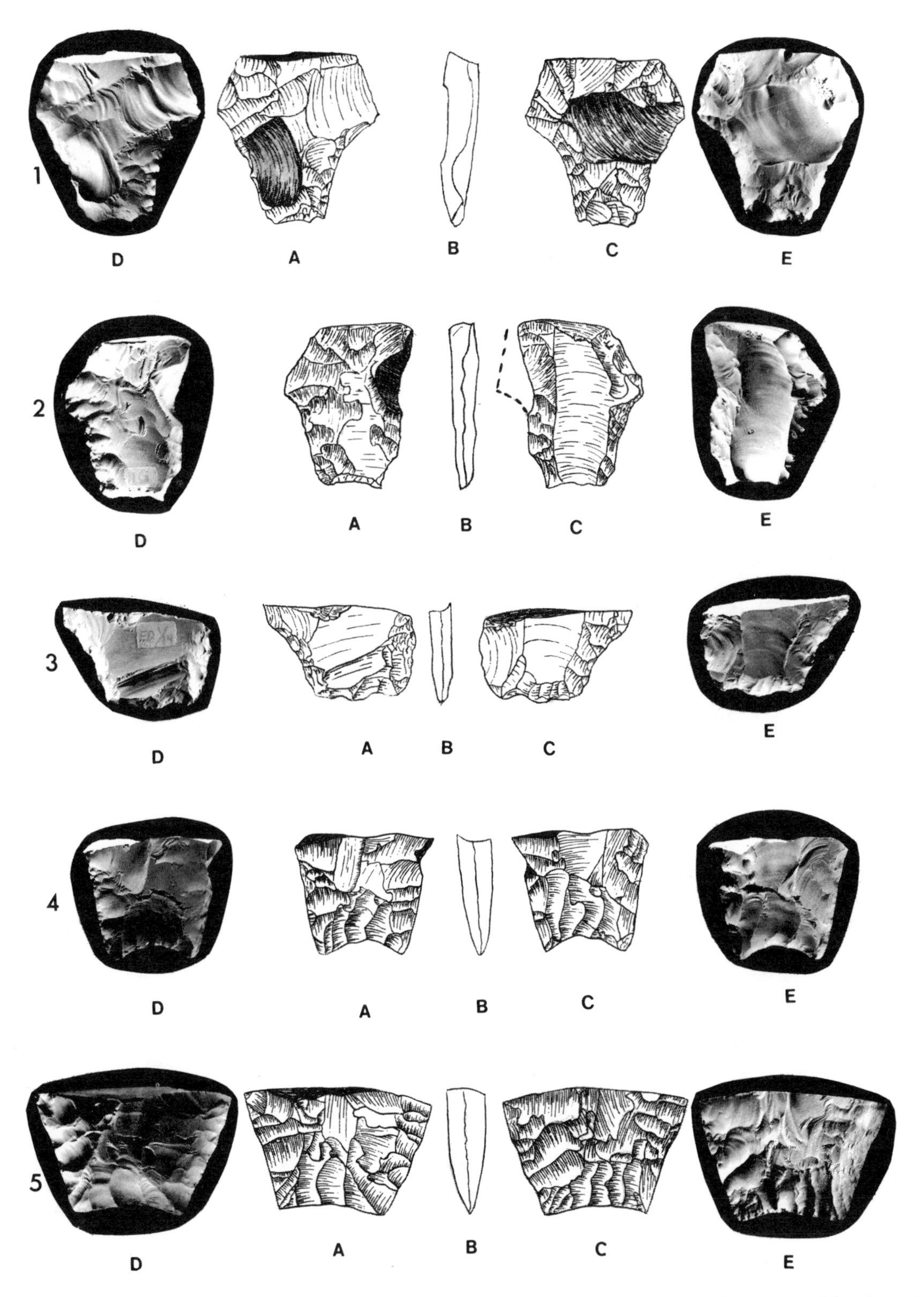

Figure 6. *El Inga Broad Stemmed* points from El Inga surface collections: 1, 2 probable preforms; 3, 4, 5 basal portions. (75% of natural size).

may be represented by many basal and stem fragments. It seems likely that large (even javelin size) versions of this type are an extreme specialization, but documentation beyond that indicated above, is lacking. Some of the basal fragments of this type would be indistinguishable from bases of the "shouldered lanceolate" type. Others are closer to a general *Clovis Lanceolate* basal form. Relationships of these types will be discussed more fully below under "comparisons."

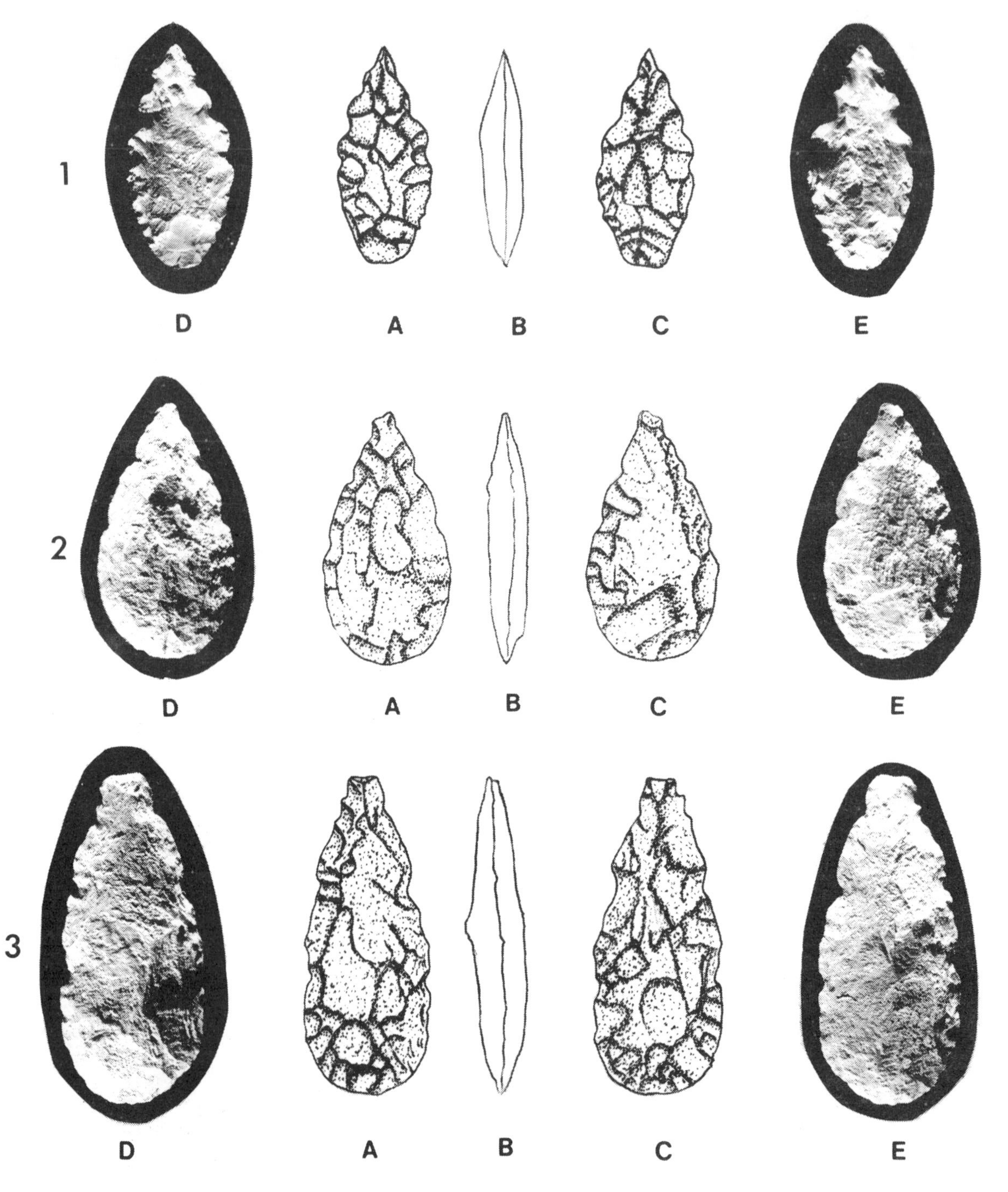

Figure 7. *Ayampitin* type points from El Inga surface collections (natural size).

Ayampitin (Figure 7)

When the El Inga collection was first examined in 1960 and during the studies done in the next few years, Bell and I differed in our interpretation of the generally ovoid or lanceolate bifaces of projectile point size. Bell most often classed these items as bifacial "knives," although a few (1965:Figures 15, 16) were identified as projectile point fragments, and even fewer (Figure 14a-e) were classed as "ovate" or "leaf-shaped" points. I was much more impressed (Mayer-Oakes 1963, 1966) with the evidence for similar "leaf-shaped" points from the Lauricocha site in Peru (Cardich 1964) and from Intihuasi Cave in Argentina (Gonzalez 1959). It is also true that prior to about 1968 the concept of "pre-form" was not very seriously considered as a functional explanation for this simple form.

From the present vantage point of thinking about lithic technology, it seems obvious that at least some of the crudely chipped "ovate biface knives" or "lanceolate" points are very likely to have functioned for their makers as preforms for projectile points that were never completed. Nonetheless, there is a set of apparently completed specimens that conform to a stereotype lanceolate plan form.

My study of the El Inga surface collection (Mayer-Oakes n.d.) resulted in the classification shown in Table 1. The *Ayampitin* style was represented by nine biface and three uniface examples to make 12% of the total point sample, second among lanceolate points only to the *El Inga Shouldered Lanceolate* style, which itself may be a preform. It seems reasonable now to expect that a northern Andean Early Man site (which may have been at a crossroads for coastal-Amazonian as well as montane traffic) should have several different kinds of Andean Early Man marker or diagnostic projectile points. The fact that both obsidian and basalt were used for the *Ayampitin* style, and the fact that both unifacial and bifacial versions of the type were made, are each interpreted as evidence supporting our contention that this form is diagnostic.

Thus we have four distinctive Early Man "marker" or diagnostic types (three of which have been documented primarily from outside Ecuador) which appear together and constitute the majority of styles at the El Inga site.

TECHNOLOGY

The Burin-blade Industry

Studies of the El Inga site recently completed (Mayer-Oakes n.d.), and those recently initiated of the San José site (Mayer-Oakes 1982c), both clearly indicate the basic nature of the Ilalo Mountain Early Man Lithic Complex. I define this as a unifacial industry dominated by blades and burins. It is characterized by classic Upper Paleolithic-like unifacial tools, both simple and complex. Although currently available chronometric evidence is inconsistent, both typology and stratigraphy suggest strongly that we have evidence for a two-stage span of development, with San José being earlier than El Inga. In any case, the lithic materials themselves provide morphological, technological and functional bases for interpreting the blade industry as "developmental," i.e., in the early stages of being established. Beyond this kind of developmentalism is the further suggestion for the development of projectile points of stone. None are associated with San José. Both uniface and biface points are associated with El Inga. In the other sections of this paper I argue that the concept of fluting as a part of the process of making stone projectile points was being invented and developed at El Inga. Most broadly then, I see the cultural processes of a developing Ilalo Mountain unifacial tool industry as being increasingly based on blade, burin and burin spall manufacturing procedures of increasing sophistication. These lithic production techniques could be transferred to the tasks of making thinned, haftable stone projectile points. We are, in effect, viewing the Ilalo Mountain lithic context as one which was particularly suitable to give rise to fluted projectile points!

In a recent review of Lynch's (1980) Guitarrero Cave monograph, Stothert (1980) points out the difficulties and inadequacies of lithic analysis when debitage (or unretouched flakes) is the predominant part of the data recovered, but left unanalyzed. As she points out:

> Standard lithic analysis has not evolved to handle the populations of unspecialized flakes that are the products of simple flaking techniques. Still, these populations may reveal aspects of the adaptation of prehistoric peoples. Simply by using the unretouched flake population as a frame of reference, Lynch might have clarified for me the significance of the 271 blades or lamellar flakes that he singled out for description. Why are these of particular importance? Do they represent a specialized blade industry? By isolating these unretouched flakes and by giving them a type number Lynch introduces the possibility that they are evidence of special flaking techniques. What would be acceptable evidence to support the existence of a blade industry? I think one would have to demonstrate that the 271 blade-like flakes could not have been produced by the same

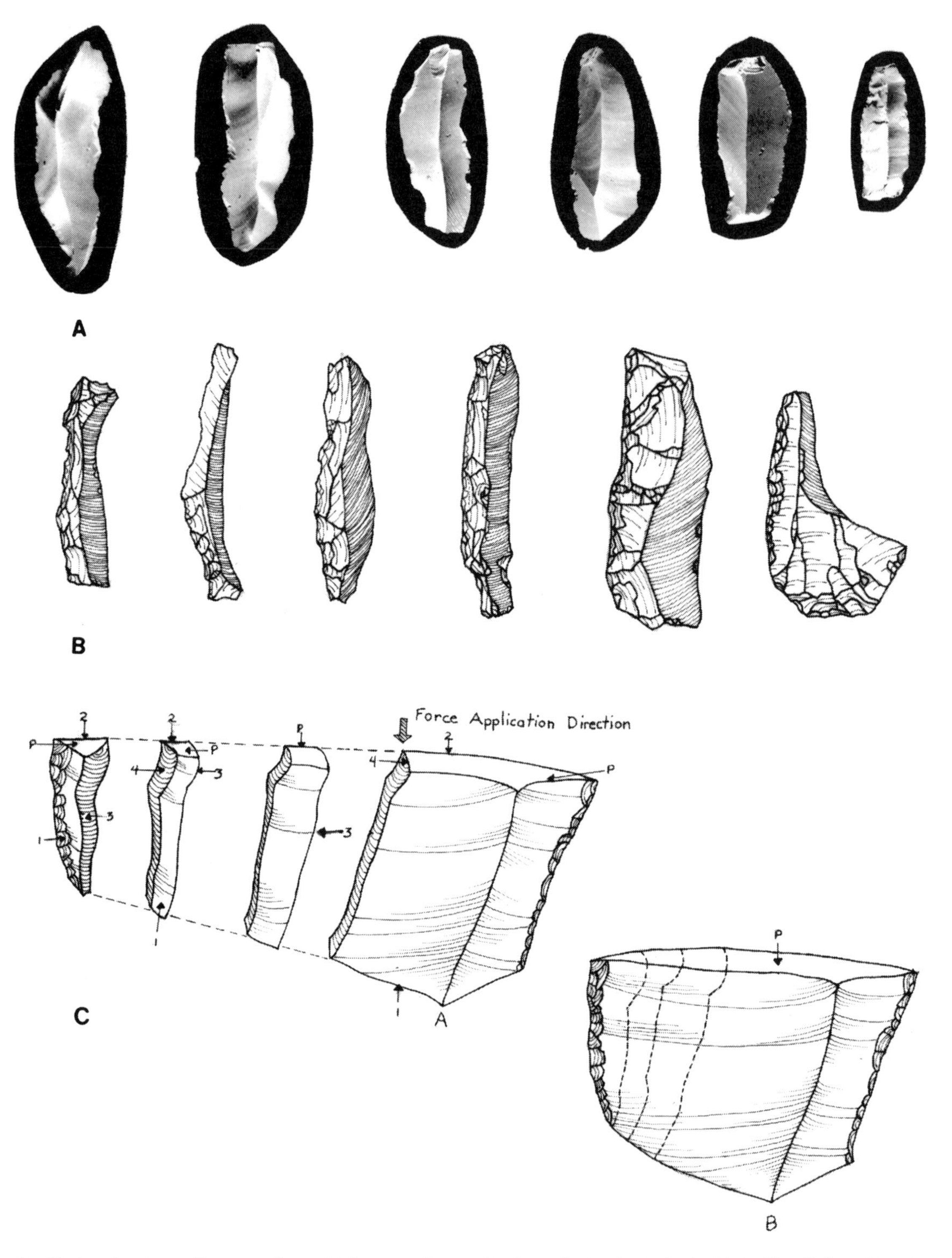

Figure 8. Blades, burin spalls, cores from El Inga, surface collections (natural size): A) blade-like flakes; B) primary burin spalls; C) broken uniface used as burin spall core (B); sequence of primary and secondary burin spall removal (A).

techniques that resulted in the rest of the flakes. I venture that this small number of lamellar flakes, scattered throughout the levels of Complex II, are part of the wide formal variation that is characteristic of all populations of simple flakes. The hypothesis can only be tested by a study of the whole flake assemblage (Stothert 1980:63).

This discussion highlights the situation at El Inga, San José, and probably other Early Man sites in the Ilalo region of highland Ecuador. The large collection of debitage from El Inga has simply been set aside for possible future study — we are talking here about some 41,381 specimens (Mayer-Oakes 1966:648) — after having been carefully sorted out on the basis of "unretouched" flake edges. The later collections (1965, 1967, 1968, 1971) of the same or larger size from both the surface and the excavations at San José have been only partially studied, the earlier studies influenced by our work at El Inga, the later studies done in the more comprehensive fashion which is now also advocated by Stothert.

But from both El Inga (Bell 1965) and San José (Mayer-Oakes 1970), the distinctive blade-like flakes, burins, and burin spalls have already been selected for special analyses. In all probability, Lynch analyzed Guitarrero Cave lithics the way he did because of a desire to make his data comparable with the trends of analysis already established for Andean lithic collections. While it is clear (Figure 8) that there are isolated forms we call blades, burins, and burin-spalls at both El Inga and San José, it is not yet demonstrated that the total lithic assemblage provides convincing evidence of special techniques necessary to interpret these forms as end products of a special production technique. In defense of this situation, it is perhaps fair to state that our discipline is only now in the first few years of an analytical capability to do what Stothert suggests. And we certainly do have the pragmatic problem of necessary time and funds to accomplish a thorough analysis. It seems clear to this writer that after a significant number of "total collection" lithic studies have been accomplished it will be possible to sample such total data collections with reasonable representativeness, instead of doing 100% "inventory" analyses. This procedure will be necessary, increasingly, as our data recovery techniques continue to improve (see Patterson 1978; n.d.).

Because both El Inga and San José (and many of the other Ilalo region Early Man sites) exhibit a strong preference for obsidian as a raw material (from 79%-99%, Salazar 1979:Table 7), we may well have a special problem in lithic technology interpretation. Boksenbaum (1978, 1980) has discussed the topic of "nodule smashing" of obsidian at Valley of Mexico Formative sites, as a complementary technology to the more refined blade-making technique. It seems clear that at least some aspects of the San José and El Inga debitage (tabular pieces, some burins, and some burin spalls) may be by-products of this kind of unspecialized stone smashing. This explanation seems particularly likely in the light of the cobble nature of much of the raw source material apparently used at San José and El Inga (Mayer-Oakes and Salazar n.d.).

On the other hand, the few "nubbin" obsidian cores remaining at El Inga and San José, the clear evidence for a polyhedral blade-producing core when done in basalt, as well as the rather specialized platform preparation and multiple channel flake removal techniques (see Figure 9) used on at least some of the *Fell's Cave Stemmed* points, all support the idea of specialized core and controlled flake (blade-like) production techniques. In addition, the highly specialized bifacial thinning by several means (including multiple channel flake removal) indicates a lithic production repertoire that included both crude and more refined techniques. As yet we have no good clues to the relationship of this lithic technology to particular kinds of environmental adaptation. Both the El Inga and San José complexes include what are conventionally interpreted as skin working and butchering tools. But the San José complex does not include the lithic projectile points necessary for large or small game killing. Perhaps they used bone or wooden projectile points.

The Fluted-facet Technique

Although Figure 9 illustrates only three of the evidences for variation in "fluting" or channel flake removal, our study of variation in the entire surface collection suggests that there are four main varieties of treatment of point bases:

1) normal (non-fluted and non-basally-thinned, see Figure 7);

2) pseudo-fluted (when a facet on an original blank assumes the role of channel flake or "flute," see Figure 6-3D, E);

3) fluted (when single or multiple "channel flake" removals and/or a platform for channel flake are present, see Figure 9);

4) basal thinned (short flake scars initiated at the proximal end of the point simply to "thin" the point, see Figure 6-4, 5).

Figure 10 shows all of the major projectile point types found in the El Inga surface collection, with an indication of which are "fluted," on the basis of categories 2 and 3 above. The illustration indicates that two different stemmed types and two different lanceolate types exhibit "fluting" or "pseudo-fluting" basal treatment. I now think of

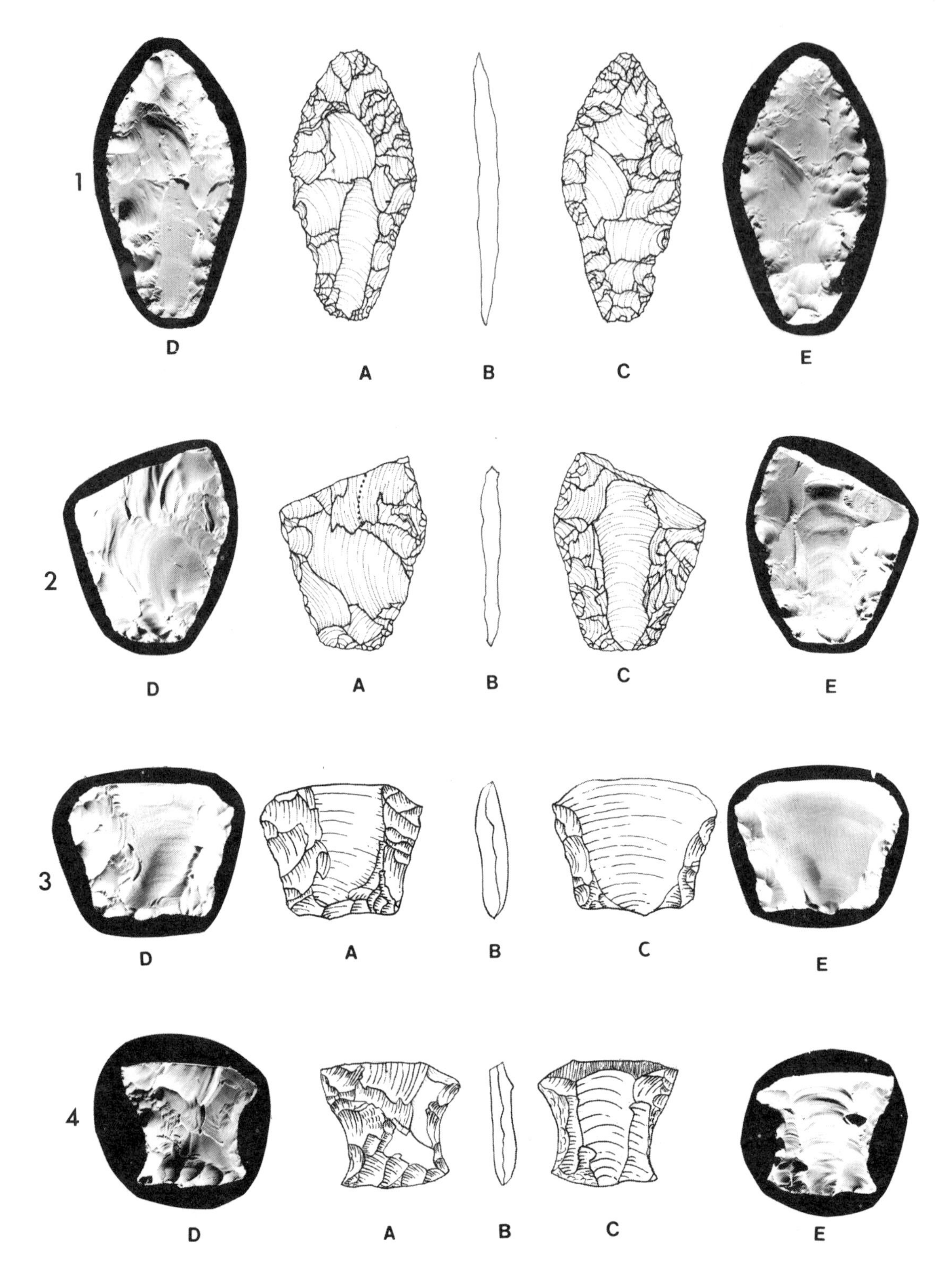

Figure 9. Examples of projectile point "fluting" from El Inga: rows 1, 2 single channel flake removal; row 3 single channel flake removal, use of "nipple" platform; row 4 multiple channel flake removal. (80% of natural size).

these two kinds of fluting as being primarily functional, designed to "thin" the proximal end of a projectile point for purposes of efficient hafting to a split or tapered shaft.

The Stemmed Point Form

Figure 10 is arranged so that the distinctions between "stemmed" and "lanceolate" point forms are emphasized. Form D, the *El Inga Shouldered Lanceolate* type, is unusual for its lateral edge profile with pronounced "shoulder." This may in fact be a preform production stage on the way to the end product shown at C, the *El Inga Broad Stemmed* type. What is most interesting is that basal (proximal) portions of these two types are essentially indistinguishable. Thus there is a real problem in classifying base-only specimens, especially as fluting (broadly considered) goes with some whole examples of each of these stemmed and lanceolate types. The two other stemmed forms seem quite distinctive, and in the El Inga version of the *Paijan* type (see Figures 4 and 5 above) we have a normal (unfluted) point that seems consistent in form. The *Fell's Cave Stemmed* type offers slight but significant variation (see Figures 2 and 3 above) when seen in complete plan form. At El Inga, the basal portion of this form seems consistently to indicate a stemmed point. However, at the Turrialba site in Costa Rica, (see Snarskis 1979:Figure 2E, F) the "waisted" lanceolate point offers a basal portion which could be compared with the base of a *Fell's Cave Stemmed*. When larger samples become available, my hunch is that the average basal width dimension will be consistently less in *Fell's Cave Stemmed*. With present small samples, however, the possibility of confusion of distinctively different types is a real problem.

The Lanceolate Point Form

At El Inga there is evidence for one very distinctive type of lanceolate point, the *Ayampitin* type shown at F in Figure 10. G in the same illustration is probably a variant of *Ayampitin*, but the one example published in some detail (Mayer-Oakes and Cameron 1971) because of its distinctive channel flake scar, may in fact represent a different form stereotype. It was suggested above that Figure 10D may be a preform. It is possible that Figure 10E is also a preform; but because most specimens of this type are rather carefully and completely chipped, it may in fact be a good candidate for function as a knife, rather than as a projectile point.

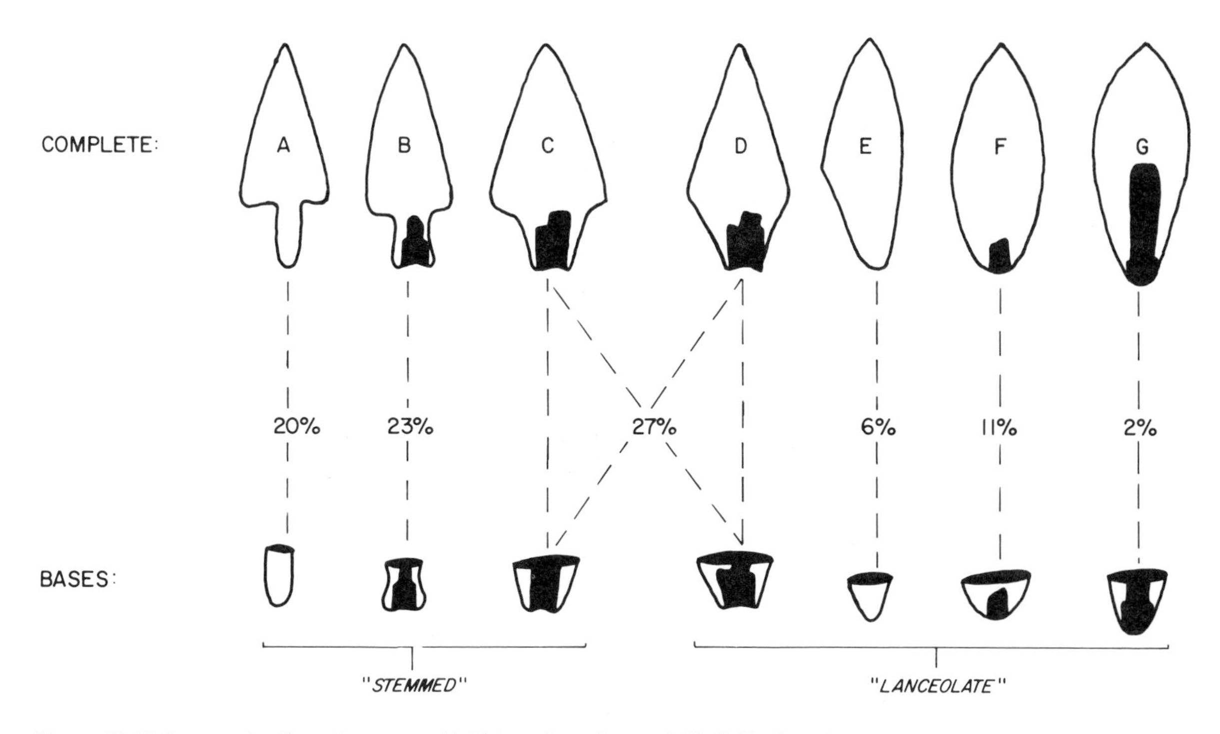

Figure 10. El Inga projectile point types: A) *El Inga Long Stemmed;* B) *Fell's Cave Stemmed;* C) *El Inga Broad Stemmed;* D) *El Inga Shouldered Lanceolate;* E) *El Inga Asymmetric Lanceolate;* F) *Ayampitin Lanceolate;* G) *El Inga Fluted Lanceolate.*

COMPARISONS

Because a number of the attributes and forms as well as techniques used at El Inga have analogs elsewhere (Chile, Peru, Argentina), additional more explicit comparisons will be made. While the present analysis has emphasized form, technological and some functional ideas have been used to help describe and thus understand the Ecuador evidence.

Figure 11 is my synthesis of the major diagnostic (or "marker") Early Man projectile points found in South and Central America. The complete specimens represent finds from seven different sites (Paijan, Peru; Fell's Cave, Chile; El Inga, Ecuador; Turrialba, Costa Rica; Los Tapiales, Guatemala; Intihuasi Cave, Argentina; and El Jobo, Venezuela). The basal examples (except for El Jobo) are all found at El Inga. The illustration (as in Figure 10) attempts to point out areas of possible confusion of attribution of type when dealing only with basal portions. In addition, the presence of a thick lens or diamond-shaped lateral cross-section in the El Jobo type suggests (following Bryan 1975; Lahren and Bonnichsen 1974) the socketed hafting style as contrasted with a split-shaft style of hafting which would favor basal thinning and/or fluting. As at El Inga, four of the continent-wide diagnostic points have a "fluting" attribute on both stemmed and lanceolate forms.

Of particular interest, now that the ideal (stereotype) point form has been schematized, is the previous usage of the term "fishtail." From the El Inga site and from Fell's Cave there are no apparent problems in identifying this characteristic stemmed style, with its concave sides (from shoulder to promixal end) and its distinctive "eared" basal termination, because of the concave curved line of the basal and stem edges. The stereotypes for Turrialba "waisted" lanceolate and *Fell's Cave Stemmed,* however, are confusing if represented only by basal portions. Similarly the *Clovis* type can be confused with the *El Inga Broad Stemmed* if represented only by basal portions. These two types share the straight line tapered (contracting towards centerline from distal end towards proximal end) basal edges and a concave

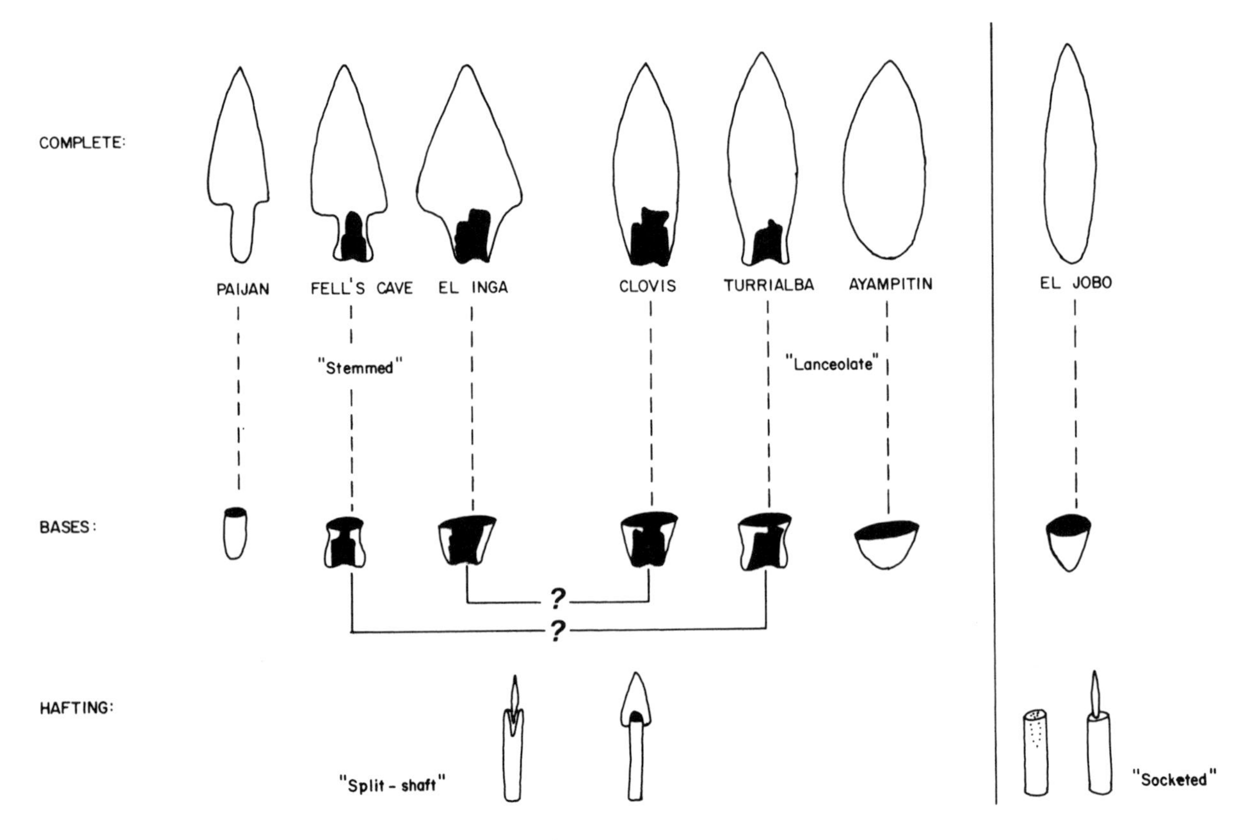

Figure 11. South American Early Man "ideal" projectile point types.

base, as well as a similar size range and lateral cross-section.

Bird and Cooke (1977, 1978) describe the presence of two types of Paleoindian points in Panama. They show the "Clovis" types (1977: Figures 3, 5, 6a) and the "fishtail" type (1977: Figures 4, 6c, 7, 8, 9, 12e, f). Snarskis (1979) describes fluted points from the Costa Rican Turrialba site in terms of preforms and three types, Type 1 being "waisted" *Clovis Lanceolate*, Type 2 being a non-waisted *Clovis* form (his Figure 2C and Figure 3B), and Type 3 being the "fishtail" stemmed point. While Snarskis does not spell out the distinction between "waisted" and "tapered" *Clovis* points, his typology generally reflects a consistent difference. Bird and Cooke, on the other hand, do not distinguish these two variants of the *Clovis Lanceolate*, although they illustrate the Central American finds which document the presence of the two different forms. (Bird and Cooke 1977:Figures 3a, d are "tapered," b and c are "waisted," I judge). Bird and Cooke also identify as the Fell's Cave "fishtail" style of stemmed point (1977:Figure 4a, d, e, 6c) four specimens which conform to the basal stereotype found on *El Inga Broad Stemmed*. That is, the concave or "waisted" contour of the stem edges is not present on these examples of the stemmed form; rather, the tapered profile is demonstrated, in my judgement.

I have not handled the Panamanian or Costa Rican specimens referred to above, but on the basis of illustrations in both Bird and Cooke (1977, 1978) reports, I conclude that the Panamanian versions of the "fishtail" point are really closer to the distinctive *El Inga Broad Stemmed* type. When considering only the basal portions of these Panama points, this conclusion is dramatically strengthened. However, the fishtail point identified by Snarskis (1979:Figures 2F and 3A) does seem to be characterized by the concave stem edges and the "eared" or waisted profile.

My conclusion, then, that the Panamanian "fishtails" are in fact not fishtails has direct implications for current interpretation of the small sample of putative Early Man points in Central and South America. It is now possible to distinguish some significant patterned variation in lanceolate forms, in stemmed forms, and in the distinctive "fluting" characteristics.

Using my model of South American Early Man projectile point stereotypes, the Panama, Costa Rica, and Guatemala sample of "Paleoindian" points presented by Bird and Cooke (1978) can be interpreted as follows:

Figure 1a. is *Clovis Fluted* (tapered) from Costa Rica
Figure 1b. is *Clovis Fluted* (waisted) from Guatemala
Figure 1c. is *Clovis Fluted* (waisted) from Panama
Figure 1d. is *Clovis Fluted* (tapered) from Guatemala
Figure 2c, d, e, f, g, are all *El Inga Broad Stemmed* from Panama.

The Costa Rican points reported by Snarskis (1979) would be interpreted, using our continental model, as follows:
Figure 2A, C and D are *Clovis Fluted* (tapered)
Figure 2F is *Fell's Cave Stemmed*
Figure 2E is *Clovis Fluted* (waisted)
Figure 4A and B are *Clovis Fluted* (tapered)

The latest additions to our universe of putative Early Man points from Central America are the "fishtails" reported by MacNeish from Belize (MacNeish et al. 1980); the fluted stemmed points reported from Belize by Hester et al. (1980); and the fluted lanceolate point from Belize reported by Hester et al. (1981). If MacNeish's "fishtails" are like the four stemmed points illustrated by Hester et al. (1980) (and Hester suggested to me that he thought they were), they are another example of a too flexible usage of the "fishtail" concept. The Belize stemmed points reported by Hester et al. (1980) are characterized by straight or expanding (to proximal end) stem edges; two of them show straight base, and two show a concave base. The latter two, incidentally, show what looks from the line drawing illustration like definite channel flake removal.

The fluted lanceolate point from Ladyville, Belize (Hester et al. 1981), is compared favorably by the authors to the Turrialba Clovis point (Snarskis 1979:Figure 2D). I agree, but would refer the interested reader to a better illustration of this Costa Rican point in Bird and Cooke (1978:Figure 5a) where the "tapered" rather than the "waisted" nature of the form is clearly evident. Because Hester et al. (1981) picked Figure 2D rather than Figure 2E (i.e., they picked the tapered rather than the waisted version) I am satisfied that my distinction between "tapered" and "waisted" is a useful one for other scholars to use. I, too, see the Belize form as a "tapered" *Clovis Fluted* rather than a "waisted" one.

The functional question to be raised once these two types (or sub-types) of Clovis are accepted is: can repeated haftings (and regrinding of the proximal portion of the lateral edges) change a "tapered" Clovis into a "waisted" one?

DISCUSSION

In this paper I have concentrated attention on the projectile point portion of the stone tool assemblage from the El Inga site. The approach has attempted to look at points with a kind of "emic" view; specifically, both the idea of a "stereotype" (or mental image in the mind of the maker) and ideas about lithic production techniques based on the current state of the art of stone chipping knowledge have been used. Contemporary flint knappers do use mental templates or stereotypes to guide their production efforts. Starting with this idea, as a means of presenting the evidence we have attempted to distinguish and utilize ideas and data from the realms of morphology, technology, and function. In this final section of the paper, an attempt is made to synthesize the information previously presented and examine it briefly in the light of known geographic distributions.

MORPHOLOGY, TECHNOLOGY AND FUNCTION

The concept of stereotype depends heavily upon morphological data, but elements of the concept also depend upon data that is primarily technological or functional in nature. Thus, the idea of a stereotype has important relationships among the three areas of substantive data: form, manufacturing processes, and function.

Morphology

The idea of "stereotypes" for projectile points depends very heavily upon the specific empirical evidence for the plan form (or outline, as used in Figures 10, 11) of the point. There are several critical components of plan form to which particular attention has been paid. The basic idea, deeply rooted in North American typological concepts, is that at El Inga there are two major and distinctly different plan forms. These we have called "stemmed" and "lanceolate." The Figure 10 depiction of our El Inga typology has summarized this morphological conclusion. However, the evidence at El Inga also indicates a third form, one intermediate between stemmed and lanceolate. *El Inga Shouldered Lanceolate* (Figure 10D) is a type that can be considered intermediate in form. Similarily, Figure 11 shows a Turrialba form which may be considered a kind of stemmed point. My personal view of such intermediate forms is strongly conditioned by the many difficulties I have experienced in attempting to allocate basal point fragments to either a stemmed or a lanceolate type.

Distinctive details or component attributes of the outline of a point have a major role to play in morphological classification. Four such characteristics have been worked with in this paper. Two of these are essentially the same attribute, namely the concepts of "eared" and "waisted." Both these terms have been used by various archaeologists in the literature on Early Man projectile points. The attribute of "concave" base is well-known, but the idea of "tapered" basal edges as I have used it in this paper has not often been isolated as diagnostic. The critical form difference between *Fell's Cave Stemmed* and *El Inga Broad Stemmed* appears to lie in the "waisted" or "eared" characteristic of the stem edges in the former and the "tapered" stem edges in the latter. Other attribute details such as "shoulder" or "barb" have not been utilized in this paper, but apparently play a similar role in defining and distinguishing El Inga point stereotypes.

Another important stereotype aspect of morphology, in addition to outline, is the complex of ideas we label "dimensions." Two major concerns relevant to dimensions bear on the problem of New World Early Man points: variations in overall dimensions such as length, width, and thickness (as denoted in both lateral and longitudinal cross-section); and the relative proportions within a point. The points used in this analysis illustrate the importance of the position of maximum width (which distinguishes *Ayampitin* from *El Inga Fluted* as lanceolate types), the proportion of stem size to face size (which distinguishes *Paijan* from *Fell's Cave Stemmed*), and the specific characteristics of the stem (which distinguishes among all three El Inga stemmed types).

A final morphological element bearing on our use of the concept of stereotype is the complex of concerns termed "problem areas," where classification difficulties, especially when the raw data are in the form of basal point fragments, are seen. The illustrations in this paper have been arranged to indicate this problem: e.g., in the types *El Inga Broad Stemmed* and *El Inga Shouldered Lanceolate*, and in the possible confusion between the Turrialba "waisted" form and *Fell's Cave Stemmed* (as well as the problem of distinguishing bases of *El Inga Broad Stemmed* from the "tapered" form of *Clovis Lanceolate*). Careful consideration of this matter should provide us with a more realistic idea of stereotypes. There are also problems to be considered with the specificity of form attributes. It has been suggested that the concept of "fishtail" is not precise enough, as these points are composed of separate formal attributes which I have split into the mutually exclusive attributes "waisted" and "tapered," thus perceiving the two distinctly different types. Another problem lies in the possibility for recognizing types that are intermediate in terms of the stereotype concept. We suggest that *El Inga Shouldered Lanceolate* is such an intermediate form, i.e., it is intermediate between a "stemmed" and a "lanceolate" form. Because, however, it is relatively frequent (19% of all points) we seem to be dealing with a stereotype form, not a sampling or other accident. Not only is the attribute "shouldered lateral profile" a problem, but the "waisted" stem or basal edge profile is a similar problem. If the edge curve is slight enough and, in a basal fragment, the amount visible is short enough, a "waisted" point can be mis-classified as a "tapered" point. A final problem area hardly touched on is that of total size. The type *El Inga Broad Stemmed* is known best from very large (long and broad) examples. These examples suggest that size ranges normally associated with projectile points (Thomas 1978) need to be left open and flexible when using the concept of stereotype. This flexibility is also necessary when there are both unifacial and bifacial examples of a given stereotype. Another conceptual flexibility lies in the lithic material used. Outline stereotypes should be expected to transcend norms of size and material, as is indicated at El Inga for both stemmed and lanceolate point stereotypes.

Technology

While it is most common perhaps to use the concept of stereotype in relation to matters of form or morphology, the concept has here been applied to technology as well. The fluting characteristics of El Inga were classified in 1970 and 1971 into four technology classes on the basis of the stereotype concepts: **normal** (unfluted), **pseudo-fluted** (utilizing one or more preform facets), **thinned base** (short basal thinning flakes), and **fluted** (removal of one or more putatively intentional long channel flakes, or by presence of a special platform). The last three of these classes can be seen to be results of a technological process aimed at creating a particular shape (grooved) or thickness (thinner than the rest of the point) for the basal portion of the point.

This series of technological classes of "fluting" suggests that at El Inga there is evidence for an early and a middle stage in a developmental sequence. Surely the idea of fluting did not spring fully formed from the mind of the first maker. It appears more likely that an idea for thinning the base occured first, followed by several unspecialized styles of thinning prior to the development of a specialized and effective technique utilizing sequential preforms and channel flake platforms. Interestingly enough, independent of the original 1972 analysis of El Inga fluting and subsequent ruminations about it, two very careful studies of Folsom fluting technique provide support for the stereotype approach. Tunnell (1975) has described the fluting process at the Adair-Steadman site as a series of stages of preform preparation with basal and edge grinding plus channel flake platform preparation as primary operational elements in a complex six-step process. Independently, and with experimental replication as a key part of the analytical process, Flenniken (1978) has developed a similar seven-step production process for the *Folsom fluted* points at Lindenmeier. The relevance of these two studies to our El Inga data, it seems to me, is in the picture they provide of technological specialization and multiple preform stages. In the light of the fluting or pseudo-fluting controversy examined recently by Rovner (1980) and Bray (1978, 1980), similarily detailed analytical studies of Clovis technology and that of other early fluted points are needed. The Tunnell and Flenniken studies are good models for such work.

If the elaborate preparation of a sequence of preforms has a developmental background, it should be expected that early and simpler forms of fluting will be represented by different and probably simpler preforming. The edge preparation by grinding (which is a characteristic activity of Folsom, as shown by both Tunnell and Flenniken) now should play a more important role in the classification and interpretation of projectile points with fluting. It is clear that several stages of grinding may be involved. Thus the observation and recording of this attribute needs to be better integrated with the overall assessment as to whether a given specimen is a preform or a final form.

Finally, the nature of an original blank for a fluted point must have been significant to the aboriginal development of the concept of fluting. This would have been of especial significance in the earliest stage of developing the concept of fluting. A blade or a flake with a concave facet that could function as the "thinning" element on the face of a point, could well have determined its acceptability to the knapper as a simple initial blank or preform.

Function

It has been difficult in this discussion to attempt to exclude concern about function, because of the complex relationships that characterize form, production process, and function on a given artifact. Mere use of the concept of "preform" involves us with a term heavily loaded with functional implications. It seems clear that we must view our fundamental past human behavior process in some kind of integrated way. That is, in order to achieve the needed aboriginal end product (a stone projectile point which will function well when hafted on a split or tapered shaft), a "thinned-base stone point" form needs to be successfully produced by a stone chipping technology in order to perform the function which has been conceived as a goal in the mind of the maker.

At such a broad level as this, function is a somewhat nebulous idea. We can, perhaps, gain a clearer or more satisfying use of the concept of function if we narrow our view. The concept of preform as used in the Folsom studies referred to seems quite clear. Each step in a lithic technological production process requires a particular form as a "blank" before applying the chipping and grinding work necessary to achieve the next stage successfully. And all of these are necessary to achieve the final stage. Thus achieving a series of technological functions, if you will, is prerequisite to achievement of the final goal of specialized implement use.

The functional requirements of size and shape seen in the socket-base as opposed to the split-base hafting option provide us with a largely unexplored area of observation and analysis. Consideration in detail of the functional and formal requirments for distinctly different styles of hafting may well throw light on the technological correlates to be anticipated. At the moment we are simply using a stereotype concept of form which seems to be a plausible way of distinguishing a possible functional difference between *El Jobo* point morphology and most other South American early point forms.

The general concept of stereotype has significance primarily in the realm of function. That is, when consciously used it has a primary role as a kind of goal statement for the user. But it also has value when considered as an element or nucleus of "style." It seems useful to employ the concept of stereotype as a prime definer of style, particularly of rather simple styles such as projectile points. This is no simple topic, however; and serious consideration is only beginning within the context of Early Man studies in the Andes (see Rick 1980:336; Lynch 1980:296-301; Aikens 1981). The questions raised by Rick and Lynch as to the geographic extent of style significance are of paramount importance to my concern with South American Early Man diagnostic point style stereotypes.

Finally, this El Inga point review has raised an issue about weaponry that has not been seriously considered yet. I refer to the large-sized examples of the *El Inga Broad Stemmed* type. It is possible that some of these projectile points are large enough to have been used as tips for thrusting spears or javelins, a possibly significant hunting specialization or adaptation.

DISTRIBUTION

While it seems likely that in some ultimate sense we must look to the Upper Paleolithic in the Old World if we seek roots for the earliest New World people (Alexander 1974), the El Inga chipped stone complex has been suggested to be distinctively Upper Paleolithic in nature (Mayer-Oakes 1972; Salazar 1979:29-31). Although it may have nothing to do with this Old World flavor, the El Inga richness in projectile point variety suggests that the site may well represent an important element in the transmission and development of ideas and forms between North and South America.

The recent information on finds in Central America (Snarskis 1979), and Bray's (1978) brief formulation of a model for understanding projectile point variety in Mesoamerica and the isthmus, have stimulated my thinking on the question of intercontinental relationships. Bird has described the presence of both "fishtail" and fluted lanceolate points from Panama in several publications. Costa Rican fluted point records began with the Hartman collection specimen from the Carnegie Museum (Swauger and Mayer-Oakes 1952), and are now augmented by the impressive Turrialba site which combines both of the North American fluted lanceolate forms with the South American "fishtail." The Guatemalan record of fluted lanceolate points indicates the presence there of both the "tapered" (Gruhn and Bryan 1977) and "waisted" (Coe 1960) forms. It seems clear that six projectile points from Belize (4 from the earliest "Lowe-ha" complex and two from the following "Sand Hill" complex) are also pertinent to the present study (MacNeish et al. 1980). One of the Lowe-ha specimens would qualify as a *Fell's Cave Stemmed* point by my criteria. In order to crystalize ideas on projectile point relationships between North and South America, I have combined the ideas gained from El Inga studies with the Central American data. The result is a model of form, technology and function stereotypes found in Figure 11.

This description of stereotypes may have some usefulness if it suggests both the complexity of Early Man projectile point styles and the apparent two-way flow of technological, morphological, and functional ideas between North and South America. It may also provide some useful target hypotheses for detailed site or small regional studies (e.g., Eisenberg 1978; Rick 1980; Lynch 1980; Mayer-Oakes n.d.). It does provide a new framework for reevaluation of specific projectile point records (e.g., Ossa 1976; Chauchat and Zevallos 1979).

Sampling Concerns

Significant quantities of data bearing on the topic of Early Man have now been collected, studied, and reported from a reasonable mixture of sites producing much data and sites that are essentially isolated finds. While the isolated finds of projectile points are weak sets of data in a quantitative sense, interpreting them by means of the concept of stereotypes as here presented can provide a strong qualitative base for planning future research directions.

CONCLUSIONS

In this paper I have attempted to present a limited amount of evidence from Ecuador and use it as a foundation for both local and broader interpretations. Looking at the Ilalo region, it appears that there is significant patterned variation in both projectile point form and "fluting" characteristics. Specifically, El Inga evidence is interpreted to support the idea that development of the fluting concept was at an early to intermediate stage, attaining the level of Clovis fluting technology and form in some cases and going beyond this towards Folsom in others. The use of a complex concept of "stereotype" has been helpful in achieving these conclusions.

Based upon broader geographic interpretations I conclude that the "fishtail" concept is a useful continental diagnostic but it must be used more rigorously than it has in the past. Although complex, the concepts of fluting and "preform" are useful, but they require more sophisticated and rigorous application in the future. The model of South American projectile point stereotypes for Early Man "works" when applied to available evidence from Panama, Costa Rica, and Belize, thus extending our ideas about North and South American relationships during Early Man times.

REFERENCES CITED

Aikens, C.M.

1981 Review of Guitarrero Cave. *American Anthropologist* 83:224-226.

Alexander, H.L.

1974 The association of Aurignacoid elements with fluted point complexes in North America. In *International conference on the prehistory and paleoecology of western North American arctic and subarctic*, edited by S. Raymond and P. Schledermann, pp. 21-31. Archaeological Association, Department of Archaeology, University of Calgary, Calgary.

Bell, R.E.

1960 Evidence of a fluted point tradition in Ecuador. *American Antiquity* 26:102-106.

1965 *Investigaciónes arqueológicas en el sitio de El Inga, Ecuador.* Casa de la Cultura Ecuatoriana, Quito.

1977 Obsidian hydration studies in highland Ecuador. *American Antiquity* 42:68-78.

Bird, J.B.

1969 A comparison of south Chilean and Ecuadorian "fishtail" projectile points. *Kroeber Anthropological Society Papers* 40:52-71.

Bird, J.B., and R. Cooke

1977 Los artefactos mas antiguos de Panama. *Revista Nacional de Cultura* 6:7-31.

1978 The occurrence in Panama of two types of Paleo-Indian projectile points. In Early Man in America from a circum-Pacific Perspective, edited by A.L. Bryan, pp. 263-272. *Department of Anthropology, University of Alberta, Occasional Papers* No. 1. Edmonton.

Boksenbaum, M.W.
1978 Lithic technology in the Basin of Mexico during the Early and Middle Preclassic. Ph.D. dissertation, City University of New York.

1980 Basic Mesoamerican stone-working: nodule smashing? *Lithic Technology* IX: 12-26.

Bonifaz, E.
1978 *Obsidianas del Paleo-Indio de la regiőn del Ilalo.* Privately printed. Quito.

Bray, W.
1978 An eighteenth century reference to a fluted point from Guatemala. *American Antiquity* 43: 457-460.

1980 Fluted points in Mesoamerica and the isthmus: a reply to Rovner. *American Antiquity* 45: 168-170.

Bryan, A.L.
1975 Paleoenvironments and cultural diversity in Late Pleistocene South America: a rejoinder to Vance Haynes and a reply to Thomas Lynch. *Quaternary Research* 5: 151-159.

Cardich, A.
1964 Lauricocha, fundamentos para una prehistoria de los Andes Centrales. *Studia Praehistorica* III. Centro Argentino de Estudios Prehistoricos. Buenos Aires.

Carluci, M.A.
1963 Puntas de proyectil, tipos, tecnica y areas de distribución en el Ecuador Andino. *Humanitas* IV: 5-56. Quito.

Chauchat, C.
1976 The Paijan complex, Pampa de Cupisnique, Peru. *Ñawpa Pacha* 13: 85-96.

1978 Additional observations on the Paijan complex. *Ñawpa Pacha* 16: 51-65.

Chauchat, C., and J. Zevallos Q.
1979 Una punta en cola de pescado procedente de la costa norte del Peru. *Ñawpa Pacha* 17: 143-147.

Coe, M.D.
1960 A fluted point from highland Guatemala. *American Antiquity* 25: 412-413.

Eisenberg, L.
1978 Paleo-Indian settlement pattern in the Hudson and Delaware river drainages. *Occasional Publications in Northeastern Anthropology* No. 4. Department of Anthropology, Franklin Pierce College, Rindge, N.H.

Flenniken, J.J.
1978 Reevaluation of the Lindenmeier Folsom: a replication experiment in lithic technology. *American Antiquity* 43: 473-480.

Gonzalez, A.R.
1959 The stratigraphy of Intihuasi cave, Argentina and its relationship to early lithic cultures of South America. Ph.D. dissertation, Columbia University.

Gruhn, R., and A.L. Bryan
1977 Los Tapiales: a Paleo-Indian campsite in the Guatemalan highlands. *American Philosophical Society, Proceedings* 121: 235-273.

Hester, T.R., T.C. Kelly, and G. Ligabue
1981 A fluted Paleo-Indian projectile point from Belize, Central America. Colha Project, *Working Papers* No. 1, Center for Archaeological Research, University of Texas at San Antonio.

Hester, T.R., H.J. Shafer, and T.C. Kelly
1980 Lithics from a preceramic site in Belize. *Lithic Technology* IX: 9-10.

Lahren, L., and R. Bonnichsen
1974 Bone foreshafts from a Clovis burial in southwestern Montana. *Science* 186: 147-150.

Lanning, E.P.
1963 A pre-agricultural occupation on the central coast of Peru. *American Antiquity* 28: 360-371.

Lynch, T.F. (editor)
1980 *Guitarrero Cave, Early Man in the Andes.* Academic Press, New York.

MacNeish, R.S., S.J.K. Wilkerson, and A. Nelken-Turner
1980 *First annual report of the Belize Archaic archaeological reconnaissance.* Robert S. Peabody Foundation for Archaeology, Phillips Academy, Andover, Massachusetts.

Mayer-Oakes, W.J.
1963 Early Man in the Andes. *Scientific American* 208(5): 116-128.

1966 El Inga projectile points — surface collections. *American Antiquity* 31(5) Part 1: 644-661.

1970 The San José site. Paper presented at XXXIX International Congress of Americanists. Lima.

1972 El Inga Obsidian Industry — an Upper Paleolithic complex from South America. Convergence, parallelism or genetic connection? VIII Congres INQUA-1969. *Etudes sur le Quaternaire dans le Monde.* pp. 995-999. Paris.

1979 Early Man research in Ecuador 1960-1980. Paper presented at XLIII International Congress of Americanists, Vancouver.

1982a Early Man in the northern Andes: problems and possibilities. In Peopling of the New World, edited by J.E. Ericson, R.E. Taylor and R. Berger. pp. 269-285. *Ballena Press Anthropological Papers* No. 23. Los Altos.

1982b Comparative archaeology: the need for closer interdisciplinary and international cooperation in the third world — an example from Ecuador. Paper presented at Transatlantic colloquium on comparative archaeology, Fort Burgwin Research Center, August 1982.

1982c Studies of the unifacial lithic assemblage from San José, Ecuador. *XLIV International Congress of Americanists, Abstracts,* p. 208. Manchester.

n.d. El Inga, a Paleo-Indian site in the Sierra of northern Ecuador, in press.

Mayer-Oakes, W.J., and R.E. Bell
1960a An early site in highland Ecuador. *Current Anthropology* 1: 429-430.

1960b Early Man site found in highland Ecuador. *Science* 131: 1805-1806.

Mayer-Oakes, W.J., and W.R. Cameron
1971 A fluted lanceolate point from El Inga, Ecuador. *Ñawpa Pacha* 7-8: 59-65.

Mayer-Oakes, W.J., and E. Salazar
n.d. Manuscript on Ecuadorian obsidian studies in possession of author.

Ossa, P.
1976 A fluted "fishtail" projectile point from La Cumbre, Moche valley, Peru. *Ñawpa Pacha* 13: 97-98.

Ossa, P., and M.E. Moseley
1972 La Cumbre; a preliminary report on research into the Early Lithic occupation of the Moche Valley, Peru. *Ñawpa Pacha* 9: 1-24.

Patterson, L.W.
1978 Lithic technology: a primer. *Special Publication,* Houston Archaeological Society.

n.d. Instruction in lithic technology in U.S. universities. Manuscript in files of Laboratory of Archaeology, Texas Tech University.

Phagan, C.J.
1970 An analysis of the Cameron collection of artifacts from highland Ecuador. M.A. thesis, Ohio State University, Columbus.

Rick, J.W.
1980 *Prehistoric hunters of the high Andes.* Academic Press, New York.

Rovner, I.
1980 Comments on Bray's "An eighteenth century reference to a fluted point from Guatemala." *American Antiquity* 45: 165-167.

Salazar, E.
1979 *El hombre temprano en la región del Ilalo, sierra del Ecuador.* Universidad de Cuenca, Cuenca.

Schobinger, J.
1969 *Prehistoria de suramerica.* Nueva Colección. Barcelona.

1973 Nuevos hallazgos de puntas "colas de pescado", y consideraciones en torno al origen y dispersion de la cultura de cazadores superiores toldense (Fell I) en Sudamérica. *Atti del XL Congresso Internazionale degli Americanisti,* Vol. 1: 33-50. Roma-Genova.

Snarskis, M.J.
1979 Turrialba: a Paleo-Indian quarry and workshop site in eastern Costa Rica. *American Antiquity* 44: 125-138.

Stothert, K.E.
1980 Review of: Guitarrero Cave: early man in the Andes. *Lithic Technology* IX: 61-63.

Swauger, J.L., and W.J. Mayer-Oakes
1952 A fluted point from Costa Rica. *American Antiquity* 17: 264-265.

Thomas, D.H.
1978 Arrowheads and atlatl darts: how the stones got the shaft. *American Antiquity* 43: 461-473.

Tunnell, C.
1975 Fluted projectile point production as revealed by lithic specimens from the Adair-Steadman site in northwest Texas. *Texas Historical Commission. Office of the State Archaeologist, Special Report* 18.

Willey, G.R.
1971 *An introduction to American archaeology.* Vol. 2, South America. Prentice-Hall, Englewood Cliffs.

Las Unidades Culturales de São Raimundo Nonato - Sudeste del Estado de Piauí - Brasil

NIÈDE GUIDON
École des Hautes Études en Sciences Sociales,
Anthropologie Préhistorique d'Amérique
44, Rue de la Tour, Paris,
FRANCE

Abstract

Since 1970, archaeological and paleoenvironmental research in the area of São Raimundo Nonato, in the caatinga zone of the southeast part of the state of Piauí in northeastern Brazil, has been carried out by a party led by Dr. Niède Guidon of the École des Hautes Études en Sciences Sociales of Paris. This region of ancient sedimentary rocks is greatly dissected; and a number of rockshelters, often with rock art, have been discovered along the walls of canyons and valleys. Excavations in two rockshelters — Toca do Sitio do Meio and Toca do Boqueirão da Pedra Furada — have exposed a sequence of human occupation levels which have yielded radiocarbon dates extending back to approximately 31,500 years ago.

The earliest series of radiocarbon dates has been obtained from Toca do Boqueirão da Pedra Furada. This site is an overhang in a high steep cliff of quartzite. The protected floor area measured approximately 40 x 15 m, of which an area approximately 20 x 15 m has been excavated to a maximum depth of approximately 3.5 m. The early deposits are largely subangular gravels, the material derived by weathering from the roof and walls of the shelter. Lithic artifacts, hearths, charcoal and ash were recovered from a series of stratified occupation levels. In general, the early lithic assemblages are characterized by flake tools and pebble cores, predominantly of quartz but some of quartzite. These artifacts are now being closely studied. More specifically, the lower occupation levels and artifact assemblages, and the associated radiocarbon dates, are as follows.

Stratum XIII. The artifact assemblage from this occupation level consists of small flake tools and small flaked pebbles, most of quartz. Of particular interest at this level was a slab of stone spalled from the rockshelter wall and bearing spots of red ocher paint, indicating the presence of rock art by this time. This slab and the lithic artifacts were directly associated with an extensive hearth, charcoal from which was dated at 17,000±400 B.P.

Strata XIV-XVIII. These occupation levels yielded small flakes and small flake pebbles of quartz and quartzite. In 1980 a sample of charcoal from Stratum XVI was dated greater than 25,000 B.P.

Stratum XIX. A large and varied assortment of tools made on small flaked pebbles of quartz or quartzite were found in association with ash and charcoal. A radiocarbon date of 26,000±600 B.P. was obtained from an extensive hearth on the west side of the shelter, and a radiocarbon date of 26,400±500 B.P. was obtained from another hearth on the east side of the shelter, while another hearth was dated 31,500±950 B.P. Four deeper levels excavated in 1984, contain similar artifacts with hearths that are being dated.

At the rockshelter site of Toca do Sitio do Meio, stratigraphy was less clear due to very large boulders incorporated in the fill, making it difficult to trace strata continuously across the area of excavation. However, radiocarbon dates of 12,000±600 B.P. and 13,900±300 B.P. were obtained on samples of charcoal in Stratum V, a stratum which yielded a "limace"; four flakes, one with lateral edges retouched; a chopper; and several pebble hammerstones. Most artifacts were of siltstone. One hearth was exposed in this stratum. Numerous artifacts, including cores, flaking debitage, retouched or utilized flakes, choppers, and hammerstones were found in lower levels, Strata VI through XXI, as yet undated.

Research in the region of São Raimundo Nonato has produced evidence for three early cultural phases. The first, extending in time from before 31,500 to 14,000 years ago, is characterized by quartz and quartzite pebble cores and utilized flakes, with very limited retouch, and likely employed as cutting tools and scrapers. The second, extending in time between 14,000 and 11,000 years ago, is characterized by well-shaped unifacial "limaces" with steep retouch and flakes of siltstone, quartz, or quartzite. The third phase, dated between 10,000 and 7000 years ago, is a quartzite industry featuring large flakes, scrapers and knives.

DESARROLLO DE LAS INVESTIGACIONES

En el año 1970 nuestro equipo inició los trabajos de campo en la región de São Raimundo Nonato. Desde entonces otras nueve misiones fueron realizadas en los años de 1973, 1974, 1975, 1978, 1979, 1980, 1981, 1982 y 1984.

Se trata de un proyecto interdisciplinario que cuenta con la colaboración de especialistas brasileños y franceses.

El tema general de la investigación es: "El poblamiento del sudeste del Piauí: los hombres y su interacción con el medio, desde la prehistoria hasta la época actual". Nuestros trabajos no se limitan a la investigación arqueológica sino que abarcan también el estudio del medio, la evolución climática, la población actual, y el registro documental de las técnicas artesanales en vías de desaparición. La prehistoria es a la vez el punto de partida de nuestros trabajos y el núcleo de coordinación de los otros aspectos de la investigación.

En la actualidad, un equipo de catorce investigadores participa en el proyecto. Estas investigaciones son financiadas por organismos franceses y brasileños.

La contribución francesa se realiza por intermedio de las siguientes instituciones:

Centre National de la Recherche Scientifique;

Ministère des Relations Extérieures;

Ecole des Hautes Etudes en Sciences Sociales;

Diréction Generale de la Recherche Scientifique et Technologique.

Los organismos que aseguran la colaboración brasileña son:

Conselho Nacional do Desenvolvimento Científico e Tecnológico;

Fundação Universidade Federal do Piauí;

Museu Paulista da Universidade de São Paulo.

Desde el punto de vista estrictamente arqueológico los resultados obtenidos hasta ahora pueden resumirse mediante las siguientes informaciones:

a) a partir de fotos aéreas, disponibles a una escala de 1:25000, se han preparado los mapas de la región sobre los cuales se ha registrado un total de 188 sitios;

b) dentro de esta cifra total pueden identificarse 172 abrigos con pinturas, de los que 121 fueron documentados en forma integral (copia de figuras sobre plástico, fotografía y cine en formato super 8) y únicamente 51 fueron foto-grafiados;

c) tres sitios de superficie fueron excavados: dos aldeas de ceramistas-agricultores y un sitio de enterramiento; los tres pertenecen al mismo período;

d) se excavó un abrigo en el que se hallaron nueve sepulturas de ceramistas-agricultores;

e) se identificaron cartográficamente 12 sitios líticos de superficie, en los que se recolectaron muestras;

f) en quince abrigos con pinturas se efectuaron excavaciones o sondeos. En tres de ellos se constató que las primeras camadas están relacionadas con los agricultores ceramistas mientras que las otras son típicas de los cazadores; todos los demás abrigos fueron únicamente frecuentados por grupos de cazadores;

g) estas excavaciones permitieron formar una colección de alrededor de 12.000 piezas líticas;

h) se recolectaron muestras para los análisis de polen y para los análisis granulométricos y químicos; éstos se encuentran en curso de realización.

LA ZONA DE SÃO RAIMUNDO NONATO

São Raimundo Nonato (Figuras 1, 2) es la ciudad más importante del sudeste del estado del Piauí, cuya población se estima alrededor de 9000 habitantes. Hasta el año 1978 los medios de comunicación eran escasos: un camino de tierra la conectaba con la ciudad de Petrolina en el estado de Pernambuco, mientras que otra ruta - también

Figura 1. Mapa de Brasil.

de tierra - permitía la comunicación con Teresina situada en el norte. Esta última fue totalmente asfaltada en el curso del año 1978 y se realizaron mejoras en la ruta a Petrolina; en la actualidad las dos carreteras disponen de un servicio regular de conservación.

Las Bases Físicas, el Medio y el Hombre

Nuestra área de trabajo ocupa una posición especial; se trata de una zona de contacto entre la cuenca sedimentaria Piauí-Maranhão (permiano-devoniano) y la depresión periférica del São Francisco medio que divide el escudo brasileño (pre-cambriano). Se pueden identificar tres formaciones geomorfológicas:

a) los altiplanos, de altura regular que varía entre 600 y 500 m. Estos altiplanos están cortados por varios valles que presentan altas paredes de arenisca, casi verticales y con relieve ruiniforme;

b) la cuesta de arenisca, en la que se encuentran valles estrechos y cañones que presentan un desnivelamiento de 200 a 250 m que forma paredes verticales. Se trata de la bajada de la depresión sedimentaria sobre el escudo primario; en esta zona se encuentran concentrados la mayoría de los abrigos pintados;

c) un pedimento, dominado por inselbergs y elementos residuales, que caracteriza la zona pre-cambriana.

La red hidrográfica es dendrítica en ciertas zonas, presentándose muy espaciada en el altiplano. Los ríos son temporarios.

La zona de São Raimundo Nonato se encuentra incluída en el Polígono de la Sequía, que se caracteriza por un clima semi-árido. La estación seca se extiende entre los meses de mayo y octubre; los niveles pluviales alcanzan una media de 644 mm presentando importantes variaciones anuales.

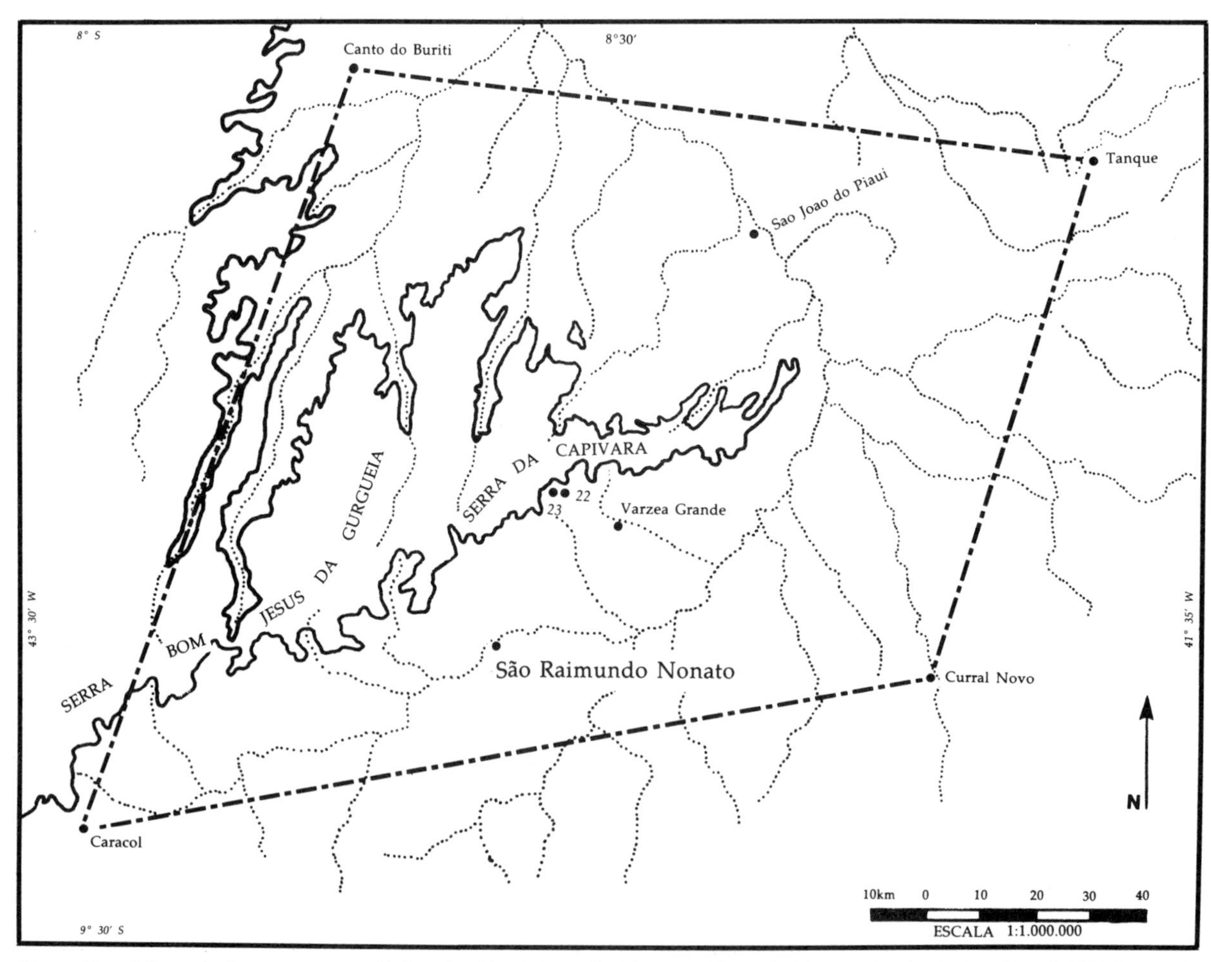

Figura 2. Mapa de la zona arqueológica de São Raimundo Nonato. Toca do Boqueirão da Pedra Furada[23], Toca do Sitio do Meio[22]).

La "caatinga" es la vegetación que domina casi totalmente en la zona, salvo algunas regiones limitadas, en las que se encuentra una vegetación de transición "caatinga-cerrado".

La fauna comprende varios tipos de felinos, venados "catingueiros", varias especies de tatus, numerosos pequeños roedores, innumerables especies de serpientes, lagartos, yacarés, aves (araras, papagayos, ñandues, etc) y una profusión de arañas, escorpiones e insectos. Casi todas las especies, que son la base de la caza, se encuentran amenazadas de extinción.

La población se concentra de manera preferencial en las ciudades y villorios; la población rural es poco numerosa, y por razones obvias, procura concentrarse en las proximidades de las fuentes de agua.

Las bases económicas; son: los pequeños rebaños (ganado y cabras), y la pequeña agricultura (frijoles, mandioca, algodón). Los latifundios son escasos y la mayor parte de las propiedades son medianas o pequeñas.

Cuando se producen sequías prolongadas, una gran parte de la población emigra hacia el sur del país, en especial, a las ciudades de São Paulo y Brasilia procurando obtener empleos que le permitan subsistir. Cuando nuevamente se producen las lluvias los campesinos regresan a sus tierras.

La red eléctrica fue instalada en 1975 suministrando energía a las principales ciudades del sudeste del estado. En ese mismo año, São Raimundo Nonato fue dotada de una red de saneamiento. La televisión llega en 1978 y el teléfono en 1981.

El Complejo Arqueológico

Además de ser una frontera geológica y ecológica, la zona de São Raimundo Nonato constituye una frontera cultural. Ha sido posible comprobar la existencia de una serie de unidades culturales que ocupan un espacio determinado; estas unidades culturales parecen sucederse en el tiempo, aunque pueden observarse también ciertos indicios de una coexistencia.

Hasta ahora nuestros esfuerzos se han concentrado en la parte central de la zona, donde aparece la mayor concentración de los sitios. Es lo que designamos como zona nuclear, y que constituye el habitat privilegiado de la unidad cultural Varzea Grande.

En la periferia de esta zona se distinguen otras unidades culturales limítrofes, observándose también ciertos indicios de incursiones rápidas de estas unidades en la zona nuclear.

La Unidad Cultural Varzea Grande

Esta unidad cultural abarca una área de alrededor de 5.000 km^2. El sustrato geológico está constituído por la cuenca sedimentaria que comprende: la Serra do Bom Jesus da Gurgueia, la Serra do Gongo, la Serra Grande, la Serra Grande da Boa Esperança, la Serra do Cumbre, la Serra Vermelha, la Serra Talhada, la Chapada da Capivara, la Chapada Dois Irmãos y la Chapada de São João.

Los sitios que integran esta unidad son principalmente:

a) abrigos que presentan manifestaciones de arte rupestre, y que habrían servido como lugar de habitación, de campamento y de entierro para los hombres prehistóricos. En sus camadas arqueológicas encontramos una gran cantidad de piezas líticas y otros vestigios.

b) sitios al aire libre, que se encuentran tanto en lo alto del altiplano como en los valles, y que se caracterizan por presentar en la superficie una abundante industria lítica.

Las características principales de la unidad cultural de Varzea Grande son:

a) un instrumental lítico típico de los grupos cazadores que presenta una evolución muy definida;

b) ciertos tipos de sepulturas;

c) un arte rupestre con una tradición mono-lítica, universal en la región y de larga duración.

Los sitios excavados que presentan dataciones más antiguas que 11.500 años son la Toca do Boqueirão da Pedra Furada, la Toca do Sitio do Meio y la Toca do Caldeirão dos Rodrigues I.

Toca do Boqueirão da Pedra Furada

En el sitio de la Toca do Boqueirão da Pedra Furada (No. 23 Figura 2) se obtuvo una serie de dataciones de una extrema importancia por ser las más antiguas de América del Sur.

La excavación realizada en este sitio en 1978 no permitió alcanzar la base rocosa del abrigo. La forma de la pared del fondo del abrigo (Figuras 3, 4) que avanzaba sobre la excavación fue lo que impidió lôgrar este objetivo. En los inicios la excavación estaba delimitada por una superficie de 7 m x 3 m; estas dimensiones se redujeron a 7 m x 0,30 m en la fase final de la excavación sin poder alcanzarse las camadas de base.

En 1980 un pequeño sondeo (1,50 m x 1,50 m) permitió alcanzar las camadas más profundas. Las dataciones que siguen fueron anunciadas en la reunión del ICPPS en México (Guidon 1981):

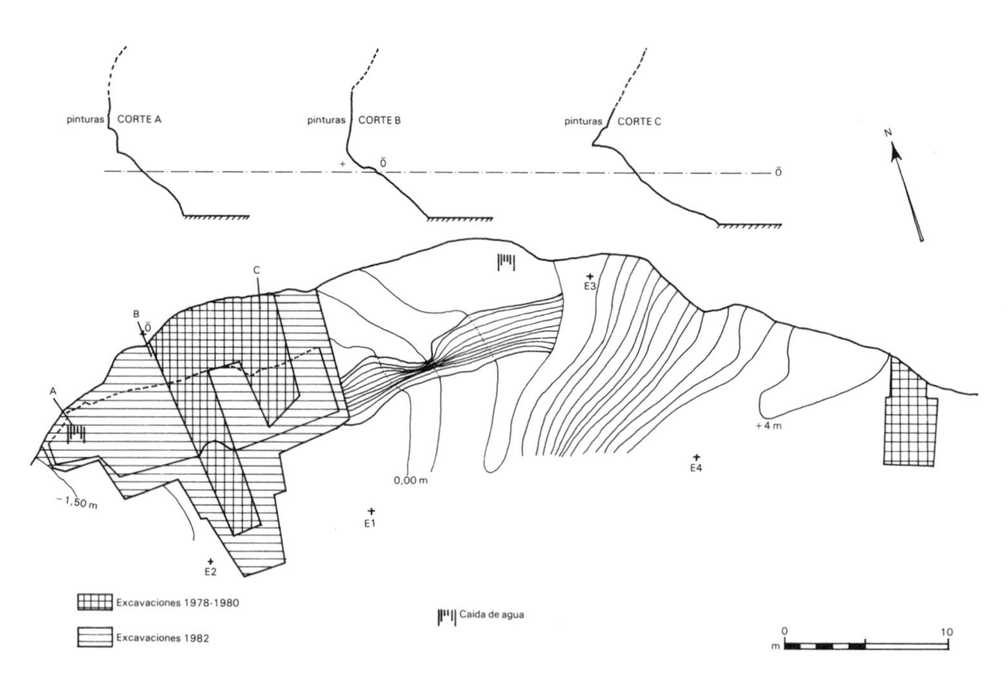

Figura 3. Plano de las excavaciones de Toca do Boqueirão da Pedra Furada.

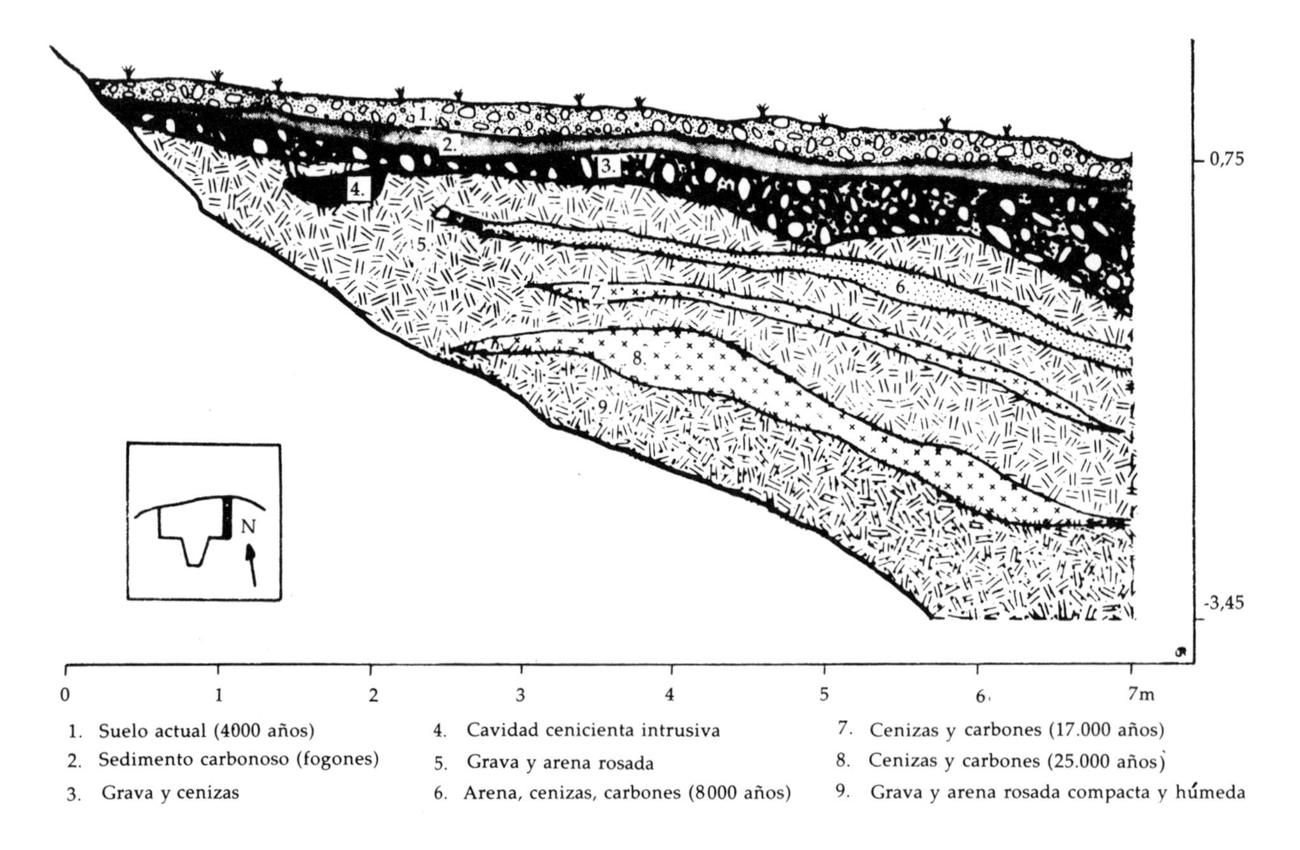

Figura 4. Sección de las camadas de Toca do Boqueirão da Pedra Furada.

6160±130 B.P. (Gif-5863) - nivel 5.
7640±140 B.P. (Gif-4928) - nivel 10;
8050±170 B.P. (Gif-4625) - nivel 12;
17.000±400 B.P. (Gif-5397) - camada de 178 a 192 cm debajo del cero; (Figuras 5, 6)

≥25.000 B.P. (Gif-5648) - camada de 192 a 203 cm debajo del cero;

≥25.000 B.P. (Gif-5398) - camada de 203 a 210 cm debajo del cero; Figura 7.

En 1982 se realizó una gran excavación en la que se alcanzaron nuevas camadas de mayor antigüedad las que suministraron una importante cantidad de muestras de carbón y de piezas líticas. (Figuras 8-11). Las Camadas XIX y XX fueron identificadas bajo el nivel que en 1980 había suministrado las muestras datadas de 25.000 B.P. Nuevos análisis han sido ya realizados en el laboratorio de Gif-sur-Yvette y los restos de carbón encontrados en la camada XIX han suministrado los seguientes fechados:

profundidad de 258 cm: 26.300±600 B.P. (Gif-5963);

profundidad de 263 cm: 26.400±500 B.P. (Gif-5962);

profundidad de 268 cm:31,500 ± 950 (Gif -6041).

A continuación suministraremos un inventario de la industria lítica encontrada en las capas cuya antigüedad supera los 11.500 B.P. (Este inventario preliminar está basado en los análisis realizados por Jacionira Silva Rocha y François Manenti).

En la camada comprendida entre 178 y 192 cm debajo del cero (datación de 17.000±400 B.P.) se encontraron 14 piezas líticas: 4 núcleos, 4 fragmentos, 5 lascas, de las cuales una es un raspador, y 1 percutor. La materia prima dominante es el cuarzo (10) y el cuarcito (4). En esta camada se encontró un "foyer" al lado del cual habían 3 pedazos de pared, uno de los cuales presentaba trazas de pintura roja.

Figura 5. Fogón con datación de 17.000±400 B.P. (Gif-5395) del nivel hubicado a 1,87 m de las superficie. Toca do Boqueirão da Pedra Furada. Sobre el piso de tierra yace una piedra larga y lisa que muestra restos de pintura roja.

Figura 6. El mismo fogón.

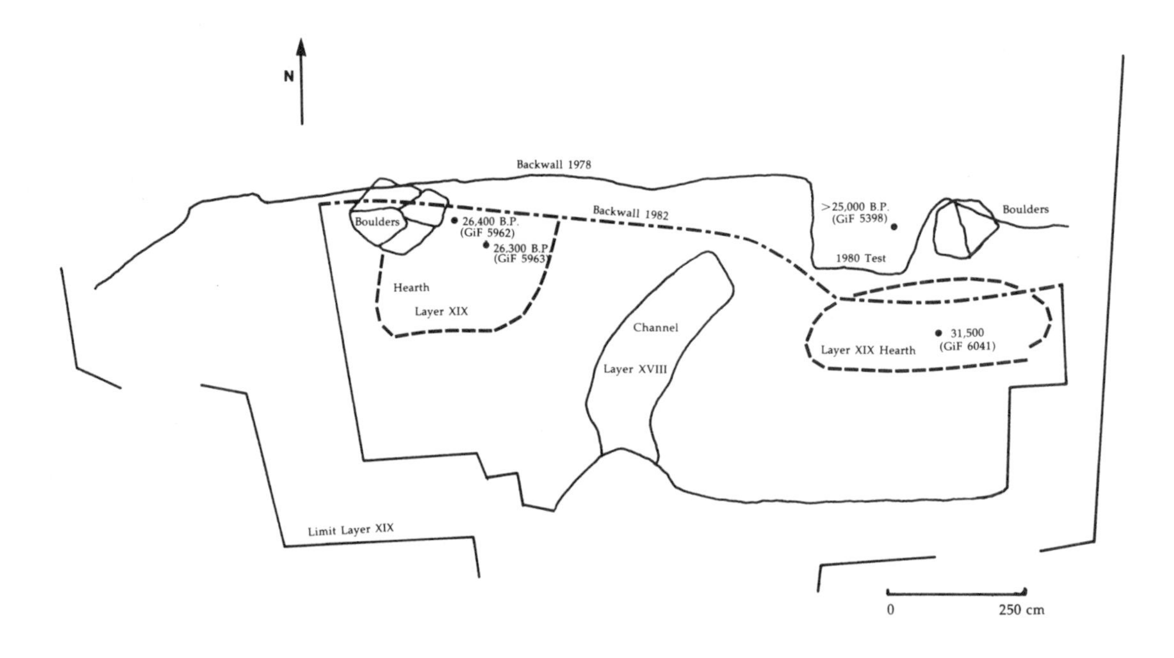

Figura 7. Plano de las piezas lascadas y fogones de camada XIX.

Figura 8. Núcleo de cuarzo de camada XIX.

Figura 9. Núcleo de cuarzo de camada XIX.

Figura 10. Núcleo de cuarzo de camada XIX.

Figura 11. Núcleo de cuarzo de camada XIX.

La camada subyascente entre 192 y 203 cm debajo del cero fue datada en ≥25.000 B.P. Se obtuvieron las siguientes piezas líticas: 4 núcleos y 3 lascas. El cuarzo domina como materia prima en seis piezas, mientras que la séptima está realizada en cuarcito. Esta camada también presentaba los restos de un "foyer".

La camada situada entre los 203 y 210 cm debajo del cero suministró un carbon que fue datado en ≥25.000 B.P. En ella fueron encontradas seis piezas líticas: 2 núcleos, 2 lascas, 1 raspador y 1 fragmento (Figuras 12, 13).

Tres de las piezas estaban realizadas en cuarzo, mientras que las otras tres eran en cuarcito.

En la camada situada a 220 cm debajo del cero se hallaron 15 piezas líticas: 5 núcleos, 6 fragmentos, 3 lascas, una de las cuales presenta trazas de utilización; 1 raspador sobre núcleo. Diez de estas piezas están realizadas en cuarzo y cinco en cuarcito. En la base de esta capa, sobre la roca, había una lasca de cuarcito con marcas de utilización.

Las fotos anexas muestran las principales piezas líticas obtenidas en la excavación de 1982.

Lo característico de estas camadas profundas es el predominio del cuarzo como materia prima. Las lascas son mayoritarias y los artefactos, son cuchillos y raspadores.

Es interesante destacar que en las camadas superiores dominaba el silex como materia prima, y las piezas más comunes eran los cuchillos, los raspadores y las láminas.

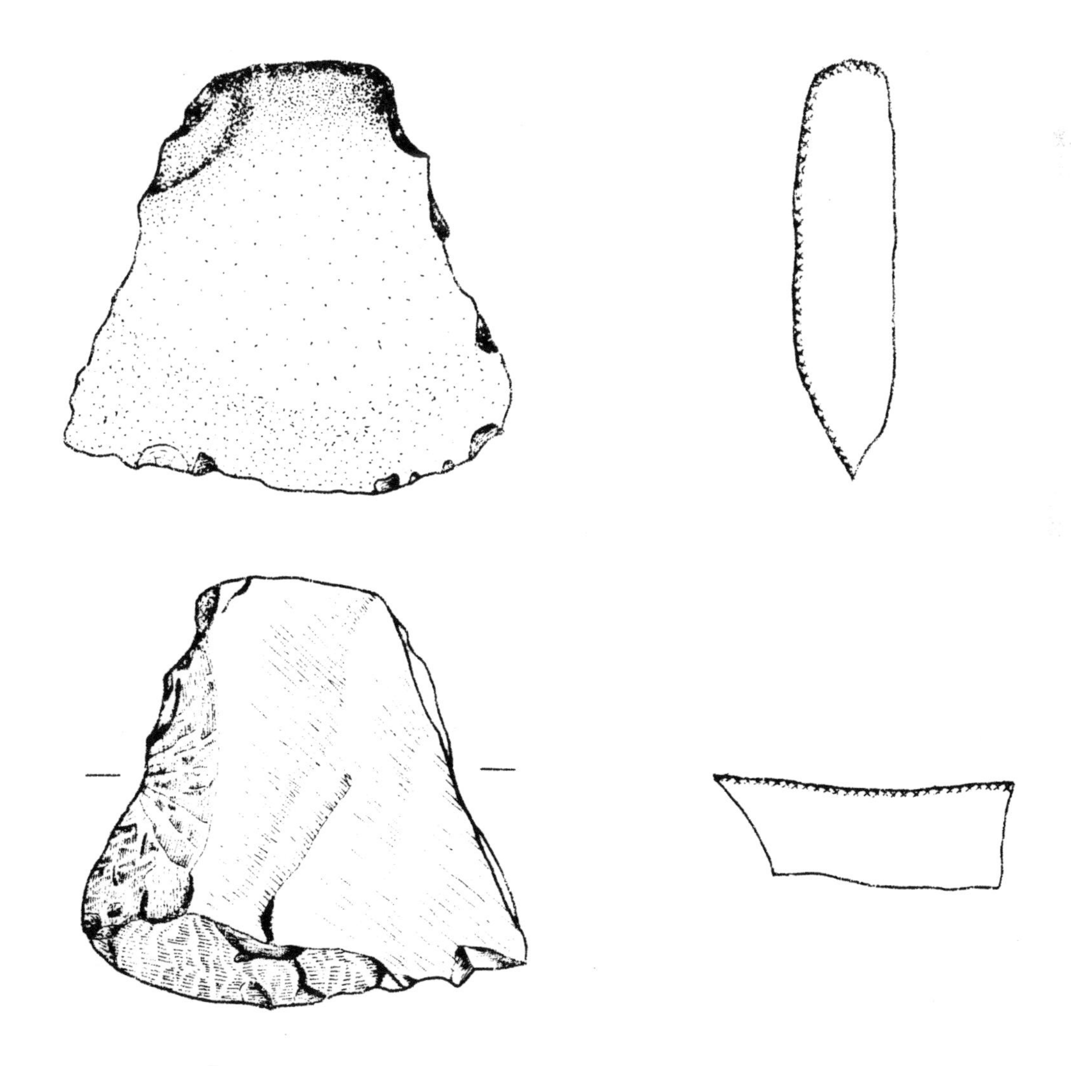

Figura 12. Raspador de cuarzo del nivel 210-220 cm. Toca do Boqueirão da Pedra Furada.

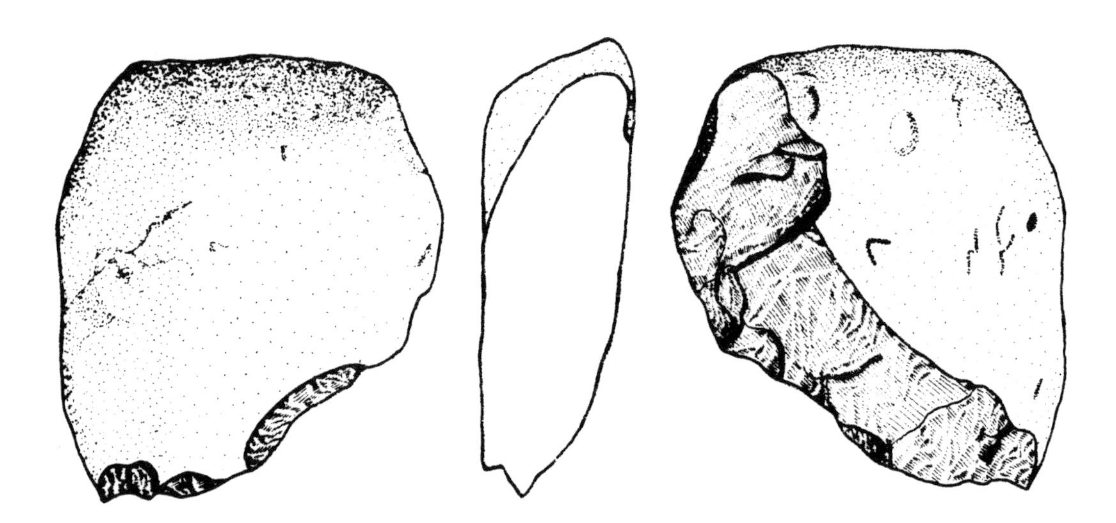

Figura 13. Núcleo de cuarzo del nivel 210-220 cm. Toca do Boqueirão da Pedra Furada.

Toca do Sitio do Meio

La Toca do Sitio do Meio (No. 22, Figura 2) cuenta también con innumerables figuras pintadas en sus paredes. En este artículo se tomarán en consideración únicamente las camadas para las que existen dataciones más antiguas que 11.500 B.P.

La excavación fue realizada en dos etapas (1978 y 1980); ciertas correlaciones entre las camadas han sido difíciles de establecer debido a la existencia de un gran número de bloques caídos, que dificultó enormemente la excavación y la manutención de los niveles naturales.

En 1978 la excavación puso al descubierto ocho camadas y niveles. A partir del nivel III-C hubo una tal cantidad de bloques caídos, que los carbones que suministraron las dataciones de 12.200±600 B.P. (Gif-4628) y 13.900±300 B.P. (Gif-4927) fueron obtenidos en una hendidura entre los bloques.

La camada X de la excavación de 1980 (primera camada de esta excavación) corresponde a la camada III-C de la excavación de 1978. Después resulta difícil establecer una correspondencia perfecta por causa de los bloques, sin embargo el inicio de la camada V (1978) debe corresponder a la camada XV de 1980. La base de la camada V (1978) puede corresponder a la camada XVII de 1980.

En la excavación de 1978, las camadas VI, VII y VIII estaban formadas por manchas esparcidas y discontinuas. Fue imposible establecer las relaciones perfectas entre ellas por hallarse todo el nivel perturbado por los bloques.

A continuación se presentan los resultados de una tipología preliminar.

En la camada V de 1978 se obtuvieron dos dataciones: una correspondiente al inicio, que da 12.200±600 B.P. y la otra en la base, de 13.900±300 B.P.

En el curso de un pequeño sondeo efectuado en esta capa V, se descubrieron 16 piezas líticas: una "limace", 4 lascas, una de las cuales es un cuchillo doble retocado y otra presenta rastros de haber sido utilizada como cuchillo, 8 fragmentos, 3 cantos rodados, dos de los cuales han sido utilizados como percutores y uno presenta un lascado que forma un filo tosco (chopper).

La materia prima dominante es el siltito (11 piezas) seguida por el cuarcito (3 piezas) habiendo igualmente dos piezas realizadas en cuarzo. En esta camada se encontró también un "foyer".

En la camada VI se encontraron únicamente grandes fragmentos de siltito, algunos de los cuales presentaban trazos e incisiones que indicaban la probabilidad que hubiesen sido utilizados como base de apoyo para la realización de otros menesteres.

La camada VII de 1978 suministró una gran cantidad de ocre, carbones situados entre los enormes bloques caídos y solamente tres piezas en siltito: un percutor, una lasca y un núcleo. Con esta camada se alcanzó el fondo de un pequeño pozo con que se terminó la excavación de 1978.

En la camada XIII de 1980 que es un poco más reciente que la capa V-1978, se encontraron catorce piezas líticas que se detallan así: 1 núcleo, 6 fragmentos, 2 astillas, 5 lascas, dos de las cuales presentan retoques de tipo de los que tienen los cuchillos.

Entre enormes bloques y a la altura de la camada XIV (1980) fueron halladas 46 piezas: 3 núcleos, 12 fragmentos, 9 cantos rodados de los cuales, uno presentaba trazas de utilización como percutor, cinco son fragmentos y uno, con un lascado que forma un filo tosco (chopper); 2 raederas sobre lasca, 1 piedra soporte, 19 lascas de las cuales 8 presentaban retoques; 5 cuchillos, 2 raspadores y tres lascas con marcas de utilización como cuchillo y un cuchillo-raspador.

La materia prima dominante es el siltito (32) mientras que el cuarzo fue utilizado para la realización de 14 piezas.

En la camada XV que corresponde a la capa V-1978 (datación de 12.440±230 B.P. (Gif-5403) se encontraron 72 piezas: 52 fragmentos naturales, 4 núcleos, 1 canto rodado, 15 lascas, de las cuales 10 presentan retoques (8 cuchillos, 2 raspadores y tres presentan trazas de utilización).

El siltito domina como materia prima (60 piezas) seguido por el cuarzo (7), el cuarcito (3) y el arenito (2).

En la camada XVI se halló un "foyer" y 14 piezas: cinco núcleos y 9 fragmentos, trece piezas habían sido realizadas en siltito y una en cuarzo.

La camada XVII suministró 80 piezas compuestas por 70 fragmentos, 9 lascas y una "limace".

La materia prima que domina es el siltito (67 piezas) seguido por el cuarzo (10), el cuarcito (2) y el arenito (1). Se hallaron también algunos carbones dispersos.

La camada XVIII suministró una datación de 14.300±400 B.P. (Gif-5399); su principal caracteríistica es la presencia de bloques.

En la camada XIX que no ha sido aun datada se encontraron 57 piezas distribuídas de la siguiente manera: 1 núcleo, 48 fragmentos, 2 raederas, 3 cuchillos, 1 raspador y 2 lascas. La materia prima de 53 piezas es el siltito, además una es de arenito y 3 en cuarzo.

La camada XX permitió encontrar 42 piezas con la siguiente composición: 2 núcleos, 32 fragmentos, 8 lascas (3 cuchillos, 3 raspadores y dos lascas que muestran trazas de utilización). De estas piezas, 42 están hechas en siltito y solamente una está realizada en arenito.

La última camada, la XXI, situada en parte sobre el fondo rocoso, suministró 34 piezas de siltito: 32 fragmentos y 2 lascas.

En resumen, es posible establecer que en estas camadas más antiguas no se observa una evolución nítida de la industria lítica.

Dominan las lascas, seguidas por los cuchillos y los raspadores. La materia prima dominante, en particular en las camadas más antiguas, es el siltito.

En las camadas más recientes se comprueba un aumento del porcentaje de cuarcito.

En las camadas superiores la industria utiliza como materia prima el silex, la calcedonia y el cuarcito. El silex no existe en el sitio, pero podría provenir de un macizo calcáreo situado a algunos kilómetros de distancia.

Toca do Caldeirão dos Rodrigues I

En la Toca do Caldeirão dos Rodrigues al pie de la pared pintada, se realizó un pequeño sondeo con una dimensión de 2 m x 2 m. Fueron identificadas nueve camadas. Dos de estas camadas suministraron las siguientes dataciones: camada VII: 9480±170 B.P. (Gif-5650); camada VIII: 18.000±600 B.P. (Gif-5406).

En esta camada VIII, sin ningún vestigio lítico, continuaba presente una misma mancha clara circular, encontrada en las camadas VI y VII. Junto a esta mancha había fragmentos de pared caídos.

No fue posible proseguir el sondeo debido a la enorme proliferación de bloques caídos que impedían el trabajo.

El reducido número de piezas halladas en las capas superiores se debe sin duda a las escasas dimensiones del sondeo. Domina el cuarcito como materia prima, y las lascas son el artefacto más frecuente.

Este abrigo no parecería presentar las condiciones propicias para haber servido de habitat debido a sus dimensiones reducidas y a su configuración que presenta las características de un corredor.

Otros Sitios

Existen otros sitios donde será probable alcanzar dataciones de alrededor de los 11.500 años. A continuación presentamos un breve resumen de los resultados de las excavaciones que se realizaron en ellos.

En el sitio de la Toca do Bojo I se obtuvieron cuatro dataciones:

a) dos en el sector Y/1/2, situado en la entrada del abrigo: para la camada 12 (8050±170 B.P. - Gif-4626) y para la camada 14 (9080±170 B.P. - Gif-4925);

b) una en el sector C/D/1, situado en la base de la pared al fondo: para la camada 13 (9700±200 B.P. - Gif-4627);

c) la última datación fue obtenida para la camada 13 del sector I/3, situado en el fondo de un divertículo, cuyo suelo actual es más elevado que el suelo en C/D/1 y Y/1/2. En estos dos sectores las camadas superiores fueron retiradas por el propietario del sitio hace unos diez años. La datación es de 7180±90 B.P. (Gif-4926).

La industria lítica ha sido realizada esencialmente sobre cuarcito y las piezas que dominan son los raspadores, los cuchillos de dorso y las lascas, en su mayoría sin retoques. También se encuentran percutores y núcleos en abundancia. La característica que parece destacarse de este conjunto es que las piezas más antiguas son de mayores dimensiones que las más recientes.

Es preciso indicar que frente al abrigo pasa un torrente, que permite la bajada de las lluvias desde lo alto de la meseta hacia el valle que se encuentra a unos 250 metros más abajo. El lecho del torrente se encuentra cubierto de cantos rodados, en especial de cuarcito.

En este mismo rango cronológico encontramos el sitio de la Toca da Boa Vista II, cuyo nivel 13 fue datado entre 9700±120 B.P. (MC-2481). La industria lítica emplea el silex, la calcedonia y el cuarcito, en tanto que las piezas características son las raederas carenadas, los cuchillos y las lascas con y sin retoque.

En el sitio de la Toca do Paraguaio, la camada 14 presenta una datación de 8670±120 B.P. (MC-2480); la industria lítica de este sitio está realizada esencialmente sobre cuarzo y cuarcito, que son materiales que abundan en la zona. Se observa una dominancia de las lascas sin retoques.

La última datación que tenemos para este período es de 7730±140 B.P. (Gif-4629) para la camada 21 del sector P-U, de la Toca da Boa Vista I. La Camada III del sector A-F fue datada en 9160±170 B.P. (Gif-5864). Se encuentra una industria sobre cuarcito que presenta las siguientes piezas: cuchillos de dorso, raspadores, lascas con y sin retoque y gran número de percutores. En el nivel 9 se encontró un fogón cuyos bordes se encontraban delimitados por un círculo formado por grandes cantos rodados.

Además de la industria lítica, se encuentran vestigios de fauna, restos vegetales y gran cantidad de restos de alimentos así como también trazas de pequeñas fogatas.

En la Toca do Paraguaio fueron excavadas dos sepulturas. La más antigua pertenece al nivel 14 y tiene una datación de 8670 años. Se trata de una mujer, que se encontraba en posición fetal dentro de una fosa circular, con grandes cantos rodados depositados sobre el esqueleto. Sobre la fosa se habría encendido una gran hoguera.

También se descubrió una sepultura más reciente (7000±100 B.P. - MC-2509) en la que el esqueleto se hallaba en posición de decubito dorsal, con los brazos a lo largo del cuerpo; la fosa de forma rectangular cavada a partir del nivel 6 presentaba el cuerpo que había sido cubierto de ramas, sobre las que se había depositado el sedimento. También sobre esta sepultura se habría encendido una gran hoguera.

Finalmente, las manifestaciones de arte rupestre son las que completan los vestigios vinculados a la unidad cultural Varzea Grande. Las dataciones obtenidas corresponden a camadas que contenían ocre, tanto en estado natural como preparado, y bloques de pared con pinturas que cayeron.

La subtradición Varzea Grande presenta algunos estilos, que aparecen ligados a las diferentes industrias encontradas. Así las manifestaciones más antiguas constituyen los variedades Serra da Capivara y Serra Talhada (Toca do Boqueirão da Pedra Furada, Sitio do Meio, Toca do Paraguaio). Los estilos Serra Branca y Salitre parecen ser más recientes, pero aun disponemos de dataciones seguras.

Las Otras Unidades Culturales

En los alrededores del área ocupada por la unidad cultural Varzea Grande y a veces dentro de su territorio, se encontraron diferentes manifestaciones de arte rupestre que indican la existencia de varias otras unidades culturales, las que se instalan frecuentemente en otros nichos ecológicos.

Sin embargo estas manifestaciones se encuentran aparentemente ligadas a diferentes grupos de cazadores (alrededor de 6000 B.P.) o a pueblos agricultores y ceramistas cuya aparición se datò alrededor de 3000 B.P.

Actualmente se están iniciando los sondeos destinados a establecer el contexto cultural de esas unidades.

CONCLUSIONES

Estas dataciones revisten una gran importancia porque demuestran una muy antigua presencia del hombre en esta región, hasta ahora desconocida por la arqueología. Son además las más antiguas para el arte rupestre de las Américas.

La inexistencia de puntas, aun en madera o hueso (esta carencia de industria en hueso o madera es un rasgo a destacar pues en estos abrigos todo material orgánico se conserva muy bien) y la representación en el arte rupestre de propulsores y dardos, pero nunca de arcos y flechas, son indicios que podrían ser característicos de estos cazadores de la región semi-árida del nordeste del Brasil.

La industria lítica presenta características bien definidas que confirman la existencia de ciertas épocas de produción de piezas típicas. El complejo lítico de Varzea Grande se encuentra muy próximo al de las industrias encontradas en Goiás, Pernambuco y Minas Gerais. La diferencia cronológica que ha podido observarse podría ser indicio de posibles vías migratorias.

En síntesis podemos indicar que existe una primera fase que se sitúa entre los 31.500 y los 14.000 años caracterizada por las lascas que son utilizadas tal cual o con algunos retoques. Son también comunes los cuchillos, en particular los cuchillos de dorso, seguidos por los raspadores.

A continuación existe otra fase, entre 14.000 y 11.000 B.P., que estaría caracterizada por "limaces" y lascas sobre siltito, cuarzo o cuarcito.

Luego entre 10.000 y 7000 años B.P. tenemos una industria en cuarcito, con lascas mayores, raspadores y cuchillos. En esta época, en algunos sitios se encuentra una hermosa industria en silex y calcedonia, cuya característica esencial es la existencia de pequeñas láminas y las raederas carenadas.

REFERENCIAS CITADAS

Guidon, N.
1981 Las unidades culturales de São Raimundo Nonato, sudeste del estado de Piaui. *Unión Internacional de Ciencias Prehistóricas y Protohistóricas, X Congreso, Comisión* XII:101-111. México, D.F.

Os Mais Antigos Vestígios Arqueológicos no Brasil Central (Estados de Minas Gerais, Goiás e Bahia)

ANDRÉ PROUS
Bolsista do CNPq brasileiro
Mission archéologique Française au Minas Gerais
Setor de Arqueologia
Museu de História Natural
Universidade Federal de Minas Gerais
Belo Horizonte, MG
BRASIL

Abstract

It is concluded that man was contemporaneous with extinct megafauna in central Brazil during the late Pleistocene and early Holocene. Bones from paleontological collections show clear traces of human activity. The ground sloth remains from Lapa Vermelha IV Cave (Minas Gerais state), including phalanges, rib fragments, pelvis, femur, and coprolites, are considered to have been redeposited laterally; and the dates of 9580±200 and 10,200±200 B.P. obtained from associated charcoal correspond closely with the animal's death. A mastodon pelvic bone found in the Lapa dos Borges Cave (Minas Gerais state) shows distinct broad man-made chopping marks (Figure 1) and signs of impact to remove bone pieces. A right humerus of ground sloth, found reworked 2.5 km inside the Brejões Cave (Bahia state), also displays clear chopping marks (6 mm deep) attributed to butchering (Figure 2).

A summary is given of the work by the Franco-Brazilian Archaeological mission in Lapa Vermelha cave. Here, the possibility of human occupation, before 25,000 B.P. (charcoal date) is considered. The oldest artifacts (quartz cores and flakes, and a unifacially retouched limestone sidescraper) are dated between 25,000 and 15,300 B.P. A human skeleton considered to be of the Lagoa Santa race is tentatively dated between 10,200 and 11,960 B.P. at the Lapa Vermelha site.

Several flakes, hearths, pigment fragments, and human bones were found in early levels in a rockshelter located in the Serra do Cipó (Minas Gerais), from which recently obtained radiocarbon dates indicate human occupation between 7900 and 11,960±250 B.P. Recent research in northern Minas Gerais state (Vale do Peruaçu) is reported. Two caves (Lapa do Boquete and Lapa da Hora) yielded lithic artifacts (Figures 3 and 4) in the lower levels related to the early Paranaíba Phase of Goias state (dated there between 9000 and 11,000). The upper levels present materials similar to the later Jatai Phase of Goias and the Lapa Pequena rockshelter farther south in Minas Gerais. No extinct fauna was found in the Peruaçu caves but at Lapa do Boquete a radiocarbon date of 11,000 years before present has been obtained for an occupation level above a stalagmitic floor, and there is clear evidence of human occupation below the stalagmitic floor.

INTRODUÇÃO

Desde as escavações de Lund em Lagoa Santa, as pesquisas sobre o povoamento mais antigo no Brasil levantam o problema da relação entre os primeiros povoadores e a fauna extinta "pleistocênica".

Em vários sítios do Rio Grande do Sul (Miller 1976) e Minas Gerais (Laming-Emperaire 1979) foram encontrados vestígios de preguiças gigantes em camadas que parecem contemporâneas da ocupação humana, ou até mais recentes que vestígios de ocupação antrópica. No entanto, nunca houve evidências claras de uma relação cultural entre os dois, como existe em outros países da América, onde há indícios certos da caça de megafauna para aproveitar a carne, e dos ossos para fabricação de instrumentos e até esculturas.

Peças oriundas de escavações paleontológicas nos parecem no entanto tirar as dúvidas sobre a caça destes animais pelo homem, enquantos elementos de datação fazem supor que parte da fauna "pleistocênica" tenha sobrevivido até um período relativamente recente.

Por outra parte, as escavações posteriores a 1970 nos Estados de Goiás e de Minas Gerais proporcionaram algumas datações pleistocênicas, outras do Holoceno mais antigo, sendo estas últimas associadas a uma indústria característica, bem diferente dos conjuntos líticos atípicos do Holoceno médio comuns no Brasil central.

OSSOS DE MEGAFAUNA NO CENTRO DO BRASIL

A maior parte dos achados aconteceu fora de sítios arqueológicos (escavação de Lund, H. Walter e do nosso Setor de Arqueologia da UFMG), e sem possibilidades de datação. Portanto, não se sabe com certeza até quando sobreviveu a megafauna. No Ceará, o centro de geocronologia da UFCE datou ossos retirados de uma cacimba de pouco mais de 5000 anos (Prof. Joaqium Torquato, comunicação pessoal); no entanto, a falta de observações precisas no ato da coleta, limita o valor desta informação.

Uma certa controversia existe a propósito de *Glossotherium* na Lapa Vermelha IV, escavada em Minas sob a orientação de A. Laming-Emperaire, razão pela qual apresentamos rapidamente os dados disponíveis atualmente, sendo que a nossa revisão do material deixado pela saudosa pesquisadora está longe de ser acabada.

Os ossos parecem ter pertencido a um mesmo exemplar de preguiça gigante, sendo várias falanges, fragmentos de costela, bacia e fêmur. Vários deles estavam nas imediações de coprólitos amarelos esféricos, cujo formato e tamanho corresponde a este tipo de animal. Os ossos estavam espalhados sobre vários metros quadrados, tendo uma profundidade entre 11,40 e 11,90 m. Esta variação vertical corresponde ao declive das camadas naturais em vários pontos do abrigo, sendo que neste caso concreto, não realizamos ainda as verificações. Na mesma superfície, havia carvões flotados, também esparsos. Deve se concluir que os ossos, desconectados, e os coprólitos, sofreram um remanejamento lateral, na matriz vermelha onde foram encontrados, provavelmente por causa da ação sazonal da água, perceptível nas verdadeiras "varvas" visíveis nestes níveis. No entanto, um problema permanece em razão da ausência de boa parte do esqueleto, apesar de ótima conservação do material coletado; seja porque o resto do esqueleto se encontra em setores ainda não escavados da Lapa, seja porque o Homem seria responsável por sua situação. Outro problema é levantado pelos paleontólogos: a diferença de mineralização entre os ossos do animal, e os restos humanos encontrados pouco mais abaixo estratigraficamente, indicaria que os ossos de fauna, apesar da sua impregnação pelo mesmo sedimento vermelho, teria sido remexido a partir dos níveis amarelos (série muito mais antiga, que foi parcialmente erodida a mais de 15.000 anos e substituída pelas argilas vermelhas). Não temos ainda resposta definitiva, mas há vários pontos de mineralização, e até formação de piso estalagmítico nos níveis vermelhos, que afetaram somente certos setores da Lapa. Antes de se obter uma datação direta dos ossos de *Glossotherium*, acreditamos que as datações de 9580±200 e 10.200±220 B.P. obtidas para carvões, uns vizinhos e outros logo inferiores correspondem de fato à morte do animal, e não a uma redeposição dos seus ossos, hipótese que deixaria inexplicada a preservação dos coprólitos na vizinhança.

Os outros elementos faunísticos encontrados na Lapa Vermelha existem até hoje, a não ser um novo representante de gastrópodo terrestre *Naesiotus* e uma bola feita por inseto fóssil *Megaphaneus* sp. o que mostra uma permanência tardia de microfauna fóssil até 6000 B.P. Por outra parte, certas espécies faunísticas, apesar das ainda existentes no Brasil, não são mais encontradas nestas latitudes (1).

Os outros achados que vamos apresentar não trazem elementos de datação, mas apoiam a tese de que os antigos moradores do Brasil caçaram a megafauna.

Em 1970, Alan Bryan, estudando a coleção paleontológica de H. Walter, depositada no Museu de História Natural da UFMG, verificou que o osso ilíaco de um mastodonte *Haplomastodon waringi* apresentava uma fratura com sinais de corte, que atribuiu a uma intervenção humana (o achado é

Bryan (1978:318; 1983:206, Figure 6). O osso foi encontrado na Lapa dos Borges, região da Lagoa Santa, a pouca distância dos sítios arqueológicos do Carroção e da Lapa Vermelha, dentro do sumidouro de uma dolina estreita sem sinal de ocupação humana. Inicialmente, nos resguardamos de emitir um diagnóstico próprio a partir da fratura, que o arqueólogo J. Tixier também considerou como de origem duvidosa. No entanto, após novo estudo, ficamos impressionados pela nitidez das incisões na base da fratura, e notamos também a presença de facetas na parte interna inferior do osso, na altura do ligamento redondo. Na face externa, na mesma altura, há vestígios de golpes violentos, que provocaram a fratura da borda óssea, enfraquecida pelos cortes; uma outra série de cortes regulares abriram o osso na região da articulação do fêmur.

A localização e as características dos trabalhos efetuados (Figura 1) fazem com que acreditamos que a única finalidade não foi desarticular a perna e retirar a carne, mas talvez obter lâminas de osso para a fabricação de instrumentos. Não tivemos, infelizmente, o tempo antes desta reunião, para proceder a experimentação de descarnamento e fratura de ossos de vaca, como tínhamos inicialmente previsto.

Um outro objeto, escavado em 1980 pelo paleontólogo C. Cartelle (Cartelle e Fonseca 1981:Figuras 5, 6), parece nos apresentar as características mais evidentes da atuação dos "paleoindios". As pesquisas foram realizadas dentro de uma galeria da gruta de Brejões, na Bahia, na entrada da qual existe um sítio arqueológico com pinturas rupestres. As enxurradas levaram, kilometros a dentro nas galerias, ossos de *Paleolama, Pampatherium, Nothrotherium*, etc; o que nos interessa é o úmero direito de *Glossotherium giganteum*, que foi levado, ainda fresco, sobre uma distância de 2,5 km, tendo sido depositado finalmente dentro de uma fenda.

A impressão de que o osso foi movimentado antes da mineralização vem do fato de que os sinais de batidas acidentais provocaram somente uma espécie de picoteamento (particularmente na epífise distal), enquanto no osso fossilizado, teriam ocasionado quebras e lascamentos típicos. Além disso, o transporte a tão grande distância teria sido muito difícil, por causa da densidade maior.

A Figura 1 mostra como a cabeça articular com o omoplata foi parcialmente eliminada, para permitir a desarticulação do braço. As marcas de corte são nítidas e limpas, mesmo no osso esponjoso, sendo também características do trabalho em osso fresco. A Figura 1 mostra também uma série de cortes profundos (atingindo 6 mm de profundidade) na parte inicial da crista deltóide, para permitir a remoção do músculo.

Todos estes vestígios não podem, em em hipótese alguma, serem atribuídos a causas naturais, ao contrário do que acontece com as marcas de dentes de roedores, que formam numerosas estrias paralelas ao final da crista. Estas últimas foram feitas pelos animais depois do abandono do osso pelo Homem, mas antes dêle ter sido levado pelas águas.

Portanto, não temos mais dúvida sobre a contemporaneidade do Homem com a megafauna no Brasil, sem que os contactos tenham que ser atribuídos *ipso facto* ao período pleistocênico. O aproveitamento alimentar nos parece inqüestionável, e a hutilização do osso para fabricar instrumentos bastante provável (*Haplomastodon* de Borges).

Figuras 1 e 2. Ilíaco de *Haplomastodon* com facetas de corte de Lapa dos Borges, Lagoa Santa (Fotos: Alan Bryan).

INDÚSTRIAS MAIS ANTIGAS DO CENTRO DE MINAS GERAIS

Lagoa Santa

Depois de publicadas as datações de cerca de 10.000 anos B.P. para o sítio de Cerca Grande, escavado por Hurt perto de Lagoa Santa, uma missão arqueológica franco-brasileira foi organizada, sob a responsabilidade científica de Mme A. Laming-Emperaire (1971-1977) para procurar novos sítios intactos que permitissem novos estudos. Realizamos uma visita preliminar ao maciço de Cerca Grande em 1971, que encontramos quase totalmente dinamitado para exploração de calcita. Abaixo dos restos da camada estalagmítica sobre a qual pararam as excavacões de Hurt, encontramos ainda carvões esparsos; o amador local Elio Diniz nos mostrou logo depois esqueletos e ossos humanos presos na parte inferior de fragmentos de piso estalagmítico, afirmando tê-los coletado nestes níveis inferiores do abrigo; neste caso, seriam evidentemente anteriores às datações de quase 10.000 anos obtidas para o nível 7 por Hurt e Blasi. Infelizmente, nenhum controle é possível em razão das destruições.

As escavações principais da Missão Franco-Brasileira foram realizadas no grande abrigo da Lapa Vermelha (município de Pedro Leopoldo), cujo sedimento foi escavado até 13 m de profundidade, sem que a base tenha sido encontrada. Mencionamos no parágrafo anterior a presença de ossos de *Glossotherium* nos sedimentos de contacto entre Pleistoceno e Holoceno.

Os sedimentos inferiores formam um conjunto estratigráfico complexo e de difícil interpretação, até que as datações radiocarbônicas permitiram escolher entre as diversas hipóteses. Os sedimentos (argilosos, arenosos e plaquetas calcáreas alternando) se acumularam num corredor de dissolução de até 2 m de largura, formando três séries paralelas, que tem de ser escavadas simultâneamente em razão do seu acentuado mergulho:

perto da parede interna, uma fenda de retração foi preenchida por material holocênico;

no centro do corredor, argilas vermelhas foram trazidas pelas águas, sofrendo, em pelo menos um ponto, uma sucção, fazendo com que este material que corresponde a transição Holoceno/Pleistoceno parece afundar, ou ter afundado em certos períodos, em direção a um ponor (sumidouro) que as escavações ainda não atingiram. Várias datações obtidas entre as cotas negativa de 11,70 m e 13,70 m forneceram uma sequência que corresponde a estratigrafia com 10.200±220, 11.680±500, 12.960±300 e 15.300±400 B.P. (Gif-3727; 3726; 3906 e 3905); esta concordância demonstra que os movimentos verticais afetaram pouco as relações estratigráficas, o que já tinha sido notado pela continuidade dos estratos argilosos. Nestes sedimentos intermediários é que foram encontrados a maior parte dos vestígios entre 11 e 13 m de profundidade: restos humanos como crânio, mandíbula, ossos longos e fragmentos de bacia, pertecendo a um exemplar feminino da raça da Lagoa Santa; alguns coprólitos; raras lascas de quartzo (material que não existe naturalmente no abrigo); carvões esparsos em todas as profundidades e vestígios de uma fogueira alimentar com ossos de um pássaro grande.

A terceira série, ocupando a parte externa do corredor, é formada por um sedimento estável, sêco, com argilas amarelas alternando com níveis de plaquetas de calcáreo. Ainda existem carvões, sempre esparsos, microfauna, raras lascas de quarzto (duas com alguns retoques) e uma raspadeira (side scraper) típica de calcáreo metamorfizado. Esta série "externa" amarela, corresponde ao mais antigo sedimento depositado no corredor, cuja parte localizada perto do paredão interno, parece ter sido erodida, descendo para o sumidouro, sendo então substituído pelo material vermelho depois da obstrução quase total do ponor. Uma datação de 25.000 B.P. foi obtida de uma amostra oriunda da parte alta da série, pouco abaixo da raspadeira mencionada. Logo abaixo, uma datação de 22.410±400 B.P. (Gif-3908) encontra-se em inversão estratigráfica com a anterior; no entanto, Gif-3908 foi realizado misturando-se quatro amostras insuficientes, inclusive uma coletada em contacto com a série sedimentar intermediária, e deve ser considerada duvidosa; no entanto confirma a idade pleistocênica dos sedimentos "amarelos". Em todo o caso, a indústria mais antiga encontrada, vem da parte superior, sendo portanto anterior a 25.000 B.P., mas pelo sistema de preenchimento descrito acima, posterior a 15.300 B.P.

Tirando-se um balanço provisório da escavação dos níveis inferiores da Lapa Vermelha IV, podemos notar em princípio que não apresentam níveis de habitação; as moradias deviam existir na parte mais alta, no talude externo do abrigo. O material encontrado no corredor deve ter caído ou ter sido jogado de lá. Não se deve portanto esperar muitos dados paleo etnológicos, a serem deduzidos de estruturas muito raras.

O interesse das escavações dos estratos profundos reside na determinação da antigüidade da ocupação humana, e da raça de Lagoa Santa.

Se for admitido o postulado que todos os carvões encontrados e datados tem origem humana, já verificamos que a colonização da região se deu há mais de 25.000 anos atrás. No entanto, não se deve *a priori* descartar a possibilidade de que parte dêles tenham se originado de fogos naturais da mata de encosta, que as enxurradas teríam levado pelos dois cones de dejeção lateral até dentro do abrigo.

Devemos portanto aceitar como maior antigüidade atualmente comprovada à dos mais antigos instrumentos inquestionáveis: entre 15.300 e 25.000 B.P.; desde período sabemos somente que os homens aproveitaram os cristais de quartzo hialino como núcleos para tirar lascas, e retocavam unifacialmente o calcáreo metamorfizado.

A idade do esqueleto da raça de Lagoa Santa ainda não é clara. Os ossos foram encontrados esparsos, flotados, quase todos dentro de uma fina formação argilo-arenosa indicando águas correntes sazonais, e a posição relativa dêles na escala vertical corresponde ao mergulho normal das camadas sedimentares. No entanto, alguns ossos de um setor bem localizado, foram encontrados mais abaixo; é o caso do crânio, encontrado dentro de um bolsão de sedimento diferente, qualificado de "marrom" nas anotações, que acreditamos inicialmente *in situ*, até que uma datação recente (Gif- 3907) tenha mostrado que os carvões eram mais recentes que os 12.000 anos esperado em razão da profundidade, comprovando que tinha sido rebaixado por um agente desconhecido, talvêz um animal, cuja cova teria sido preenchida posteriomente com o sedimento marrom. A análise morfológica dos ossos humanos tendo mostrado que todos pertencem provavelmente ao mesmo indivíduo, a idade da deposição deve ser procurada a partir da linha dos ossos superiores, dando uma idade verossimil entre 10.200 e 11.960 B.P.; no entanto, devemos ainda verificar as correlações. Em todo o caso, a presença da raça de Lagoa Santa num contêxto nitidamente pleistocênico não é ainda comprovada.

Serra do Cipó

A 60 km de Lagoa Santa, uma falha levantou o embasamento quartzítico, provocando a formação da "Serra do Cipó". Um grande abrigo foi escavado em 1976/79 pelo Setor de Arqueologia da UFMG. Os níveis do Holoceno antigo têm dataçãoes entre 7900 e 9460±110 B.P. (Gif-4508; 5087; 5088; Nuclebras/BH) correspondem a um cemitério utilizado por homens da raça de Lagoa Santa. Um nível mais abaixo ainda apresenta esqueletos, mas a má conservação dos ossos não permitiu a identificação. A camada inferior (VII) foi recentemente datada de 11.960±250 B.P. (Gif-5089), mas foi escavada sobre uma superfície reduzida. Nela foram achados somente uma grande fogueira, algumas lascas de cristal de rocha e grãos de corante. Pigmentos minerais vermelhos foram inclusive encontrados em todos os níveis; algumas pinturas enterradas feitas sobre blocos caídos, puderam ser datadas do Holoceno Médio/Recente (com datações mínimas e máximas), correspondendo a um estilo "recente" na ordem de superposições das pinturas no paredão, mas não temos atualmente possibilidade de avaliar a idade das pinturas mais antigas, já que os pigmentos foram utilizados para diversos fins, particularmente nos sepultamentos. Uma conta de colar feita de uma semente foi também achada na camada VII, mas perto de um bloco de pedra cuja parte superior tocava um sepultamento do período posterior; é portanto possível que este adorno tenha caído do componente de cima, por uma fenda de retração.

Não podemos, portanto, definir a ocupação mais antiga da Serra do Cipó, mas somente confirmar a presença do homem dentro dos abrigos. A raça de Lagoa Santa, mais uma vêz, aparece já num contêxto holocênico, depois de 11.960 B.P., mas antes de 9460 B.P.

AS INDÚSTRIAS DOS NÍVEIS INFERIORES NO LIMITE ENTRE GOIÁS E MINAS GERAIS

As pesquisas de Schmitz e de Barbosa no Vale do Paranaíba em Goiás permitiram descobrir indústrias sem pontas de projétil mas com instrumentos lascados unifacialmente, nos níveis inferiores de todos os abrigos da região, evidenciando uma ocupação sistemática do território. As datações, numerosas, vão de 10.740 B.P. até 9000 B.P. Esta indústria, diagnóstica da Fase Paranaíba, é logo substituída por outra, de lascas não retocadas (Fase Jataí). A indústria da Fase Paranaíba comporta sobretudo peças sobre lâminas estreitas e grossas, com retoque periférico marginal, existindo também raspadores planos convexos de tipo "lesma". Artefatos de osso existem, seja furadores, seja fragmentos de espátulas feitas com osso canhão de cervídeos, também encontrados nos sítios do centro de Minas Gerais, e datados com cerca de 8000 B.P. na Lapa Pequena (Bryan and Gruhn 1978).

A partir de 1980, o Setor de Arqueologia da UFMG iniciou uma série de sondagens no Vale do Peruaçú, que levou à descoberta de uma indústria aparentada, encontrada em estratigrafia em níveis profundos (IX-XII) da Lapa do Boquete existe uma datação do nivel XII de 11.000±1100 B.P. Os níveis inferiores (XIII-XV) ainda forneceram algumas peças retocadas, dentro e abaixo de um nível estalagmítico que corresponde a condições climáticas muito diversas das atuais nesta gruta, hoje totalmente seca.

Acreditamos pois, que os níveis (IX-XII) poderiam corresponder ao período entre 11.000 e 9000 B.P., sendo os inferiores algo mais antigos. No entanto, não há datações radiocarbônicas. Em outro sítio da região, a Lapa da Hora, um material que apresenta certo parentesco morfológico foi encontrado, infelizmente perturbado e superficial.

Apresentaremos rapidamente os achados líticos, acompanhados por raros instrumentos de osso, um bloco manchado de corante vermelho e uma lasca pintada (nível X).

Entre os níveis IX e XV que nos interessam aqui, pudemos observar alguma variação; mas como a escavação nestas profundidades só abrangia entre 2 e 4 metros quadrados, não podemos ter certeza se as diferenças são o resultado de mudanças culturais, ou simplesmente, da visão insuficiente que temos do espaço de cada nível de ocupação. As futuras escavações de grande superfície ajudarão a resolver o problema.

A INDÚSTRIA DO BOQUETE INFERIOR (SONDAGEM 1)

A matéria prima lítica é predominantemente de sílex local de várias qualidades, o qual aparece também nos níveis superiores. Infelizmente esta matéria tem muitas falhas internas, que dificultam o controle de lascamento. Por outra parte, muitas peças foram abandonadas em fogueiras, e foram desfiguradas pelo lascamento térmico.

O arenito silicificado, cuja a fonte encontra-se a vários kilômetros, aparece exclusivamente nos níveis IX-X, totalizando algumas dezenas de lascas de 2 a 4 cm de comprimento, que não foram retocadas.

O calcáreo foi trabalhado nos mesmos níveis, tratando-se de numerosas plaquetas espessas com retoque marginal; curiosamente, este "retoque" não

Figura 3. Úmero de *Glossotherium giganteum*, evidenciando cortes no crista deltoide e eliminacão parcial da cabeça articular com a omoplata, de Gruta de Brejões (Foto: Paulo Junqueira).

é resultado do desejo de criar um gume, mas simplesmente da limpeza dos concrecionamentos aderentes. Somente duas peças foram modificadas em forma de raspador "rabot". Aliás, o calcáreo local é altamente silicificado e responde à percussão como uma rocha frágil.

Cristais romboidais de calcita (níveis IX-X) e concrecionamentos do tipo "couve-flor" (nível XI), apareceram em número significativo, mostrando a curiosidade dos homens pré-históricos em relação à estas formas, já que não podem ter nenhuma utilidade como matéria prima, e não podem ter chegado naturalmente no local. A única matéria trazida de relativamente longe é um bloco quebrado (batedor?) de rocha básica.

Elementos para indicar as estrategias de debitagem foram obtidos a partir dos raros núcleos, com o exame da forma das lascas e das cicatrizes das suas faces externas. Parte dos núcleos tem uma tendência a ter uma forma piramidal, com um único plano de percussão, o que provoca a formação de lascas ovais e sub triangulares. O talão destas é geralmente pequeno, liso ou linear, sendo que a maior parte das lascas não ultrapassam 3 cm de comprimento. Há no entanto uma outra família de lascas, frequentemente um pouco maiores, muito mais espessas e com talão bastante inclinado; notamos que várias delas apresentavam na face externa vestígios de lascamento de preparação periféricas. Isto implica a existência de núcleos discoidais com pelo menos uma face pouco espessa, lembrando a técnica Levallois bem conhecida no Velho Mundo. Na Lapa da Hora, foi justamente encontrado um núcleo que responde a tais critérios, apresentando a característica forma em "carapaça de tartaruga".

E digno de nota que a quase totalidade das pecas retocadas pertencem a categoria de lascas maiores em cada nível, sendo que poucas lascas "grandes" foram deixadas sem aproveitamento.

Os retoques são quase sempre diretos, havendo um só caso de retoque inverso (neste ponto, entre outros, temos uma diferença com as indústrias retocadas da região vizinha de Montalvânia onde o retoque alterno é característico). Este retoque, reservado em geral a peças maiores de 4 cm, em todo o caso de 3 cm, serve seja para retocar o bulbo (casos raros, mas característicos, que não devem ser confundidos com estilhamentos importantes que ocorrem na região bulbar quando a lasca se

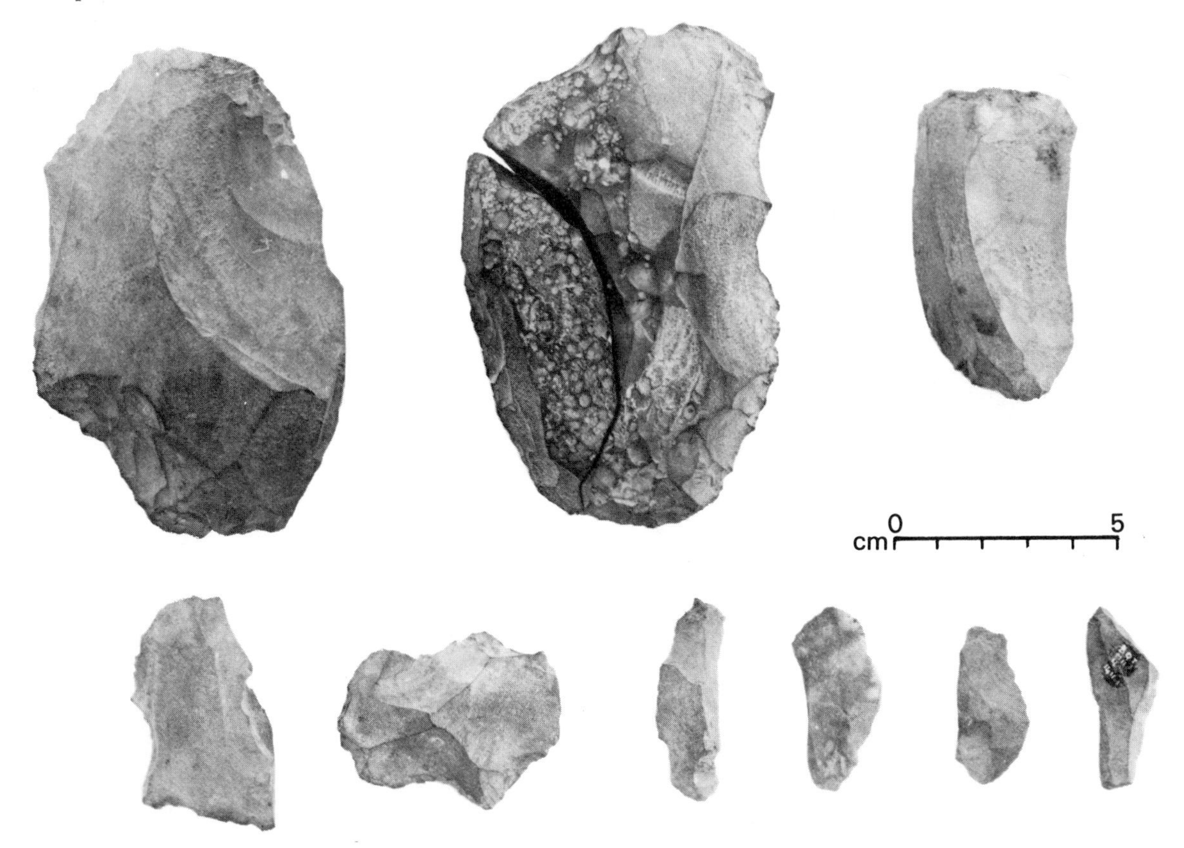

Figura 4. Lapa do Boquete. Sondagem 1, nivel 11. Raspadores concavos e convexos, lascas laminares. Observas as marcas fogo (no alto, centro).

desprende do núcleo), seja para criar gumes característicos, sendo que cada tipo se encontra mais particularmente em um ou dois níveis somente.

Na camada IX, existem grandes raspadeiras (racloir/side scraper) com mais de 7 cm em tamanho, com bulbo eventulamente retirado; há pequenas peças com retoque lateral semi abrupto na camada X, são os menores objetos retocados (3 cm), talvez facas? Algumas peças parecem buris de ângulo, sendo que alguns podem ser acidentais ("burins de Siret"), mas tem vestígios de utilização no diedro. Foram sobretudo encontrados na camada X, mas há formas atípicas na XI. Encontramos lamínulas estreitas e espessas que parecem tipicos retalhes de buril. Os raspadores (grattoirs/ end-scrapers) são numerosos nos níveis XI-XII seja plano convexos (raspadores espessos, lesma ou limace), seja raspadores menos espessos sobre lascas curtas, com frente pouco convexa; quando uma lasca apresentava um gume natural com as mesmas características, era utilizada como raspador, como demonstram os vestígios de utilização. Encontramos nestes níveis um único denticulado (XII), enquanto estes eram os instrumentos mais frequentes nos níveis superiores.

As peças de calcáreo são atípicas, a não ser uma espécie de plaina, alguns blocos com depressões "quebra côco", ou pequena mancha polida redonda, que verificamos ser o resultado de uma simples utilização como martelo de uma face, por exemplo para fincar estacas.

Os níveis inferiores, dentro e abaixo do piso estalagmítico forneceram peças na sua maior parte erodidas, com gumes gastos, ou parcialmente escondidos por concrecionamentos, estando em fase de limpeza e portanto nem sempre já identificados.

Mencionaremos 13 pequenas lascas, e uma lâmina típica. Os objetos retocados são todos espêssos, sendo um raspador duvidoso sobre plaqueta espêssa, um raspador de bico, um núcleo sub-piramidal (re utilizado como raspadeira alta em um dos lados do plano de percussão), um nódulo parcialmente lascado, com um pequeno gume côncavo. Vários grossos blocos de sílex forma trazidos das galerias vizinhas, completamente concrecionados, para servir provavelmente, de núcleos; tem partes descorticadas, que se fraturaram frequentemente.

Figura 5. Boquete, nivel 12. Raspadores frontais e lascas secundàriàs.

A indústria óssea é pobre. Encontramos duas extremidades distais do que denominamos "espátulas", normalmente feitas com o osso cannon de veados; alguns fragmentos de osso devem ainda ser examinados.

Digna de nota é a ausência de indústria sobre concha, presente nos níveis superiores.

A alimentação ainda não foi olhada em laboratório, mas as observações de campo sugerem uma modificação nítida em relação ao níveis superiores, onde dominavam coquinhos e conchas de grande gastrópodos terrestres. Nos níveis inferiores, aparecem ossos de mamíferos de porte médio, e grande quantidade de conchas de bivalvos lacustres nas fogueiras.

As numerosas estruturas de habitação dos níveis superiores, particularmente conjuntos de estacas, não foram encontradas neste contexto mais antigo.

A sondagem 1 foi abandonada por falta de condições materiais para prosseguir a descida dentro dos pisos estalagmíticos sem desbarrancar as paredes.

As outras sondagens realizadas na região não chegaram à níveis semelhantes, seja porque tenham sido interrompidas antes da base estéril, ou seja, como foi no caso da Lapa da Hora, porque os níveis antigos foram remexidos, ainda durante o período pré-histórico. Nesta última gruta porém, foram encontrados raspadores, grandes raspadeiras e núcleos, infelizmente em contêxto que não os permite datar. Juntamente apareceram artefatos de calcáreo nucleiformes com facetas planas, marcadas por pequenos círculos polidos escuros.

Concluindo esta apresentação, lembraremos que há tempo que se podia esperar indícios concretos de relacionamento entre paleo-índio e megafauna: mesmo se o significado das datações mais antigas da Lapa Vermelha (22.410 e 25.000 B.P.) é discutível, como explicamos no ano passado em Goiânia, a existência de instrumentos retocados há mais de 15.400 anos neste sítio, de várias datações também antigas, tanto em Minas como em outros estados do Brasil, fazem com que tenhamos certeza que durante os milênios de convívio, pelo menos a caça devia ter existido. Portanto, os ossos trabalhados que apresentamos não nos parecem ser revolucionários mas, pelo menos, trazem uma prova concreta.

No norte de Minas Gerais, as pesquisas em andamento, confirmam a existência do mesmo conjunto encontrado por Schmitz, Barbosa e outros em Goiás, mostrando que não se trata de uma cultura puramente local. As primeiras observações, quase todas efetuadas ainda em campo, mostram a possibilidade de se encontrarem formas originais de debitagem, e alguma evolução dentro do conjunto antigo.

AGRADECIMENTOS

Agradecemos ao Prof. C. Cartelle, que colocou o úmero de *Glossotherium* e todas as informações disponíveis à nossa disposição. Ao Dr. A. Bryan que nos mostrou em 1976 a bacia de um mastodonte trabalhada que havia encontrado nas coleções paleontológicas da UFMG.

O financiamento das escavações em Januária foi feito com verbas da Universidade de Minas Gerais e do curso de especialização em Arqueologia Pré-histórica, administrado pela FUNDEP.

A ilustração fotográfica foi realizada por Paulo Junqueira.

REFERENCIAS CITADAS

Bryan, A.L.

1978 An overview of Paleo-American prehistory from a circum-pacific perspective. In Early Man in America from a circum-Pacific perspective, edited by A.L. Bryan, pp. 306-327. *Department of Anthropology, University of Alberta, Occasional Papers* No. 1. Edmonton.

1983 Bone alteration patterns as clues for the identification of Early Man sites; or, an attempt to demythify the search for early Americans. In *Carnivores, Human Scavengers and Predators: A Question of Bone Technology*, pp. 193-217. Proceedings of the Fifteenth Annual Conference, Archaeological Association of the University of Calgary, Calgary.

Bryan, A.L., and R. Gruhn

1978 Results of a test excavation at Lapa Pequena, MG, Brazil. *Arquivos do Museu de Historia Natural* III:261-326. Belo Horizonte.

Cartelle, C., and J.S. Fonseca

1981 Espécies do gênero *Glossotherium* no Brasil. Anais do Segundo Congresso Latino Americano de Paleontologia 805-818. Porto Alegre.

Laming-Emperaire, A.

1979 Missions archéologiques franco-brésiliennes de Lagoa Santa, Minas Gerais, Brésil - Le grand abri de Lapa Vermelha (P.L.). *Revista de Prê História* I(1): 53-89. São Paulo.

Miller, E.T.

1976 Resultados preliminares das pesquisas arqueológicas paleoindigenas no Rio Grande do Sul, Brasil. *Actas del XLI Congreso Internacional de Americanistas*, Vol.3:483-491, Mexico.

Cazadores Antiguos en el Sudoeste de Goiás, Brasil

PEDRO IGNACIO SCHMITZ
Coordinador del Programa Arqueológico de Goiás.
Instituto Anchietano de Pesquisas
São Leopoldo, Rio Grande do Sul
BRASIL

Abstract

In the municipio of Serranópolis (long. 52° W, lat. 18° 20'S), southwestern Gioás State, Brazil, about 40 rockshelters are concentrated in a line extending for 25 km. At least eight of these shelters contain ancient human occupations, with radiocarbon dates ranging from 11,000 to 9000 B.P. This early occupation has been denominated the Paranaíba Phase of the Itaparica Tradition. Diagnostic Paranaíba material also is exposed along the banks of nearby creeks in a bed of gravel, believed to have been deposited during the final phase of the Pleistocene. Paranaíba material is abundant, permitting an accurate study of the lithic and bone implements, of the food and animal origin and, at least partially, of the paleoenvironmental changes.

The Paranaíba Phase represents a generalized hunting culture during the final phase of the Pleistocene and the beginning of the Holocene. The lithic industry is characterized by unifacially retouched artifacts made on thick blades (limaces). Bifacially worked artifacts are very scarce. Among these appear very rare pedunculate (contracting stemmed) arrow-heads, similar to those found at Cerca Grande, in the State of Minas Gerais, and at Alice Boër, in São Paulo.

The Itaparica Tradition, to which the Paranaíba Phase belongs, probably started around 14,000 B.P., and abruptly ended sometime after 9000 B.P. It extended over an area which nowadays is covered with tropical savanna of the "Cerrado" and "Caatinga" types. The region is marked by variable annual precipitation, with definite rainy and dry seasons. The climate during the Itaparica tradition was probably much dryer than now; evidence suggests that both precipitation and vegetation was much less abundant.

The Serranópolis Phase, which succeeded the Paranaíba Phase in the rockshelters of Goiás about 9000 B.P., is very different. Serranópolis is characterized as a generalized hunting and collecting culture. The diagnostic blade artifacts disappeared, mollusks and fruit became very abundant food remains. This abrupt change was produced in an extremely short lapse of time, and it seems to have been correlated with a very rapid change in climate.

INTRODUCCION

En el municipio de Serranópolis (long. 52°W, lat. 18°20'S), en el sudoeste de Goiás, están concentrados en un espacio de 25 km aproximadamente 40 abrigos, de los cuales al menos ocho presentan ocupaciones antiguas, cuyas dataciones van de 11.000 a 9000 años B.P., y que denominamos fase Paranaíba. El mismo material aparece también en las barrancas de lugares próximos, dentro de un nivel de cantos rodados, que pensamos debe representar el final del Pleistoceno. El material es abundante, permitiendo un buen estudio del lítico, de los implementos óseos, de los alimentos animales, y al menos parcialmente del ambiente y sus modificaciones.

EL AMBIENTE

La región de Serranópolis presenta, desde el punto de vista geológico, dos estratos importantes: el inferior, alcanzando hasta 700 m de altitud, de areniscas eólicas de la Formación Botucatu; el superior, que puede alcanzar hasta 1100 m de altitud, de basalto toleítico, de la Formación Serra Geral. En el contacto con el basalto, la arenisca fue parcialmente metamorfizada, volviéndose mucho más resistente a la erosión y de buena calidad para la producción de instrumentos lascados.

El principal responsable de la morfología actual del terreno es el río Verdinho, afluente del Paranaíba, uno de los formadores del río Paraná. Después de remover los basaltos de los estratos superiores, expuso y modeló las areniscas, creando un paisaje en el cual se suceden hoy, en espacio pequeño, altas superficies, pendientes escarpadas con abrigos, y un valle aplanado con relictos de erosión bajo la forma de mesas y morros redondeados.

Las areniscas metamórficas apenas se encuentran en una extensión de unos 25 km a lo largo del río, en altitudes que van de 550 a 700 m sobresaliendo de 100 a 200 m sobre el nivel del río.

Los abrigos se formaron por la remosión de los estratos poco resistentes, que están por debajo de la roca metamórfica. En la medida que el paredón retrocede, hay desmoronamientos de sus techos, creando taludes empinados sembrados de bloques. En los locales donde la temperatura llegó también al interior de las areniscas, surgieron torres y mesas características, que marcan el paisaje.

Los abrigos formados en los paredones, en las torres o bajo los bloques caídos en los taludes, suelen tener grandes bocas, poca profundidad y buena iluminación, con techos inclinados del frente hacia el fondo debido a las camadas inclinadas y cruzadas de las areniscas Botucatu.

Como los abrigos sólo existen en la pequeña extensión en que afloran las areniscas metamórficas, allí observamos una extraordinaria concentración de material arqueológico, no verificada en otras localidades río arriba o río abajo.

Algunas de las condiciones observadas en Serranópolis existen a lo largo de la mayor parte del valle del río Verdinho, pero otras son exclusivas: la existencia de grandes extensiones de abrigos y la abundancia de materia prima mineral, parecen ser factores decisivos para la concentración verificada y la aparente ausencia en las demás áreas.

De un modo general, en el municipio de Serranópolis, se puede observar una diversidad grande de ambientes, muy próximos debido al desnivel del terreno: en las mayores altitudes existen campos limpios; en los terrenos resultantes de la descomposición del basalto crecieron matas cerradas; en el valle arenoso y en las llanuras medias crece el "cerrado", un tipo de sabana tropical; en las áreas húmedas y pantanosas a lo largo del río se encuentra una vegetación palustre muy característica. La fauna acompaña la diversidad ambiental: en los campos hay ñandúes (*Rhea americana*); en los cerrados y matos edentados, prociónídeos, roedores, tapirídeos, aves; en los ríos peces y reptiles; en los paredones rocosos gran número de abejas melíferas y moluscos de varios géneros. Además de eso los "cerrados" producen, en la estación de las lluvias, una cantidad increíble de frutos como también fibras y esencias medicinales. Los terrenos altos son poco provistos de agua porque los cursos nacen en la pendiente, y en un curso accidentado y corto desembocan en el río. Naturalmente no podemos proyectar los recursos actuales al pasado remoto sin las debidas correcciones, y el ambiente debe haber sufrido modificaciones desde el fin del Pleistoceno, al siglo pasado, esto es de los primeros a los últimos cazadores indígenas.

El lugar más favorable para la ocupación humana es la pendiente donde se encuentran los abrigos, la materia prima, y el agua limpia; es donde se pueden dominar fácilmente todos los recursos, tanto los de las altas superficies, como los del valle y la pendiente. La costa del río no es favorable para una instalación humana porque es pantanosa, está sujeta a crecientes y es poco ventilada. El río no transporta rodados, solamente arena, y no es limpio, lo que lo convierte en lugar de poco interés.

No todos los abrigos fueron igualmente ocupados, porque no ofrecen las mismas condiciones: no son ocupados, o solamente en forma esporádica, los abrigos que no reciben sol de mañana o al medio día, que son muy húmedos,

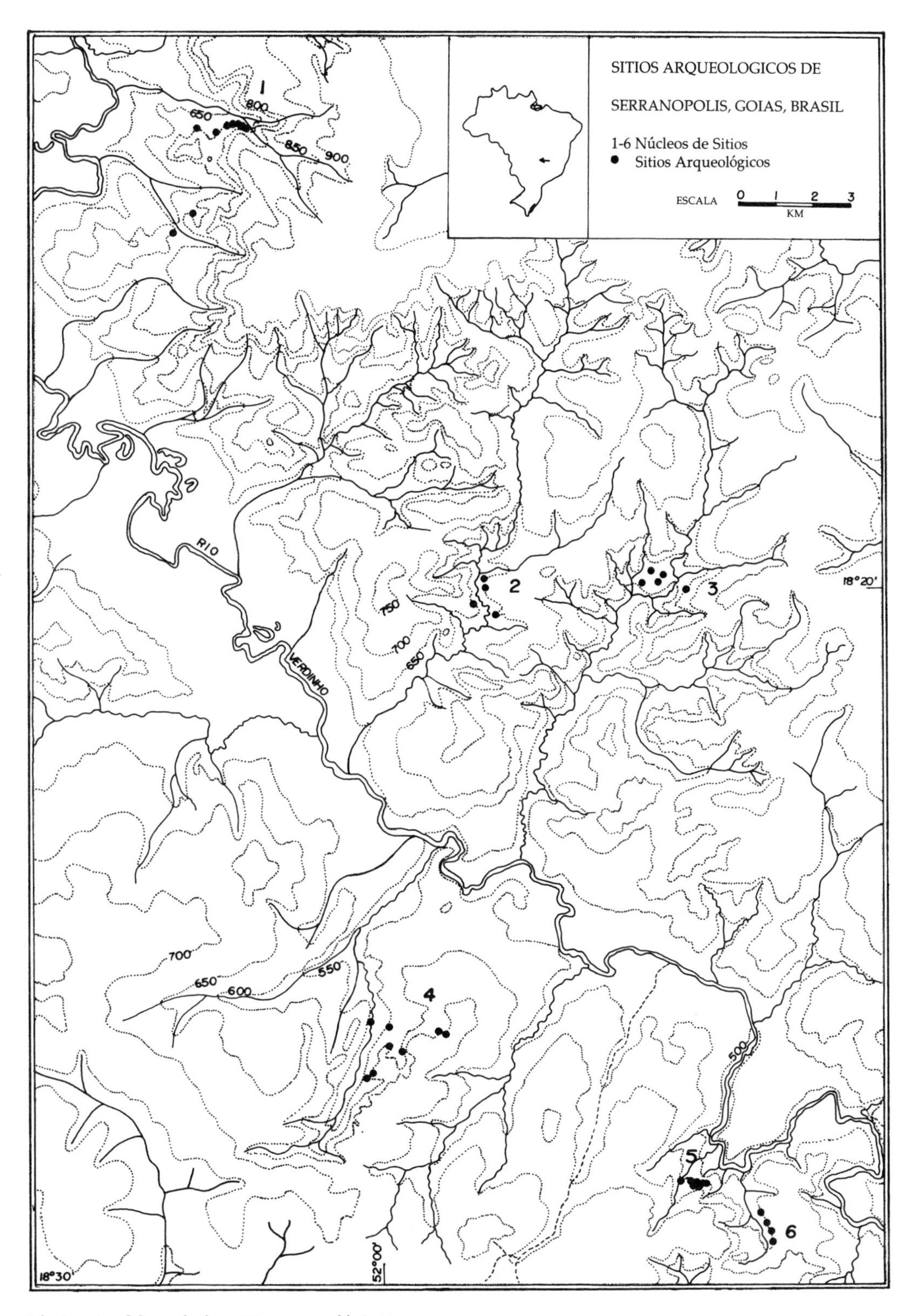

Lámina 1. Mapa de los sitios arqueológicos.

distantes del agua, presentan pisos o taludes empinados. Son ocupados tanto los abrigos pequeños, como los grandes cuando las condiciones son similares aun cuando no con la misma intensidad, presentando los abrigos grandes mucho mayor cantidad de material. Aparentemente los abrigos más apartados del río son menos procurados que los más próximos. El agua es proporcionada por pequeños cursos, que no distan más de 200 m y que mantienen un caudal regular tanto en el tiempo de sequía (hoy de mayo a setiembre), como de la lluvia (hoy de octubre hasta fines de abril).

La precipitación anual media ahora es de 1500 mm o más y la temperatura es agradable el año entero, permaneciendo entre 15°C y 30°C.

Prácticamente toda la extensión de los 25 km forma un único nicho ecológico, cuyo centro y base es el río Verdinho. Debido a los afloramientos separados de arenisca metamorfizada podemos separar este nicho en seis pequeños núcleos, donde los abrigos están concentrados (Lámina 1).

A continuación damos informaciones básicas sobre los núcleos, destacando aquellos sitios que dieron material antiguo. Generalmente los mismos presentan también materiales del período medio y reciente. Dejamos de mencionar los sitios que solamente presentan materiales recientes, y los numerosos locales de abastecimiento de materia prima superficiales, cuya cronología no fue posible determinar.

LOS SITIOS

Núcleo 1: En el margen izquierdo del río Verdinho, junto a dos pequeños cursos permanentes, donde existen al menos 10 abrigos, que distan del río de 2 a 3 km. Es probablemente el núcleo más rico debido a la gran superficie cubierta, y la proximidad del río y de los campos naturales.

Abrigo GO-JA-01: 65 m de boca, profundidad media unos 20 m, estratos fértiles de 170 cm. Pinturas y petroglifos abundantes. Fueron realizados tres cortes estratigráficos y una excavación de 40 m^2.

Abrigo GO-JA-02: 43 m de boca, profundidad media de unos 12 m, estratos fértiles de 300 cm. Regular número de petroglifos. Fue realizado un corte estratigráfico.

En los demás abrigos, cuya mayor parte presentan material superficial y uno de ellos también pinturas y petroglifos, todavía no se realizaron cortes o excavaciones.

Núcleo 2: En el margen izquierdo del río Verdinho, junto a un curso permanente donde existen cuatro abrigos, que distan del río aproximadamente 6 km. Es uno de los núcleos más pobres, porque hay poca superficie cubierta, los abrigos están dispersos y el mayor de ellos es húmedo, oscuro, y no recibe luz solar. El resto de las condiciones no serían demasiado malas.

Abrigo GO-JA-20: 70 m de boca, profundidad media unos 2 m, estratos fértiles de 230 cm. Pocas figuras pintadas. Se realizó un corte estratigráfico.

Abrigo GO-JA-22: 14 m de boca, profundidad media de unos 5 m, estratos fértiles de más de 150 cm. Muy pocas pinturas. Se hizo un corte estratigráfico.

En el abrigo oscuro fue hecho un corte estratigráfico, que resultó estéril. En el otro abrigo hay material, pinturas y petroglifos, pero no se realizó ningún trabajo.

Núcleo 3: En el margen izquierdo del río Verdinho, junto a dos cursos permanentes, donde existen 5 abrigos, que distan del río apróximadamente 8 km. Es el núcleo más pobre porque el espacio cubierto es pequeño, la mayor parte de los abrigos tiene poca iluminación solar, algunos tienen un talud empinado, el espacio útil es pequeño o excesivamente ventilado. Un abrigo tiene una pequeña pintura y otro una serie de petroglifos. Se realizaron cortes estratigráficos en tres de ellos, dando poquísimo material y reciente; los otros también tienen material, pero también poco y reciente.

Núcleo 4: En el margen derecho del río Verdinho, junto a un curso de agua permanente, con varias nacientes, junto al cual existen por lo menos 8 abrigos y dos sitios más a cielo abierto. La distancia al río es de 5 km. Es uno de los núcleos más ricos, porque hay una gran superficie cubierta y concentración de otros recursos.

Abrigo GO-JA-03: 80 m de boca, profundidad media 5 m, estratos fértiles de 270 cm. Fueron realizados 6 cortes estratigráficos.

Abrigo GO-JA-26: 8 m de boca, profundidad media de 5 m, estratos fértiles de 270 cm. Se realizó un corte estratigráfico.

Sitio abierto GO-JA-23, en la barranca de un curso, con la aparición, en un nivel profundo de rodados, en una extensión de cientos de metros, de material de la fase Paranaíba.

Sitio abierto GO-JA-29, en la barranca y en el lecho de un curso, con la aparición en el mismo nivel profundo de rodados, de materiales de la fase Paranaíba.

Los otros sitios contienen material, pinturas y o petroglifos. En uno de ellos fueron realizados 2 cortes, pero solamente apareció material reciente: el talud es muy empinado y está muy distante del agua.

Núcleo 5: En el margen derecho del río, junto a un curso permanente, cerca del cual existen por lo menos 13 abrigos de varios tamaños, con mediana abundancia de recursos. La mayor parte de los abrigos tienen materiales en superficie y diversos también pinturas y o petroglifos. Distan del río 1,5 km.

Abrigo GO-JA-13c: 20 m de boca, profundidad media 1,5 m, estratos con más de 290 cm. Fue hecho un corte estratigráfico.

Núcleo 6: En el margen derecho del río, junto a un curso ahora intermitente, junto al cual existen 4 abrigos, que distan del río 1,5 km. Es un nicho medianamente rico y en los abrigos hay problemas de exposición solar, taludes empinados y talvez dificultad para el acceso al agua en tiempo de sequía.

Abrigo GO-JA-14: 72 m de boca, profundidad media de 5 m, estratos de 110 cm. Fueron hechos dos cortes estratigráficos.

También en los otros tres abrigos fueron realizados cortes estratigráficos, pero sin alcanzar materiales antiguos.

LAS FASES CULTURALES

En los cortes estratigráficos y en la excavación aparecen tres fases culturales:

En los niveles más bajos de los abrigos, compuestos generalmente de arenas flojas, de color marrón rojizo, con bastante carbón disperso, aparece la fase lítica Paranaíba. En las barrancas de los arroyos intermitentes el material es encontrado en un nivel de rodados en profundidades bien acentuadas.

Las nueve dataciones existentes para la fase Paranaíba, la colocan entre 11.000 y 9000 B.P.

GO-JA-01 dio dataciones entre 10.580±115 B.P. (SI-3699) y 9060±65 B.P. (SI-3698).

GO-JA-02 dio dataciones de 10.120±80 B.P. (SI-3108) y 9195±74 B.P. (SI-3107).

GO-JA-03 dio una datación de 9765±75 B.P. (SI-3110).

GO-JA-14 dio una datación de 10.740±85 B.P. (SI-3111).

Otros 6 sitios con los mismos materiales no fueron (todavía) datados, en parte por no ser necesario, en parte porque los trabajos son recientes.

En los niveles intermedios de los abrigos, compuestos generalmente de paquetes de cenizas con arenas, más compactos y de color ceniza en tonalidades oscuras, con mucho carbón, aparece la fase lítica Serranópolis. Hasta ahora no fue encontrada en las barrancas de los cursos.

Las 7 dataciones existentes para la fase Serranópolis dan en su inicio 9000 B.P., sin estar demasiado claro su término.

GO-JA-01 dio las dataciones importantes para esta fase, que van de 9020±70 B.P. (SI-3697) a 6690±90 B.P. (SI-3691).

En los niveles superiores, compuestos por arenas con finos estratos de ceniza, generalmente claras, aparece la fase lito-cerámica Jataí.

Las dos dataciones de esta fase dan en su inicio, alrededor de 1000 A.D.

Las distintas fases se distinguen tanto por los sedimentos en los cuales aparecen, como por el material lítico, óseo, y los alimentos. De modo que en los cortes se percibe con bastante facilidad el pasaje de una a otra. Por lo menos entre la fase Jataí y la Serranópolis existe una discontinuidad bien visible. Entre la fase Serranópolis y la fase Paranaíba hay un cambio claro y brusco, tanto en los sedimentos, como en los implementos líticos y en los restos de alimentación.

LA FASE PARANAIBA

La industria lítica de la fase Paranaíba se caracteriza por láminas gruesas unifaciales, con buena tecnología y por raros implementos bifaciales, entre los cuales aparecen algunas puntas pedunculadas. Los artefactos más abundantes que provienen de los cortes y las excavaciones, en una primer visión, pueden se agrupados de la siguiente manera:

a) Artefactos unifaciales alargados, predominantemente simétricos, sobre láminas estrechas y gruesas, generalmente con arista dorsal, trabajadas en todo su perímetro, o con plano de lascado conservado como lado. El lascado es periférico, sin modificar el interior de la cara, el borde es regular, el ángulo del borde grande. Una de las extremidades es más delgada, la otra con el plano de lascado conservado o lascado abrupto. La cara interna lisa, raramente presenta bulbo de percusión, porque la parte correspondiente de la lámina fue removida. La cara externa a veces presenta restos de córtex. Morfológicamente son raspadores, siendo numerosas las "limaces". (Lámina 2:1-19).

Los artefactos son fabricados a partir de láminas de 10-20 cm de largo por 4-7 cm de ancho y 2-3 cm de espesor, con arista dorsal.

Muchas piezas fueron quebradas durante el uso. El desgaste se presenta de dos maneras: lascamiento fino de los bordes activos, predominantemente en la cara externa, o alisamiento de partes de bordes y de la cara interior: pudiéndose originar de la acción de raspar, las más finas de cortar. El alisamiento de todas las aristas del talón de algunas piezas sugieren enmangamiento o protección de cuero.

Algunas piezas tienen las extremidades tan estrechas y altas que deben ser tomadas por perforadores.

b) Artefactos unifaciales alargados, predominantemente asimétricos, sobre láminas estrechas y finas, con la cara superior aplanada o con dorso más bajo, trabajados en todo el perímetro, o con el plano de lascado conservado como pequeño lado. El lascado es periférico, sin afectar el interior de la cara, los bordes regulares, y el ángulo del borde grande, pero inferior al anterior; la cara interna plana. Una de las piezas presenta canalura dorsal muy marcada, pero no está claro si es intencional o producida por accidente. Morfológicamente serían raederas. (Lámina 3:1, 2, 5, 6, 6a, 9, 10).

Los artefactos son fabricados a partir de láminas con menos de 1 cm de espesor.

Hay muchas piezas quebradas y fragmentos. El desgaste se presenta como alisamiento o lascado en los bordes, en la cara superior, y pueden haberse originado en la acción de raspar y cortar.

c) Artefactos unifaciales, alargados, asimétricos, sobre lascas alargadas, con una o dos extremidades truncas, formando lados, el borde activo natural o levemente reforzado, el pasivo, próximo a la arista dorsal, cortical o rebajado, acomodado para ser usado en la mano. Lascado periférico, cara superior sin otras modificaciones, inferior lisa. Angulo del borde activo pequeño. Morfológicamente cuchillos o raederas. (Lámina 3:3, 4).

Los artefactos son fabricados a partir de láminas truncas o de lascas alargadas, obtenidas por la misma técnica de las láminas.

El desgaste se manifiesta bajo la forma de pequeñas lasquitas, eventualmente alisamiento, en los bordes activos, debiendo los artefactos haber sido usados como cuchillos para cortar, y eventualmente raspar.

d) Pequeñas lascas gruesas e irregulares con una punta saliente, producida por lascados irregulares, abruptos, que poco afectan la cara superior, natural; la cara interior es plana.

No presentan señales visibles de desgaste. Podrían ser perforadores. (Lámina 3:11, 12).

e) Tres fragmentos de puntas bifaciales: una parece ser una extremidad distal, una un fragmento medial, y la otra un fragmento proximal, mostrando aletas y un pedúnculo de base convexa. El lascado es irregular y cubre completamente las dos caras; por lo menos las dos últimas son artefactos incompletos, quebrados durante su elaboración. (Lámina 3:7, 8).

f) Una punta en arenisca alisada: gruesa, larga, con pedúnculo y aletas, como si fuera una especie de modelo de una punta de proyectil.

g) Un artefacto bifacial, con lascados irregulares en las caras y los bordes, en los cuales dos lados no pudieron ser removidos. Las aristas, de lo que podría ser el borde activo y de una de las caras, están alisadas (Lámina 3:16).

h) Discos lascados, picoteados o alisados, con aproximadamente 15 cm de diámetro y 2 a 4 cm de espesor, a veces con una de las caras levemente cóncava. También pueden ser rodados chatos, con formato conveniente.

i) Pequeñas piedras con una de las caras levemente cóncava, resultado de alisamiento o triturado.

j) Percutores pequeños, discoidales o esféricos o en arista, con muchas señales de percusión y alisamiento.

La producción de los artefactos está representada en los propios abrigos. 98% de la materia prima es arenisca recocida o cuarcita, que forma grandes porciones de los paredones de arenisca, en los cuales están los sitios.

En uno de los abrigos fue posible observar el proceso inicial de exploración de la materia prima, en un corte hecho en lo alto del talud: primero la retirada de grandes lascas corticales (hasta 30 x 20 x 7 cm); después lascas grandes de prueba de la materia prima, después láminas menores y más finas; en el mismo lugar fueron preparados y usados instrumentos. Los residuos comprenden grandes núcleos poliédricos, midiendo hasta 40 cm de diámetro; grandes y gruesas lascas corticales y no corticales, láminas y lascas grandes y finas, artefactos desechados, artefactos completos y usados, y una cantidad inmensa de lascas provenientes de la preparación de núcleos y artefactos y de su retoque.

En el interior de los abrigos fue observado solamente el proceso de acabamiento de artefactos, no apareciendo núcleos, ni grandes lascas, solamente instrumentos deshechados y una gran cantidad de lascas pequeñas, finas y alargadas, resultantes del proceso de terminación de las piezas, que deben haber sido traídas a ese lugar bajo la forma de láminas ya seleccionadas y preformadas. Naturalmente también aparece un gran número de artefactos terminados, completos o quebrados por el uso.

La industria ósea es poco abundante y variada. Los muchos huesos cortados, que se encuentran entre los restos de alimentación indican haber tenido un aprovechamiento considerable. El artefacto más común es una especie de espátula, hecha con un hueso largo de un animal de tamaño de un venado, en el cual se conserva una de las epífisis y la diáfisis es aplanada por abrasión o pulimento (Lámina 3:13, 14, 15); más raramente aparecen huesos con puntas redondeadas bien finas, y anzuelos.

Los enterramientos, correspondientes a esta población, a pesar de los numerosos cortes en abrigos, sumando 80 m^2, con material de la fase Paranaíba, hasta hoy no fueron encontrados.

Lámina 2. Material arqueológico de la fase Paranaíba: 1-19 - artefactos a).

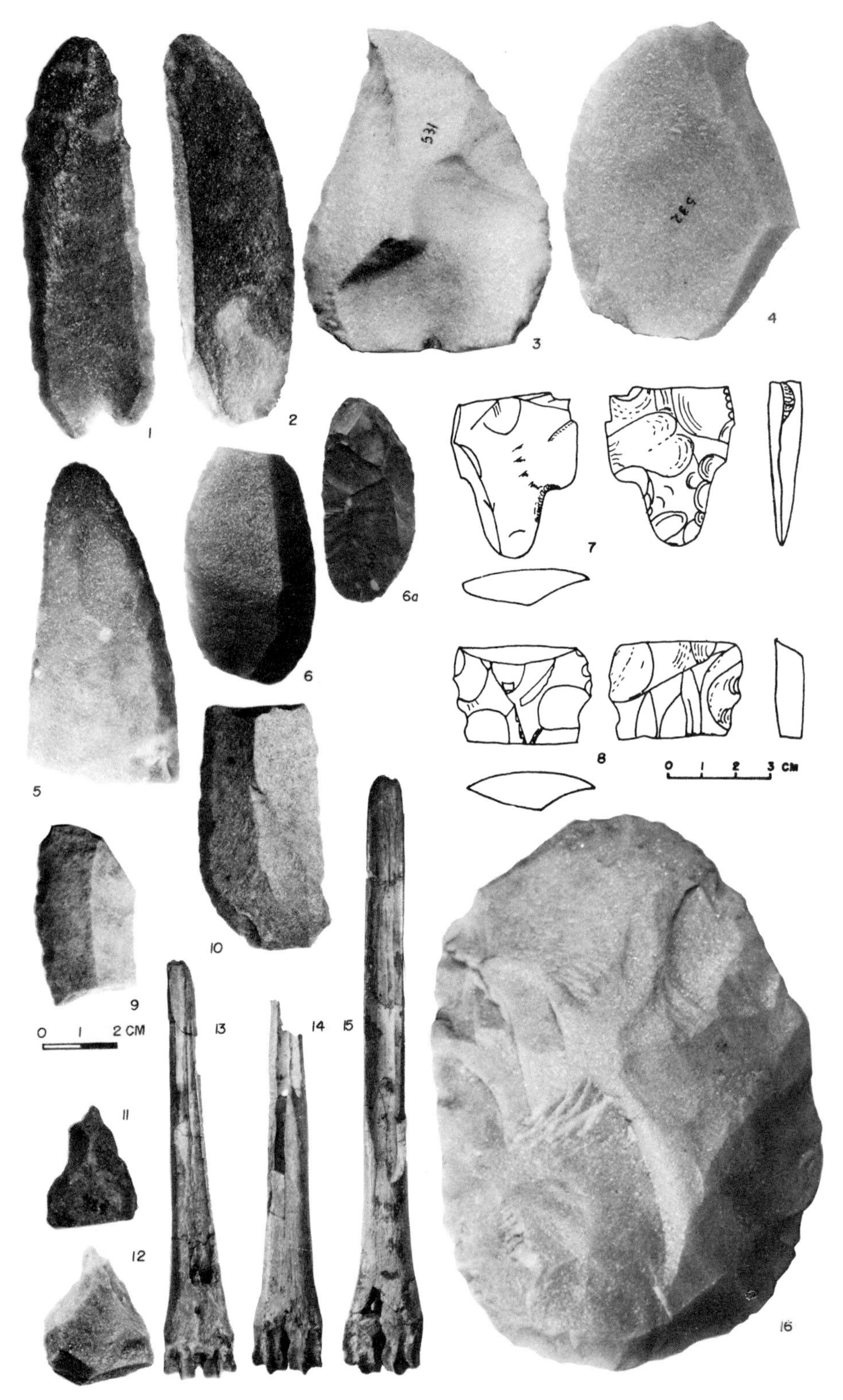

Lámina 3. Material arqueológico de la fase Paranaíba: 1, 2, 5, 6, 6a, 9, 10 - artefactos b); 3-4 - artefactos c); 11-12 - artefactos d); 7-8 - fragmentos de puntas bifaciales e); 16 - artefacto bifacial g); 13-15 - artefactos óseos.

Los restos de alimentación son muy abundantes en casi todos los abrigos, apareciendo aves, mamíferos, reptiles, peces; no moluscos. Hasta ahora no se encontró ningún hueso de animal extinguido.

Los restos conservados indican una actividad de caza intensa y generalizada, donde están presentes animales de todos los tamaños y clases, de los diversos ambientes naturales de la región, sin aparecer una tendencia en la explotación de uno de ellos en particular. La colecta se muestra menos acentuada que en las fases posteriores, encontrándose restos de frutos de palmera y otros.

LA FASE SERRANOPOLIS

La industria lítica de la fase Serranópolis es muy diferente de la anterior: los artefactos de láminas gruesas unifaciales bien acabadas desaparecen para dar orígen a una industria mal definida de lascas irregulares con gubias, picos, perforadores y raspadores pequeños. Continúan los discos y los percutores. Las lascas de deshecho, encontradas en los estratos, son mayores y más espesas, también desprendidas por percusión. El manifiesto cambio en los artefactos, es acompañado también por cambios en la materia prima, aumentando un poco la utilización de la calcedonia.

La industria ósea, todavía escasa, presenta leznas, espátulas y anzuelos.

Los enterramientos son abundantes desde el comienzo de la fase, apareciendo los esqueletos en posición fetal, decúbito lateral, generalmente izquierdo, en sepulturas. En dos enterramientos, próximo al esqueleto, aparecen cornamentas de cervídeos.

Los restos de alimentos también cambian abruptamente, apareciendo moluscos en gran cantidad y disminuyendo los huesos de caza. Son utilizados en proporciones semejantes el *Megalobulimus* sp., de tamaño grande, y los Bulimulídeos, de tamaño pequeño; raramente aparecen *Ampularia* sp., tamaño medio, y *Odontostomus* sp., tamaño pequeño. Todos ellos se multiplican en los lugares húmedos y ensombrados a lo largo de las paredes.

Los frutos recuperados son más abundantes: se registraron restos de palmeras y de numerosas plantas del "cerrado".

Se nota que a la actividad de caza generalizada, del primer período, el grupo incorpora nuevos recursos, que aparecen en la región, con los cambios climáticos, y se trasforma en un cazador recolector generalizado. Los productos cultivados recién van a aparecer en la fase siguiente, mil años después de Cristo.

CAMBIO CULTURAL Y CAMBIO AMBIENTAL

El cambio brusco de cultura observado entre la fase Paranaíba y la fase Serranópolis es difícil de explicar solamente como migración de grupos humanos, porque se produce en un ambiente grande y no sería fácil decir de donde vendría este nuevo grupo.

Existen indicios de cambios climáticos acentuados, que pueden ser la base de las diferencias culturales. En los cortes en general, y en especial en la excavación del abrigo GO-JA-01 (Schmitz 1981), podemos observar que las capas de la fase Paranaíba son de arenas flojas, de color marrón rojizo, las subyacentes (pleistocénicas) son más rojizas y compactadas, las superiores son grisáceas, compuestas de grandes paquetes de ceniza. En las barrancas de tres cursos los materiales de la fase Paranaíba están incluídos en un nivel de rodados, más o menos grandes, de acuerdo con la distancia del arrastre; el lecho de rodados forma nítidas discontinuidades tanto con los sedimentos anteriores como con los posteriores. Aparentemente el nivel de rodados está indicando el final del Pleistoceno (Ab'Sáber 1981), tanto por los datos que tenemos del material incluído, como por la caracterización general de este nivel de rodados, que estaría indicando un clima seco, de lluvias esporádicas, cortas y fuertes, y una cobertura vegetal muy abierta, inconveniente para la multiplicación de los moluscos y concentradora de los animales en las proximidades del agua. El nivel de rodados está recubierto por sedimentos arenosos de granulometría de tipo arena fina y silt, que normalmente indica un clima más húmedo, de lluvias más regulares y con una cobertura vegetal mayor. La multiplicación de los moluscos, que caracterizan la alimentación de la fase Serranópolis en oposición a la fase Paranaíba, podría así estar ligada al aumento de la humedad y de la vegetación a lo largo de los paredones y en el ambiente en general, que llevaría a la dispersión de los otros animales, resultando una mayor dificultad para su caza. El cambio de los artefactos acompañaría el cambio de la alimentación.

COMPARACIONES

Materiales semejantes a los de la fase Paranaíba y con dataciones del mismo orden aparecen sobre una amplia región del centro y nordeste del Brasil.

Dentro del mismo estado de Goiás, los sitios más próximos están a 200 km al noroeste, en Caiaponia, donde el material aparece en superfície (Schmitz et al. inédito). A 400 km más hacia el norte, en Hidrolina, fue encontrado un abrigo con una fecha de 10.750±300 B.P. (SI-2769) (Barbosa et al. 1976/77). En el municipio vecino de Niquelandia, Simonsen (1975) encontró grandes sitios superficiales. En la cuenca del Paranã, la fase Paranã también reúne materiales semejantes, aún no datados (Mendonça de Souza et al.1981).

En el estado de Minas Gerais los materiales aparecen en dos lugares: en la Lapa da Foice, datado en aproximadamente 8000 B.P. (Dias 1981) y en Peruaçu, todavía no datado (Prous y Guimarães 1981).

En el estado de Piauí, en el municipio de São Reimundo Nonato, Guidón (1981) menciona material semejante datado en 13.900±300 B.P. (Gif-4924).

En el estado de Pernambuco, el material había sido encontrado, por primera vez, en Petrolandia y datado en 7580±410 B.P. (SI-544) (Calderón 1969). Después fue encontrado en Bom Jardim, y fechado entre 11.000±250 B.P. (MC-1056) y 9520±160 B.P. (MC-1056) (Laroche et al. 1977).

Estos materiales forman la tradición Itaparica, la más extendida y mejor conocida tradición lítica del comienzo del Holoceno y final del Pleistoceno de Brasil. Es posible que ella se haya expandido dentro de un amplio ambiente de características similares, en un período aparentemente seco del final del Pleistoceno, y habría sido substituída por adaptaciones nuevas a un ambiente de cambio, alrededor del 9000 B.P.

Materiales de la misma época, pero diferentes, son encontrados en Minas Gerais, São Paulo y Rio Grande do Sul.

PALABRAS FINALES

La fase Paranaíba, de la tradición Itaparica, representa una cultura de caza generalizada en un período aparentemente del comienzo del Holoceno y final del Pleistoceno. Los sitios se encuentran tanto en abrigos, como a cielo abierto, y representan una ocupación regular de áreas de abundantes recursos. La industria lítica está caracterizada por artefactos unifaciales sobre lámina, siendo muy raros los artefactos bifaciales, donde aparecen puntas de proyectil pedunculadas, semejantes a las encontradas en Cerca Grande, Minas Gerais (Hurt 1960; Hurt y Blasi 1969) y Alice Boër, São Paulo (Beltrão 1974). Todavía no fueron encontradas puntas de proyectil en hueso.

La tradición Itaparica, cuya fecha más antigua probablemente es 13.900 B.P., y cuyo final debe estar entre 9000 e 8000 B.P., según la región, se extendió sobre un área que hoy está cubierta por "cerrado" y "caatinga" y se caracteriza por una distribución irregular de las lluvias durante el año habiendo una estación lluviosa y otra sin lluvia. El período correspondiente a la tradición Itaparica habría sido mucho más seco que el actual y la lluvia y la vegetación más escaza.

La fase Serranópolis, que sucede a la tradición Itapirica en Goiás, es muy diferente de la anterior, desapareciendo los artefactos sobre láminas y modificándose los restos de alimentación, donde se multiplican los moluscos terrestres. Este cambio se produce en un lapso de tiempo extremadamente corto y aun exije estudios más minuciosos.

RECONOCIMIENTO

Los trabajos de campo y parte de los trabajos de laboratorio fueron financiados por el Conselho Nacional de Desenvolvimento Científico e Tecnológico de Brasil, y por la Secretaria do Patrimônio Histórico e Artístico Nacional. En los trabajos de campo y laboratorio, de los cuales resultó esta síntesis, merecen destacarse Altair Sales Barbosa, Avelino Fernandes de Miranda, Irmhild Wüst, Maira Barberi Ribeiro, y una serie de investigadores tanto de la Universidad Católica de Goiás, como del Instituto Anchietano de Pesquisas, UNISINOS. Jorge Femenías tradujo el texto del original portugués.

REFERENCIAS CITADAS

Ab'Sáber, A.N.
1981 Páleo-clima e Páleo-ecología. In: P.I. Schmitz, A.S. Barbosa, e M.B. Ribeiro (Eds.) - Temas de arqueologia Brasileira 1. Páleo-índio. *Anuário de Divulgação Científica*. 5 (1978/79/80): 33-54 Instituto Goiano de Pré-história e Antropologia, Universidade Católica de Goiás, Goiânia.

Barbosa, A.S., P.I. Schmitz, e A.F. de Miranda
1976/77 Um sítio páleo-índio no médio-norte de Goiás. Novas contribuições ao estudo do páleo-índio de Goiás. *Anuário de Divulgação Científica* 3/4: 21-29. IGPA, UCG, Goiânia.

Beltrão, M. C. de M.C.
1974 Datações arqueológicas mais antigas do Brasil. *Anais da Academia Brasileira de Ciências* 46(2): 211-251.

Calderón, V.
1969 Nota prévia sobre arqueologia das regiões central e sudoeste do Estado da Bahia. PRONAPA 2, *Publicações Avulsas do Museu Paraense Emílio Goeldi* 10: 135-152. Belém. Dias, O.F. 1981 O páleo-índio em Minas Gerais. In: P.I. Schmitz, A.S. Barbosa, e M.B. Ribeiro (Eds.)- Temas de arqueologia Brasileira 1. Páleo-índio. *Anuário de Divulgação Científica* No. 5(1978/79/80): 51-54. AGPA, UCG, Goiânia.

Dias, O.F., Jr.
1981 Pesquisas arqueológicas no sudeste Brasileiro. *Boletim do Instituto de Arqueología Brasileira, Serie Especial* 2:3-22. Rio de Janeiro.

Guidon, N.
1981 O páleo-índio no Piauí. In: P.I. Schmitz, A.S. Barbosa, M.B. Ribeiro (Eds.) - Temas de arqueologia Brasileira 1. Páleo-índio. *Anuário de Divulgação Científica* No. 5(1978/79/80): 55-61. IGPA, UCG, Goiânia.

Hurt, W.R.
1960 The cultural complexes from the Lagoa Santa Region, Brazil. *American Anthropologist* 62:569-585.

Hurt, W.R., e O. Blasi
1969 O projeto arqueológico "Lagoa Santa", Minas Gerais, Brasil. (nota final). *Arquivos do Museu Paranaense*, (n.s.) *Arqueologia*, 4. Curitiba.

Laroche, A.F., A. Soares E Silva, e J.L. Rapaire
1977 *Arqueología Pernambucana. C14*. Gimnasio Pernambucano, Recife.

Mendonça de Souza, A.A.C., S.M.F. Mendonça de Souza, I. Simonsen, A.P. Oliveira, e M.A.C. Mendonça de Souza
1981 Sequência arqueológica na bacia do Paranã - I. Fases pré-cerâmicas: Cocal, Paranã e Terra Ronca. MS.

Prous, A., e C.M. Guimarães
1981 Recentes descobertas sobre os mais antigos caçadores de Minas Gerais e da Bahia. MS.

Schmitz, P.I.
1981 La evolución de la cultura en el sudoeste de Goiás. In: Contribuciones a la prehistoria de Brasil. *Pesquisas, Antropología* 32:41-83. São Leopoldo.

Simonsen, I.
1975 Alguns sítios da série Bambuí em Goiás (Nota Prévia). Museu de Antropologia, UFGO, Goiás.

Sitio Arqueológico Pleistocênico em Ambiente de Encosta: Itaboraí RJ[1]

M.C de M.C. BELTRÃO
Museu Nacional, Universidade Federal do Rio de Janeiro
BRASIL

J.R.S. de MOURA
Instituto de Geociências, UFRJ
BRASIL

W.S. de VASCONCELOS e
S.M.N. NEME
Museu Nacional, UFRJ
Rio de Janeiro
BRASIL

Abstract

The Itaboraí site is located on a slope of a limestone depression in the state of Rio de Janeiro north of Guanabara Bay (SE Brazil). It contains a succession of gravels and reddish clayey sands. Two of the gravel levels have artifacts. The upper one is overlain by a horizon radiocarbon dated as 8100±75 B.P., and therefore has an estimated age between 9000 and 12,000 B.P. The lower gravels, apparently richer in artifacts, should be considerably older (Wisconsinan II or I). Mastodon and giant ground sloth fossils have been recovered from the same gravel deposit approximately 300 m west. In addition to artifacts of the "edge trimmed tool tradition," the collections include burins on flakes, core scrapers, retouched cores, and a non-stemmed projectile point fragment. More samples will be dated by radiocarbon and thermoluminescence techniques.

INTRODUÇÃO

Os estudos realizados nos sítios arqueológicos de Alice Boër, SP (Beltrão 1966, 1968, 1969, 1973, 1974, 1978; Beltrão et al. 1981), Itaboraí, Estado de Rio de Janeiro (Beltrão 1974, 1976, 1977, 1978; Beltrão et al. 1982; Beltrão et al., this volume) e Abadiana, Goias (Beltrão et al. 1982) revelam, a partir de mútiplas linhas de evidência, que a ocupação humana no Brasil pode estar localizada cronologicamente em faixa de antigüidade superior ao Wisconsin II.

Em recente trabalho "Nota Prévia sobre a Sedimentação Neoquaternária em Alice Boër, Rio Claro, SP", Meis e Beltrão (1981) admitiram que: "Ainda que não se tenha em vista, no estágio atual da pesquisa, o estabelecimento de correlações entre depósitos estudados em Alice Boër e outros sedimentos Neoquaternários descritos na literatura, uma rápida revisão da bibliografia existentes mostra que Bigarella et al. (1965) procuram associar a ocorrência de dois episódios de deposição de rudáceos — no sul do Brasil — aos últimos eventos frios do Pleistoceno (Wisconsin I e II). Já no vale do Rio Paraíba do Sul — região de Bananal, SP — Moura e Meis (inédito) observaram estratos arenosos semelhantes aos de Alice Boër (seqüência arenosa inferior) em posição estratigráfica que denuncia idades relativamente recentes (Holoceno). Em Bananal tais depósitos encontram-se superpostos a materiais lacustres e fluviais anteriormente descritos, no vale do Rio Doce, como constituintes da porção superior do estratotipo da Formação Ipatinga (Meis e Monteiro 1979.

Assim sendo, a camada arqueológica III de Alice Boër, SP (Planalto Meridional Brasileiro) descrita por um dos autores (Beltrão 1974a) poderia tanto ser holocênica como estar incluída no Pleistoceno. Como nos parece comprovado que a camada arqueológica III de Alice Boër só é holocênica em sua parte superior (níveis 1 a 7), já que sua parte média (níveis 8 e 10) alcança antigüidades respectivas de 10.970±1020 e 14.200 ±1150 B.P. (Beltrão et al. this volume), a antigüidade da porção inferior da camada III (niveis de 18 a 20) deve se situar conseqüentemente dentro dos limites do Wisconsin II.

Os materiais arqueológicos da camada V de Alice Boër, não se encontram *"in situ"*, e suas características sugerem uma época consideravelmente mais antiga do que a da formação da camada. Isto significa que, embora a camada V se situe dentro dos limites cronológicos do Wisconsin II, os materiais que a compõem podem pertencer ao Wisconsin I.

Já o sítio arqueológico de Abadiana, GO (Planalto Central Brasileiro), apenas prospectado, situa-se numa cascalheira antiga que, pelo menos parcialmente, se encontra sob o lençol de água do rio Capivari, afluente do Corumbá (Beltrão et al. 1982).

Em conseqüência da dragagem do rio, foram retirados do local, além de "choppers" e de um artefato lascado bifacialmente em matéria prima rebelde ao lascamento conchoidal, várias lascas obtidas a partir de blocos e retocadas na extremidade, sendo que uma delas foi adaptada de modo a funcionar como plaina.

Igualmente muito antigo é o sítio arqueológico de Itaboraí, que, juntamente com os acima mencionados, permitirá, acreditamos, facilitar a revisão do quadro da entrada do homem na América.

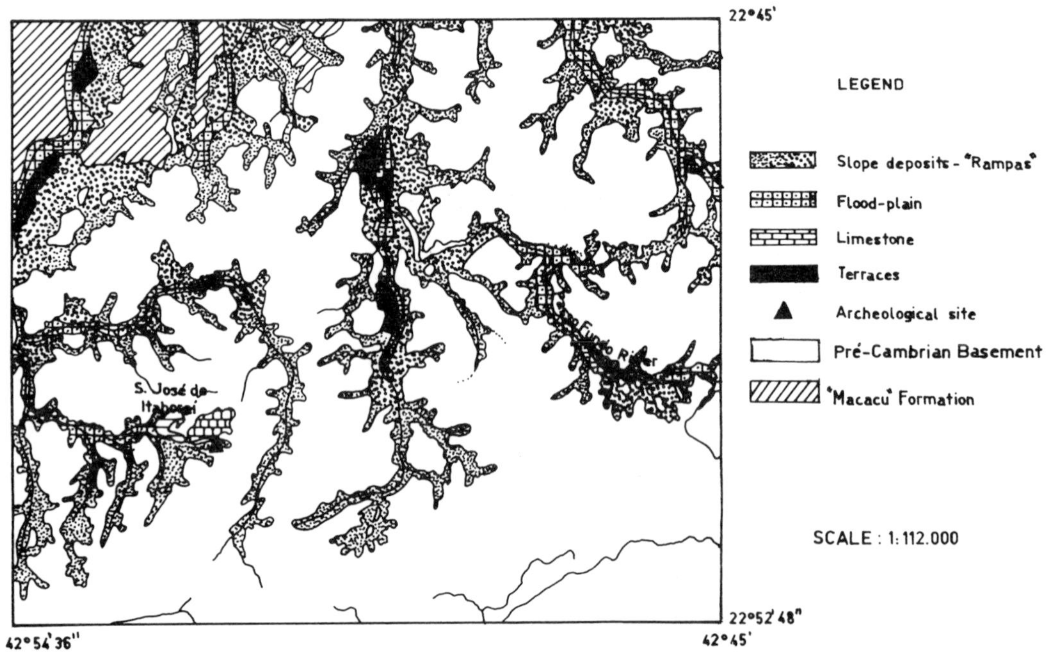

Figura 1. Sedimentacão e geomorfología Neocenozóica na região de São José de Itaboraí.

O SITIO ARQUEOLOGICO DE ITABORAI

A bacia calcária de São José de Itaboraí é uma depressão de cerca de 2000 m de comprimento por 500 m de largura e cerca de 100 m de profundidade. Está situada a setecentos metros do povoado de São José, no Município de Itaboraí, no Estado do Rio de Janeiro.

O sítio arqueológico de Itaboraí, descoberto por um dos autores (Beltrão 1974, 1976, 1977, 1978; Beltrão et al. 1982; Beltrão et al., this volume) está localizado em superficie colinosa com encostas relativamente íngremes associadas à morfologia de "Rampas". A estrutura superficial das vertentes estudadas mostra a preservação das seqüências coluviais Neoquaternárias já descritas na literatura (Moura e Meis 1981), alternadas com linhas de seixos de quartzo e/ou fragmentos de rochas subangulosas, embutidos em matriz arenosa grosseira, onde se verifica a ocorrência de artefatos. Tais cascalheiras parecem ter, na área em estudo, importante significado regional.

Na mesma depressão calcárea onde escavamos, uma cascalheira liberou exemplares da megafauna extinta. Os paleontólogos Price e Campos (1970) descreveram as características gerais do achado em uma cascalheira:

> resultante de águas pluviais em regime torrencial, em que o relevo acentuado apresentou, para sua formação, um papel preponderante. Este pacote assenta-se sobre as irregularidades do gnaisse alterado que mantém traços da xistosidade original. E encimado por um solo vermelho de espessura variável, provavelmente de formação recente. O cascalheiro consiste principalmente de um conglomerado de matacões, calhaus e seixos angulosos havendo, de uma maneira geral, predominância de calhaus. Estes são em sua maioría, formados de quartzo, havendo, também, representação de silex. A matriz é composta essencialmente de areia grosseira com uma porcentagem relativamente alta de argila. Lentes de areia grossa entremeiam o conglomerado. O conjunto apresenta aspecto de acamamento. Na parte inferior, com expessura de 70 cm, há predominância dos elementos de maior tamanho e as lentes de areia são menos desenvolvidas indicando uma maior ação de correntes. Na parte superior as lentes de areia tornam-se mais espessas e o conglomerado é constituído de calhaus pequenos e seixos. Entre as duas seqüências há, pelo menos localmente, uma pequena camada de areia ferruginosa de cor amarela viva.
>
> Os fósseis foram encontrados indistintamente em todo o depósito. De modo geral estão fragmentados e friáveis, entretanto existem ossos e dentes completos. Os ossos fragmentados, obviamente, sofreram maior transporte, sendo triturados pelos calhaus. O material mais completo, por sua vez, mostra pequeno transporte, visto o encontro de dentes completos, com as raízes de mastodontes e um grande fêmur completo de megaterídeo, mostrando que estes fósseis não sofreram praticamente transporte, indicando, ainda que o animal viveu na vizinhança do cascalheiro quando de sua formação (Price e Campos 1970:356).

Quanto à idade, Price e Campos afirmam:

> A presença de restos de *Haplomastodon* e *Eremotherium* indicam idade pleistocênica para o depósito. Quanto à idade dentro do Pleistoceno, de acordo com pesquisas recentes, há a tendência de se supor que a ligação ístmica intercontinental não tenha sido estabelecida por completo no Eo-pleistoceno e que somente no Neo-pleistoceno tornou-se possível o intercâmbio total de elementos faunísticos. Assim, torna-se provável, em vista da presença de *Haplomastodon,* que a idade Neo-pleistocênica seria a mais indicada para o cascalheiro (Price e Campos 1970:358).

Quanto ao clima concluem:

> A litologia do cascalheiro permite inferir que o mesmo formou-se sob condições torrenciais, embora os calhaus e areias não tenham sido carreados muito longe da área fonte. A faunula possivelmente indica um ambiente climático próximo ao que permite o desenvolvimento de savanas. (Price e Campos 1970:358).

Concluem os autores citados que o clima naquela época teria sido mais árido que o atual da região.

Foi destacada localmente uma alta sensibilidade da paisagem em refletir fases erosivas e de sedimentação do Quaternário Superior. Fato de grande importância, porque, não havendo no Brasil glaciações quaternárias, temos que nos apoiar em condições climáticas que possam estar relacionadas com a bem estabelecida cronologia glacial do Hemisfério Norte.

Itaboraí tem condições de se transformar, portanto, em sítio de referência, não só como apoio para determinar a idade de outros sítios com as mesmas características (a pesquisa arqueológica em rampa de coluvio encontra-se ainda incipiente no

Brasil) como para subsidiar a interpretação geocronológica das cascalheiras aí presentes, que têm sido registradas em vários pontos do Brasil Meridional. Em virtude da facilidade da preservação de carvões no local, o contexto arqueo-geológico será dentro em breve definido cronologicamente.

A pesquisa que estamos realizando no sítio arqueológico de Itaboraí tem, portanto, caráter tri-dimensional, abrangendo a arqueo-antropologia, a arqueo-geologia e a arqueo-física.

No campo de antropologia contamos com o apoio da Indiana University, EE.UU. e com a experiência de Wesley Hurt que recentemente apresentou comunicação sobre um outro sítio pleistocênico que localizamos e em que vimos trabalhando desde longa data (Beltrão 1966, 1968, 1969, 1973, 1974, 1978; Beltrão et al. 1981; Beltrão et al., this volume; Hurt, this volume).

O trabalho conjugado de arqueólogos e geomorfólogos permitiu a busca de um melhor equacionamento da dinâmica paleoambiental do Quaternário Superior na área em estudo. Para tanto, elaborou-se uma litoestratigrafia preliminar que a partir da reprodução em escala, de secções geológicas da encosta, subsidiará a localização das escavações.

Assim, as relações estratigráficas do sítio, que só foram, até agora, verificadas no topo da encosta, em um poço-piloto de 1 m de profundidade por 5 m de extensão, serão, futuramente, melhor entendidas nas partes mais baixas da rampa de coluvio onde se localiza um penhasco artificial. Nesse local a estratigrafia abrange aproximadamente 7 m de profundidade por 600 m de extensão podendo ser controlada a localização exata dos artefatos nas diferentes posições estratigráficas, quer nas cascalheiras, quer nas camadas intermediarías (Figura 1).

As camadas de 1 a 7 (Figura 2e Tabla 1) representam os testemunhos da sedimentação do Terciário Superior na área - Camadas Pré-Macacu e Formação Macacu - (Meis 1976). Os sedimentos Neoquaternários entram em contato com esta unidade delineados discordantemente por uma cascalheira de quartzo e fragmentos de rocha semi-alterada subangulosos e angulosos (camada 8).

Tabla 1. Estratigrafia do lado do Morro de Dinamite.

CRONOESTRATIGRAFIA	LITOESTRATIGRAFIA
HOLOCENO	Colúvio Castanho (Camada 13) - Material argilo-arenoso, contendo grânulos de quartzo anguloso de até 2 mm de diâmetro Colúvio Castanho Avermelhado(Camadas 10-12) - Material areno-argiloso, com poucos grânulos de até 5 mm de diâmetro (12). - Material arenoso grosseiro com baixo teor de argila de coloração castanha (11). - Seixos de Quartzo e Sílex subangulosos de até 10 cm de diâmetro, embutidos em matriz arenosa (10).
PLEISTOCENO	Colúvio Vermelho (Camada 8-9) - Material argilo-arenoso de coloração avermelhada, contendo grânulos frequentes de quartzo anguloso até 1 cm de diâmetro (9). - Seixos de Quartzo e fragmentos de rocha de até 20 cm de diâmetro, angulosos e subangulosos, butidos em matriz arenosa (8).
TERCEÁRIO	Formação MACACU (Camada 7) -Material argiloso arenoso de coloração vermelha com mosqueamento. Stone Line (Camada 6) - Linha de seixos continua predominantemente de quartzo de até 15 cm de diâmetro. Camadas Pré-Macacu (Camadas 1-5) - Materiais areno-argilosos de coloração esverdeada com diferentes níveis de osidação e altos teores em feldspato (1-3-5), alternados com linhas de seixos de quartzo e fragmentos de rocha de até 25 cm de diâmetro (2-4).

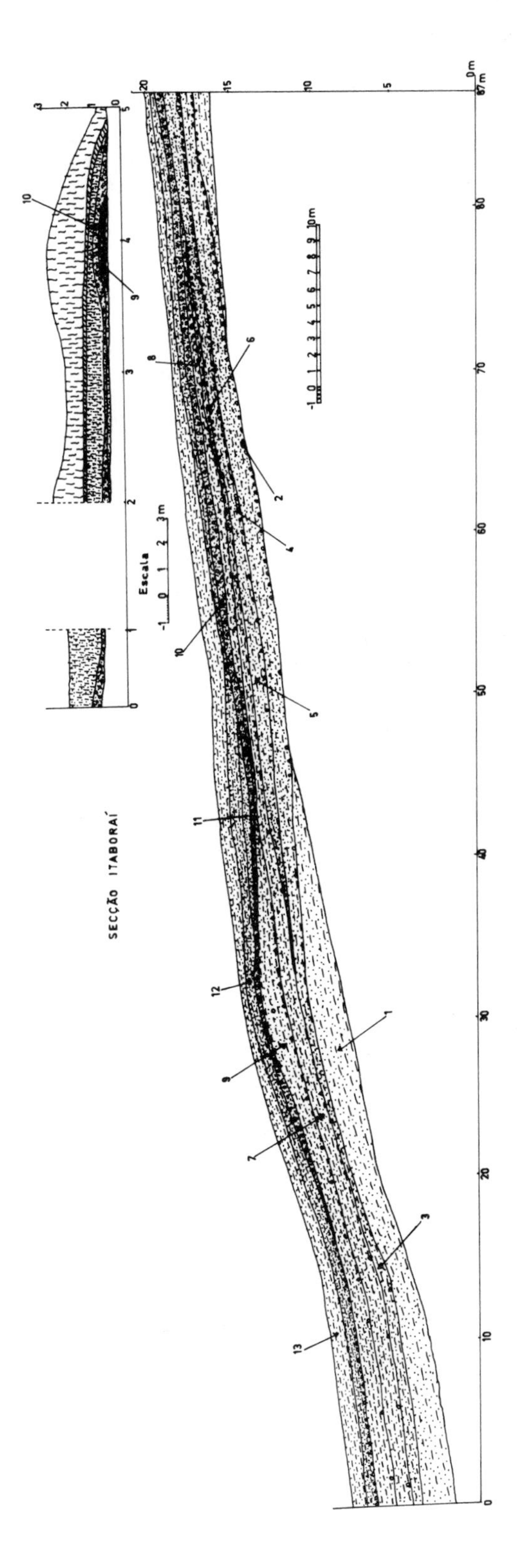

Figura 2. Secção do lado do Morro de Dinamite, São José de Itaboraí.

Um outro nível de cascalheira (camada 10) é constituído de seixos de quartzo com diâmetros variados, onde ocorre abundância de seixos de sílex. Os artefatos presentes nessa cascalheira sugerem idade situada entre 9 e 12 mil anos (edge trimmed tool tradition) o que está coerente com a datação de 8100±75 B.P. Dicarb Laboratory, EE.UU. obtida em uma camada castanho avermelhada (camada 11) em razão da ocorrência de detritos orgânicos (carvão).

A indústria da camada 10 de Itaboraí, de idade presumivelmente pleistocênica, não pode ser considerada nem pobre nem pouco característica. A julgar pela amostragem retirada do poço-piloto, embora um grupo de artefatos possa ser enquadrado dentro da chamada "edge trimmed tool tradition", outros artefatos apresentam bom nível de especialização técnica.

As vinculações tipológicas com Alice Boër estão pouco definidas, talvez porque, além do aspecto das variações culturais, há o aspecto ecológico a ser considerado. O local apresenta abundância de matéria prima, que, apesar de não ser de tão boa qualidade quanto Alice Boër, foi tratada em alguns casos com grande habilidade. Assim, por exemplo, a técnica de preparação do núcleo por lascamento circundante e retirada de lascas no sentido do comprimento pode ser encontrada não só em sílex, como até mesmo em matéria prima como o quartzo, considerada rebelde à técnica. A relação dos artefatos encontrados no topo da rampa de colúvio (camada 10), bem como em vários pontos dos cortes da estrada que correspondem à cascalheira inferior (camada 8) inclui: choppers, raspadores laterais em lasca extremamente espessa, raspador com entalhe em lasca, facas de dorso, faca-perfurador, buris em lasca, um fragmento de ponta sem pedúnculo, etc. Foram igualmente encontrados numerosos núcleos: núcleos volumosos de onde foram retiradas lascas com plano de percussão grande, geralmente raso e formando ângulo muito aberto com a superfície de lascamento; núcleos cônicos de onde foram retiradas lâminas (alguns desses núcleos foram retocados para se adaptarem à função de plaina); núcleos em que as lascas alongadas foram retiradas no sentido do comprimento do núcleo preparado. O grande número de núcleos e de artefatos apenas esboçados permitirá a reconstituição das técnicas empregadas e sua distribuição pelas diferentes camadas arqueológicas do sítio. Alguns artefatos apresentam-se manchados de vermelho, fato este já observado em sítio muito antigo localizado em um terraço fluvial em Rio Claro, Estado de São Paulo, não muito longe do sítio Alice Boër: o sítio Santo Antonio. Essas manchas tem, sua origem, ao que parece, nos sedimentos que sofreram precipitações pelos óxidos de ferro.

Alguns artefatos preservam feldspato pouco alterado; outros pátina brilhante e amarelada: outros ainda pátina incolor e escorregadia. As características da pátina, das manchas vermelhas e das concreções, poderão servir talvez, no futuro, como indicadores cronológicos.

A cascalheira inferior (camada 8), aparentemente mais rica em artefatos que a cascalheira superior (camada 10), é consideravelmente mais antiga, podendo corresponder em idade à camada V de Alice Boër, que teria sido formada segundo (Beltrão 1973, 1974; Beltrão et al. 1983) entre 20 e 40 mil anos (Wisconsin II). Contudo, os materiais aí encontrados podem, talvez, corresponder à faixa de idade do Wisconsin I.

No campo de arqueo-física realizaremos pesquisa idêntica à que foi apresentada para o sítio Alice Boër (Beltrão et al. 1981, 1983), isto é, além de datações pelo método do C-14, serão selecionados artefatos suscetíveis de datação pelo método da termoluminescência. Alguns desses artefatos podem ter sido termicamente tratados pelo homem pré-histórico, antes de trabalhados de maneira a propiciar aos materiais melhores condições de lascamento (Crabtee and Butler 1964). Outros foram ocasionalmente queimados, depois de trabalhados. Entre esses artefatos esquentados depois da feitura, aqueles que apresentarem uma coloração avermelhada total ou parcial em sua superfície externa serão selecionados especialmente para serem submetidos ao método da termoluminescência.

NOTA

1. Pesquisa financiada pelo Serviço de Patrimônio Histórico e Artístico Nacional (SPHAN), pelo Conselho de Ensino para Graduados da UFRJ (CEPG) e pelo Conselho Nacional de Desenvolvimento Científico e Tecnológico (CNPq).

REFERENCIAS CITADAS

Beltrão (antes Becker), M.C. de M.C.

1966 Quelques donnés nouvelles sur les sites préhistoriques de Rio Claro, Etat de S. Paulo. XXXVI *Congreso Internacional de Americanistas, Actas* I:445-450. Sevilla.

Beltrão, M.C. de M.C.

1968 O estágio lítico no Brasil. *Resumos da XX Reunião Anual, Sociedade Brasileira para o Progresso das Ciências* 20 (2):460.

1969 Identificação do estágio lítico superior no Brasil. Presentado a III Simpósio de Arqueologia da Area do Prata. *Pesquisas* 20:4. São Leopoldo.

1973 Datações pré-históricas mais antigas no Brasil. Resumo das Comunicações. *Academia Brasileira de Ciências,* Anais 45 (3/4): 651-652. Rio de Janeiro.

1974 Datações arqueológicas mais antigas do Brasil. *Academia Brasileira de Ciências, Anais* 46:211-251.

1976 Sítios pré-históricos e a megafauna extinta no Brasil. *Academia Brasileira de Ciências, Anais* 48 (2):355.

1977 Ocupação pré-histórica: aspectos culturais, geológicos e paleontológicos. Conveniência da abordagem interdisciplinas. *Cadernos do Museu de Arqueologia e Artes Populares da Universidade Federal do Paraná.* Paranaguá.

1978 *Prê-Histôria do Estado do Rio de Janeiro.* 276 p. Instituto Estadual do Livro e Editora Forense-Universitária. Rio de Janeiro.

Beltrão, M.C. de M.C., J. Danon, C.R. Enriquez, E. Zuleta, and G. Poupeau

1981 Thermoluminescence studies of archaeological heated cherts from the Alice Boër Site. *X Congreso, Union Internacional de Ciencas Prehistôricas y Protohistôricas, Comisiôn XII* (El Poblamiento de America), p. 96. Mexico, D.F.

Beltrão, M.C. de M.C., J. Danon, e M.M. Teles

1982 Datação pelo ^{14}C do sítio arqueológico de Itaboraí, RJ. *Academia Brasileira de Ciências, Anais* 54(1):258-259.

Bigarella, J.J., M.R. Mousinho, and J.X. Silva

1965 *Processes and environments of the Brazilian Quaternary.* 69 pp. Imprensa Universitaria. Paraná, Curitiba.

Crabtree, D.E., and B.R. Butler

1964 Notes on experiments in flint knapping: 1, heat treatment. *Tebiwa* 7:1-6.

Meis, M.R.M.

1976 Contribuição ao estudo do Terciário Superior e Quaternário da Baixada da Guanabara. Tese de Doutoramento. Universidade Federal do Rio de Janeiro.

Meis, M.R.M., e M.C. de M.C. Beltrão

1981 Nota prévia sobre a sedimentação Neoquaternária em Alice Boër, Rio Claro, SP. To be published by Comissão Técnico-Científica do Quaternário.

Meis, M.R.M., and A.M.F. Monteiro
1979 Upper Quaternary rampas: Doce river valley, southeastern Brazilian Plateau. *Zeitschrift für Geomorphologie* 23(2):132-151.

Moura, J.R.S., e M.R.M. Meis
1981 Litoestratigrafia dos depósitos de encosta — MG — RJ. *Revista Brasileira de Geociências* 10(4).

Price, L.I., e D.A. Campos
1970 Fósseis Pleistocênicos no Município de Itaboraí, Estado do Rio de Janeiro. *XXIV Congresso Brasileiro de Geologia, Anais*: 355-358. Sociedade Brasileira de Geologia. Brasilia.

Thermoluminescence Dating of Burnt Cherts From the Alice Boër Site (Brazil)

M.C. de M.C. BELTRÃO
Museu Nacional
Quinta da Boa Vista
Rio de Janeiro
BRAZIL

C.R. ENRIQUEZ, J. DANON,
E. ZULETA, and G. POUPEAU
Centro Brasileiro de Pesquisas Físicas
Av. Wenceslau Braz 71
Rio de Janiero
BRAZIL

Abstract

More than 40 culturally burnt cherts from the Alice Boër site near Rio Claro, São Paulo State, have been studied by thermoluminescence (TL). Nine of these were found to be sufficiently heated by early man to reset the TL clock to zero and thereby be suitable for TL dating.

These cherts define, for the upper half of the uppermost cultural layer (layer III), a time scale for the presence of man from about 2220 to 11,000 years. This time range is in essential agreement with the geological-paleoclimatological age estimate for the end of deposition of this layer, as well as with radiocarbon dates. In particular, it lends support to a radiocarbon age of 14,000 B.P. for a deeper level.

The overall geochronologic results are not in contradiction with earlier statements (Beltrão 1973, 1974, 1978) that the deepest cultural layer (layer V) at Alice Boër might have been deposited at least 20,000 years ago.

INTRODUCTION

The archaeological sites of Planalto Meridional in central Brazil present a rich lithic industry. In most of these sites, however, the cultural layers have a limited thickness, suggestive of short occupation times (Beltrão 1969). In this context, the Alice Boër site (São Paulo State) is of particular interest because of the wealth of its lithic industry, the great thickness of the archaeological layers and the time range they represent (at least 10,000 years), as well as the possible great antiquity of the lowermost tool-bearing level (Beltrão 1974).

Due to the low abundance of charcoal at the Alice Boër site, only a few levels from the tool-bearing layers have been radiocarbon dated (Beltrão 1974). The upper part of the cultural levels has been dated from 6000 to 14,000 radiocarbon years B.P., while the timing of deposition of the lower levels remains largely speculative.

The artifact collection from Alice Boër consists of tools and flakes of locally abundant black chert. Examination by one of us (Beltrão) revealed that before being worked most samples had been thermally pretreated a technique known to improve their flaking properties (Crabtree 1964). Laboratory simulations suggest that thermal treatment may vary, depending on flint structure, from 280°C to 500°C (Inizan et al. 1975). An annealing temperature of about 380 is sufficient to reset the thermoluminescence (TL) chronometer to zero in cherts (Wintle and Aitken 1977), thus allowing eventually a *direct* dating of stone artifacts that have been adequately heated (intentionally or not) by early man. In a previous work (Danon et al. 1980), we have shown that indeed a few cherts from Alice Boër had been sufficiently heated in the past to allow TL dating. The Archaeological Doses of eight cherts from various levels were measured, and their variation with depth in the site appeared to be compatible with the radiocarbon data.

The transformation of these archaeological doses into TL ages requires the determination of additional parameters, including site gamma-dosimetry, not available in 1980. We have now measured these parameters and present here a first approach of the TL chronology at Alice Boër.

THERMOLUMINESCENCE DATING

Principle of TL dating

Any artifact buried in a weathering horizon is continuously irradiated by a flux of ionizing particles from cosmic rays and alpha, beta and gamma rays from natural radioactivity. These radiations interact with matter mostly by ionization (i.e., by ejecting electrons from their parent atoms). In any mineral, a fraction of the ionized electrons do not recombine immediately and becomes "trapped" in crystal lattice defects. The period of most natural radioactive isotopes being long enough, the radiation dose per unit of time is constant and the number of electrons trapped is therefore proportional to time. In some traps, electrons are stable for millions of years, unless the crystal lattice is given sufficient thermal agitation by heating, so that the electrons can acquire enough energy to leave their traps. For archaeological samples, the trap-emptying temperatures of interest to TL dating are in the range from 300°C to 600°C. In practical terms, a potsherd or a hearth stone baked at such temperatures will have all their traps emptied of the previously "geologically" trapped electrons, and then the traps will progressively fill again; the number of electrons at any time being proportional to the time elapsed since their last archaeological heating. An interesting property of the trapped electrons is that, upon heating, they will progressively leave their traps and recombine to their parent atoms while emitting some light. It is this thermally induced emission of light which is called *thermoluminescence*. As the number of electrons trapped is proportional to time, so is the height of the archaeological TL emitted at a given temperature (Figure 1). The archaeological TL emitted at high temperature (say higher than 300°C) corresponds to an absorbed *Archaeological radiation Dose* (AD). The principle of TL dating is, through the measurement of a TL signal, to determine this AD (Figure 2) and deduce an age t from the relationship:

$$t = \frac{\text{Archaeological Dose}}{\text{Annual Dose}}$$

where the *annual dose* is the dose of radiation deposited per year in the mineral dated, by the radioactive impurities of the sample itself as well as by radiations originating from the embedding soil and cosmic rays. This annual dose is calculated from the U, Th and K content of the sample and its environments.

Although the principle of TL dating as summarized above is quite simple, various factors may intervene, which complicate its effective realization. For instance, the precision in the determination of the archaeological dose is limited by anomalous or saturation effects, supralinearity of the TL signal, etc...; those of the annual dose rate, by the state of disequilibrium of the U and Th series, the degree of homogeneity in the spatial distribution of the radioactive species throughout the sample, the relative efficiency of alpha and beta particles in producing TL, temporal variations of soil wetness, eventual leaching or deposition of radioactive species by running waters, etc...(Aitken and Fleming 1972; Aitken 1974). All in all, however, the best *overall* error limit (precision plus accuracy) achievable for the average date of a group of contemporaneous samples is of the order of ±7% at a 1 sigma (or 68%) confidence level. For a TL mean age of 10,000 years, this would therefore correspond to an overall error of ±700 years (Aitken and Huxtable 1980).

TL dating of cherts

TL dating of cherts was first attempted by Goksu et al. (1974). In order to eliminate spurious

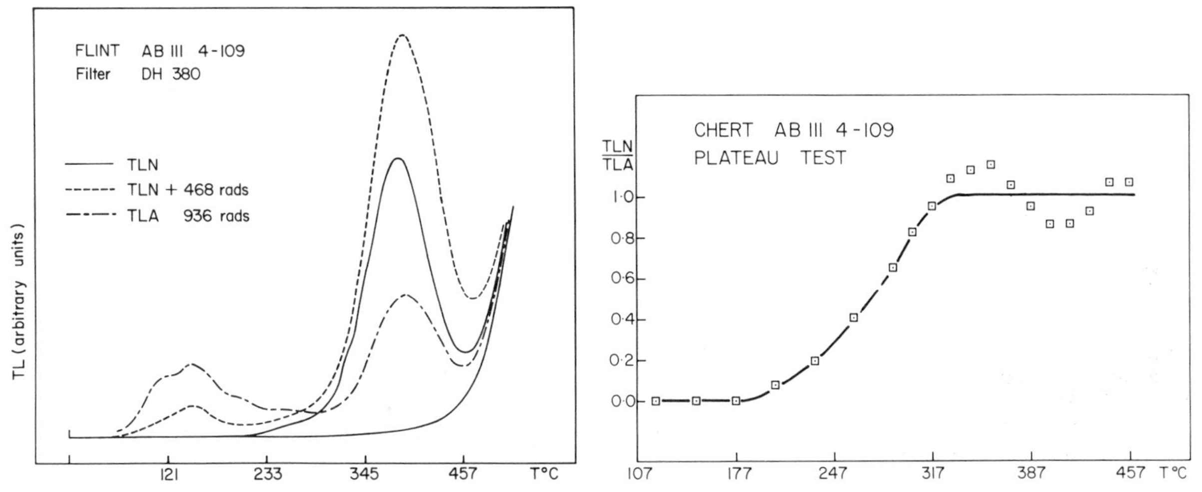

Figure 1. TL behavior of a typical unsaturated (see text) chert artifact from Alice Boër: sample 4-109

a) TL glow curve versus temperature (heating rate 10° C/second). Full line: natural TL (TLN), with a single peak centered at ~375° C; dashed line, natural TL + a laboratory β-radiation dose of 468 rads. Note the appearance of a second peak at 130° C (unseen in the TLN, due to the low thermal stability of electrons in the corresponding low-energy traps), and especially the increase of the 375° C peak; dotted line, artificial TL (TLA) released after emptying the TLN by heating and irradiating by a β-dose of 936 rads.

b) Plateau test: the ratio of TLN/TLA with temperature increases from zero at temperatures <200° C to a stable (± 15%) value above ~300° C. This suggests that above this last temperature no loss of TLN occurred since the last firing of chert 4-109 and that TL dating is feasible on this sample.

Bones fractured when dry by various natural and/or cultural agencies.

(triboluminescence and regeneration thermoluminescence) effects that had discouraged previous attempts, they used a "slice technique" for TL measurements. The same technique was then applied by Aitken and Wintle (1977) and Wintle and Aitken (1977) for the dating of a Lower Palaeolithic site. As noted by Wintle and Aitken, not too much weight was given to those ages, because there was as yet no control from other dating methods. Morever, the "slice technique" introduces an additional source of error to the TL age, of about ±6%.

This led three groups to test again the feasibility of TL dating of flint/chert materials with more conventional sample preparations using careful grinding techniques. It was thus shown independently by Danon et al. (1980), Huxtable (1980) and Valladas (1980) that crushing cherts for TL measurements did not necessarily induce adverse effects. In fact, in more than 40 samples studied by Danon et al. (1980), all prepared with a crushing technique, none showed a recognizable spurious effect. It was further shown that either the archaeological doses measured in these samples (Danon et al. 1980) or their TL ages (Valladas 1980, from two Magdelenian flint artifacts), were compatible with the radiocarbon age of the same archaeological levels in their respective sites. The results reported here confirm this tendency and thereby the ability of the TL technique to give reliable ages for archaeologically burnt flints and cherts.

STRATIGRAPHY AND TL SAMPLING AT THE ALICE BOËR SITE

The Alice Boër site, near to the town of Rio Claro (São Paulo), is situated on a flood terrace about 20 m from the bank of the Rio da Cabeça. The elevation of this terrace above running water is less than 10 m. Over an excavation depth of about four m, five lithostratigraphic units have been identified, two of which (layers III and V) yielded lithic artifacts. A detailed description of the site and the results of sedimentological studies have been reported by Beltrão (1974) and Meis and Beltrão (1981a, 1981b).

A cross-section of the excavation is given in Figure 3. Layer V, which corresponds to the older

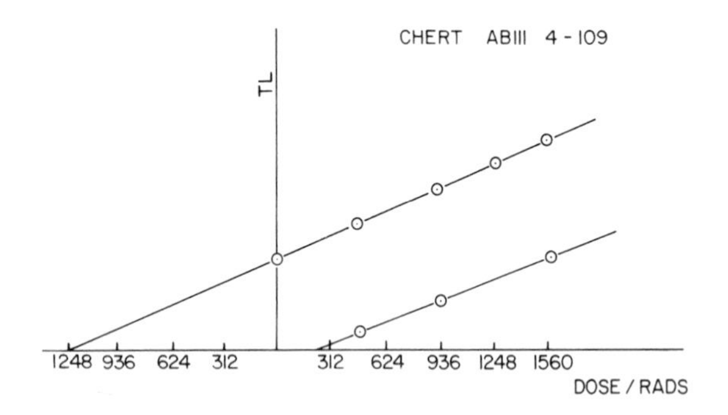

Figure 2. Determination of the archaeological dose of radiation to an archaeological sample, chert 4-109 from Alice Boër. As a result of the plateau-test (Figure 1), the TL in this sample was measured between 330°C and 400°C, with the additive radiation dose method (Aitken 1974).

Upper curve: increase of the TL level with laboratory β-doses (from 468 to 1560 rads) for aliquotes having kept their TLN.
The linearity of TL increase with dose indicate that we are still far from saturation. Extrapolation backwards gives the "equivalent β-dose" to which the sample was submitted since last firing ("first glow growth curve").

Lower curve: The artificial TL of aliquotes reirradiated after TLN + dose emptying is again linear with dose. Extrapolation backward to the zero TL level allows to determine a "supra linear" effect. The sum of the β-equivalent dose and supra linearity dose effect constitute the Archaeological Dose of equation 1 ("second glow growth curve").

bed of the Rio da Cabeça, lies directly on bedrock composed of more or less altered silts. This layer, rich in gravels and blocs, contains a primitive tool industry of two different traditions. One of these consists of elaborated artifacts of elongated flakes, not laminates, probably obtained by indirect percussion and carefully retouched. The second tradition is represented by more rudimentary lithic tools, including one "chopping-tool," voluminous nuclei and poorly retouched massive flakes used as scrapers.

Layer IV above is a sterile level of river sands deposited in a torrential regime and lacking internal stratification. Layer III, with a total thickness of about 2 m, is the richest archaeological layer. It is composed of clayey sands characteristic of a humid climatic phase. Sedimentological evidence suggests that the sedimentation rates toward the base of this layer may have been considerably slower than in the upper levels. The limit between layers III and IV corresponds to an unconformity, the top of layer IV being an erosional surface.

Again, no clear stratigraphy is discernible within layer III. As a consequence, excavation was carried out by horizontal levels of an arbitrary thickness of 10 cm each. Nineteen levels were defined, all of them bearing chert artifacts. Several hundred artifacts were recovered, most of them in levels 7 and 8. The description and illustrations of the most typical objects have been given in Beltrão (1974). Down to level 10, both bifacially flaked stone projectile points and biface tools are present. The last biface fragment was discovered at level 16. Below, only unifacially flaked artifacts were found. Whether the absence of bifacially flaked points below level 10 and bifaces below level 16 are a statistical effect or genuine is not clear because very few tools were found in the lower levels.

The two upper layers of the site are culturally sterile. Layer II is composed of 1.5 m of colluvial deposits, capped by the present soil (layer I).

Radiocarbon dates were obtained on charcoal for four levels of layer III. Charcoal from levels 3, 5 and 8 gave similar ages (Table 2) which average 6090±43 B.P. One small charcoal sample from level 10 gave a radiocarbon age of 14,200±1150 B.P. (SI-1208). From regional considerations, it seems that the sedimentation of layer III might have started a little

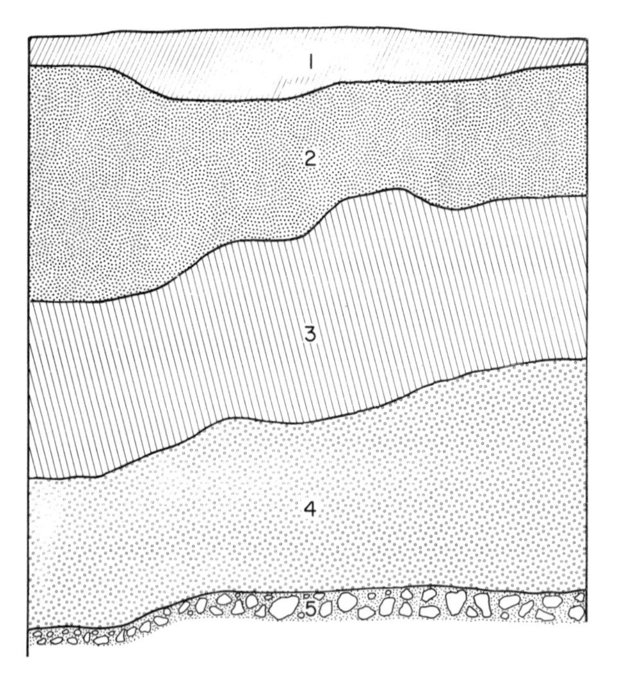

Figure 3. A typical stratigraphic column at Alice Boër. Layer I is the present soil. Layers II to V are described in the text. All contacts between layers II to V are erosional surfaces. Note the inclination of contact lines between layers. From field observations it seems that the sedimentation itself was more or less parallel to these inclined contacts. The total thickness of the outcrop is of the order of 4 to 5 m.

Table 1[1]. Classification of Alice Boër Cherts According to the Results From TL Measurements.

Saturated	Fading	Inhomogeneous	Anomalous	Prehistoric
AB III 1-10	AB III 8-132	AB III 2-53	AB III 8-62	heat-treated
AB III 7-502	AB III 8-213	AB III 4-47	AB III 12-5	AB III 1-1
AB III 8-200	AB III 10-37	AB III 5-78		AB III 1-11
AB III 9-6	AB III 13-37	AB III 7-37		AB III 1-22
AB III 9-13	AB III 15-3	AB III 7-387		AB III 4-53
AB III 9-40	AB III 16-4	AB III 8-8		AB III 4-109
AB III 10	AB III 19-1	AB III 13-16		AB III 7-212
AB III 10-1		AB III 14-4		AB III 8-339
AB III 10-36				AB III 8-340
AB III 11-4				
AB III 11-9				
AB III 11-34				
AB III 13-2				
AB III 14-11				
AB III 18-4				
AB III 18-13				
AB III 18-8				
AB III 19-2				

Saturated –TLN is saturated or near saturation. The equivalent dose is typically 10^4 to 10^5 rads.
Fading –Do not fulfill the plateau test of TL storage stability over archaeological time.
Inhomogeneous –Very poor reproducibility of TL measurements. Possibly partially heated samples.
Anomalous –Unexpected behavior of TL measurements.
Prehistoric Heat-Treated –The equivalent dose is in the range of 10^2 to 10^3 rads; fulfill the plateau test and can be dated.

[1] Modified from Danon et al. (1980).

Table 2. TL Ages of Burnt Cherts From the Upper Part of Alice Boër Layer III.

Stratigraphy	Sample	Archaeological[1] Dose Rads	TL age[2,3] yrs	Remarks[3]
Layer II				Sedimentation of layer II starts 3500 yrs
Layer III				C-14 ages
level 1	1-1	499±43	2200±280	
	1-11	533±3	2370±220	
	1-22	452±16	2000±200	
2				
3				6050±100(SI-1205)
4	4-53	646±87	2870±450	
	4-109	1435±29	3400±200	
5				6135±160(SI-1206)
6				
7	7-212	1534±150	6350±1220	
8	8-339	2468±2	10,970±1020	6085±160(SI-1207)
	8-340	2464±6	10,950±1020	
9				
10				14,200±1150(SI-1208)

[1] error calculated from regression line (see Figure 2) calculation.

[2] includes error on annual dose calculation as given in text. This does not take into account a possible variation with time of the external dose due to variable soil water content.

[3] C-14 and TL ages are given ±1 sigma.

less than 3500 years ago (Beltrão 1974). The beginning of the deposition of layer V was estimated by the same author, from sedimentological arguments, to be as early as about 20,000 years.

Burnt chert artifacts were present at all levels in layer III. On the contrary, no evidence of fire treatment was found on the tools from layer V. We have therefore concentrated our efforts on the layer III artifacts. Only chert artifacts showing signs of archaeological fire treatment were selected for TL dating. Special attention in the sampling procedure was given to samples from level 10 and below (i.e., to levels presumably older than ca. 14,000 years). Finally, cherts representative of all 19 levels (except levels 6 and 17) were studied. As far as we could judge from hand specimens, the chert material was homogenous through all levels and apparently similar to the chert blocs now available in the nearby Rio da Cabeça bed, presumably the source of the archaeological fragments. Among the archaeologically burnt specimens, those presenting a reddish color on some to all of their external surface (an indication of strong fire treatment), were especially selected. Our results are given in the next section.

TL AGE DETERMINATIONS

Forty-three chert artifacts showing evidence of fire treatment were selected for this study, and their TL measured with a coarse grain technique (Danon et al. 1980). The TLN of all cherts represent a single peak, centered in our heating conditions (10C/s) at 375°C. The TLA presents another peak centered at 130°C.

Of the 43 cherts, 34 were rejected for TL dating, either because of saturation (18 samples) and/or presence of anomalous fading (7 samples), non-reproducible TL signal (8 samples) or anomalous TL behavior (2 samples). The cherts with saturated TL (Figure 4) correspond to samples archaeologically unheated. They were found at all levels, especially from level 7 downwards (Table 1) and correspond to samples that were never archaeologically heated enough to remove their geological TL. In effect, their "beta-equivalent dose" at high temperature corresponds to saturation values of 10^4 to no more than 10^5 rads (Figure 5), which would require irradiation times in the Alice Boër site conditions of 60,000 to 500,000 years, which would contradict both geological and radiocarbon data.

Eight cherts are unsaturated and present a good plateau at temperatures greater than 300°C. Fading tests showed no anomalous loss of TL over a period of three months. The archaeological doses

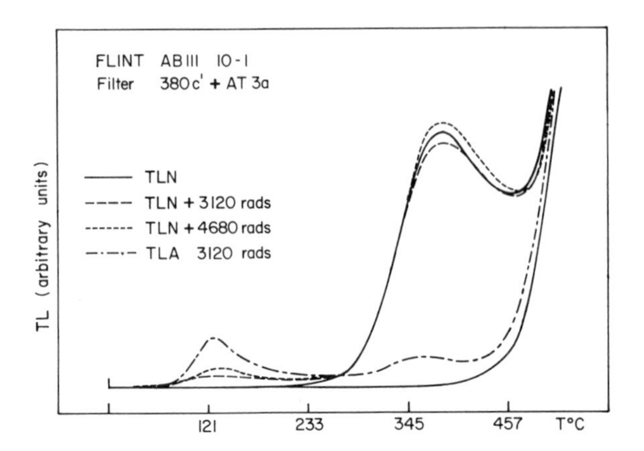

Figure 4. A typical saturated chert from Alice Boër: even the addition of strong β-radiation doses to the natural TL (up to nearly 5000 rads) do not modify the TL answer. All high temperature electron traps were already occupied. For comparison the TLA produced by a 3120 rads laboratory dose is shown.

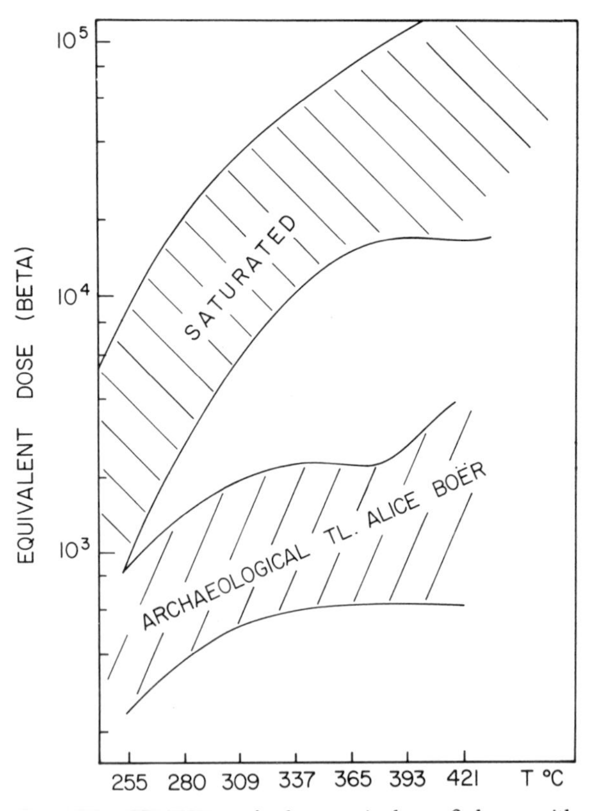

Figure 5. Variation of the equivalent ß-dose with temperature for the 26 Alice Boër cherts of columns 1 and 5 of Table 1. For temperatures above 250°C, the saturated and unsaturated cherts fall within two discrete areas of this diagram. Similar results for a population of cherts from various origins were obtained also by Melcher and Zimmerman (1977).

and annual dose rates of these samples were determined with conventional procedures to ascertain their TL age.

Archaeological Dose

The Archaeological Dose rates of the unsaturated cherts were determined from the TL lecture of powder aliquotes (100-150 μm grain size) of the samples with the additive dose method. Supra-linearity effects are negligible for most samples.

The Archaeological Dose received by the samples (Table 2) is the sum of the beta-equivalent dose + supra-linearity dose effect. The major source of error in the AD comes from the supra-linear dose evaluation. This is why the older the sample and the smaller its supra-linearity, the better the precision on the AD. It may also be seen from Table 2 that the AD increases as expected with depth below surface. An apparent anomaly occurs at level 4, where two samples show widely differing AD. It will be shown below that this is simply a consequence of their widely different internal radioactivity.

Annual Dose rates

The annual radiation dose received by the cherts comes from their own radioactivity and the environmental flux of ionizing particles. The latter, composed of cosmic ray particles plus gamma rays of the soil radioactive isotopes, was measured from TL dosimeters ($Ca\ F_2$) buried for about one year in various places in the Alice Boër excavation. No significant variation of the dose from place to place was observed and we have adopted for the environmental dose an annual value of 200 $\pm$ 25 mrads. This value is also the average value determined from many soils in São Paulo State by Watanabe (personal communication)

The internal dose in individual cherts has to be calculated from their U, Th and K contents. However, due to the small size and low radioactivity of our samples, individual determinations in our gamma-counting system were not possible. Therefore, groups of samples (up to 7) were measured together. Thorium and potassium were constantly below detection levels, setting upper limits for these elements of $<$0.1 ppm and $<$100 ppm, respectively. Uranium was on the average of 0.6 ppm, but fission-track uranium content determinations revealed large variations from chert to chert on the range of .2 ppm to 1 ppm. Uranium fission-track mapping also showed that the distribution of uranium within cherts was very homogeneous, fission stars (corresponding to ponctual concentrations of uranium) having been observed only rarely (and only in a few cherts). The U content determinations by the fission-track method were made on small fragments and it is not certain they are representative of the whole cherts. Therefore, the annual dose rates were calculated as the average of two extremes, assuming a content of radioactive species of either (i) 0.2 ppm of μ, with no Th or K; (ii) 1 ppm U + 0.1 ppm Th +100 ppm K.

Proper determination of the annual dose requires that the relative efficiency of alpha and beta particles for producing TL is known. Huxtable (1980) found, for flints from four different sites, a k-alpha relative efficiency factor of 0.094 ± 0.03 (but with individual values extending from 0.057 to 0.14). Similar results are given in Valladas (1980). The k-alpha factor for one of our cherts was determined by Steve Sutton at Washington University to be 0.11. We have thus adopted a value of 0.10 for the calculation of internal radiation doses, using Bell's (1979) tabulation for U, Th and K annual dose rate data. Assuming that both U and Th radioactive series are at equilibrium, without loss of radon or thoron, we calculated an annual internal dose rate of 25 ± 10 mrads.

The total annual dose rate received by the Alice Boër cherts amounts therefore to an average value of 225 ± 21 mrads/year. The very large (90%) contribution of the environmental dose, the large uncertainties of the internal dose, and in a certain measure the K-alpha factor, do not affect critically the total annual dose. The average value arrived at above was used for all cherts except sample 4-109. In effect, sample 4-109 was found, from fission-track analysis and gamma-spectrometry, to contain significantly more uranium than others. With 5.2 ± 0.5 ppm U, and Th and K respectively, less than 0.05 ppm and less than 80 ppm, the dose rate of the sample was taken at 422 ± 23 mrads/year.

TL ages

The TL ages, calculated from equation (1) for the eight dated cherts, are reported in Table 2. The TL ages range from an average of 2190 ± 185 B.P. at level 1 to about 11,000 B.P. at level 8. For those layers where two or three samples could be dated, the TL ages of cherts are, in general, remarkably concordant, as expected for samples having had a similar environmental and cosmic ray exposure history. (Remember that in the Alice Boër cherts, the external dose accounts for ~90% of the Archaeological Dose). Finally, it can be noted that the increase of TL age is correlative with stratigraphic depth.

Still, only the upper half of layer III could be dated. In effect, it is remarkable that only cherts from levels 1 to 8 were found to be convenient for TL dating: nearly half of the 20 samples from these levels were categorized as "saturated" (Table 1). On the contrary, the 23 cherts from levels 9 to 19 (the deepest level in layer III) were mostly (15 samples) in the latter group, and none was burnt to temperatures high enough to cure totally the geological TL. As throughout the whole stratigraphic column we have kept constant our sampling criteria (from visual observation), we are led to conclude that below level 8 (i.e., before about 11,000 years ago), none to possibly a very few chert artifacts were strongly heated in hearths at the excavated site. Earlier, we have (Danon et al. 1980) suggested that among the possible explanations to this observation can be included either a change in fire technology or a change in utilization of the site. Whatever the actual answer, it is a fact that heating of cherts at temperatures above 380°C was absent before 11,000 B.P.

DISCUSSION

Chronology of the Alice Boër Site

In order to discuss the overall chronology of the Alice Boër site, we have reported in Table 2, along with our TL measurements, all previously available data. From the comparison of columns 4 and 5 from Table 2, it turns out that the time-scale defined by thermoluminescence dating of the upper levels of layer III is in general agreement with other (geological and radiocarbon) scales. Still, some discrepancies occur between TL and radiocarbon ages at levels 3-4 and 8.

In agreement with regional geological data, Beltrão had conservatively estimated that the beginning of sedimentation of layer III had begun less than 3500 years ago. A lower limit to the age of this sedimentation can be set by a drastic climatic change toward an arid regime that took place some time between 2400 and 2700 years ago. The dating of chert artifacts from the top level of layer III at 2190±185 B.P. (error standard on the mean age of three cherts) is in satisfying agreement with the above boundaries. Similarly, the TL ages of 6350±1220 B.P. for level 7 is concordant with the average radiocarbon age of 6090±43 B.P. for levels 3 to 8. Finally, the two remarkably concordant TL ages of 11,000 B.P. for level 8 are significantly older than the above dates and younger than the radiocarbon age of the underlying level 10.

The only TL data apparently incompatible with the radiocarbon ages are the two concordant TL ages for level 4, apparently "too young" by about 2000 years. To understand this discrepancy, it is important to remember that the archaeological excavation levels were defined purely geometrically as horizontal 10 cm thick layers, while contacts between layers suggest a slightly inclined stratigraphy (Figure 3). Moreover, some parts of the site could not be worked conveniently and were not investigated due to obvious stratigraphic disturbance by burrowing animals. In the absence of other evidence, we tend to believe that the slight but significant differences between radiocarbon and TL ages at level 4 might result from either the obliquity or irregularity of stratigraphy over the "geometric" definition of sampling or sediment disturbance by nearby (unseen) animal reworking.

In this respect, it is interesting that in a subsequent excavation at Alice Boër, a few meters from the one described here, one burnt chert artifact, sample 54, from a topographic level (nivel 1.30) roughly equivalent to level 5 of layer III yielded a TL age of 3800±350 B.P. quite compatible with those of chert 4-109, of 3400±200 B.P.

All the TL ages have been calculated by assuming that the present level of soil humidity was representative of average conditions since the beginning of burial of the chert artifacts. The agreement of TL ages with other ages (i) at the top of layer III; (ii) at levels 5-7 with radiocarbon ages, suggest that this hypothesis is not an unreasonable one and that the overall error in our TL ages (including radioactive source calibration and other systematic errors) is of the order of 10%.

The TL chronology at Alice Boër appears therefore to be in *essential agreement* with the radiocarbon time scale, and suggests that the deposition of the earliest flaked artifacts at Alice Boër at the bottom of layer III and especially in layer V may have started significantly earlier than 14,000 years ago. In fact, from geological evidence, Beltrão (1973, 1974, 1978) suggested that the deposition of layer V occured more than 20,000 years ago and possibly as long as 40,000 years ago. From the coarseness of sedimentation products (including rocky blocs), it may be inferred that the deposition was made under a high energy regime. As a consequence, the tools in this layer cannot be considered as *in situ*, and therefore the estimated age of 20,000 to 40,000 years must be considered only as a *minimum* age for the two different artifact cultural traditions found in layer V.

Comparison with other sites

Both leaf-shaped and contracted-stemmed points are found in the lithic industry at Alice Boër at various depths in layer III, from levels 1 to 10 for the stemmed points and 4 to 9 for the leaf-shaped points. Bifacially flaked stone projectile points are known in a number of places in South America.

Following Hurt (this volume), similar contracted stemmed points occur by about 8000 radiocarbon years B.P. at El Inga (Ecuador) and at Las Casitas (Venezuela). The results at Alice Boër indicate that both types of points were present in central Brazil by about 14,000 years ago and that their production without major stylistic change continued up to about 2000 years ago.

Another kind of tool was found in layer III. From layer V to layer III (up to level 3), Beltrão (1978) identified a series of burins (identification confirmed by J. Tixier), of which the more elaborate ones were found in the upper levels. There is therefore a suggestion here that, as in Europe, the stylistic evolution through time of burins could serve to establish a relative chronology of sites from central Brazil. Accordingly, careful typological studies of burins and further absolute dating at Alice Boër might be of paramount importance.

Another site, Itaboraí (state of Rio de Janeiro) being excavated by one of us (see report in this volume by M.C. de M.C. Beltrão), might also shed some light on the early occupation of southeastern Brazil. Within a stratigraphic section of about 7 m, this site has two archaeological layers with a lithic industry (typologically different from that at Alice Boër), interbedded between sterile deposits. The stratigraphy at Itaboraí is reminiscent of that at Alice Boër, in that it suggests that here too some archaeological materials might have been deposited more than 14,000 years ago. Intensive work on this site is in progress, and excavated burnt stones associated with charcoal fragments will be analysed by radiocarbon and TL methods.

Evidence for the early presence of man in northeastern Brazil is presented by Guidon in this volume. Three sites in the state of Piaui have been radiocarbon dated 17,000 or more years, and one of these has three dates greater than 25,000 B.P. Lapa Vermelha, near Belo Horizonte in Minas Gerais State, has yielded quartz flakes in levels dated between 15,000 and 25,000 B.P. (Prous this volume). The possibility that the Alice Boër site was occuppied by 20,000 B.P. and that occupants made bifacial projectile points by 14,000 B.P. seems reasonable given this supporting evidence for the early occupation of Brazil.

CONCLUSION

Ten years ago, evidence for an early (more than 12,000 years ago) arrival of man in Brazil was very faint. The Alice Boër site and its radiocarbon dating suggested strongly that early man was present in central Brazil by 14,000 years ago, and probably significantly earlier. The TL chronology established from the dating of archaeologically burnt chert artifacts for the upper part of the major cultural layer (layer III) confirms the radiocarbon time scale. These results demonstrate clearly that a lithic industry characterized by contracted-stemmed points was already in place by about 14,000 B.P., and persisted for more than 11,000 years. Rather inexplicably, no chert artifacts were found sufficiently reheated in the deepest half of layer III to allow TL dating below the levels attributed to a radiocarbon age of $14{,}200 \pm 1000$ years. Whether this unfortunate (for TL dating purposes) lack of reheating of the most deeply buried chert artifacts is a consequence of some change in fire technology or for some other reason is still unresolved.

Other sites (Itaboraí in the state of Rio de Janeiro, Lapa Vermelha in Minas Gerais, and three recently radiocarbon dated sites in Piaui) support the evidence from Alice Boër that early man was present in Brazil significantly earlier than 12,000 B.P. Although much archaeological and geochronological work remains to be done at these sites, there are now at least five sites in Brazil at which evidence exists in favor of the presence of man more than 17,000 years ago.

Precisely dating these and similar sites in Brazil, and elsewhere in America, where now a number of quite ancient localities is suspected (cf. Bryan 1978) will be a major task of the next few years. Recent developments of existing techniques, as radiocarbon (with the use of accelerator dating) and TL (with the use of new datable materials, as calcite and chert), as well as the emergence of new methods (as electron spin resonance dating), open fascinating possibilities in this respect.

ACKNOWLEDGEMENTS

The authors wish to thank Steve Sutton for his help in determining the relative alpha and beta efficiency in the production of TL, and Dr. A. Rivera, who kindly took some time from an already overburdened schedule to determine by gammametry the U, Th and K abundances of our chert collection. They are indebted to A.L. Bryan, whose careful reading of the manuscript and suggestions led to significant improvements of this article. One of us (G.P.) thanks the Centre National de la Recherche Scientifique (France) for partial financial support.

REFERENCES CITED

Aitken, M.J.
1974 *Physics and archaeology.* Clarendon Press, Oxford.

Aitken, M.J., and S.I. Fleming
1972 Thermoluminescence dosimetry in archaeological dating. In *Topics in radiation dosimetry,* edited by F.H. Attix, pp. 1-78. Academic Press, New York.

Aitken, M.J., and J. Huxtable
1980 Accuracy of thermoluminescence dates. *Nature* 286: 911.

Aitken, M.J., and A.G. Wintle
1977 Thermoluminescence dosimetry in archaeological dating. In *Topics in radiation dosimetry,* edited by F.H. Attix, pp. 1-78. Academic Press, New York.

Bell, W.T.
1979 Thermoluminescence dating: radiation dose-rate data. *Archaeometry* 21: 243-245.

Beltrão, M.C. de M.C.
1969 Identifiçao do estágro litico superior no Brasil, *Anais do Terceiro Simpôsio de Arqueologia da ârea do Prata, São Leopoldo*: 4.

1973 Dataçoes pre-historicas mais antigas do Brasil, Academia Brasileira de Ciencias, FUNAI, *Boletim Informativo* II(7): 58-62.

1974 Dataçoes arqueologicas mais antigas do Brasil, *Anais da Academia Brasileira Ciencias* 46: 211-251.

1978 *Pre-historia do estado do Rio de Janeiro.* Ed. Forense Universitaria/SEEC, Rio de Janeiro.

Bryan, A.L. (editor)
1978 Early man in America from a circum-Pacific perspective. *University of Alberta, Department of Anthropology, Occasional Papers* No. 1, Edmonton.

Crabtree, D.E.
1964 Notes on experiments in flint knapping: 1, heat treatment of silica materials. *Tebiwa* 7: 1-6.

Danon, J., C.R. Enriquez, E. Zuleta, M.C. de M.C. Beltrão and G. Poupeau
1980 Thermoluminescence dating of archaeologically heated cherts. A case study: the Alice Boër site. *PACT Journal.*

Goksu, H.Y., J.H. Fremlin, H.T. Irwin, and R. Fryxell
1974 Age determination of burned flint by a thermoluminescence method. *Science* 181: 651-654.

Huxtable, J.
1980 Fine grain thermoluminescent (TL) techniques applied to flint dating. *PACT Journal.*

Inizan, M.L., H. Roche, and J. Tixier
1975 Avantages d'un traitement thermique pour la taille des roches siliceuses. *Quaternaria* 19: 1-18.

Meis, M.R.M, and M.C. de M.C. Beltrão
1981a Sedimentacão fluvial neoquaternaria em Rio Claro, SP: Sitio arqueologico de Alice Boër. IV Symposio Quaternario no Brasil, julha 1981, *Publicacão Especial* 1: 27-31. Abstract. Rio de Janeiro.

1981b Nota previa sobre a sedimentação neoquaternaria em Alice Boër, Rio Claro, SP. To be published by Comissão Tecnico-Cientifica do Quaternario.

Melcher, C.L., and D.W. Zimmerman
1977 Thermoluminescent determination of prehistoric heat treatment of chert artifacts. *Science* 197:1359-1362.

Valladas, H.
1980 La thermoluminescence de pierres de foyers prehistoriques: essai de chronologie. These 3ème cycle, pp. 151. Universite Paris I, Paris.

Wintle, A.G., and M.J. Aitken
1977 Thermoluminescence dating of burnt flint: application to a Lower Paleolithic site, Terra Amata. *Archaeometry* 19: 111-130.

The Cultural Relationships of the Alice Boër Site, State of São Paulo, Brazil

WESLEY R. HURT
Department of Anthropology
Indiana University
Bloomington, Indiana 47405
U.S.A.

Abstract

At the Alice Boër site, Rio Claro Basin, State of São Paulo, Brazil, there is evidence of a cultural complex comprised of simple, unifaced, core and flake artifacts that in their technique of manufacture resemble those of the Edge-Trimmed Tool Tradition of northwest South America. Although this assemblage is undated at the Alice Boër site, dates from higher levels indicate that it may be more than 14,000 years old. This industry is separated by a sterile stratum (Bed IV) from Bed III, which contains in its lower half a continuation of the unifacial industry. In the upper half of Bed III, beginning with Level 10, there are bifacial pressure retouched contracted stemmed projectile points as well as leaf-shaped points that resemble points from El Inga II types from Ecuador, and Las Casitas and Canaima complexes of Venezuela. The contracting stemmed points from the Alice Boër site appear to be older than the estimated dates from these other cultural complexes, for a radiocarbon date from Level 10 produced an age of ca. 14,000 years; while in Level 8, 20 cm above, a thermoluminescent date from fire-hardened chert produced a date of 11,000 years.

INTRODUCTION

One of the oldest radiocarbon dated sites of southcentral Brazil which has been described in the literature is the Alice Boër site, State of São Paulo. The site was discovered in 1961 and a report on excavations was presented for the first time in the XXXVI Congreso Nacional de Americanistas, 1964 (Beltrão 1966). Bryan and Beltrão (1978) published a summary report on this site giving further details. Although the lithic industries of this multi-component site have unique features, they also have traits shared with other Paleoindian sites, not only in eastern Brazil, but in western South America. The geological significance and techniques for dating this site have been given in other papers (Meis and Beltrão 1981; Beltrão et al., this volume). This report will discuss primarily the cultural relationships, and more particularly the observations I made in 1980-81 on the collections of lithic material housed in the Museu Nacional in Rio de Janeiro.

ARTIFACTS FROM BASAL UNIT (BED V)

In a publication by Beltrão (1974), descriptions and photographs were given of a series of lithic objects, considered to be tools, from the Alice Boër site. These objects came from two distinct stratigraphic units, Bed III and the surface of Bed V. Those objects from the most recent unit, Bed III, are unquestionable artifacts, for they contain universally recognized types such as contracting stemmed (tanged or pedunculated) and leaf-shaped projectile points. Extreme caution, however, must be maintained in attempting to segregate the much simpler, possibly man-made tools from naturally-formed objects resulting from accidentally battering chert cobbles as they were transported by water in stratigraphic Unit V, the bed of an ancient river. Nevertheless, some of the latter lithic objects meet more than one criterion of man-made tools. Obviously, the more of these criteria that are present in a particular object, the more certain that they were made by man; and in this paper only those objects which meet all or most of these criteria will be discussed.

I do not accept all of the specimens from Bed V, considered to be lithic artifacts by Beltrão (e.g., Beltrão 1974: Figures 36, 39). However, I consider several specimens from Unit V to be man-made tools because they have the following characteristics: (1) *Repetition of morphology* — three of the specimens from Bed V and one from Bed III are ovoid flakes which have a rounded projection on one side resulting in a rough keyhole or diamond shape (Figures 1, 2; Beltrão 1974: Figures 33, 37, 40); (2) *Multiple and secondary flaking* — on the specimens considered as man-made tools described above, as well as on a blade scraper (Beltrão 1974: Figure 38; Bryan and Beltrão 1978: Figure 9) the dorsal sides of which exhibit large percussion flaking while the edges have retouching or secondary flaking; (3) *Sharp projections along the edges* — when cobbles batter each other in a river bed it is to be expected that the edges would be rounded rather than have sharp projections as do some of the objects considered by me to be artifacts (e.g., a triangular flake from Bed V that has at the tip a typical flake removed in the style of a burin; in addition, some of the waterborne cobbles should have ground edges, a trait not present in specimens considered as artifacts); (4) *Bulbs of percussion and striking platforms* — on complete man-made flake and blade tools one or both characteristics are present, for example on the blade scraper from Bed V (Beltrão 1974: Figure 38).

COMPARISONS WITH OTHER SOUTH AMERICAN ASSEMBLAGES

Beltrão estimates the age of Unit V as ca. 20,000 - 40,000 years on the basis of dates from the more recent Unit III. In Brazil, no stone industry has as yet been found identical with that of the surface of Unit V; but in techniques of manufacture, there are similarities to the stone industries of the Itaboraí site in the State of Rio de Janeiro and with the Lapa Pequena site in Minas Gerais (Bryan and Gruhn 1978). These techniques include: (1) large flaking confined to one side, although there may be small occasional flaking or retouching on the opposite side; (2) flaking confined mainly to the working edge of the artifact; (3) minimum alteration of the basic form of the flake or core with the result that there is no overall flaking of the artifact; and (4) absence of stone projectile points. In addition, at the Lapa Pequena site, the ovoid flakes with the

Figure 1. Keyhold-shaped artifact from Bed IV.

Figure 2. Broken ovoid from Bed V.

rounded projection are represented by several specimens (Bryan and Gruhn 1978:Plate IV). At the Itaboraí site, the stone instruments show the same charactertistic but the exact forms are different. Absent at this latter site are the T-shaped tools and blade tools. On the other hand, steep scraper-planes and spokeshaves, characteristic of the Edge-trimmed Tool Tradition of northwest South America (Hurt et al, 1976: 20), are present at Itaboraí but have not been found at Alice Boër.

Assuming that Beltrão's (1978) estimate of a minimum age of 20,000 years, with the possibility of 40,000 years, is correct, a simple unifacial stone technology of the type described above must have endured for a long period of time in certain areas of eastern Brazil. At the Lapa Pequena Rockshelter, the major period of occupation occurred between 8240±160 and 7590±100 years (Bryan and Gruhn 1978:268). These dates are similar to that of 8100±75 B.P. (DIC-2272) taken from a hearth in the Itaboraí site (discovered by Beltrão, 1974:219; 1978: 131, 209-212) in the 1981 field project of the Museu Nacional and the Indiana University Museum. This age, however, should be considered a minimum age, for the hearth lay above a colluvial stone line with which the artifacts seem to have been associated.

The 12,500 B.P. age assigned to the closely-related Edge-Trimmed Tool Tradition of northwest South America (Hurt et al. 1976:20) also supports the theory that the Tradition began in the Late Pleistocene. In the oldest manifestation in the El Abra Rockshelters of the Sabana de Bogotá, radiocarbon dates indicate an age greater than 12,500 years. At these rockshelters, simple types of tools persisted until at least 1640 years ago when maize and ceramics of the Chibcha Indians were already present. Toward the end of the sequence, ground stone tools were added. In specific tool types, the El Abra Industry contained steep-nosed scraper planes and spokeshaves, a trait found at the Itaboraí site but not at the Alice Boër site. The presence of rectangular, percussion-flaked blade tools with edge flaking is a trait shared with the Tangurapa Industry of northwest Uruguay, surveyed in 1980 by the Museo de Historia Natural of Salto and Indiana University. Other types of uniface tools, such as large hook-shaped blade tools and large expanded-based drills, occur in the Tangurapa Industry but not at Alice Boër. Because the Tangurapa Industry is confined to the highest terraces of the northwestern tributaries of the Uruguay River, it seems to be older than an unnamed pressure-flaked biface industry, confined to the floodplains, that has been radiocarbon dated to a maximum of 11,000 years (Klaus Hilbert, personal communication, 1980). I also observed a T-shaped tool of the Alice Boër type in a collection of artifacts I made on a high terrace of the Loa Valley in the Atacama Desert of northern Chile.

ARTIFACTS FROM UPPER UNIT (BED III)

Bed IV, a sterile sand, separates Bed V of the Alice Boër site from Bed III, which contains artifacts. Bed III consists of about two meters of colluvium and river flood deposits, that change from fine grains at the base to coarser grains in the upper meter. A climatic interpretation derived from grain size is that the basal deposits indicate a regime with a general, well-distributed rainfall pattern which changes to torrential conditions in the upper layers. This interpretation, as well as the dates given below, support Bombin's (1976) conclusion that the climate during his *Pre-Atlantic Phase* (13,950 - 11,950 B.P.) of southern Brazil was characterized by a torrential rainfall regime. A charcoal sample obtained from the 90-100 cm level of Unit III produced a date of 14,200±1150 B.P., although the small size of the sample suggested that the probable error could be ±3000 years. As yet, no fire-altered stone artifacts suitable for thermoluminescence dating have been found in the level from which the radiocarbon date was obtained, but an artifact from the 80 cm depth (Level 8) produced a TL date of ca. 11,000 years (Beltrão et al., this volume). According to G. Poupeau (personal communication 1982), the precision of the TL dating is about ±10% at 1 sigma (i.e., 68% confidence level); in other words it could be about 1000 years too young. Further support that the 14,000 year date for the middle of Bed III is plausible is indicated by the 13,900 and 12,200 dates for lithic sites excavated by Guidon (1981) in the State of Piauí, northeast Brazil. (Older dates are announced by Guidon this volume — Editor).

As yet, the artifacts from Unit III of the Alice Boër site have not been described in detail; and the remarks that follow are based upon Beltrão's publication and upon personal observation. In the lower meter of Unit III were found a series of uniface flake scrapers and knives (Beltrão 1974: Figures 20, 30, 32, and 34), types which occur on so many sites of various ages that they cannot be considered diagnostic. Beltrão illustrates a single T-shaped tool (of the type from Unit V) from Unit III (Beltrão 1974: Figure 33). In the upper meter of Unit III, there was also the addition of bifacial pressure-flaked tools, including contracting stemmed points, which are not only of aid in tracing cultural relationships but also in establishing a local chronology. One tool, locally known as a "lesma" (Beltrão 1974: Figure 28), is made of an elongated blade, with triangular cross section, a pointed end, and a large amount of invasive flaking confined to the upper surface plus retouched edges. Judging by use marks on one side of one extremity, the instrument was held diagonally in the hand for use as a cutting instrument. To produce such a long blade, a prepared polyhedral core probably was necessary. This tool is also characteristic of the Paranaíba Phase of southeast Goias, dated between 10,750 and 9000 B.P. (Schmitz 1980: 192, this volume). In the upper half of Bed III, Beltrão also illustrates a probable burin from Bed V (Schmitz 1980: Figure 35).

DISCUSSION OF EARLY BIFACIAL PROJECTILE POINTS IN SOUTH AMERICA

Equally diagnostic of the Unit III level 9 (90-100 cm), where the C-14 date of ca. 14,000 years was obtained, are two widespread types of bifacially flaked projectile points, one with contracting stems, and the other a leaf-shaped form (Schmitz: Figures 1, 18). On the basis of known dates, this is the earliest occurrence of the contracting stemmed (tanged) point in South America. In the Cerca Grande Complex farther northeast in the State of Minas Gerais, the contracting stem type apparently has a maximum age of 10,000 years (Hurt and Blasi 1969). On the other hand, narrow lanceolate points have been found in the Taima-taima mastodon kill in northern Venezuela, with several radiocarbon dates ranging from 14,000±300 B.P. to 12,980±83 B.P. (Bryan et al. 1978). The El Jobo points at this site, however, are much thicker and narrower than the broad leaf-shaped points from the Alice Boër site. Both leaf-shaped and contracting stemmed points occur in the El Inga II Complex of Ecuador. If Willey's (1971) estimate for El Inga of 6000 B.C. (7950 B.P.) is correct, the examples from the Alice Boër site appear to be much older. The contracting stemmed point is also present in the Las Casitas and Canaima complexes of Venuezuela, estimated by Rouse and Cruxent (1963) to have an age of ca. 5500 B.C.

SUMMARY AND CONCLUSIONS

The Alice Boër site, São Paulo, Brazil, contains two components that apparently existed during the terminal phases of the Pleistocene, extending into the Holocene. The most ancient culture, associated with the surface of stratigraphic Unit V, seems to be older than 14,000 years and contains unifacial flake and blade tools. In the simplicity of their manufacture, these tools resemble those of the Edge-trimmed Tool Tradition of northwestern South America; that is, the flaking was usually confined to the working edge. In specific types of tools, however, there were great differences.

The unifacial industry continued in the lower half of stratigraphic unit III; while bifacial pressure-flaked tools were added in the upper half of Unit III. Major tool types include lanceolate and constricted-stemmed points that resemble those from sites in western South America.

REFERENCES CITED

Beltrão, M.C. de M.C. (formerly Becker)

1966 Quelques donnés nouvelles sur les sites préhistoriques de Rio Claro, Etat de São Paulo. *Congreso Internacional de Americanistas,* Actas I: 445-450. Sevilla.

Beltrão, M.C. de M.C.

1974 Datacões arqueologicas mais antigas do Brasil. *Anais da Academia Brasileira de Ciencias* 46(2): 212-251.

1978 *Pre-Historia do estado do Rio de Janeiro.* Instituto Estadual do Livro e Editora Forense-Universitaria, Rio de Janeiro.

Bombin, M.
1976 Modelo paleoecológico evolutivo para o Neoquaternário da região da Campanha-oeste do Rio Grande do Sul (Brasil) — A Formação Touro Passo, seu Conteúdo Fossilífero da Pedogénese Pos-deposicional. *Communições do Museu de Ci'encias da PUCRS* 15: 1-90. Porto Alegre.

Bryan, A.L., and M.C. de M.C. Beltrão
1978 An early stratified sequence near Rio Claro, east central São Paulo State, Brazil. In: Early Man from a circum-Pacific perspective, edited by A.L. Bryan. *Department of Anthropology, University of Alberta, Occasional Papers* 1: 303-305. Edmonton.

Bryan, A.L., and R. Gruhn
1978 Results of a Test Excavation at Lapa Pequena, MG, Brazil. UFMG, *Arquivos do Museu de Historia Natural* III: 261-320. Belo Horizonte.

Bryan, A.L., R. Casamiquela, J.M. Cruxent, R. Gruhn, and C. Ochsenius
1978 An El Jobo mastodon kill at Taima-taima, Venezuela. *Science* 200:1275-1277.

Guidon, N.
1981 Las unidades culturales de São Raimundo Nonato, Sudeste del Estado de Piaui. *Union Internacional de Ciencias Prehistoricas y Protohistoricas, X Congreso, Comisión* XII: 101-111. Mexico.

Hurt, W.R., and O. Blasi
1969 O Projeto Arqueologico "Lagoa Santa" — Minas Gerais, Brasil. *Arquivos do Museu Paranaense, Nova Serie, Arqueologia No. 4.* Curitiba.

Hurt, W., T. van der Hammen, and G. Correal Urrego
1976 The El Abra Rockshelters, Sabana de Bogotá, Colombia, South America. *Indiana University Museum Occasional Papers and Monographs,* No. 2. Bloomington.

Meis, M.R.M., and M.C. de M.C. Beltrão
1981 The Alice Boër site: lithostratigraphic background. *Union Internacional de Ciencias Prehistoricas y Protohistoricas, X Congreso, Comisión* XII: 99-100. Mexico.

Rouse, I., and J.M. Cruxent
1963 *Venezuelan Archaeology.* Yale University Carribean Series No. 6. Yale University Press, New Haven.

Schmitz, P.I.
1980 A Evolucão de Cultura no Sudoeste de Goias, Brasil. *Pesquisas, Antropologia* no. 31, Instituto Anchietano de Pesquisas, São Leopoldo.

Willey, G.R.
1971 *An Introduction to American Archaeology,* Vol 2. South America. Prentice Hall, Englewood Cliffs.

Investigaciones Arqueológicas en el Sitio 2 de Arroyo Seco (Pdo. de Tres Arroyos — Pcia. de Buenos Aires — República Argentina)

FRANCISCO FIDALGO
Profesor Titular de Geomorfología
Facultad de Ciencias Naturales y Museo
1900 La Plata
ARGENTINA

LUIS M. MEO GUZMAN
Director del Museo Municipal "José A. Mulazzi"
7500 Tres Arroyos

GUSTAVO G. POLITIS
Becario del CONICET. Facultad de Ciencias Naturales y Museo
1900 La Plata

MONICA C. SALEMME
División Paleontología Vertebrados
Facultad de Ciencias Naturales y Museo
1900 La Plata

EDUARDO P. TONNI
Miembro de la Carrera del Investigador CIC Pcia. de Buenos Aires Facultad de Ciencias Naturales y Museo
1900 La Plata
Con la colaboración de:

JORGE E. CARBONARI
GABRIEL J. GOMEZ
ROBERTO A. HUARTE
Consejo Nacional de Investigaciones Científicas y Técnicas.

ANIBAL J. FIGINI
Facultad de Ciencias Naturales y Museo
1900 La Plata

Abstract

The archaeological locality of Arroyo Seco is situated near the city of Tres Arroyos, Buenos Aires Province, Argentina, at long. 60° 14′ 39″ W and lat. 38° 21′ 38″ S. Site 2 is an open-air, multicomponent site located on a small hill at the right bank of the Primer Brazo de los Tres Arroyos, also called Arroyo Seco.

Several partial hypotheses are postulated based upon evidence from different disciplines (geology, paleoethnozoology, archaeology, isotopic geology). These are then put together into general statements at different levels.

1. The geological and pedological stratigraphy (Figure 3) is subjected to interpretation.

Figure 4, Profile 1 represents the most probable interpretation of the stratigraphy because it appears to be vertified by the faunal and archaeological evidence, as well as by the radiocarbon dates.

Stratigraphic units A and B are composed of sediments deposited under more arid conditions than those existing today in the same area. This interpretation has been verified not only by the faunal analysis (presence of Patagonic and Central Argentinian elements) but also by the geological record (dominance of eolic sedimentation).

A similar situation to Profile 1 has been observed and described in other places. For example at Lobería (Fidalgo and Tonni 1981), the extinct fauna has been found only below the uncomformity plane, that is, in the lower lithostratigraphic unit, corresponding to the lower part of Unit Y.

2. The Arroyo Seco site was initially occupied by hunters, probably hunter-gatherers, who used the tools which form the lower component, found in the lower part of Unit Y and Unit S (Figure 3). These tools are mostly quartzite flakes with unifacial marginal retouch. The lithic remains were found associated with bones of modern animals and extinct Pampean megamammals. A primary adaptation to guanaco and deer hunting as the economic basis for their subsistence can be inferred from the great abundance of remains of these animals. Probably some extinct species were also eaten, although they were only occasionally hunted.

3. At the present time, we think that at least some of the human skeletal remains from the top of Unit Z belonged to the hunters who made the flake tools of the lower component, and hunted deer, guanaco, extinct horse (*Equus*), and giant sloth (*Megatherium*). This hypothesis is based upon the fact that the burials and the lower component have common elements which were not registered in the other component, and no stratigraphic intrusion was detected.

Small portions of hematite were found within the lower component; some of them show one smooth side. Abundant powdered ochre was observed within the sediments which cover the human bones in burials 4, 6, 7 and 8.

A fragment of a marine mollusk shell, *Adelomelon (Pachycymbiola) brasiliana*, has been found within the lower component. The ornaments of burials 1, 4, 6, 7, and 8 were composed of some beads made of mollusk shells. Two articulated shells of *Amiantis* sp. were also registered in connection with one burial.

Abundant remains of extinct Pampean megamammals were recognized within the lower component. Within burial No. 8 a dermal plate of *Glyptodon* sp. was found.

All Pleistocenic faunal remains were found in Unit S and at the base of Unit Y (with the exception of the dermal plate of *Glyptodon* sp.), even in some grids without burials. Faunal remains and the lithic materials lying above the burials were indistinguishable from those belonging to the other sectors where human bones have not been recorded.

If the burials would have been made during the occupation represented by the middle or upper component, it should be expected that the burials would have disturbed the spatial distribution of the materials within the lower component. However, this situation has not yet been recognized.

Although some of the remains may not be *in situ*, evidence of significant disturbance was not detected, except for that produced by human occupation itself. No evidence was found to support the idea that human burials were intruded through the zone containing the megafauna. Most of the human skeletons are articulated and the horse metapodial which has been assigned to *Equus (Amerhippus)* sp. is also articulated with some of the autopodial bones.

4. The occupation represented by the lower component had elaborate funerary practices, as evidenced by the discovery of child burials with ceremonial ornaments (perforated canines, beads of mollusk shells, powdered ochre around the cranium); adult burials without ornaments; and multiple burials of two or three individuals.

5. Tools are also present in the upper part of Unit Y, the middle component. These tools are principally worked on quartzite, mostly with marginal unifacial retouch, but bifacial triangular projectile points are also present, as well as a larger percentage of tools associated with grinding activities (fragments of grinding stones and "manos").

The faunal remains associated with this level are mostly guanaco, deer and "ñandú" (*Rhea*); no remains of extinct megamammals, or European fauna have been found in this zone.

6. The economy of the aborigines who occupied the site after extinction of the megafauna was based on hunting and gathering as evidenced by the finding of fragments of grinding stones and "manos." No human skeletal remains were definitely associated with these later hunters and gatherers.

No suitable materials for radiocarbon dating have been found in this zone. Nevertheless, if the interpretation proposed in Profile 1 is the most likely, the tools from the lower and upper part of zone Y would be separated by a considerable time interval. The upper part of zone Y occupation must have preceded initial European contact.

7. Tools were also found in Unit X, the upper component. Incised brown pottery and tools worked on opal were found here in the humus which suffered a major disturbance due to ploughing in recent years. Surface collecting has further reduced the value of this last aboriginal assemblage. Post-hispanic contact goods were also found. It is difficult to determine if there is evidence of Hispanic and aboriginal contact, or, whether there is an artificial association of two diachronic settlements, one aboriginal and the other European.

8. The lithic tools found at the site do not represent great morphological variation. Basically, the tools are quartzite flakes of medium size with unifacial marginal retouch; other elements were added through time. Some types are very frequently found from the lower to the upper levels at other sites in southern Buenos Aires Province. These types include: single side scrapers, double side scrapers, double convergent-side scrapers, end scrapers with an extended active edge, and end and side scrapers. Triangular bifacial projectile points were added to this initial assemblage and the frequency of grinding tools was increased. Later on, pottery and the use of opal tools were added.

INTRODUCCION

El Sitio 2 de la Localidad Arqueológica Arroyo Seco presentó desde las primeras etapas de su investigación características excepcionales que motivaron un estudio amplio de sus componentes.

En efecto, desde el comienzo de las excavaciones la problemática complexiva evidenciada por la información que se iba exhumando, motivó la formación de un equipo interdisciplinario de investigación que permitiese concluir en un estudio integral del sitio.

La investigación se estructuró de manera tal que las disciplinas concurrentes tuviesen un estrecho intercambio de información, pero que a la vez sus hipótesis fuesen formuladas independientemente. Luego de confrontadas éstas, se reunieron en hipótesis generales.

En el presente informe se efectúa una exposición detallada y objetiva de la información reunida en el Sitio 2. Asimismo, se realiza una primera interpretación y se propone las hipótesis más ampliamente contrastadas.

Las investigaciones en el Sitio 2 de la Localidad Arqueológica Arroyo Seco están en pleno desarrollo y obviamente nuevos datos podrán confirmar o disconfirmar lo aquí expuesto.

En tal sentido, Dr. Alberto Marcellino está actualmente realizando el estudio del material óseo humano, cuyos resultados serán seguramente de singular importancia para esta investigación.

ANTECEDENTES

La Localidad Arqueológica Arroyo Seco fue visitada hace aproximadamente 50 años por el Sr. José Mulazzi; el material recolectado fue incorporado a su colección con la indicación: "Paradero No. 1."

En 1972, un grupo de entusiastas aficionados, los Sres. A. Elgart, A. Morán y J. Móttola, excavararon los sitios 1 y 2 mediante el método de cuadrículas y capas artificiales de 0,20 m. De esta manera abrieron aproximadamente 20 cuadrículas en cada sitio y obtuvieron gran cantidad de material arqueológico y faunístico. También exhumaron tres esqueletos humanos en el Sitio 1 y otros dos en el Sitio 2. Entre el material faunístico recuperado hallaron restos óseos de *Toxodon* y *Antifer*.

Debido a las características excepcionales que presentaba el sitio, las personas citadas decidieron solicitar asesoramiento al Dr. Alberto Rex Gonzalez, quien comenzó las excavaciones con una campaña efectuada en enero de 1977.

En estos trabajos de campo participó uno de los autores (GGP) y posteriormente el nombramiento de otro de los autores (LMG) como director del Museo Municipal de Tres Arroyos, dio origen a las tareas de investigación sistemática en el lugar.

Ante la necesidad de un estudio integral se conformó el equipo multidisciplinario en el que intervinieron los demás autores del presente informe.

En la provincia de Buenos Aires sólo se conoce un registro de asociación íntima de restos culturales y fauna pleistocénica: el yacimiento "Estancia La Moderna" en el partido de Azul (Palanca et al. 1972; Zetti et al. 1972; Palanca y Politis 1979). En el caso del yacimiento de Los Flamencos II, en el partido de Saavedra (Austral 1972), la asociación se registró en términos estratigráficos, lo cual no necesariamente implica coexistencia entre el artefacto y los restos de fauna extinguida.

Los hallazgos realizados por F. Ameghino y otros investigadores a fines del siglo pasado y a principios del actual (i.e.: zona costera de los partidos de Gral. Pueyrredón, Gral. Alvarado, Necochea) fueron objeto de importantes y en ocasiones bien fundadas revisiones, por lo tanto sólo poseen un interés histórico, ya que la información brindada es incontrastable y altamente aleatoria. Un resumen de éstos ha sido expuesto por Daino (1979).

Asimismo, los numerosos restos culturales hallados en la provincia de Buenos Aires, especialmente aquellos que provienen del área al sur del río Salado, fueron estudiados sistemáticamente a partir de 1950. En ese año se publicó el trabajo de Menghin y Bórmida, donde los autores postulan un esquema para el desarrollo cultural de la Región Pampeana Bonaerense. Posteriormente, los trabajos de Bórmida (s/f, 1960, 1962) y Sanguinetti de Bórmida (1966, 1970) continuaron esta línea de investigación. Austral (1965, 1971) y Madrazo (1968, 1972, 1973) propusieron diferentes esquemas interpretativos para la arqueología pampeana. Recientemente Orquera (1981) ha realizado un análisis crítico de la bibliografía existente y expone sus opiniones en

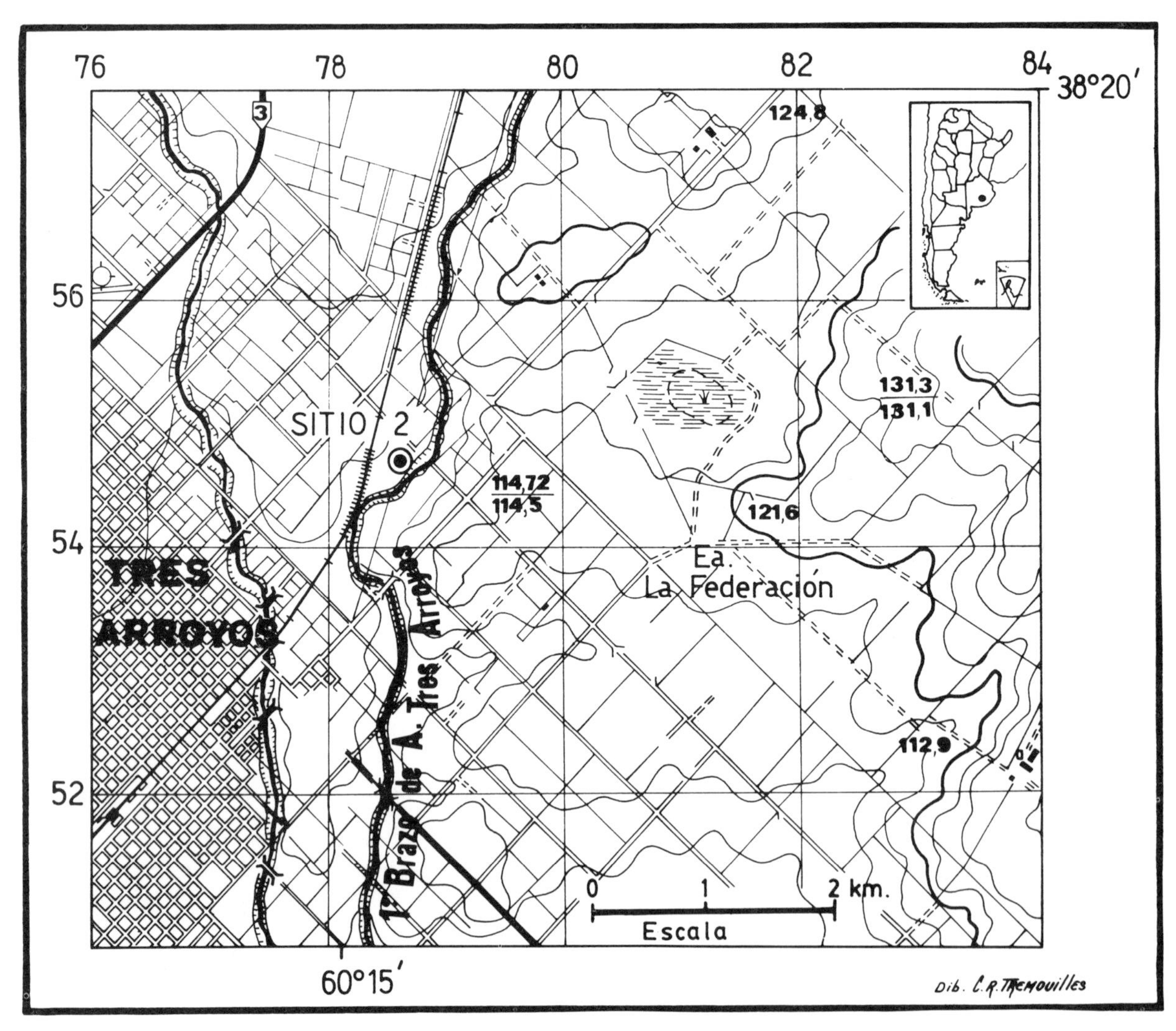

Figura 1. Mapa de ubicación.

algunos casos discordantes con las de los autores citados.

En suma, obviamente no hay consenso general acerca de la interpretación de los procesos culturales en la Región Pampeana Bonaerense.

AGRADECIMIENTOS

El Dr. A.R. González informó a los autores sobre la importancia de los hallazgos y realizó las primeras tareas sistemáticas de investigación en el Sitio.

El Dr. Rosendo Pascual y el Licenciado Carlos Aschero aportaron valiosos comentarios y observaciones. Al Licenciado Néstor Porro se deben los datos climáticos del área.

En los trabajos de campaña colaboraron las siguientes personas: María E. Albeck, Teresa Acedo, Marcelo Alvarez, Juan C. Benítez, Adriana Boedo, Laura Cáceres, Juana Campodónico, Graciela Canale, Aldo Cangiani, Flavia Carballo, Lili S. de Carbonari, Mónica Carminatti, Marcela Cid de la Paz, Sergio Echeverry, Nora Flegenheimer, Laura Gachón, Pablo Garcé, Jorge García, Inés Gordillo, Laura Güerci, Estela Mansur, Gastón Martínez, Alfredo Morán, Claudio Morán, Alejandrina Nieto, María L. Obregozo, Darío Olmo, Daniel Olivera, Cristina Scattolín, Hernán Vidal y Luis Zunino.

El personal técnico de la División Paleontología Vertebrados del Museo de La Plata, Gerardo Fabris, José Laza, Víctor Melemenis, Omar Molina y Juan Moly, realizó la preparación de la mayor parte del material óseo recuperado.

Luis Ferreyra y Carlos Tremouilles, del Museo de La Plata, se encargaron de la realización de las fotografías y dibujos, respectivamente.

Los trabajos de campaña fueron financiados por la Comisión de Investigaciones Científicas de la provincia de Buenos Aires y por la Municipalidad de Tres Arroyos.

A todos ellos el agradecimiento de los autores, exclusivos responsables de lo expresado en la presente contribución.

METODOLOGIA

El Sitio 2 fue excavado utilizando el método de cuadrículas de 2 m de lado, con una profundidad no superior a 1,50 m, límite hasta donde se detectaron los materiales culturales y faunísticos. Las cuadrículas se abrieron al Este del terreno excavado por Móttola, Elgart y Morán (Figura 2). Entre una y otra se dejaron testigos de 0,20 m de ancho.

Se excavó en capas artificiales de 0,05 m tomando las coordenadas tridimensionales de cada hallazgo. Cada elemento recuperado se guardó en bolsas donde se consignó: Localidad, Sitio, cuadrículas, unidad litoestratigráfica, tipo de material, nivel artificial y las coordenadas a las paredes N y E de la cuadrícula. La profundidad se tomó con respecto a un nivel 0 (cero) determinado por el punto más alto del sitio.

Los restos óseos, cuando estaban mal preservados, se limpiaron con pincel y se recubrieron con una capa de laca a la piroxilina, excepto aquellos utilizados para fechados radiocarbónicos. Todo el material óseo fue preparado para su estudio por los técnicos de la División Palentología Vertebrados del Museo de La Plata.

Los restos de infantes fueron cubiertos con vendas enyesadas y luego extraídos en bloque. Posteriormente se trataron en el laboratorio de preparación. El material óseo humano correspondiente a individuos adultos fue preparado en el Instituto de Antropología de la Universidad de Córdoba.

Los primeros 0,25 m de sedimento fueron extraídos a pala y pasados por zaranda de malla de 0,005 m, debido a que la parte superior del terreno estaba disturbada por sucesivos laboreos llevados a cabo años anteriores.

Para el análisis específico de cada uno de los aspectos de la investigación se utilizaron técnicas y métodos particulares que son explicitados en los correspondientes apartados de este informe.

UBICACION

La Localidad Arqueológica Arroyo Seco está ubicada en el partido de Tres Arroyos a los 60°14'39" de long. Oeste y 38°21'38" de lat. Sur. (Figura 1).

Esta localidad se encuentra en las vecindades del éjido urbano de Tres Arroyos en terrenos pertenecientes al Municipio y en la actualidad parcialmente a cargo del Tiro Federal de esa ciudad.

Los tres sitios ubicados están localizados en las cercanías del Primer Brazo de los Tres Arroyos o Arroyo Seco, en terrenos altos, próximos al curso de agua.

El Sitio 1 se encuentra 200 m al sur del grupo de construcciones del Tiro Federal, sobre la margen izquierda del arroyo citado.

El Sitio 2, motivo de la presente comunicación, está ubicado sobre una "lomada" 2000 m al oeste del grupo de construcciones del Tiro Federal, sobre la margen derecha.

El Sitio 3, se localiza en frente del Sitio 2, separado de éste por una pequeña laguna actual-

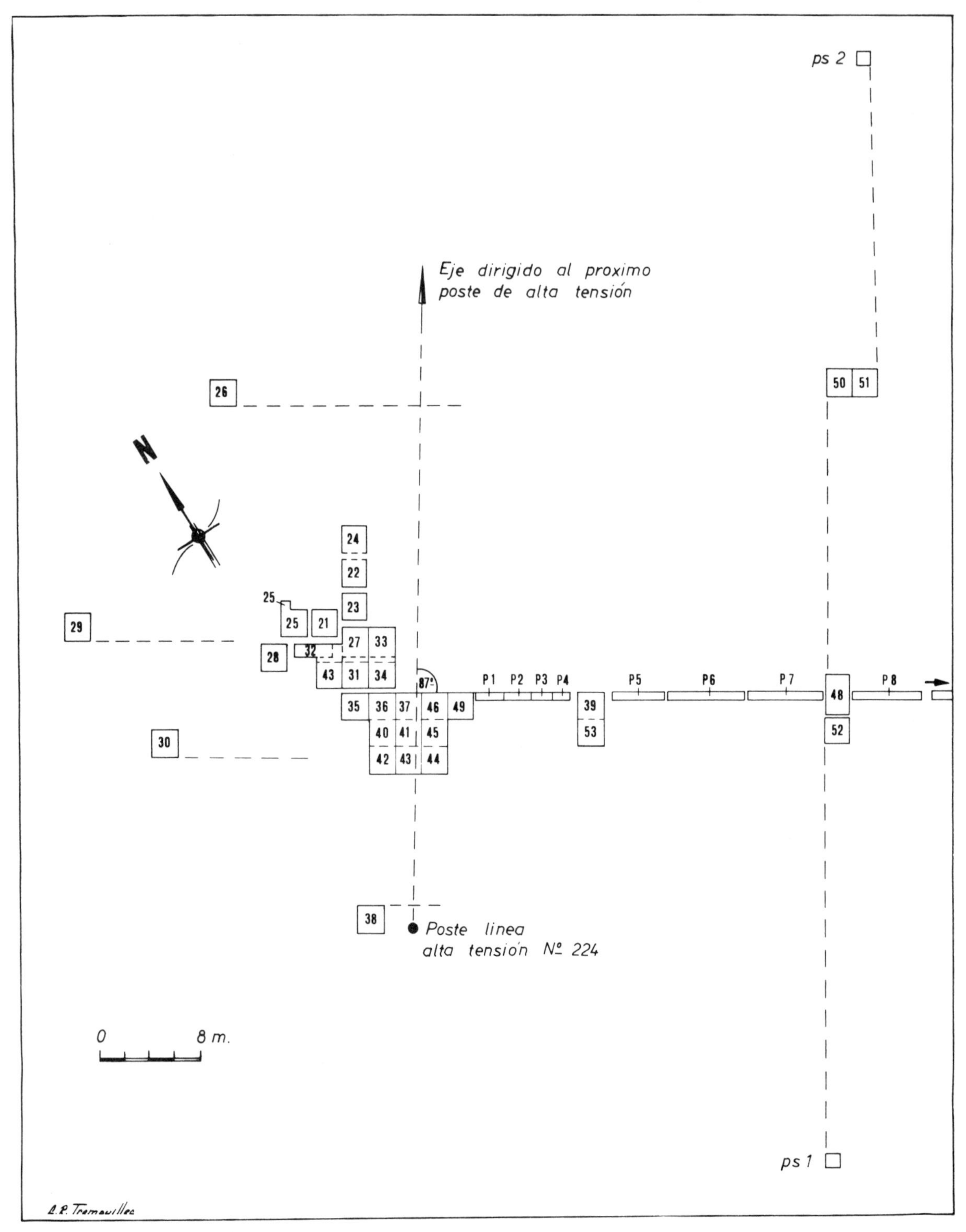

Figura 2. Planimetría del Sitio 2.

mente casi seca. Este sitio está en terrenos pertenecientes al Sr. Luis Fanucci.

La zona ocupada por estos sitios se encuentra en la Hoja Topográfica Estancia Tres Lagunas (3960-9-2), Escala 1:50.000 publicada por el Insituto Geográfico Militar.

CLIMA

Los datos meteorológicos corresponden a la estación Tres Arroyos (38°23'S y 60°16'W) para el período 1901-1950.

La temperatura media anual es de 13,99°C con un promedio para el invierno de 7,63°C y de 20,63°C para el verano. La precipitación media anual es de 693 mm, registrándose los máximos en los meses de otoño y de verano (75 mm en marzo y 74 mm en diciembre).

El clima, según el método de Thornwaite, corresponde al tipo C2 B'2 r a', es decir subhúmedo húmedo, mesotérmico con poco déficit y concentración estival de la eficiencia térmica $<48\%$ (N. Porro, comunicación personal).

BIOGEOGRAFIA

a) Zoogeografía: no se conocen datos precisos sobre la actual fauna de vertebrados del Partido de Tres Arroyos. Se han tomado como referencia los trabajos generales de Ringuelet y Arámburu (1957) y de Cabrera (1957/1960), incluyéndose asimismo datos de algunas especies en particular (i.e.: las citadas de Contreras y Reig 1965) y escasas observaciones visuales efectuadas durante el transcurso de las tareas de campaña.

El área se encuentra en el Dominio Pampásico de la Subregión Guayano-Brasileña (Ringuelet 1955, 1961), muy próxima al límite del Dominio Central con influencia marcada en la Sierra de la Ventana, pero cuya extensión al pie de sierra y llanura circundante no está determinada.

El territorio del partido de Tres Arroyos está ubicado en el Area Interserrana donde dominan en forma casi total las aves y mamíferos de zonas abiertas. Entre las primeras se observan aun algunos ejemplares de *Rhea americana* (Linné 1758) aunque ya muy reducidos fundamentalmente por acción antrópica. *Nothura maculosa* Salvadori 1895 y *Eudromia elegans* D'Orbigny y Geoffroy 1832, son dos tinámidos relativamente frecuentes; *Caracara plancus* (J.F. Miller 1877) y *Milvago chimango* (Vieillot 1816) son las más frecuentes rapaces diurnas. En los ambientes lóticos y lénticos con hidrofitia abundante, se encuentran varias especies de Anatidae (i.e.: *Anas flavirostris* Vieillot 1816; *Anas georgica* Gmelin 1789; *Dendrocygna bicolor* (Vieillot 1816)) y la ubicua *Fulica leucoptera* Vieillot 1817. En invierno se registra por lo menos una especie que anida en la región patagónica: *Chloephaga picta* (Gmelin 1789).

Entre los reptiles se encuentran numerosos ejemplares de *Tupinambis teguixin* (Linné 1758), en especial en las márgenes vegetadas de los arroyos. *Bothrops alternata* (Dumeril y Bibrom 1854), si bien es típica de ambientes rupestres, también frecuenta la llanura (Gallardo 1977).

Como ocurre en todas las áreas con intensa explotación agropecuaria, muchos mamíferos silvestres se encuentran reducidos o totalmente extinguidos en épocas recientes. Ejemplo de esto son *Canis* (*Pseudalopex*) *gymnocercus* (Fischer 1814) y *Lagostomus maximus* (Blainville 1817) de la que sólo pueden observarse algunas "vizcacheras" en las márgenes del río Quequén Salado. Asimismo, sobre las márgenes de este río se encuentran las poblaciones de *Ctenomys talarum* (Thomas 1898) que penetran más hacia el interior; *Ctenomys australis* (Rusconi 1934) es simpátrica con las especies anteriores en la zona del balneario Claromecó (Contreras y Reig 1965).

En lo que respecta a los micromamíferos - y fundamentalmente a los cricétidos - que resultan buenos indicadores zoogeográficos y ambientales, no hay datos precisos para el área por lo que las menciones generales para la provincia de Buenos Aires son de escasa utilidad.

b) Fitogeografía: el área comprendida por el partido de Tres Arroyos está incluida en la Provincia Pampeana del Dominio Chaqueño (Cabrera 1971). Dentro de la Provincia Pampeana pertenece al Distrito Austral, caracterizado por la estepa de gramíneas "formada por grandes matas del género *Stipa*, entre las cuales el suelo queda desnudo durante gran parte del año" (Cabrera 1971: 28). En la zona la especie dominante es *Stipa caudata* Trinius 1830, conspicua principalmente en los campos poco modificados o en descanso por un largo período. En las márgenes de los arroyos se forman extensos "cardales" de *Eryngium eburneum* Decne., mientras que en las áreas bajas inundables predominan los "duraznillales" de *Solanum glaucum* Dun.

En toda esta parte del Distrito Austral no se encuentra vegetación arbórea natural ni tampoco arbustos leñosos de cierto porte.

GEOLOGIA

Ubicación

La localidad Arqueológica Arroyo Seco se encuentra en las vecindades del ejido urbano de la ciudad de Tres Arroyos, sobre la margen derecha del Primer Brazo del arroyo homónimo.

La zona reconocida, mucho más amplia, abarca parte de las hojas topográficas en escala de 1:50.000 publicadas por el Instituto Geográfico Militar, que a continuación se mencionan:

1. Estancia Tres Lagunas (3960-9-2);
2. Tres Arroyos (3960-9-1);
3. Estancia San Bautista (3960-9-3);
4. Estancia Santa Rita (3960-9-4);
5. San Francisco de Bellocq (3960-15-2) y
6. Claromecó (3960-15-4).

En el ángulo NO de la primera de las hojas mencionadas se encuentra la Localidad Arqueológica, ubicada a los 60°14′ 39″ de longitud Oeste y a los 38°21′38″ de latitud Sur.

Fisiografía

a) Rasgos Regionales

El paisaje de la zona responde a características bonaerense, casi una planicie, con muy suaves ondulaciones originadas en relación con cauces fluviales y depresiones, de drenaje frecuentemente centrípeto. En este último caso pueden tener agua en forma permanente o transitoria y es común la vegetación característica de la región en estas zonas de relieve negativo.

Los cursos más destacados de la zona son el Primero, Segundo y Tercer Brazo del denominado Tres Arroyos. Los dos mencionados en primer término se unen al sureste dentro del éjido urbano de la localidad homónima, constituyendo un cauce único que al encontrarse con el Tercer Brazo forman el Arroyo Claromecó o Tres Arroyos.

Este último presenta un valle más amplio que los dos anteriores y va a desembocar en la costa atlántica después de atravesar una cadena medanosa de unos dos kilómetros de ancho.

Dentro de las hojas topográficas mencionadas precedentemente el Primer Brazo de los Tres Arroyos tiene un recorrido de unos 8 km y un ancho del valle que va de 30 a 200 m. El Segundo y Tercer Brazo tienen dimensiones similares, aunque algo mayores este último. Finalmente el Arroyo Claromecó tiene una longitud de unos 45 km y un ancho del valle que oscila entre los 500 m y los 2 a 3 km.

En la zona se encuentran también depresiones de variadas dimensiones aunque no mayores a 2 km de diámetro. Pueden tener agua temporaria o permanentemente, constituyendo en general pequeñas lagunas muy típicas como las denominadas Tres Lagunas en la hoja topográfica homónima.

Vinculados con estas depresiones o con ciertos sectores de los valles, se encuentran en sus vecindades pequeñas lomadas que en casos excepcionales recuerdan a formas típicamente medanosas muy degradadas con suelos frecuentemente zonales, bien desarrollados.

En las vecindades de la costa atlántica una cadena medanosa de unos 2 km de ancho y alturas promedio de 5 a 8 m se disponen en forma paralela a la playa, contituyendo una barrera natural a las aguas que vienen del continente y que es sólo atravesada por el Arroyo Claromecó o Tres Arroyos en la zona estudiada. Las ondulaciones, en este caso, son mucho más pronunciadas que en el resto del paisaje y cuando falta la vegetación las acumulaciones sedimentarias pueden integrar formas linguoides y/o barkanoides.

b) Rasgos Locales

Son los diferenciados en la misma zona del yacimiento que presenta divisorias y zonas bajas, muy típicas en la región en estudio.

El lugar donde se ubica el Sitio 2, constituye una pequeña lomada que tiene suave pendiente hacia el Este y Sureste en dirección al cauce del Primer Brazo de los Tres Arroyos, pero algo más pronunciada hacia el Oeste donde se encuentra una pequeña depresión que funciona temporariamente como laguna.

Si a esto sumamos una depresión tributaria al arroyo principal, es decir al cauce del Primer Brazo por el norte y noreste, es claro que resulta una pequeña elevación casi aislada por la presencia de agua en condiciones favorables.

En la parte más alta de esta elevación, cuya altura máxima en relación con el cauce y la laguna temporaria no supera los 3 m, se han encontrado hasta ahora los elementos faunísticos y culturales descriptos en el presente trabajo.

Geología

a) Características Regionales

El paisaje de la zona responde a características geológicas vinculadas con procesos desarrollados bajo diferentes condiciones climáticas en los distintos momentos de su evolución.

La geología entonces comprende escencialmente una roca de base o sustrato sobre el cual evolucionó el paisaje y un conjunto de unidades litoestratigráficas y edafoestratigráficas vinculadas con dicha evolución.

Formación Pampeana

El sustrato está constituido por la Formación Pampeana en el sentido Fidalgo et al. (1973) tratándose de un limo arcilloso a limo arenoso castaño amarillento a castaño rojizo, a veces con tonalidades blanquecinas en relación con la presencia de carbonato de calcio presente en muy distintas formas y variadas características.

Esta roca se ubica generalmente en las partes altas del relieve integrando tanto las divisorias principales como secundarias. En esa posición y observando los cortes producidos para la construcción de caminos o aun en los lugares donde se han abierto canteras con el fin de usar estos sedimentos como subrasantes de excelente calidad, es posible percibir diferentes características en el carbonato de calcio que presentan, conocido bajo la antigua y difundida denominación de "tosca" en nuestro país y particularmente en la Región Pampeana.

Unos 4 km al sureste del cruce de la ruta 228 con el Primer Brazo de los Tres Arroyos se encuentra una de estas divisorias que muestra varios afloramientos artificiales que permiten observar distintos aspectos en cuanto a las características y distribución del carbonato de calcio, que comúnmente se concentra en las partes más elevadas de estos relieves. Una de las formas más frecuentes es cuando se presenta en "venas" horizontales y verticales aproximadamente de 1 ó 2 cm de espesor, pero con continuidad lateral de varios metros, a veces decenas de metros, integrando un enrejado que separa los paralelepípedos de distintas dimensiones de "sedimentos pampeanos." Estas formas geométricas comúnmente tienen su eje mayor en sentido horizontal con dimensiones promedio que oscilan entre 5 y 10 cm de longitud y eje menor en posición vertical con dimensiones promedio que van de 1 a 3 cm como máximo.

El espesor del conjunto carbonatado comúnmente puede tener de 0,50 a 1,50 m como promedio, pero en oportunidades estas formas se ven complicadas por la presencia de capas de carbonato de calcio de espesores mayores a las "venas", pero de desarrollo único, es decir individual en la secuencia como representando la zona de concentración máxima. El espesor de estas últimas puede alcanzar de 0,20 a 0,30 m.

Al menos parcialmente, algunas de estas formas carbonatadas corresponden, según creemos, a horizontes K de suelos que han sido decapitados constituyendo la superficie de erosión una verdadera y clara discordancia de erosión.

Estos rasgos pueden observarse repetidas veces en los cortes de los caminos que cruzan las divisorias y se dirigen en forma perpendicular a los arroyos de la zona, donde además se nota con claridad en la parte baja el desarrollo de la planicie de inundación donde distintas unidades más jóvenes sobreyace al sustrato.

En la Formación Pampeana del área no se han hallado fósiles, pero pocos kilómetros al oeste, en las inmediaciones del Río Quequén Salado, la fauna encontrada parece ser de una antigüedad Plioceno (Montehermosense), particularmente Plioceno medio a superior.

Formación Luján

Esta unidad se encuentra ampliamente distribuída en los arroyos y lagunas de la provincia de Buenos Aires. Se trata de sedimentos lacustres y/o fluviales típicamente continentales representados también en esta región.

En la planicie de inundación próxima al cauce del Primer Brazo de los Tres Arroyos realizamos dos perforaciones en las cercanías del Sitio 2.

La primera de las perforaciones alcanzó una profundidad de 2,50 m. Los primeros 0,70 m son de una arena fina castaña donde el suelo actual tiene un desarrollo que ocupa la mitad del dicho espesor. Se trata de sedimentos esencialmente fluviales que en parte se asemejan a los sedimentos eólicos más jóvenes de la región y constituyen el aluvio. Continúan unos 0,80 m de sedimentos limo arcillosos gris a gris verdosos con manchas oscuras de posible materia orgánica.

Los sedimentos anteriores pasan a un limo arcilloso verde, muy plástico con un espesor aproximado a los 0,50 m que en los siguientes 0,30 m inferiores presenta rodados de tosca de 0,005 a 0,001 m de diámetro, algo redondeados.

Finalmente se pasa a sedimentos arenosos castaños donde continúan los clastos de tosca que aumentan de diámetro hasta 0,02 a 0,03 m haciéndose finalmente imposible avanzar con el barreno debido quizás a la presencia de la Formación Pampeana, más resistente.

Algo más cerca del cauce y más alejado de la divisoria vecina hicimos una nueva perforación con barreno que superó levemente los 3 m, alejada algo menos de 50 m de la anterior.

Los primeros 1,50 m a partir de la superficie están integrado por una arena fina a limo castaño los primeros 0,70 m, pasando a limo arcilloso gris con algo de materia orgánica diseminada en los 0,80 m restantes.

Por debajo continúan 0,50 a 0,70 m de una arcilla algo limosa, sumamente plástica y con gran proporción de materia orgánica lo que le da, sobre todo en húmedo, un color negro oscuro pronunciado.

Continúa hacia abajo un limo arcilloso con algo de arena de color verde a verde amarillento típico en un espesor de 0,60 m, que pasa a sedimentos verde grisáceos y enseguida a castaño amarillento con pequeños rodados de tosca no mayores de 0,01 m de diámetro y cuyo espesor no parece ser mayor de 0,50 m pues el barreno llega a un sedimento compactado en el cual no se puede seguir perforando.

Finalmente en una cuadrícula abandonada perforamos 4,50 m con barreno apareciendo el agua a los 3,80 m y en los últimos 0,30 m ó 0,40 m un sedimento limo arenoso a limo arcilloso castaño verdoso a verde que puede ser similar a aquéllos encontrados en los perfiles anteriores.

Un kilómetro al sur de la unión del Primero y Segundo Brazo de Los Tres Arroyos se ve un cauce de poco más de una decena de metros de ancho con un rápido que tiene unos 40 cm de altura desarrollado sobre los sedimentos de la Formación Pampeana ya descriptos precedentemente. En discordancia de erosión sobre los sedimentos mencionados se ven 0,50 a 0,60 m de un limo arenoso de color verde a verde amarillento que en parte se hace arcilloso y que hacia la base pasa una grava muy fina con clastos no mayores de 0,002 a 0,005 m de color gris a castaño.

Se sobrepone a los sedimentos anteriores, también en discordancia de erosión, un depósito de color gris a gris blanquecino integrado por una arena fina a limo grueso con cantidades a veces subordinadas de arcilla, pero con abundante presencia de *Littoridina* sp. *y Biomphalaria* sp. También se observan sobre todo en la parte inferior de este depósito unos 0,50 m de espesor lentes con concentración de materia orgánica negra redepositada que a veces forman capitas de 3 a 10 mm de espesor término medio.

Rematando el perfil, sedimentos de textura arenosa a limosa de color castaño de unos 0,30 a 0,40 m de espesor, con un principio de edafización y cierta abundancia de materia orgánica en un espesor de 0,05 a 0,10 m adquiriendo por ese motivo un color negro pronunciado.

La unión del cauce anterior con el Tercer Brazo de los Tres Arroyos da lugar a la presencia de un cauce y una planicie de inundación mucho mayores y que integran el Arroyo Claromecó o Tres Arroyos. Unos 100 m aguas abajo de la confluencia se ven unos pequeños rápidos sobre, posiblemente, sedimentos de la Formación Pampeana que se encuentran por debajo de la superficie del agua. Lateralmente y por encima, unos 4 a 4,5 m de un sedimento verdoso a amarillento hasta la mitad de dicho espesor, constituído por arena a limo esencialmente arcilloso, que hacia la parte superior en los 2 m restantes, pasan a un limo castaño a castaño grisáceo. En este último se ve un nivel de unos 0,15 a 0,20 m de espesor en la zona próxima a la superficie caracterizada por formas tubulares, ramificadas en parte, de 1 a 3 mm dediámetro y varios centímetros de longitud donde predominan los fenómenos de oxidación que se advierten por la presencia de hematita y limonita. Se trata de huecos que fueron y en parte son ocupados por raíces y restos de ellas en cuya relación parecen haberse producido los procesos de meteorización química mencionados.

En la desembocadura del Arroyo Claromecó se pueden observar unos 2 a 2,5 m de espesor de un depósito similar, sobre todo en su parte inferior, a los sedimentos limo arenosos a areno limoso de color verde a verde amarillento ya descriptos en las vecindades de los cauces actuales. Por encima hay unos 0,70 a 1,00 m de sedimentos arenosos gris oscuros a negros con gran cantidad de materia orgánica y olor fétido donde hallamos restos de valvas de moluscos rotas en parte, pero donde en un caso se pudo determinar *Tagelus* sp. También se hallaron restos de *Panochthus* sp., lo cual indica que tales sedimentos (de mezcla) corresponden a lagunas costeras.

Unos quinientos metros aguas arriba se ven cerca del nuevo puente en construcción que cruzará el arroyo, un afloramiento a partir del pelo de agua de color verde similar al anterior que remata a su vez en un depósito de 0,20 a 0,30 m de color negro pronunciado pero correspondiente posiblemente a un suelo enterrado a su vez por sedimentos arenosos grises a gris blanquecino de origen dudoso.

También cerca de la desembocadura del arroyo Claromecó se ve un sedimento gris integrado por una arenisca gruesa hasta conglomerado muy fino con estratificación entrecruzada bien pronunciada, posiblemente en cubeta y con inclinación hacia el oeste en general. Se trata de un sedimento bien cementado cuya proporción abundante de carbonato de calcio hace que se la pueda considerar una carbonatita arenosa cuyas relaciones no tenemos claras todavía. En este depósito se ven también concentraciones de *Littoridina* sp. y *Biomphalaria* sp.

En la hoja San Francisco Bellocq y en las vecindades del Vivero Forestal y la Escuela Agrícola abandonada, se pueden recorrer extensos afloramientos en sentido horizontal en la barrancas del arroyo Claromecó, con exposiciones que oscilan entre 2 y 4 m de potencia. En la parte inferior suele verse que el cauce del arroyo al menos en parte está elaborado sobre la Formación Pampeana y en discordancia de erosión sobre ella se encuentran sedimentos fluviales limo arenoso a limo arcilloso en un espesor que varía de 1,50 a 2 m. En su base el color puede ser castaño a gris amarillento pasando el resto a tener un color verde a verde amarillento pronunciado. Por encima se pasa transicionalmente a veces a sedimentos de similar textura, gris a gris blanquecinos, con capitas de 1 cm de espesor o menores de materia orgánica redepositada de color gris a negro muy pronunciado, especialmente en la parte inferior del depósito que en conjunto tiene 1,50 m de espesor de término medio. Estos sedimentos en su parte superior están en contacto con limos arenosos castaño amarillentos de espesor variable, pero casi siempre menores a 2 m de sedimentos eólicos. A veces en ese contacto se ven remanentes de un sedimento oscuro con abundante materia orgánica de no más de 2 ó 3 cm de espesor que podría ser un remanente de un suelo enterrado.

Sedimentos eólicos

Recorriendo las Hojas Estancia Tres Lagunas, Estancia Santa Rita, San Francisco Bellocq y Claromecó en las vecindades del arroyo colector de la zona se ve que, salvo en las áreas correspondientes a la planicie de inundación, predominan los sedimentos de la Formación Pampeano y los sedimentos eólicos que con variable espesor culminan las formas de relieve.

En términos generales, parecería que el espesor de estos sedimentos eólicos disminuye en forma muy generalizada desde los cuerpos de agua dulce actuales en dirección al este, o más específicamente desde las planicies de inundación y playas de las lagunas de variadas dimensiones, en dicha dirección.

Desde la curva más pronunciada del camino asfaltado que de Tres Arroyos se dirige a Necochea, se encuentra dentro de la Hoja Estancia Tres Lagunas, un camino que va en dirección al suroeste y un kilómetro antes de llegar al arroyo Claromecó cambia de rumbo hacia sureste, en dirección al "Boliche El Descanso." Allí se observa un perfil sobre sedimentos eólicos típicos que representan dos unidades litoestratigráficas claramente separadas por una discordancia de erosión. En ambos casos se trata de depósitos constituídos esencialmente por un limo grueso a arena fina con cantidades muy subordinadas de arcilla, pero con una evidente proporción de carbonato de calcio pulverulento diseminado en los depósitos, en proporción aparentemente mayor en la unidad inferior, al menos en observación megascópica. El color en ambos casos es castaño a castaño amarillento presentando tonalidades blanquecinas la unidad inferior, además de mostrar manchas blanco mate de 2 a 3 mm de diámetro término medio.

En las distintas descripciones realizadas en el acápite Formación Luján además hemos visto cómo los perfiles a veces rematan en sedimentos eólicos que son los que aquí nos competen.

En las vecindades del Sitio 2 y cerca de 700 m al Norte del mismo, se ve el corte del camino de acceso al lugar un afloramiento de unos 2,30 m de exposición donde puede observarse desde la superficie hacia abajo 1 a 1,20 m de sedimentos de color castaño rojizo a castaño amarillento fácilmente disgregables donde hay un suelo actual bastante bien desarrollado, zonal, de no menos de 70 a 80 cm de espesor. Por debajo, un sedimento de características similares al anterior, pero de color algo más claro, blanquecino debido a la presencia mayor de carbonato de calcio pulverulento, a veces con concentraciones de pocos milímetros de diámetro, pero fácilmente disgregables formando a veces punteados o lunares muy diseminados. El espesor de la vista de la unidad inferior, alcanza a 1 m aproximadamente de exposición.

Interpretación y correlación

De las localidades descriptas precedentemente es claro que podemos reconocer, como se indica en el cuadro adjunto, algunas de las unidades que en la Pampa Deprimida han sido diferenciadas en Fidalgo et al. (1973), Fidalgo (1979), y en la zona interserrana Tonni y Fidalgo (1978), Fidalgo y Tonni (1981).

Con ello reconocemos regionalmente en las vecindades de la Cuenca de los Tres Arroyos las unidades que a continuación se detallan en orden decreciente de edad:

Formación Pampeana: La mencionamos en el sentido de Fidalgo et al. (1975) existiendo por las razones citadas precedentemente la posibilidad que la edad sea Plioceno Medio-Superior.

Después de un gran hiatus se depositaron los sedimentos correspondientes a la Formación Luján de la forma descripta en Fidalgo et al. (1973) y dividiéndola en dos miembros como en la Pampa Deprimida, es decir Miembro Guerrero y Miembro Río Salado con características muy similares a aquella región. La diferencia existente con la última región mencionada es aquella correspondiente a la restringida zona donde se observa pronunciada oxidación a través de la existencia de colores rojos y amarillos, por la presencia quizás de hematita y muy especialmente limonita.

En el caso de una de las perforaciones realizadas en la vecindad del sitio arqueológico y en observaciones en las vencindades de la desembocadura del arroyo Claromecó, sobre el Miembro Guerrero de la Formación Luján hay desarrollado un suelo que ha sido enterrado por los depósitos correspondientes al Miembro Río Salado y que podría ser equivalente al suelo Puesto Callejón Viejo descripto en esa posición en la Pampa Deprimida.

Además, sobre el Miembro Río Salado de la Formación Luján en las inmediaciones del Vivero Forestal sobre las barrancas del arroyo Claromecó se ven remanentes de un suelo enterrado que puede ser equivalente al Suelo Puesto Berrondo (Fidalgo et al. 1973).

Finalmente, se encuentran los sedimentos eólicos que en la Pampa Deprimida describimos bajo la denominación de Formación La Postrera como única unidad y que en la zona interserrana comprendida entre entre las sierras Tandil y Ventana hemos diferenciado como dos unidades diferentes y en oportunidades, tres (Fidalgo y Tonni 1981).

En la zona de Tres Arroyos y sus vecindades, distinguimos regionalmente dos de estas unidades de origen esencialmente eólico y con ellas creemos que está relacionado el interesante Sitio 2 de Arroyo Seco.

Texturalmente los dos depósitos son de características similares tanto macroscópicamente como a través de análisis de laboratorio. El color es también muy semejante, castaño amarillento en general, pero la unidad inferior parece presentar una proporción mayor de carbonato de calcio, que le suministra cuando la distribución es masiva, un color pronunciadamente blanquecino en seco. Otras veces se ve un punteado o muy pequeños lunares con carbonato de calcio cementado que la da un aspecto moteado y que en oportunidades, pero más raramente, aparecen en la unidad superior.

b) Características Locales.

En la pequeña lomada donde se ubica el Sitio 2 de Arroyo Seco hemos estudiado diferentes cortes correspondientes a las paredes de cuadrículas que se realizaban durante la exploración. Con tal motivo hemos podido objetivamente diferenciar en espesores de 0,80 a 1,20 m los siguientes elementos (Figura 3).

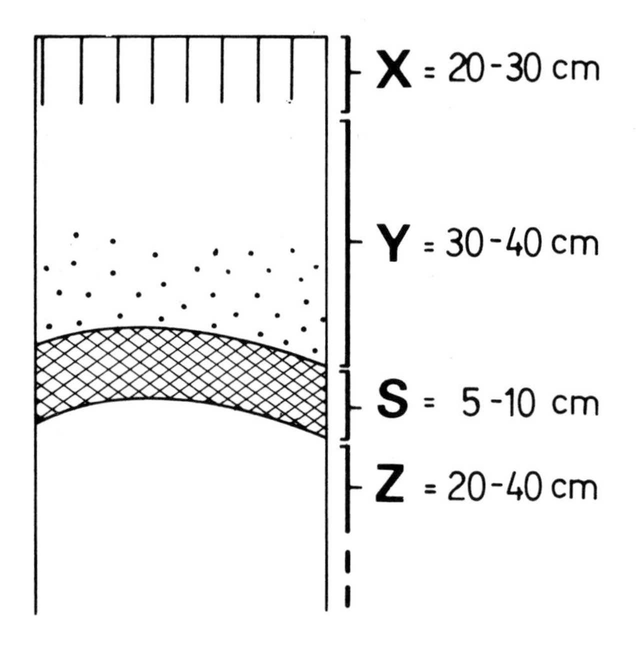

Figura 3. Perfil geológico del Sitio 2.

X: Suelo actual en parte muy afectado por acción antrópica. Unos 20 a 30 cm de espesor, con abundante materia orgánica que le da un color negro oscuro en húmedo y grisáceo en seco. Una mayor concentración de arcilla en los 10 a 15 cm inferiores de este suelo, disminuyendo luego hacia abajo. También en esa dirección se hace menor rápidamente la presencia de materia orgánica, motivo por el cual se pasa rápidamente a los colores de la roca madre Y.

Y: 30 a 40 cm de espesor de un limo grueso a arena fina con cantidades muy subordinadas de arcilla. El color es castaño amarillento y aproximadamente en su mitad inferior suele presentar aunque en forma difusa un moteado muy débil integrado por pequeñas concentraciones de carbonato de calcio en forma de lunares de un diámetro individual no mayor a los 2 ó 3 mm.

S: Es una zona de unos 5 a 10 cm de espesor donde se nota en parte una clara concentración de carbonato de calcio en forma de una capa que generalmente muestra un límite inferior más neto, más definido que el límite superior que a veces es algo más difuso.

Z: Se encuentra en la parte inferior de los perfiles observados en el yacimiento y es también como Y un limo grueso a arena fina con cantidades subordinadas de arcilla de color castaño amarillento que en parte puede tener tonalidades grisáceas. Su espesor a la vista varió de 20 a 40 cm dependiendo de la profundidad a que se trabajó la cuadrícula.

Las observaciones sobre los perfiles expuestos en las cuadrículas fueron realizadas en distintos meses de los años 1979 y 1980 especialmente en primavera, verano y otoño. Por esta razón el grado de humedad y su distribución en el perfil fue diferente en distintos momentos influenciando evidentemente las observaciones.

Como las características descriptas precedentemente no concordaban con las observaciones regionales realizadas, se resolvió, después de oir la opinión de distintos especialistas que se reunieron en la zona, con quienes se intercambió ideas, realizar una extensa trinchera que saliendo del yacimiento se prolongara hasta las vecindades del cauce del Primer Brazo en una extensión de 82 m.

Las observaciones en la trinchera nos permitieron deferenciar las dos unidades litoestratigráficas de origen eólico que venimos diferenciando regionalmente en la zona interserrana.

Las dos unidades litoestratigráficas son: (Figura 4, perfil 4).

A: sedimentos integrados por un limo grueso a arena fina de color castaño amarillento, con un espesor de 40 a 50 cm en cuyos 20 a 30 cm superiores se encuentra desarrollado un suelo con las características similares a las del perfil en el yacimiento y afectado en gran parte por acción antrópica. También resulta evidente en esta unidad la acción de animales cavadores de distinto tipo, desde vizcachas a escarabajos, en todo su espesor.

B: constituye un depósito de sedimentos con características similares al que lo sobreyace, tanto en textura como en color. Presenta un moteado integrado por pequeñas concentraciones de carbonato de calcio de 2 a 3 mm de espesor término medio y que se observa en un espesor de 10 a 15 cm desapareciendo hacia la parte inferior.

El límite superior cuando la presencia del moteado es definida resulta claro, no así cuando el carbonato de calcio no se observa como sucede en trechos reducidos. En esta zona de límite, pero siempre en B, también se observan concentraciones de carbonato de calcio junto con sedimentos integrando formas esféricas o al menos marcadamente redondeadas, de diámetros que oscilan entre 5 y 10 cm y otras cilíndricas o poliédricas de eje mayor similar mientras los diámetros basales pueden alcanzar de 2 a 4 cm.

A medida que nos desplazamos por la trinchera en dirección al arroyo la presencia de las pequeñas concentraciones de carbonato de calcio aumentan su diámetro individualmente y de 2 a 3 mm en la parte más alta pasan a 2 ó 3 cm en la parte media de la pendiente. En esta última y ya cerca del arroyo también se observan rodados de tosca de 1 a 2 cm de diámetro redepositados por la acción fluvial.

Discusión

Los perfiles descriptos tanto regionales como locales, sobre todo estos últimos, pueden ser interpretados de diferentes formas.

En sentido regional se han reconocido dos unidades de origen eólico descriptas y fundamentadas desde el punto de vista geológico y faunístico por Fidalgo y Tonni (1981) y incluso en los alrededores de la ciudad de Miramar, determinados perfiles muestran la posibilidad de distinguir tres unidades litoestratigráficas de origen eólico.

Los perfiles observados en el Sitio Arqueológico en cambio pueden dar lugar a variadas interpretaciones motivo por el cual, y para tratar de ser más claros, graficaremos las posibilidades factibles de la siguiente forma:

Los perfiles 1, 2 y 3 contituyen interpretaciones que pueden hacerse del perfil generalizado para la mayor parte de la zona de excavación y descripto en las páginas precedentes bajo las letras X, Y, S y Z (Figuras 3 y 4).

En cambio las letras A, B y C en los perfiles se refieren a las unidades litoestratigráficas factibles de considerar. La letra "d" indica la o las discordancias en cada perfil a lo que habría que agregar como cuarta, quinta y sexta posibilidad la presencia de otra discordancia en la base de S en cada uno de los perfiles 1, 2 y 3 representados.

Cualquiera de las seis posibilidades planteadas u otras que puedan plantearse, nos enfrentan a un mayor o menor número de interrogantes en algunos casos de difícil o imposible explicación.

Relacionando estos perfiles con las características vistas para las unidades geológicas de origen eólico en las observaciones de tipo regional, es claro que la capa S de mayor concentración de carbonato de calcio aquí, es un fenómeno estrictamente local y relacionada con la mayor

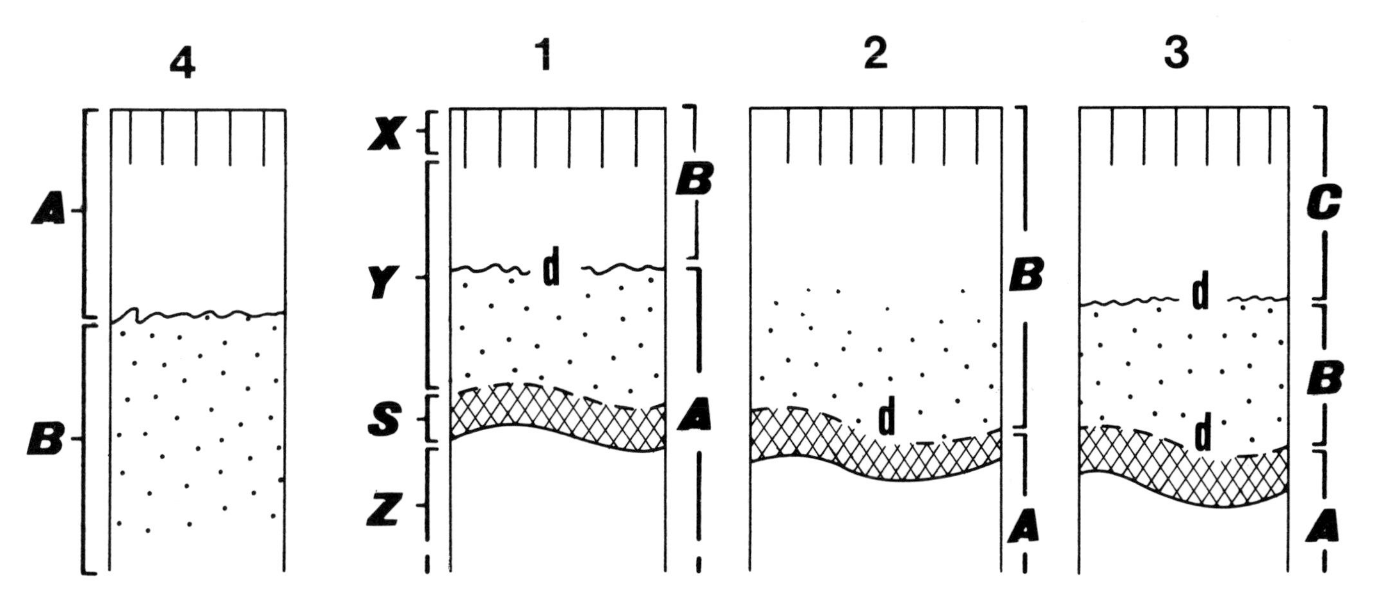

Figura 4. Perfiles geológicos. 1, 2, 3: Distintas interpretaciones del perfil representado en la figura 3. 4: Perfil regional.

parte de las cuadrículas excavadas, pero ni aun con todas ellas.

Cuáles pueden ser las causas de la mencionada concentración de carbonato de calcio? Tentativamente se sugiere, después del cambio de ideas mencionado precedentemente, que podría deberse a:

1. El movimiento de tierra al efectuar las excavaciones para los enterratorios habría facilitado la lixiviación y posterior concentración del carbonato de calcio en la forma descripta.

2. Siendo un lugar temporalmente habitado y utilizado para el despostamiento durante ese tiempo es probable que la vegetación desapareciera superficialmente y se viera así facilitada la infiltración de agua de lluvia con el consiguiente perfeccionamiento de la lixiviación y precipitación del carbonato de calcio a profundidad donde el sedimento está más compactado o con más carbonato de calcio o ambas.

Los aspectos planteados en 1 y 2 han tenido lugar sin duda, sea individualmente o aun complementándose, pero cuál fue la influencia sobre la formación de la capa S es lo que no podemos cuantificar.

En base a las características y distribución de la fauna descripta en el capítulo correspondiente, así como en base a los elementos culturales encontrados en las cuadrículas exploradas creemos que el perfil 1 es el que tiene mayores posibilidades de ser el perfil real entre las seis alternativas planteadas.

PALEOETNOZOOLOGIA

Métodos

Los restos faunísticos fueron extraídos teniendo como control una estratigrafía artificial en capas de 5 cm y atendiendo a su ubicación tridimensional. Para su análisis bioestratigráfico se tomó en consideración ésta y la estratigrafía natural.

El análisis faunístico se basa en el material recuperado de 32 cuadrículas. Se determinó el número total de piezas, el número mínimo de individuos (NMI) y los percentajes respectivos, correspondientes a cada taxón.

Los restos exhumados de las nueve cuadrículas donde se detectaron enterratorios fueron posteriormente analizadas en forma independiente del total. Ello se efectuó para verificar la probable existencia de un comportamiento diferente en dicho universo de muestra.

El material analizado corresponde en todos los casos a aquellos restos extraídos a partir de una profundidad promedio de 40 cm, desde el nivel 0 (cero) del sitio, dado que hasta dicha profundidad el sedimento está fuertemente disturbado por la actividad humana. Este hecho queda evidenciado por la presencia de trozos de ladrillos, vidrios, alambres, etc., junto a restos de mamíferos introducidos desde Europa (vaca, caballo, oveja).

En el caso de las cuadrículas 22, 23 y 24 (alineadas en sentido NE-SO y abarcando 12 m^2), el espesor de sedimentos alterados es mayor, alcanzando en la cuadrícula 24 hasta los 65 cm y disminuyendo hasta 40 cm en dirección sur (cuadrícula 23) (Figura 2).

En la cuadrícula 28, un pozo de alrededor de 30 cm de diámetro y de 110 cm de profundidad, es otra muestra de la actividad humana, puesto que allí se extrajeron restos óseos quemados de fauna introducida. El resto de la cuadrícula presentó escasa disturbación, por lo tanto se tomaron en cuenta para el análisis aquellos materiales faunísticos hallados por debajo de los 34 cm.

La máxima profundidad alcanzada fue de 150 cm y la mínima de 75 cm, en sólo tres casos. El promedio aproximado fue de 100 cm.

El número mínimo de individuos, la edad relativa de los mismos y el cálculo porcentual se obtuvieron en base a los métodos ya explicitados en otros trabajos (Tonni y Laza 1976; Raffino et al. 1977). En el cálculo porcentual se consideraron, únicamente, los taxa reconocidos a nivel genérico y/o específico.

El NMI fue considerado en conjunto, puesto que no fue posible observar en el campo límites físicos que permitieran una segura subdivisión en grupos de los materiales recuperados. Como se verá, estos límites han sido detectados a través del análisis bioestratigráfico.

Es necesario aclarar que: a) Para *Lama glama guanicoe* y *Blastoceros bezoarticus* se tomaron en consideración, para determinar el NMI, los astrágalos y escafoides por ser estos los elementos mejor representados; b) para *Rhea americana* se contabilizaron los extremos distales de tarso-metatarso y se los asignó a individuos juveniles o adultos. En cuanto a los restos de cáscara de huevo de Rheidae, se asignaron a un solo ejemplar; pero no se tuvo en cuenta en el cálculo porcentual por hallarse representado un nivel específico; c) en los armadillos (*Chaetophractus, Zaedyus*) los elementos mejor representados son las placas dérmicas. Por tal razón, a pesar de ser numerosas, se contabilizaron como pertenecientes a un único individuo.

Sistemática

Se incluye aquí un listado de los distintos taxa registrados en el Sitio 2 de la Localidad Arqueológica de Arroyo Seco.

BIVALVIA, HETERODONTA, VENERIDAE: *Amiantis* sp.

GASTROPODA, CAENOGASTROPODA, VOLUTIDAE: *Adelomelon (Pachycymbiola) brasiliana* (Solander 1786)
AVES, RHEIFORMES, RHEIDAE: *Rhea americana* (Linné 1758) TINAMIFORMES, TINAMIDAE: *Eudromia* sp. (cf. *E. elegans*) (D'Orbigny y Geoffroy 1832); *Nothura* sp. (cf. *N. maculosa* Salvadori 1895) STRIGIFORMES, STRIGIDAE: *Speotyto cunicularia* (Molina 1782)
MAMMALIA, CARNIVORA, MUSTELIDAE: *Lyncodon* sp. (cf. *L. patagonicus* (Blainville 1842); CANIDAE: Canidae *gen. et sp. indet.*; RODENTIA, CAVIIDAE: *Dolichotis patagonum* (Zimmermann 1780); OCTODONTIDAE: *Ctenomys* sp.; CRICETIDAE: Cricetidae *gen. et sp. indet.*; CHINCHILLIDAE: *Lagostomus* sp. (cf. *L. maximus* (Desmarest 1877); EDENTATA, DASYPODIDAE: Dasypodidae *gen. et sp. indet.*; *Chaetophractus villosus* (Desmarest 1804); *Zaedyus pichiy* (Desmarest 1804); *Eutatus seguini* Gervais 1867; GLYPTODONTIDAE: *Glyptodon* sp.; MEGATHERIIDAE: *Megatherium americanum* Cuvier 1796; MYLODONTIDAE: Mylodontinae *gen. et sp. indet.* (cf. *Mylodon* sp.), *Glossotherium (Glossotherium) robustum* (Owen 1840); LITOPTERNA, MACRAUCHENIIDAE: *Macrauchenia* sp. (cf. *M. patachonica* Owen 1840); ARTIODACTYLA, CAMELIDAE: *Lama glama guanicoe* (Muller 1776); *Palaeolama* P. Gervais 1867 (cf. *P. wedelli* (P. Gervais 1855)); CERVIDAE: *Blastoceros bezoarticus* Fitzinger 1860; PERISSODACTYLA; EQUIDAE: *Equus* sp. (cf. *E. (Amerhippus)* sp.); "Hippidiforme" indet. (*Onohippidium* sp. - *Hippidion* sp.).

CARACTERISTICAS DESTACABLES DE ALGUNAS DE LAS ESPECIES REGISTRADAS

Rhea americana

El género *Rhea* involucra en la actualidad a una sola especie *Rhea americana*, que se distribuye desde Río Negro en Argentina, hacia el norte a través de Uruguay, Paraguay, Bolivia y sur de Brasil.

Está representado por una especie extinguida en el Pleistoceno medio de la provincia de Buenos Aires (*vide* Ameghino y Rusconi 1932; Brodkorb 1963), mientras que a la especie viviente se la conoce a partir del Pleistoceno tardío en Brasil y Argentina (Brodkorb 1963; Tonni 1980).

Las dos especies que comprende actualmente la Familia Rheidae se caracterizan por ser aves corredoras terrestres que frecuentan áreas abiertas como estepas y sabanas, o áreas abiertas en zonas boscosas xerófilas.

Rhea americana fue muy abundante en el territorio bonaerense hasta fines del siglo pasado (Armaignac 1976) y aun hasta comienzos del presente. Su retracción se ha debido, por lo menos en parte, a la intensa explotación agropecuaria.

Chaetophractus villosus

Este dasipódido habita en todo el centro de la Argentina, en los Dominios Pampásico, Norte del Patagónico y parte del Central (*sensu* Ringuelet 1961). También habita en Uruguay.

Restos indudablemente referibles a alguna especie del género, han sido registrados en sedimentos asignables al Pleistoceno medio en la provincia de Buenos Aires (ejemplar MLP no. 69-VIII-1-2) y quizá anteriores (Pleistoceno temprano?, ejemplar MLP no. 76-VII-2-22). De cualquier manera, debe señalarse que placas con una esculturación muy similar a la de la especie viviente, se encuentran en sedimentos referibles a las Edades Montehermosense y Uquiense (Ameghino 1889; Pascual et al. 1965).

Zaedyus pichiy

Cabrera y Yepes señalan que (1960, T.II:68): "Es una especie preferentemente patagónica, llegando por el litoral hasta la provincia de Buenos Aires."

Cabrera (1957/60:217) da como distribución de la subespecie típica de este dasipódido "...desde el sur de San Luis y el centro de Buenos Aires hasta la cuenca del río Santa Cruz."

Ringuelet y Arámburu (1957) lo mencionan para Bahía Blanca, Sierra de la Ventana y Bonifacio.

Contreras (1973) lo incluye en el elenco de mamíferos de la zona de Laguna Chasicó.

No hay registros que certifiquen la presencia actual de esta especie en el área de Tres Arroyos. Si bien todos los restos consisten en placas dérmicas, por lo cual no es posible determinar un número mínimo probable, sus restos son más numerosos que los de *Chaetophractus villosus* (véase Tabla I), especie que actualmente domina en la zona.

Glossotherium (Glossotherium) robustum

Este edentado piloso es muy frecuente en los sedimentos del Pleistoceno tardío de la provincia de Buenos Aires. Si bien el material de Arroyo Seco es poco significativo (una tibia), su morfología es inseparable de la de *G. g. robustum* Owen 1840 de la Formación Pampeana de la provincia de Buenos Aires y del Pleistoceno del sur de Brasil.

Los primeros registros incuestionables para el género en el territorio argentino corresponden al Pleistoceno medio (Cattoi 1966).

Glossotherium (Glossotherium) robustum fue muy probablemente un herbívoro de áreas abiertas áridas o semiáridas, a juzgar fundamentalmente por las asociaciones en que se encuentra.

Megatherium americanum

Las especies del género *Megatherium* tienen una amplia distribución geográfica, principalmente en los sedimentos pleistocénicos de la llanura Chacobonaerense. También han sido registrados en el Pleistoceno de Brasil (Paula Couto 1970) y de Bolivia (Ortega 1970).

Megatherium americanum está restringida, por lo menos en el territorio argentino, al Pleistoceno tardío. Sus restos han sido hallados en repetidas oportunidades en sedimentos de origen eólico

(Kraglievich in Greslebin 1924; Frenguelli 1931; Zetti 1964).

Ctenomys sp.

Los octodóntidos de este género están representados en el territorio argentino por numerosas especies aun no adecuadamente estudiadas. Se las encuentra en todas las áreas abiertas no inundables del territorio, desde Jujuy por el norte hasta los canales fueguinos por el sur. Habita además en Chile, Bolivia, Perú, Paraguay, Brasil y Uruguay. Sus poblaciones son muy numerosas en las regiones áridas y semiáridas. En la provincia de Buenos Aires se encuentran en áreas con suelos sueltos, marcadamente arenosos, de la costa atlántica y parte de la rioplatense, franja medanosa del Oeste y partidos de Villarino y Patagones.

El género ha sido registrado en la Argentina, por lo menos a partir del Pleistoceno medio y quizá desde el Pleistoceno temprano. Sus restos son muy abundantes en sedimentos referibles al Pleistoceno tardío de distintas localidades de la provincia de Buenos Aires.

Las características del material aquí estudiado (tamaño, morfología dentaria, etc.), son coincidentes con las de las especies de menor tamaño del género (*C. talarum, C. mendocinus, C. magellanicus*). Por tal motivo, y hasta tanto se obtenga un mejor conocimiento de las especies vivientes, no es posible con material óseo fragmentario arribar a su determinación específica.

En el partido de Tres Arroyos se hallan actualmente poblaciones de dos especies de *Ctenomys: C. australis* y *C. talarum. C. australis* (una de las especies de mayor tamaño del género) se encuentra en el área del balneario Claromecó, mientras que *C. talarum* habita en áreas vecinas a ambas márgenes del río Quequén Salado hasta unos 20 km aguas arriba de su desembocadura (Contreras y Reig 1965).

Ninguna especie de *Ctenomys* habita actualmente en el área del sitio estudiado y lo mismo parece verificarse en épocas recientes, de acuerdo con los datos proporcionados por antiguos pobladores.

Lagostomus sp.

El material referido es escaso y poco representativo como para atribuirlo a alguna especie determinada.

La única especie viviente del género, *L. maximus*, es un roedor característico de áreas abiertas no inundables. Está diferenciada en tres razas geográficas (*fide* Cabrera 1957/60) que habitan en el territorio argentino desde las provincias del Norte hasta la zona septentrional de Chubut.

Alguna especie del género quizá ya estaba presente en sedimentos asignables al Pleistoceno temprano (Pascual 1966). La especie actual ha sido registrada en el Pleistoceno tardío de la provincia de Buenos Aires (materiales MLP no. 76-IV-27-16). *L. debilis* Ameghino 1889, del Pleistoceno tardío (Edad Lujanense) de la provincia de Buenos Aires, muy probablemente constituya un sinónimo de *L. maximus* (Desmarest 1817), pues los datos merísticos y morfológicos proporcionados por Ameghino se encuentran dentro de los límites de la especie citada.

Dolichotis patagonum

Actualmente extinguida en el área y probablemente en toda la provincia, excepto en la zona de Chasicó (Contreras 1973). Para 1839, Darwin cita como distribución septentrional máxima los 37° 30′ latitud Sur (*fide* Cabrera 1953), aunque en un viaje de Crawford realizado en 1871 (1974) menciona a este cávido en la zona medanosa del partido de Lincoln (NO de la provincia de Bs. As.).

Lama glama guanicoe

No hay registros para esta especie con posterioridad al contacto europeo. Sus restos en el área son prehispánicos (Tonni y Politis 1980). Respecto a sus requerimientos ecológicos y aspectos taxonómicos véase Tonni y Politis (1980) y la bibliografía allí citada.

Blastoceros bezoarticus

Seguramente se encontraba en el área por lo menos hasta fines del siglo pasado. Armaignac (1976) cita numerosas manadas en las proximidades de Lobería y Necochea en la segunda mitad del siglo XIX.

Equus (Amerhippus) sp.

Distintas especies de équidos fósiles americanos han sido referidas al subgénero *Amerhippus* Hoffstetter 1950. El registro incluye al Pleistoceno tardío de Argentina, Chile, Brasil, Colombia, Ecuador, Perú y Bolivia y al Pleistoceno del Sur de Estados Unidos de América del Norte (California).

Como bien lo señalara Hoffstetter (1950, 1952) las diferencias fundamentales entre los subgéneros *Equus* y *Amerhippus* radican en la presencia, en el primero, de pozos de esmalte en los incisivos inferiores y su ausencia en los segundos. Otras características morfológicas esqueletarias se encuentran dentro del rango de variación de las distintas especies de *Equus* o aun en las razas domésticas de *Equus caballus*. En el sitio de Arroyo Seco no se han recuperado elementos que permitan una asignación indudable de los mismos a *Equus (Amerhippus)*, pero la asociación con fauna extinguida y autóctona y la ausencia de fauna exótica permiten referir dichos materiales con alto grado de probabilidad al subgénero extinguido. Más aun, los elementos del esqueleto apendicular son morfológicamente inseparables de los *E. (Amerhippus) curvidens* (Owen 1844), especie frecuente en los sedimentos del Pleistoceno tardío de la provincia de Buenos Aires.

Complejo *Hippidion-Onohippidium*

Los restos recuperados del sitio de Arroyo Seco corresponden a elementos del esqueleto apendicular de alguna especie del grupo de los Hippidiformes (*sensu* Hoffstetter 1950), más concretamente de los géneros *Hippidion* Owen 1870, u *Onohippidium* Moreno 1891. Los elementos del esqueleto apendicular de las especies de estos géneros - caracterizados por su robustez respecto a las especies de *Equus* - son muy semejantes y no poseen caracteres constantes - al menos en muestras pequeñas — que permitan su asignación a una u otra. Restos referibles a especies de estos géneros se encuentran en todo el Pleistoceno de la provincia de Buenos Aires.

Macrauchenia cf. patachonica

La única especie conocida de Macrauchenia es *M. patachonica* Owen 1839. Sus restos han sido hallados en sedimentos asignables al Pleistoceno tardío de Argentina, Bolivia, Brasil y Uruguay. La presencia del género en el Pleistoceno medio de Argentina no está suficientemente documentada.

La dentición hipsodonte, selenodonte, con cemento, señala características propias de un herbívoro pastador. La fusión de elementos en el esqueleto apendicular indica que *M. patachonica* era un buen corredor. Por ello puede inferirse que era una especie adaptada a áreas abiertas, estépicas, como lo ha sugerido Hoffstetter (*vide* Paula Couto 1979:375) y parece verificarse también por las características de algunos sedimentos portadores (v.g.: "Médano invasor de Sayape, San Luis, citados por Frenguelli 1931").

Análisis del Material

Se expondrán aquí una serie de enunciados observacionales vinculados al material faunístico recuperado. Gran parte de estos enunciados se basan en la formulación previa de hipótesis sistemáticas de distinto grado, las cuales han sido ampliamente contrastadas.

1 - Los restos de megamamíferos pampeanos extinguidos corresponden a *Macrauchenia patachonica, Palaeolama* cf. *wedelli, Glossotherium (Glossotherium) robustum, Megatherium americanum, Equus (Amerhippus)* sp., Equidae sp. indet. del complejo *Onohippidium* o *Hippidion* sp., Mylodontinae indet., *Eutatus seguini.*

2 - En adición, asociada al enterratorio de un párvulo, se registró una placa dérmica de *Glyptodon* sp.

3 - La mayor parte de los restos de especies extinguidas consisten en elementos del esqueleto apendicular y corresponden a individuos adultos.

4 - Se registra un elemento exótico al paleoecosistema: el molusco gastrópodo *Adelomelon (Pachycymbiola) brasiliana.* Este molusco habita actualmente en la latitud del sitio arqueológico, en fondos arenosos de la zona intercotidal y litoral.

5 - Integrando el ajuar funerario de varios enterratorios, se registró un conjunto de caninos de cánidos y una valva de *Amiantis* sp. articulada.

6 - El porcentaje mayor de los elementos faunísticos registrados corresponden a especies que integran la fauna indígena actual de la provincia de Buenos Aires.

6a - Algunas neoespecies no se encuentran actualmente en el área del sitio Arroyo Seco. Tal es el caso de *Blastoceros bezoarticus, Lama g. guanicoe, Dolichotis patagonum, Ctenomys* sp. y probablemente *Zaedyus pichiy* y *Lyncodon* cf. *patagonicus.*

7 - El NMI para el guanaco es altamente significativo. Este camélido no habita actualmente el área ni la habitó en tiempos históricos.

8 - La moda del registro faunístico se centraliza en *L. g. guanicoe, Ctenomys* sp. y *Blastoceros bezoarticus,* tanto en NMI como en la cantidad de restos óseos recuperados

9 - Se registra una gran proporción de restos óseos fragmentarios e identificables.

10 - La mayor parte de los restos óseos atribuibles a neoespecies, al igual que lo verificado con las especies extinguidas, corresponden a elementos del esqueleto apendicular, fundamentalmente del autopodio.

11 - Son escasos los elementos del esqueleto axial (vértebras y costillas) y los craneanos (bullas timpánicas, trozos de maxilar con dientes). Sólo en el caso de *Ctenomys* sp. están bien representados los fragmentos craneanos y mandibulares.

12 - Se registran numerosos fragmentos de cáscara de huevo de Rheidae, (probablemente *Rhea americana*), algunos con idicios de fuego.

13 - Se registran algunos fragmentos óseos con indicios de fuego.

14 - Al igual que para las especies extinguidas, la mayor parte de los restos óseos de neoespecies corresponden a individuos adultos. Sólo unos pocos restos de guanaco y venado son asignables a individuos juveniles.

Aspectos Bioestratigráficos

La nomenclatura geológica (X, Y, S, y Z) es la utilizada en el capítulo correspondiente.

Desde el punto de vista bioestratigráfico, pueden formularse los siguientes enunciados observacionales:

1 - Todos los restos recuperados son asignables a taxa que, por lo menos a nivel genérico, están representados a partir del Pleistoceno medio y tardío en la provincia de Buenos Aires.

2 - No se ha detectado la presencia de fauna introducida en el momento del contacto europeo o con posterioridad a él.

3 - En los sedimentos que se encuentran desde unos 12 cm (base de Y) por encima de la zona de concentración de carbonato (S), la diversidad específica es menor que en los situados dentro y por debajo de esa zona. Con respecto a esta situación debe tenerse en cuenta que aquellos sedimentos están considerablemente disturbados, al menos en sus términos superiores (X y parte de Y). Por tal razón, es altamente probable que las diferencias en la diversidad específica sean el resultado de esa situación.

4 - Por debajo de la zona de concentración de carbonato (techo de Z), se registra una especie exótica al paleoecosistema. Este molusco actual ha sido también registrado en sedimentos marinos referibles al Pleistoceno tardío ("Belgranense", Formación Pascua) y al Holoceno (Fm. Las Escobas) (*vide* Camacho 1966).

5 - Inmediatamente por encima y dentro de la zona de concentración de carbonato (base de Y y S), se registran restos asignables a varias especies de megamamíferos pampaeanos extinguidos, cuya diversidad específica es mayor dentro de la zona de concentración de carbonato (S) que por encima de la misma (base de Y).

6 - *Megatherium americanum* y *Macrauchenia patachonica* son dos especies características de la Edad Lujanense (*sensu* Pascual et al. 1965), referida al Pleistoceno tardío.

En base a las observaciones de campo y a la metodología utilizada en la extracción de los restos faunísticos, desde el punto de vista bioestratigráfico se verifica la existencia de dos agregados:

I) El procedente de la zona de concentración de carbonato (S) y hasta unos 12 cm por encima del techo de la misma (base de Y). Este agregado está compuesto por especies de la megafauna pampeana extinguida y especies de la fauna autóctona actual.

II) El procedente de los sedimentos situados a partir de los 12 cm por encima del techo de la zona de concentración de carbonato (techo de Y), hasta la zona de disturbación antrópica reciente y/o actual (X). Este agregado está compuesto exclusivamente por especies de la fauna autóctona actual.

Teniendo en cuenta lo expresado, el agregado I) es referible a la Edad Mamífero Lujanense (*sensu* Pascual et al. 1965). El agregado II) corresponde al Holoceno o Reciente, anterior al siglo XVI.

Desde el punto de vista tafonómico, es importante destacar que los restos se hallan en gran parte fragmentados, por lo menos en lo que respecta a los huesos largos. En adición, los restos de los megamamíferos extinguidos y algunos asignables a neoespecies, están fuertemente fracturados, aunque con las distintas partes muy próximas o aun contactando entre sí.

Como se desprende de lo expresado, los restos se presentan desarticulados, a excepción de parte de un autopodio asignado a *Equus (Amerhippus)* sp., que fue hallado en la base de Y.

Aspectos Paleoambientales y Paleozoogeográficos

Los restos óseos extraídos del Sitio 2 de Arroyo Seco son cuantitativa y cualitativamente significativos para ser utilizados en la interpretación del paleoecositema, al menos desde el punto de vista de la componente de los megamamíferos.

En muy pocas oportunidades puede obtenerse una muestra representativa de parte de la fauna integrante de un paleoecositema, que haya coexistido en un área reducida y durante un intervalo de tiempo de magnitud poco significativa. Sin embargo, las interpretaciones realizadas a través del análisis de este cuantioso material faunístico, presentan una apreciable fuente de error: la selectividad de origen antrópico.

Es correcto inferir que si bien una parte del material faunístico fue incorporada al sedimento en forma natural, el mayor porcentaje corresponde a especies seleccionadas por los habitantes del sitio, de acuerdo con sus requerimientos económicos.

Por tal razón, la muestra seguramente estará hiperdimensionada en aquellas especies de particular importancia económica, esto es, volumen de carne proporcionada, cueros, etc. Por el mismo motivo, los micromamíferos -de gran utilidad para las interpretaciones paleoambientales- pueden no estar representados, o estarlo en un porcentaje menor al normal; en muchos casos aun, los micromamíferos pueden haber sido incorporados al sedimento sin la intervención humana, en especial en las especies fosoriales, de hecho representadas en el sitio.

Lo expuesto implica claramente que en base a una muestra seleccionada, sólo pueden realizarse interpretaciones paleoambientales generales. Ciertamente, en estos casos las pruebas negativas (ausencia o falta de registro) no tienen el mismo valor que para muestras obtenidas de sitios sin selección antrópica.

Desde el punto de vista paleozoogeográfico, la muestra es útil, pues las especies presentes en el sitio -aunque seleccionadas y alteradas en su frecuencia- serán las que habitan en el área o estaban muy próximas al alcance de los habitantes del sitio.

Todas las neoespecies de mamíferos registradas corresponden a ambientes llanos, de vegetación abierta, graminosa. Otro tanto se verifica con las escasas aves recuperadas del yacimiento, y pueden inferirse requerimientos similares para los megamamíferos extinguidos.

Todas las neoespecies y géneros con especies vivientes registrados, están actualmente representados en el ámbito del Dominio Pampásico (*sensu* Ringuelet 1961). Sin embargo, se detectaron algunas frecuencias significativas desde el punto de vista paleozoogeográfico.

El guanaco (*Lama g. guanicoe*) es el mamífero más frecuente registrado en el Sitio 2 de Arroyo Seco. En la provincia de Buenos Aires sólo habita en el área de Sierra de la Ventana y esta situación actual, a partir de una distribución que incluía el norte del territorio bonaerense (i.e.: proximidades de Luján), fue explicada por causas fundamentales climáticas (Tonni y Politis 1980).

Los tuco-tuco (*Ctenomys* sp.) tampoco se encuentran actualmente en el área, aunque sí se hallan poblaciones bien establecidas en la costa atlántica y en las márgenes del río Quequén Salado (*vide* Contreras y Reig 1965).

La mara (*Dolichotis patagonum*) no habita actualmente en el área ni en áreas próximas, pero seguramente estuvo representada hasta fines del siglo pasado. Lo mismo ocurre con *Blastoceros bezoarticus*.

En cuanto al huroncito patagónico (*Lyncodon patagonicus*) y al pichi (*Zaedyus pichiy*), es probable que se encuentren actualmente en el área del sitio o en sus proximidades, por ello no está documentado.

En suma, de las especies citadas, sólo el guanaco no está ni lo estuvo por lo menos en tiempos históricos, presente en el área. Las otras especies pudieron estar presentes con poblaciones bien establecidas, cuando la densidad demográfica era menor o no incidían otros factores de alteración y deterioro ambiental. Sin embargo, respecto a esto último debe señalarse que tal explicación no es la única y más correcta, sino sólo la más simple y fácilmente contrastable.

En el Hemisferio Norte, donde se poseen datos múltiples acerca de fluctuaciones climáticas en tiempos recientes, se han realizado numerosos trabajos sobre los cambios en la distribución de distintas especies de mamíferos y aves como respuesta a dichas fluctuaciones climáticas (vase como ejemplo los aportes de Harris 1964; Moreau 1969; Schultz 1972).

Si bien en la Argentina hay escasos trabajos , y muy recientes, que incorporen al campo observacional la existencia de posibles fluctuaciones climáticas, este factor debe ser considerado para la explicación de ciertas corologías actuales. En tal caso, la presencia y frecuencia de algunas especies, hasta ahora aleatoriamente consideradas, tendrán un alto valor en las investigaciones paleoambientales y paleozoogeográficas.

Retomando el tema central de este Apartado, es posible inferir que las condiciones ambientales durante la depositación de los sedimentos portadores del material faunístico analizado, fueron algo distintas a las actuales en la misma zona. Ciertamente, esto es en cuanto a la información que proporcionan los restos óseos recuperados.

El partido de Tres Arroyos tiene actualmente un clima C_2B' r a', es decir subhúmedo húmedo, mesotérmico, con escaso déficit de agua y concentración estival de la eficiencia térmica menor al 48%. Bajo este régimen climático se desarrolla una vegetación de gramíneas indígenas con predominio de especies de los géneros *Stipa*, *Pictochaetium* y *Poa* y gran diversidad de gramíneas espontáneas.

La presencia y frecuencia de mamíferos cuyas especies, aunque habitan o habitaron recientemente en el Dominio Pampásico, también se encuentran en el Norte del Patagónico, permite inferir condiciones algo más áridas que las actuales. Al respecto es importante señalarse que el Dominio Pampásico, como lo han destacado los zoogeógrafos, es un área de engranaje con los Dominios Subtropical, Patagónico y por el Oeste con el Central. En las condiciones actuales se verifica una penetración hacia el sur de especies de marcada "estirpe" subtropical, por lo menos hasta la latitud del Partido de Necochea, por el Este (*vide* Reig 1964, 1965). No hay datos referentes al Partido de Tres Arroyos que permitan la confirmación o disconfirmación de una situación similar.

Es evidente, empero, que en la teriofauna rescatada del Sitio 2 de Arroyo Seco, se registran especies de "estirpe" patagónica y ninguna decididamente subtropical. Más aun, una de ellas, el guanaco, es indicadora de condiciones ambientales más áridas (*vide* Tonni y Politis 1980).

En suma, el actual territorio bonaerense es un área de engranaje entre distintos ámbitos zoogeográficos, sin endemismos conspicuos, por lo cual no es de esperar cambios marcados en la composición faunística como respuesta a fluctuciones climáticas, sino sólo variaciones en la distribución. Estas variaciones serán más o menos marcadas dependiendo de la situación geográfica del sitio prospectado. En el Sitio 2 de Arroyo Seco se verifica un de las citadas variaciones en la distribución de ciertos mamíferos, cuya explicación más parsimoniosa radica en las fluctuaciones climáticas.

Utilización del Recurso Fauna

De acuerdo con el análisis efectudo a través del estudio de los restos faunísticos del sitio 2 de Arroyo Seco, se desprende lo siguiente:

La presencia de ejemplares adultos, tanto de especies extinguidas como de fauna indígena puede interpretarse como: A) el resultado de la selectividad en la caza; o B) la época del año que se efectúa la caza.

Sin embargo, la segunda posibilidad no sería la más probable, dado que en los megamamíferos el período de maduración es largo, por lo cual los individuos jóvenes deberían estar presentes todo el año. Por lo tanto, es más correcto explicar la frecuencia de individuos adultos a través de un criterio selectivo en la caza.

La caza de megamamíferos extinguidos no fue una actividad habitual de los habitantes del sitio.

Probablemente esta situación se debió a la menor frecuencia de estos animales en la zona, con respecto a otros megamamíferos que, seguramente, se presentaban en mayor proporción.

Asimismo, la baja densidad demográfica de las bandas de cazadores haría muy dificultosa la captura de estos animales, los que, por otro lado, no podrían ser aprovechados en su totalidad dado el gran volumen de carne suministrado. Además, teniendo la posibilidad de obtener con mayor facilidad y más rápidamente animales de tamaño menor, tales como el guanaco y el venado, que de la misma manera les aportaran una cantidad de carne suficiente, es evidente que dichos megamamíferos extinguidos deben haber sido cazados sólo en ocasiones circunstanciales (*vide* Palanca y Politis 1979).

La alta proporción de elementos del autopodio, correspondientes a neoespecies, podría indicar que dicho paraje fue, probablemente el sitio donde la actividad principal fue el despostamiento secundario.

La presencia de extremos distales de los huesos largos y no así de sus diáfisis -que en general están sumamente fragmentadas- puede deberse a factores tafonómicos; pero, más probablemente, sean el producto de la rotura intencional para la extracción de la médula y/o la confección de instrumentos.

La existencia de varios fragmentos óseos inidentificables y de cáscara de huevo con indicios de fuego, podrían indicar algún tipo de cocción de los alimentos.

El alto porcentaje de restos de roedores puede haber sido incorporado al sedimento en forma natural, dado que la selección humana seguramente recayó sobre animales con mayor volumen de carne.

En cuanto a la presencia de los enterratorios, debe señalarse que no se ha observado ninguna situación diferente en lo que a fauna respecta, en aquellas cuadrículas donde se recuperaron restos óseos humanos. Por el contrario, se verificó un comportamiento faunístico semejante al del resto del área excavada.

Es importante destacar que tres elementos faunísticos - una placa de *Glyptodon* sp., una valva de *Amiantis* sp. articulada y un conjunto de caninos de Canidae- no se presentan en el registro con la frecuencia de los demás taxa, sino que están formando parte exclusivamente del ajuar funerario.

En lo que respecta a los caninos, varios esqueletos humanos articulados presentaban como ajuar funerario collares con pendientes elaborados con caninos superiores e inferiores de Cánidos.

Los caninos de los Cánidos (se sigue aquí la clasificación de los Canidae propuesta por Van Gelder 1978), no son buenos elementos diagnósticos, por lo cual no fue posible precisar las hipótesis sistemáticas más allá del nivel familiar. Sin embargo, por el tamaño y morfología es posible separar dos categorías de caninos: 1) Caninos de tamaño y morfología similar a los inferiores y superiores de *Canis (Pseudalopex) gymnocercus* (Fischer 1814). 2) Caninos de tamaño similar a los de *Chrysocyon brachyurus* (Illiger 1815) o a los de *Canis (Pseudalopex) culpaeus* (Molina 1782). Algunos caninos de esta categoría son -además- morfológicamente inseparables de los de *Chrysocyon brachyurus*.

Es necesario destacar que en el sitio sólo se han hallado dos restos esqueletarios (una vértebra y una bulla timpánica) aparte de los caninos atribuibles a cánidos. Por otra parte, actualmente en el área sólo se encuentra un cánido silvestre: *Canis (Pseudalopex) gymnocercus* (*fide* Cabrera 1957/60), aunque en el Pleistoceno tardío probablemente se encontrara otra especie, *Dusycion avus* (Burmeister 1884), cuyos caninos son de tamaño y morfología similar a algunos de los del grupo 2).

Como se señaló al comienzo, los caninos no son buenos elementos diagnósticos, al menos considerados aisladamente y más aun cuando han sido alterados parcialmente para la confección de pendientes. Por ello, aparte de las expuestas, existen varias otras posibilidades respecto a la asignación del material:

a) Corresponden a algún otro cánido silvestre con características de tamaño y morfología similares a los considerados. Estos cánidos pueden ser especies aun existentes que actualmente no habitan el área.

b) Corresponden a *Canis (Canis) familiaris*, dentro de cuya amplia variabilidad hay morfologías similares a las observadas.

A pesar que los caninos usados como pendientes hallados en este sitio son muy numerosos, no son elementos cualitativamente importantes como para confirmar alguna de las proposiciones esbozadas. De cualquier forma, la asignación a *C. (P) gymnocercus* y a *Chrysocyon brachyurus* parecen ser las más probables.

Si se considera que excepto los caninos, sólo hay registros de otros dos elementos esqueletarios atribuibles a Cánidos, es claro que los ejemplares no fueron capturados en el lugar. Por esta causa surgen otras variables en cuanto a la interpretación de los materiales, lo que se vincula directamente con las explicaciones arqueológicas (véase apartado Arqueologia).

Características de la Ocupación

Los aspectos paleoetnozoológicos explicitados en los distintos apartados permiten formular algunas consideraciones respecto a la funcionalidad del Sitio 2 de Arroyo Seco.

Para ello debe considerarse que el sitio no fue aun excavado totalmente. Esto, por sí mismo, plantea una limitación en la asignación del mismo a una categoría, dado que el sector trabajado puede ser una muestra representativa que permita referirlo a una categoría definida, o sólo puede representar un área de actividad dentro de un contexto más amplio.

En consecuencia, parece más correcto asumir que representa un área de actividad. En tal sentido, la actividad predominante en el sitio parece haber sido el despostamiento secundario.

Las evidencias que apoyan esta hipótesis son: 1) alta proporción de elementos del autopodio; 2) escaso porcentaje de huesos largos, elementos craneanos (a excepción de algunos dientes) y del esqueleto axial (vértebras y costillas); 3) huesos con incisiones producidas por instrumentos cortantes.

No obstante, resulta dificultoso establecer si los habitantes del sitio consumían o no allí los alimentos. Se registran elementos óseos y cáscaras de huevo fragmentados con indicios de fuego, que implicarían algún tipo de cocción de los mismos. Sin embargo, en caso de haber consumido en el lugar, debería registrarse mayor cantidad de huesos largos, además de otros elementos que también suministran un gran volumen de carne.

Respecto al estado de conservación de los huesos largos, es necesario considerar, por un lado, los procesos tafonómicos y por el otro, la rotura intencional para la extracción de la médula y/o la confección de instrumentos. Esta última posibilidad conduce a considerar otro tipo de actividad, además del despostamiento secundario.

Como ya se expresó anteriormente, los huesos largos son escasos y en su mayoría corresponden a elementos del esqueleto apendicular de megamamíferos extinguidos.

Con respecto a esto, es probable que dichos animales hayan sido capturados ocasionalmente en las inmediaciones del sitio, abandonando allí los elementos óseos de gran tamaño, previa separación de la masa muscular útil. Una evidencia de esto la constituyen las numerosas marcas que se observan en la cresta cnemial de la tibia de *Megatherium americanum* (*vide* Tonni et al. 1982) y en las diáfisis de fémur de *Equus (Amerhippus)* sp., que seguramente fueron producidas con instrumentos cortantes empleados para la separación de la masa muscular.

Asimismo, debe tenerse en cuenta que no es posible precisar desde el punto de vista faunístico, si todos los elementos óseos recuperados fueron utilizados por un único grupo en el mismo momento, o por varios grupos en momentos distintos. Para ello debe complementarse la información con los resultados del registro arqueológico. Esta es otra de las razones por la cual resulta altamente aleatorio ubicar al Sitio en una categoría definida.

En suma, la actividad principal desarrollada en el momento de ocupación del Sitio 2 probablememte haya sido el despostamiento secundario. Obviamente, otras actividades pueden haberse efectuado simultáneamente, pero ellas no han dejado evidencias arqueológicas.

ARQUEOLOGIA

El material arqueológico descrito en este apartado ha sido recuperado en los trabajos de campaña realizados en 1979 y 1980.

Estas tareas se distribuyeron de la siguiente manera: febrero, mayo y noviembre de 1979 y marzo-abril de 1980, totalizando 110 días de trabajo sobre el terreno. Diariamente se mantuvo un promedio de 10 personas en las excavaciones.

La metodología de la excavación ya fue expuesta en el apartado correspondiente.

La ubicación de las cuadrículas y de la trinchera se registró en un plano general del sitio (Figura 2) que se confeccionó de acuerdo a datos provenientes de la fotografía aérea del I.G.M., datos altimétricos y distancias acotadas por el ingeniero Ricardo Di Rocco. El mapeo del área excavada fue efectuado por el Arq. Luis Zunino y los colaboradores Lic. M. Alvarez, Srta. Teresa Acedo y Lic. Mónica Salemme.

Las primeras cuadrículas se abrieron en terrenos cercanos a los excavados por los señores Móttola, Elgart y Morán. Las cuadrículas 21, 22, 23, 24 y 25 se dispusieron en forma de "L", limitando por el Sur y por el Este con una excavación de 3 x 3 m, que habían abierto los aficionados citados y en donde se registró la presencia de un enterratorio secundario (cuadrícula A 20, en la anotación de Móttola, Elgart y Morán).

Las primeras veinte cuadrículas se concentraron en un sector del sitio, denominado sector A y posteriormente en sectores alejados del núcleo de las excavaciones, con el fin de detectar los límites de la ocupación.

Durante el final de la 3ra. campaña, en noviembre de 1979, invitados por el Museo

Municipal José A. Mulazzi, se reunieron en el lugar excavado algunos investigadores, cuyos temas de estudio están relacionados con los hallazgos de Arroyo Seco. Esta reunión contó con la presencia de: Lic. Carlos Aschero, Prof. Antonio Austral, Lic. Luis Borrero, Dr. Rodolfo Casamiquela, Lic. Néstor Kristkautzky, Dr. Alberto Marcellino y Lic. Abel Orquera.

Del fluído intercambio de opiniones, surgieron interesantes propuestas que fueron incorporadas al posterior trabajo de campo y gabinete. Los autores agradecen a los investigadores asistentes su participación y el valioso aporte al estudio del sitio.

Una de las propuestas surgidas de esta reunión fue la de realizar una trinchera que cortara transversalmente la "lomada" en la cual se localiza el sitio, con la finalidad de ampliar las observaciones geológicas hasta el mento. Teniendo en cuenta que los objetivos principales de este trabajo no eran la recuperación de material, es que se utilizó una metodología de excavación distinta.

Figura 5. Vista de varias cuadrículas y de la trinchera del Sitio 2.

La trinchera fue excavada a pala, en capas artificiales de 0,20 m, todo el sedimento extraido fue pasado por zaranda y tuvo una longitud de 80 m por un ancho de 0,60 m. Cada 6 m se conservaron testigos transversales de 0,20 m de ancho. A lo largo de la trinchera se ubicaron dos cuadrículas, 39 y 48, para controlar detalladamente la estratigrafía y su vinculación con los elementos culturales y faunísticos.

De esta forma la superficie excavada a través de cuadrículas de aproximadamente 132 m^2 más la superficie que ocupó la trinchera de 51 m^2.

Algunos factores de alteración deben ser considerados, pues probablemente modificaron la disposición original de algunos materiales.

Las cuadrículas 22, 23 y 24 tenían los primeros 0,65 m superiores alterados totalmente en los cuales había mezcla de carbón, huesos quemados, vidrio hierro, artefactos líticos etc. Probablemente esta disturbación sea el resultado de la excavación y posterior relleno de un gran pozo pues se tiene conocimiento que ha principios de siglo existió en las inmediaciones una vivienda y vinculado a ella puede estar relacionado el origen del pozo mencionado.

Asimismo se registraron cuevas de mamíferos fosoriales en varias cuadrículas y en distintos niveles. La característica forma de galería, el sedimento más oscuro y flojo que el circundante y, en algunos casos, los restos óseos de los organismos cavadores permitieron la identificación de los límites del sedimento disturbado.

Las modificaciones que produjeron antiguos pozos realizados por aficionados fueron determinadas por presentar líneas de contacto rectas entre el sedimento oscuro y flojo de relleno y el circundante, más claro y compacto.

En S y Z fueron registrados restos óseos de *Ctenomys* sp., pero resultó difícil determinar los restos de las cuevas. La alteración debida a estos organismos no ha sido muy importante de acuerdo

Figura 6a. Base de la Unidad Y (vease perfil de la Figura 3) donde se observan restos dispersos de *Megatherium americanum* e instrumentos líticos. La flecha indica la pieza 41/30.

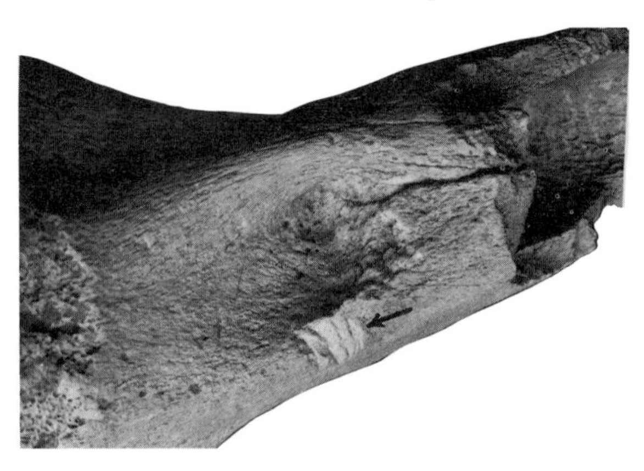

Figura 6b. Mitad distal de fémur de *Equus* (*Amerihippus*) sp. La flecha señala probables huellas de descarne.

a la posición articulada de casi todos los restos humanos hallados en Z, pero igualmente ha sido considerada como un posible factor de alteración en la ubicación de los materiales más pequeños de S y Z. En Y las cuevas fueron facilmente identificadas y el material hallado con ellas no fue considerado en el presente trabajo.

Los 0,20-0,30 m superiores, a causa de las continuas aradas, se presentaban disturbadas. El material aquí recuperado estaría correctamente ubicado dentro de la unidad, pero no conservaría la posición original en el suelo.

Material Lítico

Para la confección del presente informe se han analizado 252 piezas, las cuales presentan evidencias de poseer retalla, retoque o microtoque intencional o de haber sido modificadas por pulimento, alisado o "picado." Hay una excepción que debe acotarse: 4 guijarros basálticos que no presentan evidencias de modificaciones de origen antrópico. Su inclusión en la lista descriptiva se debe a las características de la materia prima y a la distribución estratigráfica restringida.

En esta etapa de la investigación sólo se realizó un listado de los instrumentos en el cual se menciona la materia prima, las medidas, los caracteres morfológicos relevantes y la posición estratigráfica. No se han tratado las connotaciones funcionales vinculadas a la morfología de los instrumentos, pues esto se complementará en el futuro con otros análisis y con mayor número de registro.

Para la descripción técnico-morfológica se ha tomado como base el trabajo de Carlos Aschero "Ensayo para una clasificación morfológica de artefactos líticos aplicada a estudios tipológicos comparativos" (MS. Informe presentado al CONICET). En el mismo se definen los términos descriptivos que utilizan en el presente informe. Debido a que el trabajo de Aschero no ha sido publicado, se reproducirán en este informe algunas de las definiciones de los términos más frecuentemente usados.

"Caracteres morfológicos referentes a la forma-base:

Denominamos forma-base al artefacto o lito en estado natural sobre el que se elaboró un instrumento. En el primer caso, hablamos de lascas, láminas, hojas u otras variedades morfológicas de estos tres grupos mayores de artefactos; en el segundo caso, hablamos de guijarros, lajas, clastos, etc.; es decir de distintas formas de las rocas en estado natural."

"6. Grupo de los "cepillos" ("rabots"):

Instrumentos de filo normal irregular o regular (muy poco frecuente) con aristas regulares o sinuosas regulares. Bisel asimétrico; extensión del filo: perimetral o extendido; módulo de espesor: grueso.

Serie técnica: instrumentos de talla extendida, instrumentos de talla extendida y retoque marginal.

Formas-base: nódulos, lascas nodulares, lascas gruesas.

"7. Grupo de los raspadores:

Instrumentos de filo normal regular, con aristas regulares. Bisel asimétrico; extensión del filo: restringido, corto, extendido o perimetral; módulo de espesor: variable.

Serie técnica: instrumentos de retoque marginal, instrumentos de talla extendida y retoque marginal, instrumentos de retoque extendido.

Formas-base: lascas, láminas hojas, lascas gruesas.

"9. Grupo de las raederas:

Instrumentos de filo normal regular, de arista regular. Bisel asimétrico; extensión del filo: largo; módulo de espesor: delgado o mediano.

Serie técnica: instrumentos de retoque marginal, instrumentos de retoque profundo, instrumentos de talla extendida y retoque marginal, instrumentos de talla bipolar y retoque marginal"

Formas-base: lascas, láminas, hojas, nódulos tabulares delgados.

Los materiales han sido, ordenados en el listado en base a su posición estratigráfica (Vease: *Geología*, pp. 230-236).

La organización de los datos en la descripción es la siguiente:

ej.: 27/7/ Cuarcita 20x30x16 fb: Lasca. Raedera doble convergente.

En donde: 27 es el número de la cuadrícula en la cual fue hallado el instrumento y 7 el número de catálogo, independiente para cada cuadrícula. 20x30x16 son las medidas del instrumento expresadas en milímetros que corresponden al largo, el ancho y el espesor respectivamente. fb.: Lasca. Es la forma base sobre la cual fue confeccionado el instrumento. Raedera doble convergente: son las características morfológicas más relevantes del instrumento.

En los instrumentos donde se consigna "fragmento" se refiere a aquellos en los cuales existen evidencias de haberse fragmentado con posterioridad a la confección de la pieza, ya sea durante su uso o por factores tafonómicos.

Con excepción de la cuarcita, la calcedonia y el basalto, el resto de la materia prima fue determinada megascópicamente por el Dr. Mario Teruggi, al cual los autores le agradecen su colaboración.

La orientación de las piezas se realizó de acuerdo con el eje morfológico.

Descripción del Material Lítico

21/5 Cuarcita 73x35x15 fb.:lámina. Raedera doble lateral convergente en punta roma con retoque alterno. La base está adelgazada. Filos festoneados irregulares.

21/3 Basalto 95x42x33 Fragmento con una cara pulida.

21/ Basalto (negro) 57x41x31 Nódulo sin evidencias de utilización y/o modificación de origen antrópico.

24/7 Toba Silicificada 31x25x9 fb.:lasca. Punta bifacial en forma de triángulo isósceles, con filos aserrados y base ligeramente cóncava y adelgazada. Retalla y retoque extendidos sobre las dos caras.

25/2 Cuarcita 42x28x10 fb.:lasca. Raspador distal con filo lateral complementario en bisel asimétrico oblicuo. Con retalla extendida sobre la cara dorsal.

25 bis/1 Opalo ocráceo 15x26x11 fb.:lasca secundaria. Fragmento de lasca con retoque lateral bifacial más retoque lateral unifacial. Evidencias de técnica bipolar.

25 bis/3 Toba 32x19x6 fb.:Nódulo. Fragmento con microretoque bifacial sobre un borde formando un filo sinuoso regular sobre arista.

26/4 Cuarcita 67x26x14 fb.:Lámina. Raedera doble lateral convergente con una punta destacada triédrica con retoque unifacial. Retalla extendida sobre cara ventral (Figura 7).

26/5 Fangolita silicificada 88x86x83 Fragmento hemiesferoide con una superficie regularizada por percusión.

27/1 Cuarcita 67x47x15 fb.:lasca. Raedera doble lateral alterna.

27/2 Toba silicificada 29x30x19 fb.:lasca secundaria. Raspador frontal + filo complementario lateral sobre cara ventral en bisel asimétrico abrupto + filo complementario proximal sobre cara ventral en bisel asimétrico abrupto.

29/1 Cuarcita 68x47x18 fb.:lasca. Raedera doble lateral con retoque alterno.

33/1 Pórfido 92x67x41 Fragmento, Lito con una cara pulida y borde alisado.

33/3 Cuarcita 35x30x5 fb.:lasca. Lasca con un filo lateral corto en bisel asimétrico abrupto + un filo complementario en bisel asimétrico oblicuo lateral.

(de 33/4 a 33/17, en X y en parte superior de Y)

33/4 Calcedonia 25x15x10 fb.:lasca. Raspador frontal + filo complementario lateral en bisel asimétrico abrupto.

33/7 Opalo ocráceo 21x11x15 fb.:lasca. Lasca con retoque en el extremo proximal y distal. Evidencias de técnica bipolar.

33/15 Cuarcita 65x47x9 fb.:lasca. Lasca con retoque alterno que forma un filo dentado irregular en los bordes.

33/17 Basalto 35x30x5 fb.:lasca. Lasca con retoque discontinuo bifacial sobre un borde.

35/1 Cuarcita 57x24x11 fb.:lámina. Raedera doble lateral convergente en punta en el extremo distal. En el extremo proximal los filos convergen en punta roma. La cara dorsal tiene retalla extendida.

36/2 Basalto 37x26x17 Guijarro sin modificaciones de origen antrópico.

36/3 Cuarcita (grano grueso) 66x52x21 fb.:lasca. Raspador con filo extendido.

36/4 Cuarcita 19x20x4 fb.:lasca. Fragmento con microretoque alterno.

36/6 Toba silicificada 47x40x20 fb.:lasca. Lasca con retalla sobre la cara dorsal.

36/7 Cuarcita 40x51x8 fb.:lasca. Lasca con retalla bifacial y retoques discontinuos que forman un filo en bisel asimétrico oblicuo.

36/13 Cuarcita 20x16x8 fb.:lasca. Lasca con retalla. Evidencias de técnica bipolar.

37/1 bis Cuarcita 35x18x17 fb.:lasca. Lasca con retoque lateral que forman un filo en bisel asimétrico oblicuo.

37/2 bis Cuarcita 32x34x11 fb.:lasca. Fragmento de lasca con retoques en los bordes laterales sobre cara ventral.

38/1 Toba silicificada 56x62x20 fb.:lasca. Raspador con filo extendido.

38/8 Cuarcita 30x26x14 fb.:lasca. Lasca con filo en bisel asimétrico oblicuo.

41/1 Toba silicificada 23x21x7 fb.:lasca. Raspador frontal + un filo complementario lateral con bisel asimétrico oblicuo + punta destacada en extremo proximal fragmentado (Figura 7).

41/2 Cuarcita 18x32x6 fb.:lasca/fragmento. Sección proximal de punta bifacial con base recta y adelgazada y bordes subparalelos. El ancho máximo no es el de la base.

41/3 Cuarcita 23x11x7 fb.: Fragmento con dos filos laterales con bisel asimétrico, uno abrupto y otro oblicuo. Retalla y retoque extendidos sobre cara dorsal.

41/5 Cuarcita 36x27x11 fb.:lasca. Raedera simple lateral con punta no destacada en el extremo proximal.

41/7 Riolita 73x41x12 fb.:fragmento. Lito tabular con una cara alisada.

41/10 Cuarcita 27x22x18 fb.:lasca. Lasca con retoques en borde proximal y distal. Técnica bipolar.

41/13 Cuarcita 29x15x11 fb.:lasca. Fragmento de raedera simple lateral.

42/1 Cuarcita 17x16x4 fb.:lasca. Raspador con filo extendido.

42/2 Cuarcita 69x40x12 fb.:lámina. Fragmento. Raedera simple lateral recta.

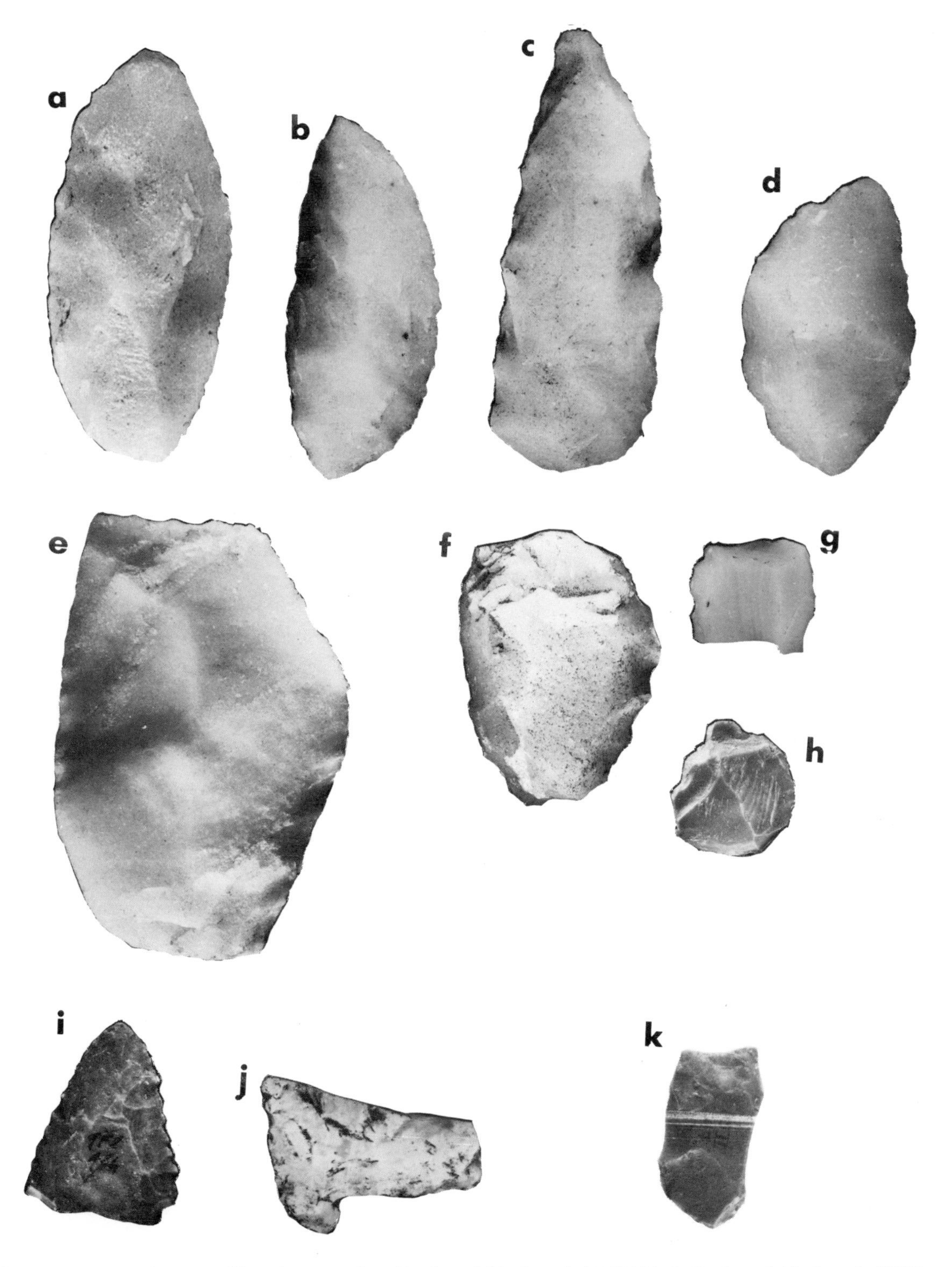

Figura 7. Componente Superior. (Tamaño natural). a.-Raedera doble lateral. (44/2 bis); b.-Raedera doble lateral. (35/1); c.-Raedera doble lateral. (26/4); d.-Raedera doble lateral. (45/7); e.-Raedera doble lateral. (27/1); f.-Raspador con filo extendido. (44/1 bis); g.-Fragmento. Raspador distal. (48/2); h.-Raspador frontal. (41/1); i.-Punta triangular bifacial. (24/7); j.-Fragmento. Base de punta bifacial. (53/25); k.-Lasca con retoque bifacial. (47/7).

42/3 Opalo ocráceo 16x20x5 fb.:lasca. Fragmento con retoque y retalla bifacial extendido. Posible pedúnculo de punta de proyectil.

42/8 Cuarcita 20x26x10 fb.:lasca. Fragmento con retoque y retalla sobre la totalidad de una cara.

42/2 bis Cuarcita 29x38x15 fb.:lasca. Lasca con retoque sumario sobre el borde distal.

44/1 bis Cuarcita 44x34x10 fb.:lasca. Raspador con filo extendido.

44/2 bis Cuarcita 64x27x15 fb.:lámina. Raedera doble lateral convergente en punta roma en el extremo distal y proximal. Retalla extendida sobre cara dorsal (Figura 7).

44/4 Toba silicificada 80x64x26 fb.:nódulo. Lito con lascados en los extremos y las caras alisadas.

44/6 bis Sílice tipo Jaspe 26x26x7 fb.:lasca secundaria. Lasca con retoque lateral.

44/10 Calcedonia 22x16x10 fb.:lasca secundaria. Raspador frontal.

44/11 Toba silicificada 31x25x12 fb.:lasca. Lasca con retoque lateral que forma un filo en bisel asimétrico abrupto.

(de 45/1 a 45/7 en X y parte sup. de Y)

45/1 Cuarcita 37x45x15 fb.:lasca. Lasca con retoques.

45/2 Cuarcita 47x27x7 fb.:lasca. Lasca con retoque lateral, formando un ápice en el extremo distal.

45/3 Toba silicificada 52x24x9 fb.:lámina secundaria. Raedera doble lateral convergente en punta, con retoque alterno.

45/5 Cuarcita 30x26x5 fb.:lasca. Lasca con microretoque lateral.

45/7 Cuarcita 46x37x13 fb.:lasca. Raedera doble lateral convergente en punta roma en el extremo distal y proximal (Figura 7).

47/1 Cuarcita 60x53x12 fb.:lasca. Lasca con microretoques ultramarginales.

47/2 Cuarcita 36x16x13 fb. :lámina. Raedera simple lateral.

47/4 Basalto 32x24x6 fb.:nódulo. Raspador frontal + filo complementario lateral en bisel asimétrico oblicuo.

47/14 Cuarcita 40x26x10 fb.: lasca. Raspador distal + raedera simple lateral.

48/2 Calcedonia 21x17x11 fb.:lasca. Fragmento de raspador distal.

49/1 Opalo ocráceo 18x15x6 fb.:lasca secundaria. Lasca con retoques que forman un filo en bisel asimétrico oblicuo.

50/2 Cuarcita 25x18x6 fb.:lasca. Lasca con retoque lateral que forma un filo en bisel asimétrico oblicuo + retoque lateral sobre cara ventral que forma un filo en bisel asimétrico oblicuo.

50/3 Basalto 63x29x20 Guijarro sin modificaciones de origen antrópico,

50/1 Cuarcita 44x21x8 fb.:lámina. Fragmento de raedera doble convergente en punta roma.

53/34 Basalto 56x42x30 fb.:nódulo. Guijarro partido transversalmente. Presenta "puntos de picado" en un extremo.

53/33 Basalto 67x37x30 Fragmento. Lito con dos caras alisadas.

53/31 Cuarcita 20x20x10 Fragmento fb. Lasca. Fragmento con filo en bisel asimétrico abrupto + filo complementario alterno en bisel asimétrico oblicuo.

53/24 Cuarcita 27x20x6 fb.:lasca. Lasca con un filo lateral en bisel simétrico oblicuo.

53/22 Cuarcita 18x20x4 fb.:lasca. Raspador perimetral.

53/25 Silex 26x34x7 fb.:lasca. Fragmento de base de punta bifacial con lados paralelos. Base irregular.

53/32 Indet. 55x31x15 fb.:lasca secundaria. Lasca con retoque lateral.

Parte Superior de Y

21/14 Calcedonia 20x33x10 fb.:lasca. Raspador con filo extendido (Figura 8).

21/15 Cuarcita 36x36x11 fb.:lasca. Raedera doble lateral con retoque alterno y muesca en extremo distal.

21/62 Cuarcita 28x18x14 fb.:lasca. Lasca con retalla abrupta perimetral.

25/5 Cuarcita 61x31x11 fb.:lasca. Raedera doble lateral convergente en punta (Figura 8).

25/6 Cuarcita 49x43x10 fb.:lasca. Lasca con microretoque ultramarginal.

26/10 Cuarcita 52x25x14 fb.:lámina. Raedera doble lateral convergente en punta. Retalla y retoque extendidos sobre cara ventral.

27/26 Cuarcita 41x28x6 fb.:lasca. Lasca con retoque continuo lateral que forma un filo en bisel asimétrico oblicuo.

27/27 Toba 62x40x15 fb.:lasca. Raedera simple lateral + retoque alternante sobrel el borde opuesto.

29/2 Cuarcita 37x44x11 fb.:lasca. Raedera transversal + filo complementario lateral en bisel asimétrico abrupto.

29/10 Cuarcita 41x40x25 fb.:lasca secundaria. Raspador distal + denticulado irregular formando una punta no destacada.

31/1 Lutita silicificada 52x36x11 fb.:lasca. Lasca con retalla.

32/1 Cuarcita 39x30x7 fb.:lasca. Lasca con retoque lateral + muesca lateral.

32/6 Toba silicificada 64x29x9 fb.:lámina secundaria. Raedera simple lateral.

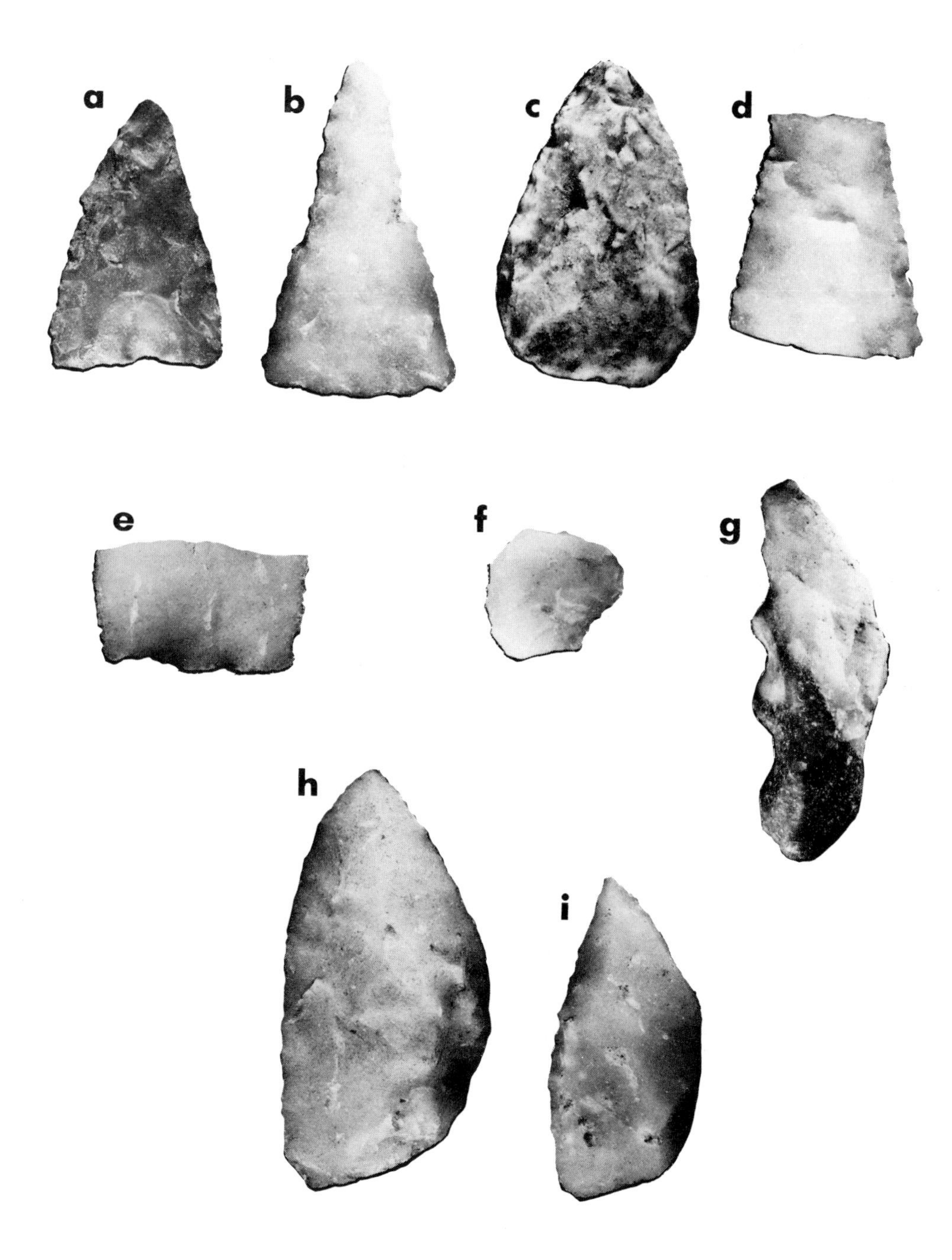

Figura 8. Componente Medio. (Tamaño natural). a.-Punta triangular bifacial. (44/3); b.-Punta triangular bifacial. (53/14); c.-Punta triangular bifacial. (Ps/1); d.-Fragmento. Punta triangular bifacial. (34/2); e.-Fragmento. Base de punta de proyectil bifacial. (38/4); f.-Raspador con filo extendido. (21/14); g.-Lámina con filo en bisel asimétrico abrupto. (34/5); h.-Raedera doble lateral. (25/5); i.-Raedera doble lateral. (47/33).

32/10 Cuarcita 31x31x15 fb.:lasca. Raspador distal con retoque alterno extendido.

32/11 Basalto 21x21x5 fb.:lasca secundaria. Raedera doble-proximal y distal- con retoque bifacial en los extremos. Evidencias de técnica bipolar.

32/15 Cuarcita 14x15x4 fb.:lasca. Raspador perimetral.

32/7 Ignimbrita 111x72x49 Lito con las caras pulidas y un extremo muy pulido. El extremo opuesto presenta "puntos de picado." (Figura 9).

35/2 Cuarcita 37x25x6 fb.:lasca. Lasca con microretoque en los bordes.

38/2 Granito 137x104x15 Fragmento tabular con dos caras pulidas y un borde alisado.

39/1 Cuarcita 35x22x12 fb: lasca con muesca lateral y retoque distal que forma un filo en bisel asimétrico oblicuo.

39/2 Cuarcita 47x32x19 fb.:lasca primaria. Raspador distal + un filo dentado irregular en bisel asimétrico abrupto.

39/3 Cuarcita 31x17x7 fb.:lasca. Lasca con retoque lateral.

39/4 Basalto 62x48x29 fb.:nódulo. Guijarro partido transversalmente con un negativo de lascado sobre una de sus caras.

39/6 Cuarcita 29x20x11 fb.:lasca. Fragmento. Raedera doble lateral convergente en punta con retoque alterno. Retalla extendida sobre cara ventral.

39/8 Cuarcita 42x25x5 fb.:lasca. Lasca con retoque continuo.

39/10 Cuarcita 36x20x7 fb.:lasca. Lasca con retalla extendida sobre cara dorsal.

39/9 Cuarcita 37x24x14 fb.:lasca. Lasca con microretoque sumario.

39/13 Basalto 62x33x28 fb.:nódulo. Guijarro sin modificaciones de origen antrópico.

39/12 Vulcanita 68x61x38 Fragmento. Lito con los extremos distal y proximal planos y pulidos

46/1 Cuarcita 35x28x11 fb.:lasca. Fragmento. Raspador con filo extendido.

46/2 Toba silicificada 56x46x17 fb.:lasca secundaria. Lasca con retoque sumario.

46/3 Cuarcita 38x24x11 fb.:lasca. Lasca con retoque lateral.

47/16 Traquita - Riolita de grano fino 98x120x30 fb.:nódulo o lasca nodular. Lito tabular con retoque frontal que forma un filo en bisel asimétrico abrupto. La cara ventral es totalmente plana y pulida. La cara dorsal tiene reservas de corteza en aproximadamente el 70% de la superficie.

47/18 Toba silicificada 18x20x14 Lasca con retalla y retoque.

47/19 Cuarcita 19x17x5 fb.:lasca. Raspador frontal.

Figura 9. Componente Medio. a.-Cepillo con filo perimetral. Vista superior (39/40). Tamaño natural. b.-Cepillo con filo perimetral. Vista lateral (39/40). Tamaño natural. c.-Lito con caras pulidas y un extremo muy pulido (32/7).

47/17 Cuarcita 49x25x11 fb.:lasca. Raedera doble lateral convergente

47/21 y 47/25 Cuarcita 94x67x30 (dos fragmentos que se unen entre sí). Núcleo chato con evidencias de rupturas frescas.

47/22 Cuarcita 31x26x9 fb.:lasca. Lasca con retoque bifacial en el extremo distal y proximal con un filo en bisel oblicuo. Evidencias de técnica bipolar.

47/23 Cuarcita 64x62x21 fb.:lasca. Lasca con retoques continuos en el borde proximal.

47/24 Cuarcita 37x34x15 fb.:lasca. Raedera simple lateral.

47/26 Pórfido 32x29x13 Fragmento. Lito con una superficie pulida.

47/27 Cuarcita 38x38x13 fb.:lasca. Fragmento con microretoque alterno.

47/30 Cuarcita 42x31x10 fb.:lasca secundaria. Lasca con microretoque continuo en los dos bordes laterales, sobre cara ventral.

47/33 Cuarcita 44x22x10 fb.:lámina secundaria. Raedera doble lateral convergente en punta (Figura 8).

47/31 Cuarcita 46x26x9 fb.:lasca. Lasca con microretoques continuos sobre cara ventral.

48/3 Micacita o esquisto micáseo 77x46x12 Fragmento con una cara alisada.

51/1 Cuarcita 59x52x27 fb.:nódulo. Fragmento. Lito con dos caras pulidas y puntos de picado en el centro de una de las caras.

51/3 Cuarcita 35x26x9 fb.:lasca. Fragmento. Raedera simple lateral + filo complementario alterno en bisel asimétrico oblicuo. Ambos filos convergen en punta no destacada.

52/5 Cuarcita 29x15x5 fb.:lasca. Lasca con retoque extendido que forma un filo en bisel asimétrico oblicuo.

52/3 Cuarcita 18x21x6 fb.:lasca. Fragmento. Lasca con retoque lateral bifacial que forma un filo en bisel simétrico oblicuo.

53/1 Cuarcita 27x33x9 fb.:lasca. Raedera doble lateral convergente en punta roma.

53/2 Cuarcita 27x33x9 fb.:lasca. Fragmento. Lasca con retoque lateral sobre cara ventral que forma un filo en bisel simétrico oblicuo.

53/3 Cuarcita 30x22x11 fb.:lasca. Fragmento de raedera simple lateral.

53/4 Cuarcita 25x23x5 fb.:lasca. Fragmento con retoque lateral continuo que forma un filo en bisel asimétrico oblicuo.

53/6 Cuarcita 47x30x20 fb.:lasca secundaria. Lasca con retalla y retoque sumario.

53/8 Cuarcita 34x17x9 fb.:lámina. Fragmento. Raedera doble lateral convergente.

Ps/1 Indet. 46x30x9 fb.:lasca secundaria. Punta bifacial en forma de triángulo isosceles con base recta y adelgazada. Bordes laterales ligeramente cóncavos. Retalla extendida sobre ambas caras (Figura 8).

21/22 Toba silicificada 41x55x17 fb.:lasca secundaria. Lasca con retoque sumario.

28/9 Cuarcita 23x25x9 fb.:lasca. Lasca con retoque en el extremo distal y proximal. Evidencias de técnica bipolar.

28/8 Indet. 39x56x11 fb.:lasca. Raedera simple lateral.

28/11 Cuarcita 33x17x8 fb.:lasca. Raedera doble lateral convergente.

32/22 Cuarcita 47x27x12 fb.:lasca. Fragmento. Raedera doble lateral con retoque alterno. La cara dorsal tiene retalla extendida.

32/43 Riolita 106x86x32 Lito subretangular con dos caras alisadas. En los extremos presenta algunos negativos de lascado.

33/16 Riolita o Pórfido cuarcífero 64x45x19 Fragmento. Lito con una cara muy pulida y un borde alisado.

34/2 Cuarcita 33x28x7 fb.:lasca. Fragmento de punta bifacial en forma de triángulo isósceles. Base recta con los bordes laterales rectos, levemente dentados. Le falta el tercio distal (Figura 8).

34/3 Cuarcita 21x19x6 fb.:lasca. Fragmento. Lasca con retoques que forman un filo en bisel simétrico oblicuo.

34/4 Cuarcita 25x25x10 fb.:lasca. Lasca con retoque en el extremo proximal y distal. Talla bipolar.

35/4 Pórfido 75x73x54 Lito con la superficie parcialmente pulida. Presenta una depresión subcónica en una de sus caras y en la opuesta está fragmentada con evidencias de una depresión similar a la anterior.

36/16 Cuarcita 21x21x26 fb.:lasca. Lasca con retalla o probable núcleo agotado.

37/1 Cuarcita 37x35x4 fb.:lasca. Fragmento de lasca con punta no destacada entre muescas.

37/2 Arenisca de grano fino 66x64x45 Fragmento. Lito con una depresión pronunciada subcónica, en el centro de una cara. Huellas de golpes en ambas caras.

37/4 Toba silicificada 49x26x25 fb.:nódulo o lasca nodular. Núcleo agotado.

37/3 Cuarcita 21x19x7 fb.:lasca. Lasca con retoque bifacial en el borde distal + un filo complementario lateral en bisel simétrico oblicuo.

38/4 Cuarcita 20x13x7 fb.:lasca. Fragmento. Sección proximal de punta de proyectil bifacial con base recta y adelgazada. El ancho mayor no es el de la base (Figura 8).

38/5 Cuarcita 50x28x6 fb.:lasca. Raedera doble lateral convergente en punta roma con retoque alterno.

39/14 Cuarcita 32x45x16 fb.:lasca. Lasca con retoque que forma un filo en bisel asimétrico abrupto + un filo en bisel asimétrico oblicuo.

39/17 Cuarcita 25x17x7 fb.:lasca. Raspador distal.

39/15 Basalto 51x49x30 fb.:nódulo. Fragmento partido longitudinalmente y con "puntos de picado" en uno de los extremos.

39/16 Cuarcita 44x22x11 fb.:lámina. Raedera doble lateral convergente en punta. La base está adelgazada con retoques en la cara ventral. Retalla extendida totalmente sobre la cara dorsal.

39/20 Cuarcita 41x23x9 fb.:lasca. Raedera simple lateral + filo complementario lateral con retoques sobre cara ventral.

39/40 Toba silicificada 57x50x37 fb.:núcleo o lasca nodular. Cepillo con filo perimetral muy abrupto (Figura 9).

40/2 Cuarcita 31x29x14 fb.:lasca. Lasca con retoque lateral.

41/21 Cuarcita 51x21x12 fb.:lámina. Raedera doble lateral convergente en punta roma (triédrica?).

41/23 Cuarcita 32x30x23 fb.:lasca secundaria. Raspador frontal.

41/27 Cuarcita 38x34x10 fb.:lasca. Raedera simple transversal.

42/5 Basalto 25x24x15 fb.:lasca. Lasca con retoque bifacial en el extremo distal + retoques en el extremo proximal. Evidencias de técnica bipolar.

42/6 Cuarcita 22x24x5 fb.:lasca. Lasca con retoque lateral.

42/7 Cuarcita 16x46x15 fb.:lasca. Fragmento de raedera doble lateral.

43/3 Cuarcita 48x35x17 fb.:nódulo. Lito rodado con "puntos de picado" en un extremo.

44/3 Toba silicificada 38x25x6 fb.:lasca. Punta bifacial con forma de triángulo isósceles con base levemente cóncava y bordes laterales ligeramente convexos. Base adelgazada y retalla extendida sobre ambas caras (Figura 8).

45/8 Cuarcita 30x26x7 fb.:lasca. Lasca con retoque lateral alternante.

45/9 Cuarcita 15x22x10 fb.:lasca. Fragmento con retoque lateral.

46/4 Cuarcita 31x37x7 fb.:lasca. Fragmento. Lasca con microretoque alterno.

46/5 Cuarcita 55x40x12 fb.:lasca. Lasca con retalla.

48/4 Calcedonia 36x20x8 fb.:lasca. Lasca con retoque discontinuo alterno.

49/3 Cuarcita 30x20x9 fb.:lasca. Fragmento. Raedera doble lateral convergente en punta destacada.

49/4 Cuarcita 37x31x15 fb.:lasca. Fragmento de raedera simple lateral.

50/4 Cuarcita 31x12x2 fb.:lámina. Lámina triangular con retalla y retoque total sobre cara dorsal.

50/6 Basalto 59x43x20 fb.:nódulo. Guijarro con "puntos de picado" en un extremo.

53/9 Cuarcita 36x26x9 fb.:lasca. Raspador con filo extendido.

53/13 Cuarcita 8x14x15 fb.:lasca. Lasca con microretoques. Evidencias de técnica bipolar.

53/14 Cuarcita 47x28x10 fb.:lasca. Punta en forma de triángulo isósceles bifacial, sin pedúnculo, con base ligeramente convexa. En la parte media del limbo hay un leve estrechamiento (Figura 8).

Parte Media de Y

28/25 Basalto 57x57x33 fb.:nódulo. Lito con una cara pulida, bordes alisados, con "puntos de picado" en sectores del borde y lascados sobre una de las caras.

31/3 Basalto 36x30x12 fb.:nódulo. Guijarro con retoque bifacial en un extremo formando un filo con bisel asimétrico oblicuo más "puntos de picado" en el extremo opuesto.

34/5 Cuarcita 52x20x17 fb.:lámina. Lámina con un filo en bisel asimétrico abrupto formando un dentado irregular (Figura 8).

34/6 Cuarcita 35x24x5 fb.:lasca. Lasca con retoque continuo sobre la cara ventral.

Parte Inferior de Y

21/30 Cuarcita 74x30x23 fb.:lámina. Lámina con un borde con retoque continuo y otro borde con retoque abrupto recto (Figura 10).

21/32 Cuarcita 47x26x28 fb.:núcelo. Cepillo con filo perimetral muy abrupto (Figura 11).

21/34 Toba silicificada 32x47x8 fb.:lasca primaria. Raedera simple transversal.

21/36 Diabasa 52x52x27 Hemiesfera pulida y fragmentada con una cara plana. Se trata de la mitad de una boleadora con surco.

21/48 Indet. 67x35x12 fb.:lasca. Raedera doble lateral convergente en punta roma con retalla parcialmente extendida (Figura 10).

21/26 Limolita 340x160x35 fb.:nódulo. Lito subrectangular con una cara alisada con una depresión poco profunda. En la otra cara presenta dos depresiones subcónicas.

25/18 Cuarcita 22x26x11 fb.:lasca. Fragmento con retalla bifacial y retoque lateral unifacial sobre un borde formando un filo en bisel simétrico agudo.

26/16 Toba 44x43x19 Fragmento con una cara pulida.

28/28 Toba endurecida (estratificación compacta) 78x44x30 Fragmento. Lito con dos caras pulidas y bordes alisados.

28/41 Cuarcita 31x25x7 fb.:lasca. Lasca con retoque distal formando un filo con bisel simétrico oblicuo + un filo lateral complementario con retoque continuo. Probable técnica bipolar.

31/4 Cuarcita 30x24x10 fb.:lasca. Raedera simple lateral.

31/5 Toba silicificada 46x29x10 fb.:lasca. Lasca con retoque lateral.

32/25 Calcedonia 34x20x6 fb.:lasca secundaria. Lasca con retoque bifacial en un extremo que forma

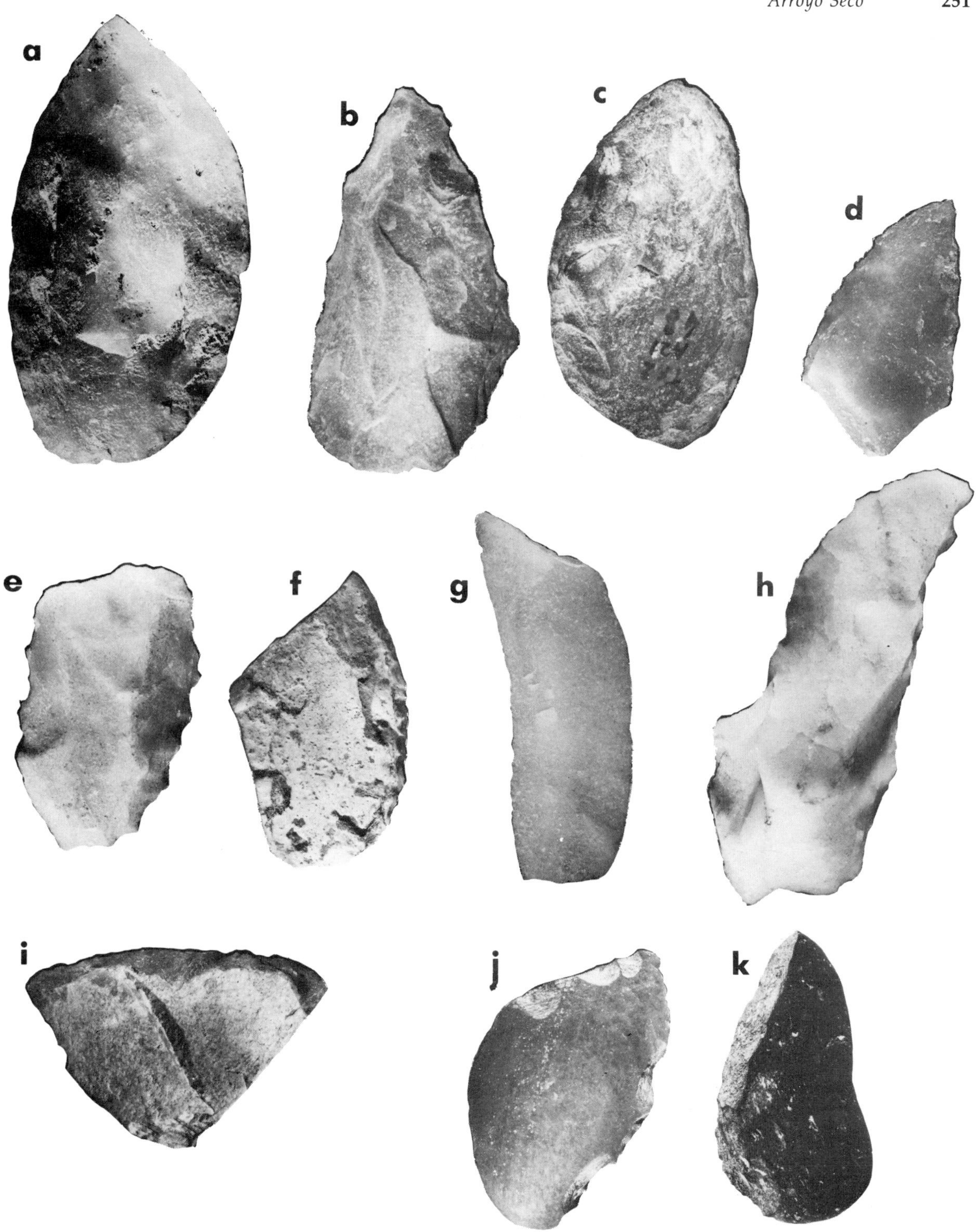

Figura 10. Componente Inferior. (Tamaño natural). a.-Raedera doble lateral (48/6); b.-Raedera doble lateral. (32/36); c.-Raedera doble lateral. (21/48); d.-Fragmento. Raedera simple lateral. (27/25); e.-Lasca con filo extendido dentado irregular. (23/18); f.-Fragmento con microretoque bifacial. (28/44); g.-Lámina con microretoque continuo. (53/19); h.-Lámina con retoque. (21/30); i.-Raedera simple transversal. (21/34); j.-Lasca con retoque lateral. (41/25); k.-Lito con retoque lateral. (47/35).

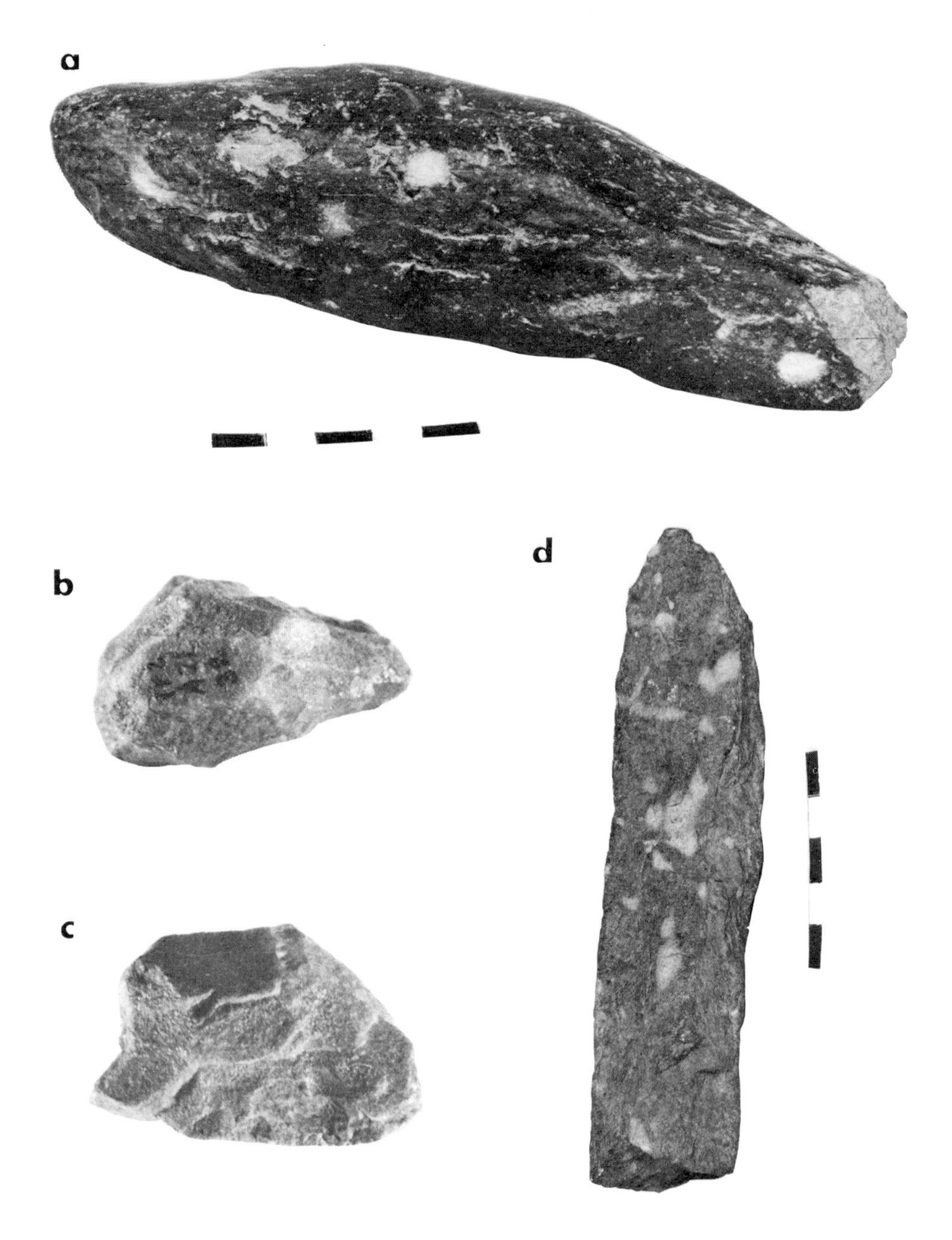

Figura 11. Componente Inferior. a.-Lito con puntos de picado en un extremo (41/30); b.-Cepillo con filo perimetral (21/32). Vista superior. Tamaño natural; c.-Cepillo con filo perimetral (21/32). Vista lateral. Tamaño natural; d.-Lito subrectangular con extremo en punta roma (37/8).

un filo en bisel simétrico oblicuo. Probable técnica bipolar.

32/26 Cuarcita 23x24x10 fb.:lasca. Lasca con retoque.

32/28 Cuarcita 20x16x4 fb.:lasca. Raspador frontal.

32/33 Cuarcita 45x14x7 fb.:lámina con microretoque sumario.

32/34 Cuarcita 32x11x4 Fragmento. Fb:lámina. Lámina con retoque.

32/35 Cuarcita 30x17x5 fb.:lasa. Lasca con retoque que forma una punta no destacada.

33/28 Cuarcita 26x24x15 fb.:nódulo. Núcleo agotado.

34/7 Calcedonia 38x23x10 fb.:lasca. Fragmento. Lasca con un filo en bisel asimétrico abrupto.

34/9 Cuarcita 25x20x4 fb.:lasca. Fragmento de raedera doble lateral. Retalla extendida sobre cara dorsal.

36/17 Cuarcita 32x16x11 Fragmento. Fb.:lasca. Lasca con retoque sumario que forma un filo en bisel simétrico oblicuo.

36/18 Cuarcita 28x16x7 fb.:lasca. Raedera doble convergente en punta roma.

37/7 Cuarcita 59x27x12 fb.:lámina. Lámina con un filo cóncavo lateral en bisel simétrico oblicuo. (Figura 10).

37/8 Fangolita-Lutita 159x41x22 fb.:lasca nodulalar. Lito subrectangular con un extremo en punta roma y retalla discontinua sobre ambas caras. (Figura 11).

38/7 Cuarcita 25x21x7 fb.:lasca. Lasca con retoque distal que forma un filo asimétrico en bisel oblicuo más un filo complementario lateral en bisel simétrico oblicuo. Evidencias de técnica bipolar.

41/25 Basalto 51x33x11 fb.:lasca. Lasca con retoque lateral continuo sobre los dos bordes laterales (Figura 10).

41/29 Cuarcita 31x15x6 fb.:lasca. Lasca con microretoque sumario.

41/30 Riolita 178x74x47 fb.:nódulo. Lito con "puntos de picado" en un extremo y negativos de lascados en el opuesto (Figura 11).

42/13 Cuarcita 27x21x5 fb.:lasca. Lasca con microretoque discontinuo. Evidencias de técnica bipolar.

45/12 Cuarcita 25x23x5 fb.:lasca. Lasca con retoque lateral.

46/6 Cuarcita 44x28x15 fb.:lasca. Raspador frontal + filo complementario lateral en bisel asimétrico oblicuo.

49/5 Cuarcita 36x30x29 fb.:núcleo o lasca nodular. Núcleo agotado.

53/15 Cuarcita 29x16x5 fb.:lasca. Lasca con retalla bifacial total que forma un filo perimetral en bisel simétrico oblicuo.

27/32/1 Cuarcita 27x22x7 fb.:lasca. Lasca con microretoque alternante.

S y parte Superior de Z

33/32 Cuarcita 32x22x7 fb.:lasca. Lasca con retoque sumario

33/33 Basalto 45x30x24 fb.:nódulo. Fragmento de guijarro con retalla discontinua.

41/36 Cuarcita 37x18x6 fb.:lasca. Lasca con microretoque.

42/10 Cuarcita 31x26x13 fb.:lasca. Lasca con retoque sobre un borde.

34/11 Basalto Guijarro esferoidal con un diámetro que oscila entre 26 y 27 mm.

21/64 Arenisca micácea estratificada 57x51x16 Fragmento tabular con las dos caras pulidas y un borde alisado.

21/55 Calcedonia 12x13x6 fb.:lasca. Raspador distal.

23/15 Cuarcita 27x20x6 fb.:lasca. Lasca con retoque sumario.

23/18 Cuarcita 47x30x12 fb.:lasca. Lasca con filo extendido dentado irregular. Bisel asimétrico. (Figura 10).

24/16 Diabasa 55x20x16 Fragmento con una cara alisada.

24/14 Calcedonia 14x11x6 fb.:lasca. Fragmento de raspador frontal.

25 bis/6 Cuarcita 31x10x11 fb.:lasca. Lasca con retalla.

25 bis/4 Toba 26x39x8 fb.:nódulo. Lasca con retoque en un borde.

28/44 Toba silicificada 47x28x6 fb.:nódulo. Fragmento con microretoque bifaciales sobre un borde determinando un filo sobre arista sinuoso regular y en bisel simétrico oblicuo (Figura 10).

32/36 Toba silicificada 62x35x14 fb.:lasca. Raedera doble lateral convergente en punta roma. (Figura 10).

32/37 Cuarcita 22x31x9 fb.:lasca. Lasca con filo lateral en bisel simétrico oblicuo.

36/21 Calcedonia 32x13x16 fb.:lámina. Lámina con microretoques.

37/10 Basalto 40x16x7 fb.:lámina primaria. Lámina con retoque en el extremo distal. Evidencias de técnica bipolar.

39/24 Cuarcita 22x28x7 fb.:lasca. Raedera doble transversal con retoque bifacial sobre el borde proximal y distal. Técnica bipolar.

47/35 Basalto 46x25x11 fb.:nódulo. Lito con retoque lateral que forma un filo en bisel asimétrico abrupto (Figura 10).

47/36 Cuarcita 17x15x8 fb.:lasca. Lasca con retoque continuo en dos bordes. Talla bipolar.

27/32/3 Cuarcita 28x32x9 fb.:lasca secundaria. Lasca con un filo en bisel asimétrico abrupto más una muesca complementaria.

27/32/4 Arenisca micácea 74x60x12 Fragmento tabular con una cara probablemente alisada.

21/51 Cuarcita 32x30x10 fb.:lasca. Lasca con

microretoque alterno. Evidencias de técnica bipolar.

24/17 Cuarcita 58x30x16 fb.:lasca. Lasca con microretoques. En el extremo distal tiene una punta triédrica no destacada.

24/18 Cuarcita 22x27x11 fb.:lasca. Fragmento de raedera doble.

24/10 Cuarcita 21x27x11 fb.:lasca. Fragmento con un filo en bisel asimétrico agudo + filo complementario en bisel asimétrico abrupto + muesca con microretoques. Retalla y retoques extendido sobre cara dorsal.

25/24 Cuarcita 20x16x4 fb.:lasca. Lasca con microretoques sumarios.

25/20 Basalto 18x20x10 fb.:lasca secundaria. Lasca secundaria con filo en bisel asimétrico abrupto.

25/26 Basalto 52x13x9 fb.:lasca primaria. Lasca con retoques discontinuos alternantes.

26/17 Toba silicificada 44x52x10 fb.:lasca. Lasca con retoque sumario.

27/22 Cuarcita 30x20x11 fb.:lasca. Fragmento. Lasca con retoque continuo.

27/24 Calcedonia 24x14x33 fb.:lasca. Fragmento con una cara con retoque y retalla extendida.

27/25 Cuarcita 42x24x11 fb.:lasca. Fragmento de raedera simple lateral con microretoque continuo que forma un filo primario dentado regular (Figura 10).

29/20 Cuarcita 39x29x10 fb.:lasca. Lasca con retoque.

32/40 Cuarcita 26x30x7 fb.:lasca. Fragmento con retoques que forman un filo en bisel simétrico oblicuo.

32/42 Cuarcita 34x10x6 fb.:lasca. Lasca con retoque sumario.

33/30 Cuarcita 19x16x15 fb.:lasca. Lasca con retoque sobre dos bordes.

33/31 Hematita 16x10x7 Fragmento con un pulimento leve.

41/37 Cuarcita 37x27x8 fb.:lasca. Lasca con retoque.

41/31 Cuarcita 28x23x21 fb.:núcleo. Raspador con filo extendido.

45/14 Cuarcita 36x26x8 fb.:lasca. Lasca con retoque alterno. Evidencias de técnica bipolar.

45/15 Cuarcita 31x24x9 fb.:lasca. Lasca con retoque continuo en los dos bordes.

48/6 Cuarcita 72x40x14 fb.:lasca. Raedera doble lateral convergente en punta con un filo dentado regular. Retalla y retoque totalmente extendido sobre cara dorsal (Figura 10).

52/6 Cuarcita 33x20x12 fb.:lasca. Lasca con microretoque lateral alternante.

53/19 Cuarcita 62x21x6 fb.:lámina. Lámina con microretoque continuo que forma un filo en bisel simétrico oblicuo. Punta no destacada en el extremo distal.

27/32/5 Basalto 29x24x10 fb.:lasca secundaria. Fragmento con un filo en bisel asimétrico abrupto.

29/25 Cuarcita 30x27x22 fb.:lasca nodular. Raspador frontal.

50/8 Cuarcita 29x18x5 fb.:lasca. Lasca con retoque continuo.

DESCRIPCION DE LA ALFARERIA

Los términos usados en la descripcíon del material de alfarería son los definidos en la "Primera Convención Nacional de Antropología." Todos los fragmentos, 15 en total, fueron hallados en X exclusivamente.

36/20 Sup. externa: color ante con manchas oscuras, tratamiento alisado. Sup. interna: color marrón oscuro, tratamiento alisado. Color de la pasta: negro, grosor 10 mm.

35/10 Sup. externa: color ante con manchas oscuras, tratamiento alisado. Sup. interna: color marrón oscuro con manchas negras, tratamiento alisado. Color de la pasta: negro, grosor 9 mm. Borde con labio levemente evertido. La superficie externa presenta decoración formada por incisiones poco profundas que forman un escalonado. Presenta restos de pintura roja sobre la superficie externa (Figura 12).

37/1 bis Sup. externa: color ante, tratamiento alisado. Sup. interna: color ante más oscuro, tratamiento alisado. Color de la pasta: ante claro, grosor 6 mm.

37/2 bis Sup. externa: color ante, tratamiento alisado. Sup. interna: color ante oscuro, tratamiento alisado. Color de la pasta: negro, grosor 8 mm.

37/3 bis Sup. externa: color marrón oscuro, tratamiento alisado. Sup. interna: color marrón oscuro, tratamiento alisado. Color de la pasta: negro, grosor 9 mm.

37/4 bis Sup. externa: color negro, tratamiento alisado. Sup. interna: color marrón oscuro, tratamiento alisado. Color de la pasta: negro, grosor 8 mm. Borde con labio adelgazado (Figura 12).

40/3 Sup. externa: color ante oscuro, tratamiento alisado. Sup. interna: color ante oscuro, tratamiento alisado. Color de la pasta: negro, grosor 6mm.

40/4 Sup. externa: color ante, tratamiento alisado. Sup. interna: color ante oscuro, tratamiento alisado. Color de la pasta: marrón claro, grosor 10 mm.

40/5 Sup. externa: color marrón claro, tratamiento alisado. Sup. interna: color marrón claro, tratamiento alisado. Color de la pasta: marrón claro, grosor 7 mm.

40/6 Sup. externa: color marrón oscuro, tratamiento alisado. Sup. interna: no se observa color, tratamiento alisado. Color de la pasta: negro, grosor 7 mm.

40/7 Sup. externa: color ante rojizo, tratamiento alisado. Sup. interna: color marrón oscuro, tratamiento alisado. Color de la pasta: negro, grosor 9 mm.

44/9 Sup. externa: color ante, tratamiento alisado. Sup. interna: color ante oscuro, tratamiento alisado. Color de la pasta: negro, grosor 9 mm.

52/23 Sup. externa: color marrón oscuro, tratamiento alisado-pulido. Sup. interna: color ante oscuro, tratamiento alisado. Color de la pasta: negro, grosor 10 mm. Borde directo. En la sup. interna hay un surco que queda por la implantación del "rodete" superior de la pieza (Figura 12).

53/26 Sup. externa: color ante, tratamiento alisado. Sup. interna: color ante, tratamiento alisado. Color de la pasta: marrón claro, grosor 7 mm.

53/27 Sup. externa: color ante, tratamiento alisado. Sup. interna: color marrón claro, tratamiento alisado. Color de la pasta: negro, grosor 10 mm.

En general, los fragmentos de alfarería presentan una cocción deficiente, con la pasta negra, mientras que la superficie externa e interna conserva un color marrón claro a marrón oscuro.

No se ha observado antiplástico en los tiestos recuperados, con excepción del 53/27 que tiene algunos granos de cuarzo y del 40/6 que tiene algunos pequeños tiestos molidos que han sido incorporados a la pasta.

Algunas Observaciones Acerca del Registro Arqueológico

A lo largo de la descripción del material lítico, se realizaron algunas observaciones, las cuales se resumen en los siguientes puntos:

La materia prima más utilizada en todos los niveles fue la cuarcita. El porcentaje de este material se mantiene más o menos constante.

En segundo término fueron utilizados el basalto, la toba silicificada y la calcedonia. Las variaciones que existe en el porcentaje de un nivel a otro, en el registro de estos tres materiales, no son significativas debido al tamaño de la muestra.

El ópalo ocráceo aparece exclusivamente en X, y de los instrumentos confeccionados en este material, cuatro presentan retoque bifacial.

El resto de los materiales utilizados (que representan aproximadamente el 10% del total) se encuentran distribuídos irregularmente en los distintos niveles. Debido al tamaño de la muestra, la aparición de algunos de estos materiales en determinados niveles, sólo indica presencia o ausencia en el registro y no frecuencia de uso.

Toda la materia prima utilizada es alóctona en el sedimento en el que se la registró. Los afloramientos naturales más cercanos se encuentran a varias decenas de kilómetros. En el caso de los rodados basálticos, probablemente hayan sido obtenidos en el litoral Atlántico Bonaerense, distante 50 km aproximadamente.

La mayoría de los instrumentos líticos fueron confeccionados sobre lascas.

La forma base lámina ha sido utilizada en su mayor parte para la elaboración de raederas laterales doble convergente. Una proporción menor de láminas presenta retoque y microretoque lateral.

En general los instrumentos cuya forma base es núcleo, pertenecen a dos grupos distintos. Uno, los confeccionados sobre guijarros basálticos y otros, los confeccionados sobre núcleos de cuarcita agotados, los que en algunos casos fueron reutilizados como forma base para raspadores.

En todos los niveles se registran claras evidencias de retalla y retoque por percusión y

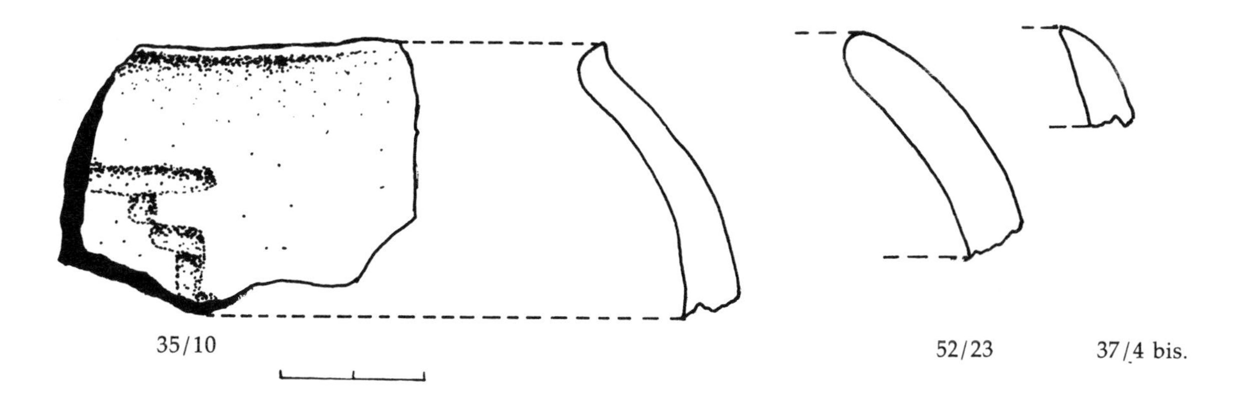

Figura 12. Perfiles de borde de recipientes de alfarería. 35/10 Vista frontal de borde con decoración incisa. (Tamaño natural).

presión, talla bipolar y alisamiento y pulido que producen caras planas o levemente cóncavas o redondeadas.

En todos lo niveles la retalla y el retoque han sido preferentemente unifaciales. El trabajo bifacial sólo está representado en el 9% de los artefactos. La frecuencia de este rasgo es muy baja en los niveles inferiores. Sólo se ha registrado una lasca con retoque lateral bifacial en Z, dos con retalla bifacial y una con retoque bifacial en la parte inferior de Y. Desde la parte superior de Y, hasta X, el trabajo sobre las dos caras se hace más frecuente, vinculado a la presencia de puntas bifaciales triangulares y a las bases de puntas lanceoladas.

La Alfarería se halla exclusivamente en X, en donde se registraron 13 tiestos y 3 bordes distintos. (Figura 12).

Las puntas bifaciales pueden ser separadas en dos grupos distintos: a) En forma de triángulo isósceles, sin pedúnculo, de base recta (levemente cóncava o concexa). b) Lanceoladas de base recta (el ancho máximo del instrumento no es el de la base).

Las puntas triangulares tienen los lados rectos o levemente convexos, con excepción del ejemplar 24/7 cuyos bordes presentan un dentado irregular. De los cinco ejemplares recuperados, cuatro están completos, y uno (34/2) tiene fragmentado el tercio distal.

Sólo se han registrado tres ejemplares fragmentados de puntas lanceoladas de los cuales se conserva el tercio proximal.

Es importante señalar que las puntas bifaciales, hasta el presente no se han registrado en la parte inferior de Y, ni tampoco en S y en Z. Se ha verificado la presencia de artefactos descriptos como raederas doble lateral convergente, en todos los niveles. Este grupo reúne varios atributos morfológicos, los que se repiten dando formas más o menos similares. De acuerdo al registro actual este grupo de raederas aumenta la frecuencia de registro desde Z hasta el techo de Y, decreciendo en X. Pero aquí se debe tener en cuenta las numerosas recolecciones superficiales que se han realizado durante años en el sitio, las que han alterado la frecuencia de registro en el nivel superior. Asímismo este factor puede haber contribuído a la ausencia de raederas transversales en X.

En todos los niveles se registran lascas con retoque marginal unifacial. En algunos casos podría tratarse de fragmentos de instrumentos definidos morfológicamente como "raederas" pero debido a lo incompleto de la pieza se optó por describirlas como "lascas con filo en bisel asimétrico oblicuo." También se encontraron fragmentos de probables artefactos definidos con el mismo criterio como "raspadores", los cuales fueron descriptos como "lascas con filo en bisel asimétrico abrupto" por los mismos motivos. Entre una y otra forma se encuentran gran cantidad de variantes y formas intermedias, fragmentadas y enteras. En todos los nivels se han hallado raspadores. Estos instrumentos presentan el filo ubicado en distinta posición: distal (frontal), extendido, perimetral.

Dos pequeños raspadores frontales de calcedonia (21/55 y 24/14), que se registraron en S y Z, pueden ser denominados como "microraspadores." Estos son los más pequeños encontrados en el sitio, pues en los niveles superiores, los raspadores de calcedonia hallados son de mayor tamaño. Las láminas con retoque se registran solamente en la parte inferior de Y (32/33, 22/34, 38/17 y 32/7) y en S y Z (36/21, 37/10 y 53/19). En la cuadrícula 41 los niveles superiores estaban alterados de forma tal, que no se podía distinguir la separación entre X y Y; dentro de este sedimento se registró una lámina con retoque. Los instrumentos 39/40 y 21/32 fueron descriptos como "cepillo con filo perimetral muy abrupto." Ambos ejemplares son marcadamente espesos, y el bisel del filo presenta frecuentes charnelas y resaltos.

En el trabajo utilizado como base para la descripción morfológica descriptiva de los materiales del sitio 2 de Arroyo Seco (Aschero m/s), en algunos casos, el grupo de los raspadores y el de los cepillos podrían confundirse. En efecto, los raspadores nucleiformes o los atípicos no se pueden diferenciar claramente, en los materiales de la Región Pampeana, de los cepillos de filo regular frontal o los de filo perimetral.

La distinción clara entre este tipo de raspadores y los cepillos, tampoco ha sido objetivamente expresada en la Primera Convención Nacional de Antropología (1966) donde como sinónimo de cepillo se consigna a "raspador grueso, grosero" (página 59).

Por lo tanto, si bien se ha consignado a los dos ejemplares referidos como cepillo, esto no implica que mediante micro-análisis complementarios y mayores estudios tipológicos, se pueda considerar la posibilidad de que se trate de raspadores.

Si bien no se ha realizado un estudio detallado de los deshechos de talla, se puede expresar que en su mayoría, están formadas por lascas de tamaño mediano, pequeño y un gran número de esquirlas. También se han hallado algunas láminas. Prácticamente no se han registrado lascas grandes.

HIPOTESIS

El tamaño de la muestra descrita no es lo suficientemente significativo como para arribar a resultados concluyentes. En efecto, en algunos casos la presencia o ausencia de determinados instrumentos, en determinados niveles se puede deber a una situación aleatoria en el registro más que una situación real dentro del sitio.

Asimismo, la frecuencia de registro de un instrumento, se podría ver sensiblemente modificada en el hallazgo de 2 ó 3 piezas más.

En términos generales, los rasgos que están presentes en todos los niveles son:

Preferencia de uso de la cuarcita como materia prima.

Utilización en segundo término de basalto, toba silicificada y calcedonia.

Preferencia de retalla unifacial y retoque marginal.

Retalla y retoque bifacial escasamente representada.

Talla bipolar escasamente representada.

Raederas simples, dobles laterales y dobles laterales convergentes.

Raspadores frontales, con filo extendido y perimetrales.

Litos pulidos o alisados, que representan caras planas, levemente cóncavas o redondeadas.

Asimismo se ha verificado que otros rasgos muestran una distribución estratigráfica restringida o que varían la frecuencia de registros en los distintos niveles. Estos elementos conducen a la postulación de tres componentes, cuya caracterización es la siguiente:

Componente Superior

Distribución estratigráfica: X

Presencia frecuente (5% o más) de: Alfarería lisa, con superficie exterior e interior alisada, marrón oscura.

Guijarros basálticos sin modificaciones de origen antrópico o partidos transversalmente.

Instrumentos con retoque bifacial confeccionados sobre ópalo ocráceo.

Raspadores frontales y extendidos.

Raederas doble lateral convergente.

Instrumentos pulidos o alisados con caras planas, levemente cóncavas o redondeadas.

Presencia (más del 1%) de:

Punta bifacial triangular.

Base de punta lanceolada.

Componente Medio

Distribución estratigráfica: Parte superior de Y.

Presencia frecuente de:

Puntas bifaciales triangulares.

Litos alisados o pulidos, con caras planas, levemente cóncavas o redondeadas.

Raspadores frontales.

Raederas doble lateral convergente y simple lateral.

Presencia de:

Raspadores con filo extendido y perimetrales.

Raederas transversales y dobles laterales con retoque alterno.

Núcleos agotados de cuarcita.

Base de punta lanceolada.

Ausencia de:

Alfarería e instrumentos bifaciales sobre ópalo ocráceo.

Componente Inferior

Distribución estratigráfica: Parte Inferior de Y, S y techo de Z

Presencia frecuente de:

Láminas con retoque lateral.

Raspadores frontales (dos microraspadores).

Raederas doble lateral convergente.

Presencia de:

Raederas simple lateral, doble lateral y transversal.

Raspadores perimetral y con filo extendido.

Núcleos agotados.

Nódulo y lasca nodular de gran tamaño, alargadas con lascado en el extremo.

Esfera y hemisfera pulida. (mitad de boleadora).

Litos alisados o pulidos con caras planas o levemente cóncavas.

Ausencia de:

Alfarería, instrumentos bifaciales sobre ópalo ocráceo y puntas bifaciales.

Debe mencionarse que un buen número de los instrumentos líticos hallados en el Sitio 2, presentan similitud morfológica con materiales descritos por diferentes investigadores (vease por ej. Bórmida s/f., 1962; Austral 1965; Madrazo 1972) para la Región Pampeana. La gran mayoría de estos materiales no han sido hallados en contextos estratigráficamente asociados.

A causa de esto y de que es necesario profundizar el estudio del Sitio 2, es que se prefiere no realizar comparaciones o correlaciones basadas exclusivamente en algunas semejanzas morfológicas.

Probablemente, con mayor información proveniente no sólo de Arroyo Seco sino de toda la región, se podrá efectuar una integración mayor de datos, que permita aumentar el conocimiento sobre el desarrollo cultural prehispánico de la Región Pampeana.

Enterratorios Humanos

En el área excavada hasta el presente se han registrado 14 esqueletos humanos. Estos esqueletos formaban parte de enterratorios primarios, múltiples e individuales. Una sola excepción podría ser el enterratorio No. 7 en el cual la disposición de los huesos podría sugerir la posibilidad de que se trate de un enterratorio secundario.

Los esqueletos 1, 2, y 3 formaban el enterratorio múltiple No. 1 (Figura 13a). Los esqueletos 4, 5 y 6 componían el enterratorio múltiple No. 2 (Figura 13b), y los esqueletos 7 y 8, el enterratorio múltiple No. 4. El resto de los esqueletos fueron enterrados individualmente.

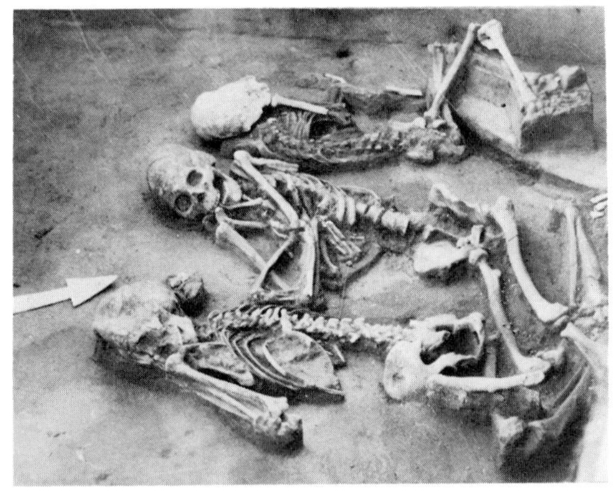

Figura 13a. Enterratorio múltiple N° 1 formado por los esqueletos 1, 2 y 3.

Figura 13b. Enterratorio múltiple N° 2 formado por los esqueletos 4, 5 y 6.

Descripción

ENTERRATORIO No. 1 (Figura 13a).

Cantidad de esqueletos: tres.

Esqueleto No. 1: Estado de conservación: Bueno. Totalmente articulado. Edad: Adulto Sexo: Masculino. Posición: Decúbito ventral, en direción O-E, con los pies y las piernas formando un ángulo recto con el resto del cuerpo y los brazos extendidos a lo largo del mismo. Ajuar: Alrededor del cráneo se disponían seis caninos de cánido, perforados en la parte radicular. Ubicación estratigráfica: Techo de Z, en contacto con S. Cuadrícula: No. 21.

Esqueleto No. 2: Estado de conservación: Bueno. Totalmente articulado. Edad: Adulto. Sexo: Femenino. Posición: Decúbito lateral derecho, en dirección SO-NE, con las manos debajo de la cabeza y las piernas semiflexionadas. Ajuar: Alrededor del cráneo se disponían numerosas conchillas circulares de unos 3 mm de diámetro, con una perforación en el centro. Ubicación estratigráfica: Techo de Z. Cuadrícula: No. 21.

Esqueleto No. 3: Estado de conservación: Bueno. Totalmente articulado. Los huesos del cráneo están muy aplastados. Edad: Infantil. Sexo: ? Posición: Decúbito dorsal extendido en dirección E-O con los brazos extendidos a los costados del cuerpo. Ajuar: No poseía.

ENTERRATORIO No. 2 (Figura 13b).

Cantidad de esqueletos: tres.

Esqueleto No. 4: Estado de conservación: Bueno. El cráneo sumamente aplastado en sentido transversal. Totalmente articulado. Edad: Adulto. Sexo: Femenino. Posición: Decúbito ventro lateral derecho, en dirección SO-NE, manos en la cara, que está mirando hacia el este. Pierna derecha ligeramente flexionada. Ajuar: No poseía. Ubicación estratigráfica: Techo de Z, en contacto con S. Cuadrícula: No. 27.

Esqueleto No. 5: Estado de conservación: Bueno. Totalmente articulado. Edad: Adulto. Sexo: Masculino. Posición: Decúbito lateral derecho. En dirección SO-NE. Piernas totalmente flexionadas a nivel de rodillas y semiflexionadas con respecto al eje del tronco. Posición de los miembros superiores semejante a los "brazos en jarra" con el dorso de la mano derecha apoyado sobre la espalda del esqueleto No. 4. Ajuar: No poseía. Ubicación estratigráfica: Techo de Z, en contacto con S. Cuadrícula: No. 27.

Esqueleto No. 6: Estado de conservación: Bueno. Cráneo sumamente aplastado. Totalmente articulado. Edad: Adulto. Sexo: Femenino. Posición: Decúbito lateral izquierdo, en dirección SO-NE, con torsión de tórax hacia posición ventral con brazo derecho totalmente flexionado sobre este último. Piernas semiflexionadas. La izquierda en ángulo recto con el eje del tronco y la derecha en angulo obtuso con respecto del mismo, y semiflexión de ambas rodillas. Ajuar: No poseía. Ubicación estratigráfica: Techo de Z, en contacto con S. Cuadrícula: No. 27.

ENTERRATORIO No. 3

Cantidad de esqueletos: Dos.

Esqueleto No. 7: Estado de conservación: Bueno. Con el 70% de los huesos articulados, el resto se encontraba disperso y faltaban. Edad: Adulto. Sexo: Masculino. Posición: Decúbito lateral derecho y en dirección NE-SO. El antebrazo izquierdo forma ángulo recto con el brazo y la mano apoya sobre la pelvis. El brazo derecho está extendido paralelo al eje de la columna vertebral y el antebrazo forma ángulo recto con el mismo. La pierna derecha está flexionada a nivel de rodilla. Ajuar: No poseía. Ubicación estratigráfica: Techo de Z y en contacto con S. Cuadrícula: No. 33 y 34.

Esqueleto No.8: Estado de Conservación: Bueno. Con el 40% de los huesos articulados, el resto faltaba. Edad: Adulto. Sexo: Masculino. Posición: Decúbito lateral derecho y en dirección NE-SO. Ajuar: No poseía. Ubicación estratigráfica: Techo de Z, en contacto con S. Cuadrícula: No. 33 y 34.

ENTERRATORIO No. 4 (Figura 14b).

Cantidad de esqueletos: Uno.

Esqueleto No. 9: Estado de conservación: Cráneo sumamente aplastado. Faltaban algunos huesos de los pies. Edad:Infantil Sexo:? Posición: Decúbito dorsal, extendido, con la pierna derecha flexionada a nivel de rodilla y la pierna izquierda flexionada formando ángulo recto con el eje de la columna vertebral, habiendo una superposición de tibias y peronés. Los brazos extendidos a los costados del cuerpo. Ajuar: Lo componían más de ciento cincuenta caninos de cánidos perforados en la parte radicular y numerosas cuentas circulares con perforación central realizadas en conchillas. Los mismos estaban dispuestos alrededor del cráneo, de los huesos del tobillo y de la muñeca. Sobre el cráneo y alrededor del mismo, el sedimento contenía gran cantidad de ocre. Ubicación estratigráfica: Techo de Z. Cuadrícula: No. 34.

ENTERRATORIO No. 5

Cantidad de esqueletos: Uno.

Esqueleto No. 10: Estado de conservación: Bueno. Solamente había algunos huesos del cráneo, parte de la columna vertebral y un húmero. Todos estaban articulados. Edad: Infantil. Sexo: ? Posición: ? Ajuar: Lo componían 10 caninos de cánidos con perforación en el extremo radicular, distribuídos sin ningún orden sobre los huesos. Ubicación estratigráfica: Techo de Z. Cuadrícula: No. 43.

ENTERRATORIO No. 6 (Figura 14a).

Cantidad de esqueletos: Uno.

Esqueleto No. 11: Estado de conservación: Regular. Totalmente articulado, presentando aplastamiento de los huesos del cráneo. Edad: Infantil (aproximadamente entre 2 y 3 años). Sexo: ? Posición: Decúbito dorsal extendido en dirección SE-NE. Los brazos extendidos a lo largo del cuerpo. Ajuar: Lo formaban numerosas cuentas de conchillas marinas, de forma rectangular, con una perforación en la parte central dispuestas alrededor del cráneo, las muñecas y los tobillos. Todo estaba cubierto con ocre. Ubicación estratigráfica: Techo de Z. Cuadrícula: No. 48.

ENTERRATORIO NO. 7

Cantidad de esqueletos: Uno.

Esqueleto No. 12 Estado de conservación: Regular. Compuesto solamente por huesos largos, totalmente desarticulados. Edad: Infantil Sexo: ? Posición: Huesos largos totalmente desarticulados. Ajuar: Junto a los huesos que estaban cubiertos con ocre se disponían cuentas de conchillas marinas de

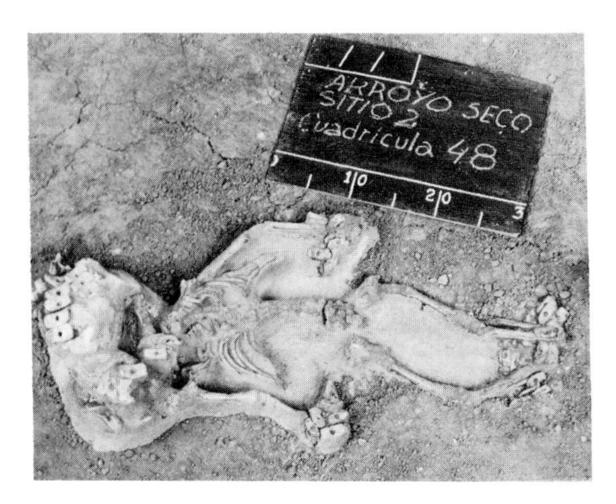

Figura 14a. Enterratorio N° 6 formado por el esqueleto 11. Se observan las cuentas rectangulares de conchillas marinas en la greión del cráneo, manos y pies.

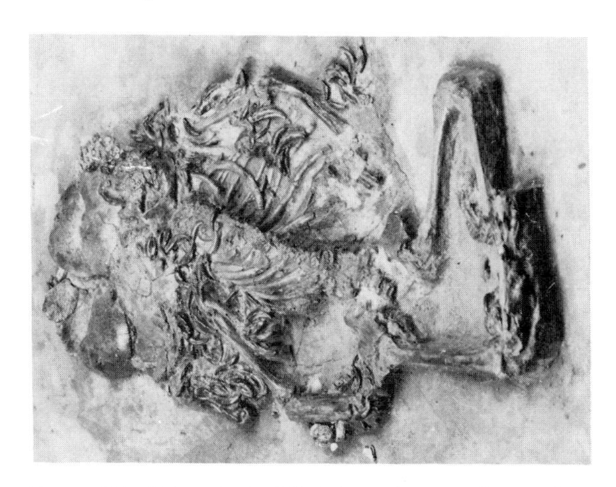

Figura 14b. Enterratorio N° 4 formado por el esqueleto 9. Se observan los caninos de cánido dispuestos alrededor del craneo.

forma rectangular con una perforación central. Ubicación estratigráfica: Techo de Z. Cuadrícula: No. 52.

ENTERRATORIO No. 8

Cantidad de esqueletos: Uno.

Esqueleto No. 13: Estado de conservación: Regular. Compuesto por fragmentos de cráneo y parte de la columna vertebral. Edad: Infantil. Sexo: ? Posición: ? Dirección SE-NO. Ajuar: Estaba compuesto por cuentas con una perforación central, ubicadas junto a los fragmentos craneales. El estado de conservación de las conchillas no es tan bueno como el de los enterratorios 6 y 7. Al lado de los huesos craneales, en dirección Oeste, se encontraban una placa de *Glyptodon* sp. y otros huesos indeterminados. Todo este conjunto estaba cubierto por ocre. Ubicación estratigráfica: Techo de Z. Cuadrícula: No. 48

ENTERRATORIO No. 9

Cantidad de esqueletos: Uno.

Esqueleto No. 14: Estado de conservación: Bueno. Faltan los huesos de la pelvis y algunas partes posteriores del craneo. El resto está totalmente articulado. Edad: Adulto. Sexo: ? Posición Decúbito lateral derecho, en dirección)-E, con las piernas semiflexionadas. Ajuar: No poseía. Ubicación estratigráfica: Sector "A." Unidad Inferior. Cuadrícula: 42 y 43. Es importante señalar la presencia de algunas piedras de "tosca" sobre este enterratorio a nivel de la parte superior de la unidad Y. Asimismo una piedra de este tipo se ubicaba debajo del cráneo.

Los enterratorios descritos, se han hallado en dos sectores definidos y separados dentro del sitio. En el sector A se ubicaron los enterratorios No. 1, 2, 3, 4, 5 y 9 y en el sector B, los enterratorios No. 6, 7 y 8.

La posición articulada de casi todas las piezas óseas, evidencian sin duda que los esqueletos han sido enterrados inmediatamente después de la muerte del individuo, pero sólo en un caso, el enterratorio múltiple No. 1, se han podido identificar las marcas del pozo. Estas se evidencian solamente a nivel de los esqueletos, es decir en Z y no en los sedimentos suprayacentes.

En el sector A, todos los esqueletos se encuentran en el techo de Z, y en algunos casos en contacto con S, pero siempre se hallan debajo de los materiales del componente inferior.

Teniendo en cuenta que se encuentran en la misma posición estratigráfica y con similares caracteristicas inhumatorios, es muy probable que todos los esqueletos del sector A correspondan a un mismo grupo. La única excepción podria ser el esqueleto No. 14, debido a la presencia de un rasgo inhumatorio distinto (piedras de "tosca").

Los enterratorios del sector B, si bien tienen una posición estratigráfica similar a la del sector A, presentan características particulares y pueden o no estar vinculados a los otros enterratorios. Debido a que se trata de individuos de muy corta edad y en estado de conservación regular, no se ha podido registrar ningún tipo de características raciales que lo relacionen con los demás esqueletos.

Algunas Consideraciones Acerca de las Actividades que se Desarrollaron en el Sitio 2

De acuerdo al registro del material lítico y alfarero, se podrían plantear varias hipótesis sobre las actividades que se desarrollaron en el Sitio 2 durante los momentos de ocupación.

No obstante, hay una serie de requisitos que este sitio no cumple y que desde nuestro punto de vista son importantes para la formulación de hipótesis sobre este tema.

La excavación parcial de este sitio a cielo abierto y con límites poco precisos, permite expresar que no hay evidencias para desechar la posibilidad de otras actividades en áreas todavía no excavadas.

En base a la información proporcionada por la trinchera y los pozos de sondeo es probable que el sitio tenga una extensión mucho mayor que la del área excavada, la que representaría aproximadamente el 20% del lugar ocupado.

Por otro lado, no se han podido identificar verdaderos "pisos de ocupación" los que permitan garantizar la estricta contemporaneidad de los materiales de cada uno de los componentes definidos. Sí se puede postular que los portadores del componente inferior ocuparon el sitio primero, luego lo hicieron los del componente medio y posteriormente los del componente superior. Pero dentro de cada componente, no es posible determinar si el sitio fue ocupado una o varias veces.

Por lo tanto, dentro del componente inferior, por ejemplo, pueden estar representadas varias ocupaciones distintas, con características industriales similares. El registro arqueológico puede entonces representar diferentes actividades en el mismo lugar, no estrictamente contemporaneas.

La única excepción podría ser el conjunto de huesos de *Megatherium,* probablemente de un mismo individuo (Tonni et al. en prensa) cuya disposición sugiere una contemporaneidad absoluta entre ellos y algunos materiales líticos asociados.

Habría que tener en cuenta en este tipo de análisis, también, la preservación diferencial del material arqueológico. En efecto, algunas actividades efectuadas con asiduidad en el sitio pueden no haber dejado evidencias (v. gr. trabajos en madera, recolección y preparación de vegetales que no necesiten ser molidos para su ingestión etc.).

No obstante, analizando el conjunto lítico estudiado, surgen algunas consideraciones interesantes. Se descarta el componente superior por los problemas que ya se han mencionado.

En los otros dos componentes se observa una gran variedad de formas y tecnologías, que implicarían una confección para distintos usos (ver Descripción de materiales). Podrían existir, asimismo, algunas diferencias en las actividades efectuadas por los portadores de cada componente, por esto sólo se podrá determinar com análisis complementarios más minuciosos.

Los litos con caras pulidas y/o alisadas, se vinculan, en base a innumerables datos arqueológicos y etnográficos, a las actividades de molienda. Algunos han funcionado como elemento pasivo en la molienda por fricción y quizás por percusión (21/26) o como elemento activo en la molienda por percusión (32/7 y 39/12). No obstante, los elementos activos en la molienda por fricción ("manos"), pueden confundirse con los "sobadores" de cueros, ya que no se conoce ningún criterio morfológico para separarlos.

La amplia variación en los ángulos del bisel de los filos, y la diversidad de forma y tamaño de los instrumentos, estaría indicando a su vez, usos diferentes.

La presencia de microraspadores sugeriría el uso de hueso o madera, pues probablemente sería necesario enmangarlos para su utilización.

Asimismo se han observado numerosos huesos con huellas de corte probablemente producidas en las tareas de descarne. Estas marcas son particularmente notorias en algunos huesos largos de megamemíferos pampeanos extinguidos (Figura 6b).

Por otro lado, hay dos sectores del sitio que fueron utilizados con frecuencia para enterrar individuos.

También se registraron otros elementos que indicarían actividades diversas: puntas de proyectil, núcleos agotados, mitad de boleadora etc.

En suma, es correcto expresar que la gran variedad de elementos reforzaría la hipótesis que en el Sitio 2, en los diferentes momentos de ocupación, se han realizado múltiples actividades.

EDADES RADIOCARBONICAS DE RESTOS ÓSEOS[1]

Parte de la información contenida en este capítulo fue expuesta en el VII Congreso Nacional de Arqueología, Colonia del Sacramento - Uruguay, 1980.

Introducción

Las dataciones radiocarbónicas se efectuaron en el Laboratorio de Tritio y Radiocarbono - LATYR - de La Plata (LP) Argentina, con muestras suministradas por el Dr. E. Tonni, los Lic. L.M. Guzmán y G. Politis, consistentes en restos óseos pertenecientes a un megamamífero extinguido, *Megatherium americanum*, y a un esqueleto humano (no. 5), provenientes de la localidad arqueológica de Arroyo Seco (lat. 38° 21′ 38″ S, long. 60° 14′ 39″ O) Argentina.

Los objetivos de este capítulo consisten en:

a) informar los resultados radiocarbónicos obtenidos, considerando las características de la fracción de hueso utilizada para datación;

b) analizar esas edades radiocarbónicas respecto a su concordancia con las edades de muerte de los especímenes, a través de una evaluación de los posibles contaminantes presentes en las muestras;

c) consideraciones sobre la relación de las edades radiocarbónicas obtenidas, con las estimadas mediante aspectos bioestratigráficos y geológico estratigráficos.

Características de las Muestras

Los restos óseos en general, están constituídos por una fracción mineral compuesta por una mezcla de hidroxiapatita y carbonato de calcio y una fracción orgánica en donde prevalece una escleroproteína denominada colágeno.

La fracción mineral (moléculas de naturaleza iónica) puede sufrir contaminación natural, por introducción de carbono exógeno de sustancias presentes en el medio ambiente de depositación a través de procesos de intercambio iónico, isotópico y recristalización (Hassan et al. 1977).

El colágeno es una sustancia orgánica de alto peso molecular (polímero) con uniones químicas de tipo covalente, que le confieren alta estabilidad y escasa reactividad frente a otras sustancias eventualmente contaminantes, presentes en el medio natural donde se conservó el resto.

Estado de Preservación de las Muestras

Se efectuó un análisis macroscópico de ambas muestras, con el fin de determinar el grado de alteración física de cada una de ellas.

1) Fragmento de fémur de *Megatherium americanum*: profundas incisiones alteraron la continuidad del hueso compacto, el hueso esponjoso se encontraba prácticamente reemplazado por sedimento y algunos canalículos estaban ocupados por pequeñas raíces. Presentaba ligera carbonatación superficial.

2) Fragmentos de costillas, vértebras y huesos de la mano de un esqueleto humano (No. 5): a pesar del estado fragmentario de los huesos largos, tanto el hueso compacto como el esponjoso se encontraban en buen estado de preservación;

presentaban escaso sedimento dentro de su masa y ausencia de raices. Exhibía escasa carbonatación superficial.

Medio en Contacto con la Muestra

Esta información ha sido extraída de la "Ficha registro de muestras radiocarbónicas" suministrada conjuntamente con los restos óseos objeto de datación, y por observación directa de perfiles en el sitio arqueológico, por personal del LATYR.

Se describen las características (macroscópicas) del sedimento donde se encontraban las muestras, con el objeto de determinar la posible presencia de sustancias que por su naturaleza puedan intervenir en los procesos de contaminación radiocarbónica.

El sedimento que contenía a los restos óseos es de origen eólico, de color castaño-amarillento, de textura limo-arenosa.

Los restos del megamamífero extinguido descansaban sobre una capa de carbonato (base unidad Y).

El esqueleto humano se encontraba inmediatamente por debajo de esa capa (techo unidad Z).

No se observó infiltración húmica que afectara directamente a los restos en cuestión, aunque en la porción superior del perfil y a 0,60 m por encima de los mismos, se desarrollaba el suelo actual. La presencia de raíces alrededor de estos restos era escasa.

Pretratamiento y Tratamiento de los Restos Óseos Analizados

El pretratamiento empleado tuvo la finalidad de separar las fracciones constituyentes del hueso (mineral y orgánica) y contempló simultáneamente la eliminación de sustancias contaminantes (carbonatos) observadas en distinta magnitud en los perfiles estratigráficos del sitio estudiado.

El tratamiento consistió en extraer el colágeno y separarlo de los contaminantes remanentes (ácidos húmicos, raíces) utilizando una metodología basada en los principios formulados por Longin (1971).

La proteína obtenida fue quemada en un tubo de combustión y el gas liberado - dióxido de carbono - fue purificado por procesos físicos y químicos, en una línea de alto vacío. La determinación de la actividad del C-14 fue efectuada en fase gaseosa por medio de un contador proporcional (Figini et al. 1977).

Resultados Obtenidos

Las edades son expresadas en años radiocarbónicos anteriores al presente B.P. (año 1950). Se utilizó 5568 años como período de semidesintegración del Carbono-14 y se adoptó como estandar contemporáneo de referencia el 95% de la actividad del ácido oxálico N B S. El error con que se informan los fechados radiocarbónicos es de una desviación estandar (± 1 sigma) que incluye las variaciones estatísticas de la medición de la muestra, del estandar de referencia y del background. No se efectuaron correcciones por fraccionamento isotópico.

LP-53 *Megatherium americanum* (colágeno) 8390 ± 140 B.P.

LP-55 Esqueleto humano (colágeno) 8558 ± 316 B.P.

Interpretación y Conclusiones

Si consideramos el estado de preservación de ambas muestras podemos señalar que el fragmento de fémur de *Megatherium americanum* presenta una mayor alteración física. Esto podría evidenciar un incremento en el grado de contaminación de este resto óseo respecto a los del esqueleto humano, haciendo por ello menos confiable a la muestra de megamamífero extinguido para determinar la cronología absoluta del evento en estudio.

Pero si consideramos la naturaleza de las sustancias contaminantes presentes en el sedimento que conservó a los restos óseos, y los niveles inferidos en el mismo, podemos estimar que es altamente probable que el método utilizado permita eliminar los contaminantes que pudieran alterar los resultados.

A través de las consideraciones efectuadas, es probable que las edades obtenidas correspondan a las edades de muerte de los especímenes (en años radiocarbónicos); siendo mayor esa probabilidad en el esqueleto humano No. 5.

Desde un punto de vista estadístico hay concordancia entre las dos dataciones radiocarbónicas obtenidas. Esas edades absolutas corresponden a un mismo evento; este se encontraría dentro de un ámbito definido como Edad Mamífero Lujanense (Pascual et al. 1965) y Epoca de deposición de la Formación La Postrera (Fidalgo et al 1973; Fidalgo et al. 1975; Fidalgo y Tonni 1981).

La aplicación de estudios isotópicos, bioestratigráficos y geológico estratigráficos dieron resultados concordantes, desde el punto de vista cronológico.

CONCLUSIONES

En los informes expuestos precedentemente se han formulado una serie de hipótesis parciales referentes a cada aspecto de la investigación. Estas pueden ser reunidas en enunciados generales de distinto grado.

1) El sitio fue ocupado inicialmente por cazadores, quizá también recolectores, que portaban los instrumentos del Componente Inferior.

Estos instrumentos están confeccionados básicamente sobre cuarcita con retalla y retoque marginal unifacial.

El material lítico se halló asociado con restos óseos de especies de la fauna indígena y de megamamíferos pampeanos extinguidos. La abundancia de restos de guanaco y venado permite inferir una gran adaptación a la caza de estos mamíferos como base de la economía. Probablemente algunas especies de megamamíferos extinguidos fueron utilizadas para complementar la dieta, aunque su captura fue ocasional debido en parte a su escaso número o a la carencia de recursos técnicos necesarios. Tal situación se verifica con *Megatherium americanum* y especialmente con *Equus* sp. cf. *E. (Amerhippus)* sp.

2) Los esqueletos humanos hallados en el techo de la unidad Z (probablemente con excepción del No. 14) corresponden a los portadores del componente inferior.

Esta hipótesis se fundamenta en que los enterratorios y el componente inferior presentan algunos elementos similares, que no se registran en los otros dos componentes, a saber:

En el Componente Inferior se hallaron pequeños panes de hematita, en algunos casos con una cara alisada y en los enterratorios No. 4, No. 6, No. 7 y No. 8 se conservaba abundante hematita en polvo en el sedimento que cubría los huesos humanos.

En el Componente Inferior se halló parte de un molusco marino: *Adelomelon (P.) brasiliana* y en los enterrratorios No. 1, No. 4, No. 6, No. 7 y No. 8 el ajuar estaba compuesto en parte por diversas cuentas confeccionadas sobre valvas, que en los casos determinados corresponden a moluscos marinos. También se registró en vinculación con un enterratorio las dos valvas de *Amiantis* sp.

En el Componente Inferior se verificó la presencia de abundantes restos de megamamíferos pampeanos y en el enterratorio No. 8 se halló una placa dérmica de *Glyptodon* sp.

Asimismo es importante destacar que los restos de fauna pleistocénica fueron hallados en S y en la parte inferior de Y (con excepción de la placa de *Glyptodon*), en cuadrículas con enterratorios y sin enterratorios. Por sobre estos la disposición del material lítico y óseo no difería de aquellos sectores donde no se registraron huesos humanos.

Si los enterratorios se hubieran efectuado por los ocupantes del sitio representados por el componente Medio o Superior, se debería esperar que la inhumación hubiese afectado la disposición del material en el componente inferior y esta situación no ha sido observada.

Se dispone de dos fechados radiocarbónicos realizados sobre muestras procedentes de los niveles inferiores. Uno corresponde a un trozo de fémur de *M. americanum* (base de la unidad Y) y el otro a

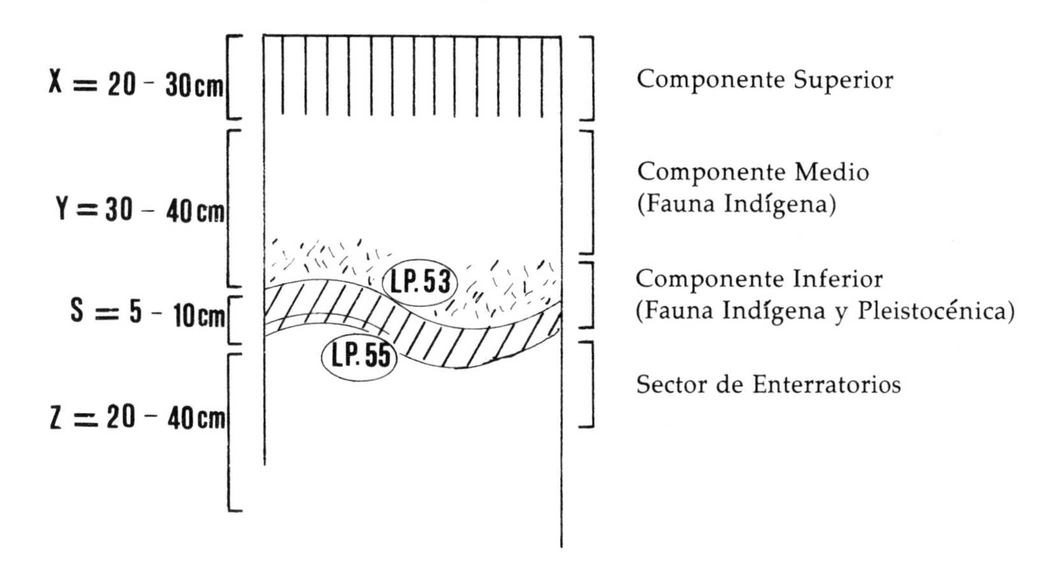

Figura 15. Ubicación de los componentes. Los restos faunísticos y las muestras fechadas por radiocarbono en el perfil del Sitio 2.

parte de un esqueleto humano (techo de la unidad Z). Ambas dataciones ubicarían cronológicamente al nivel inferior de ocupación en una antiguedad aproximada de 8500 años radiocarbono B.P. (8390±240 B.P. (LP-53) y 8558±316 B.P. (LP-55)).

Es probable que algunos de los elementos no se encuentren *in situ* aunque en estos niveles no se detectaron indicadores de disturbaciones importantes, excepto las provocadas por la ocupación. Como pruebas pueden citarse la posición articulada de casi todos los esqueletos humanos, y el metapodio articulado con algunos huesos del autopodio que corresponden probablemente a *E. (Amerhippus)* sp.

3) Los partadores del Componente Inferior desarrollaron un elaborado culto funerario.

Esto se verifica a través del hallazgo de entierros de párvulos con ajuar (caninos de cánidos perforados, cuentas de valvas de moluscos, ocre espolvoreado alrededor del cráneo), de adultos sin ajuar y enterratorios múltiples de dos o tres individuos. Este enunciado observacional se complementa con la hipótesis que atribuye tal compartamiento a los partadores del Componente Inferior

4) Desde el punto de vista geológico, entre las distintas alternativas planteadas, la graficada en el perfil 1 (véase Figura 4), aparece como la más probable.

Tal situación es verificable a través de las evidencias faunísticas y arqueológicas, ya que geológicamente no pudo ser detectado localmente ningún plano de erosión o discontinuidad en la sedimentación. Sin embargo, una situación similar a la planteada en el perfil 1 es la que se observa y describe para otras áreas (i.e.: Lobería, véase Fidalgo y Tonni 1981), donde la fauna extinguida se encuentra exclusivamente por debajo del plano de discordancia, es decir en la unidad litoestratigráfica inferior.

5) Los instrumentos que integran el Componente Medio se encuentran en la parte superior de la unidad Y.

Estos instrumentos están confeccionados fundamentalmente sobre cuarcita con predominio de retalla y retoque marginal unifacial, pero con la presencia de puntas de proyectil triangulares bifaciales y de una mayor proporción de elementos vinculados a la molienda (fragmentos de molinos y manos).

Los restos faunísticos asociados en este nivel están compuestos principalmente por guanaco, venado y ñandú; no se ha registrado ningún resto de megamamíferos extinguidos ni de fauna europea.

6) La economía de los portadores del Componente Medio se basó en la caza de especies de la fauna indígena y en la recolección.

Esto último se verifica a través del hallazgo de fragmentos de molinos y manos.

En las investigaciones efectuadas por los autores no se halló ningún resto humano asignable con cierto grado de probabilidad a los portadores de este componente.

No se poseen fechados radiocarbónicos para estos niveles. Sin embargo, si la interpretación ofrecida en el perfil 1 es la más probable, un lapso importante separaría los artefactos del Componente Inferior de los de Medio. Este último es, además, anterior al siglo XVI, a juzgar por la ausencia de elementos faunísticos y culturales de origen europeo.

7) En la unidad X se registraron los instrumentos del Componente Superior.

Este componente presenta como característica sobresaliente, la incorporación de la alfarería marrón lisa e incisa y de instrumentos realizados en ópalo ocráceo.

Los materiales fueron hallados dentro de sedimentos afectados por el arado, lo que constituye un factor importante de disturbación. Asimismo, las numerosas recolecciones superficiales alteraron la real frecuencia de los tipos. Por lo tanto, este componente es el más pobremente caracterizaado.

Los restos culturales y faunísticos pos-hispánicos hallados en la unidad X pueden ser evidencias del contacto hispano-indígena, o el resultado de la asociación artificial de dos ocupaciones diacrónicas, una indígena y la otra europea.

8) Los instrumentos líticos hallados en el sitio no presentan variaciones morfológicas notables.

En base a instrumentos confeccionados sobre lascas de cuarcita de mediano tamaño con retoque y retalla marginal unifacial, se incorporaron nuevos elementos en el transcurso del tiempo. Desde los niveles inferiores a los superiores se registran algunos tipos muy frecuentes en los yacimientos del área Interserrana Bonaerense tales como raederas simples laterales, dobles laterales y dobles laterales convergentes y raspadores frontales con filo extendido y perimetrales. Sobre este "stock" inicial se anexaron las puntas de proyectil triangulares bifaciales y aumentó la utilización de elementos vinculados a la molienda. Posteriormente se incorporaron la alfarería y los instrumentos sobre ópalo ocráceo.

9) Los sedimentos que integran las unidades Z e Y se depositaron bajo condiciones más áridas que las actuales en la misma zona. Esto se verifica no sólo a través del análisis faunístico (presencia de elementos patagónicos y centrales), sino también del geológico (predominio de la sedimentación eólica).

POST SCRIPTUM: Luego de haber sido enviado este trabajo para su publicación se recibieron algunos nuevos resultados de análisis radiocarbónicos. The Smithsonian Environmental Research Center entregó los siguientes fechados:

SI-5481 *Megatherium americanum* (femur) colageno 1800±110 B.P.

SI-5482 Esqueleto humano 7 (enterratorio 3) colageno 5910±55 B.P.

El Dr. R. Stuckenrath, en el análisis de los resultados obtenidos señala que: "Both samples were heavily mineralized, and collagen yield was both low and unsatisfactory. While normal fractions ready for combustion are dark brown, glossy and exhibit conchoidal fracture, these two were gritty, grey, and burned only with difficulty at high temperatures and 0_2 flow. I did not like them and so do not believe the ages produced." En concecuencia, ambos fechados deben ser en principio descartados.

Por otra parte el Laboratorio de Tritio y Radiocarbono de la Universidad Nacional de La Plata (LATYR) efectuó dos fechados sobre carbonato pedogenético proveniente de la base y del techo de la unidad S cuyos resultados fueron: LP-93 = 5740 ± 120 y LP-94 = 5700 ± 120 B.P. respectivamente. Un tercer fechado fue llevado a cabo sobre $CaCO_3$ pedogénetico que se se presentaba como un débil moteado en la base de la Unidad Y; este análisis dio como resultado LP-92 = 1890±80 B.P.

Se postula que:

a) La concordancia de las edades de las dos muestras de carbonato que corresponden a diferentes profundidades dentro de la unidad S (LP-93 y LP-94) indican que ese horizonte corresponde a un mismo proceso de meteorización.

b) La secuencia cronológica presenta un orden acorde a su posición estratigráfica. El salto cronológico entre los fechados de los carbonatos podría indicar una variación de las condiciones climáticas que dan origen a la secuencia estratigráfica en el sitio o una discordancia de erosión.

c) En el sitio se ha observado un suelo actual y humedad en el perfil en determinadas épocar del año. Por tales motivos, y debido a que el carbonato no es un material "ideal" para determinar edades absolutas, las edades radiocarbónicas obtenidas en los horizontes de carbonatos de las unidades S y Y serían más "jóvenes" que sus edades verdaderas. Esta situación estaría causada por la posible contaminación por carbonatos "jóvenes" disueltos en aguas meteóricas que se infiltraron en el perfil.

Estas conclusiones estarían apoyando la hipótesis de que los esqueletos que se encuentran debajo de S serían más antiguos que ca. 5700 años B.P. Esto a su vez estaría en concordancia con la hipótesis que atribuye los esqueletos hallados en el techo de Z al Componente Inferior.

NOTA

[1]Los autores de esta sección son J. Carbonari, G. Gomez, R. Huarte, y A. Figiní.

REFERENCIAS CITADAS

Ameghino, F.
1889 *Contribución al conocimiento de los mamíferos fósiles de la República Argentina.* CONI, Buenos Aires.

Ameghino, C., y C. Rusconi
1932 Nuevas subespecies de avestruz fósil del Plioceno de Buenos Aires. *Anales de la Sociedad Científica de Argentina* 114: 38-42.

Armaignac, H.
1976 Viaje por las pampas argentinas. Editora. EUDEBA, Buenos Aires.

Austral, A.
1965 Investigaciones Prehistóricas en el curso inferior del río Sauce Grande. (Partido de Coronel de Marina Leonardo Rosales, Provincia de Buenos Aires. República Argentina).*Trabajos de Prehistoria del Seminario de Historia Primitiva del Hombre, de la Universidad de Madrid y del Instituto Español de Prehistoria del Consejo Superior de Investigaciones Científicas.* Vol. XIX. Madrid.

1971 El yacimiento arqueológico Vallejo en el noroeste de la Provincia de La Pampa. Contribución a la sistematización de la prehistoria y arqueología de la región Pampeana. *Relaciones de la Sociedad Argentina de Antropología* V(2): 49-70.

1972 El yacimiento de Los Flamencos II. La coexistencia del hombre con fauna extinguida en la Región Pampeana. *Relaciones de la Sociedad Argentina de Antropología* VI (n.s.):203-209.

Bormida, M.
s/f *Prolegómenos para una Arqueología de la Pampa Bonaerense.* Edición Oficial de la Pcia. de Bs. As., Dirección de Museo y Archivo Histórico, La Plata.

1960 Investigaciones paleontológicas en la región de Bolívar. (Pcia. de Buenos Aires). *Anales de la Comisión de Investigaciones Científica* Vol. I (1960). La Plata, 1961: 196-283.

1962 El Epiprotolítico de La Pampa Bonaerense: la industria de La Montura. (Pdo. de Bolívar, Pcia. de Buenos Aires). *Jornadas Internacionales de Arqueología y Etnografía* 2:113-132.

Brodkorb, P.
1963 Catalogue of fossil birds. Part. 1: *Florida State Museum, Biological Sciences Bulletin* 7(4): 179-293.

Cabrera, A.
1953 Los roedores argentinos de la familia "Caviidae." *Facultad de Agronomía y Veterinaria de la Universidad de Buenos Aires, Publicación* 3.

1957/60 Catálogo de los mamíferos de América del Sur. *Revista del Museo Argentino de Ciencias Naturales Bernardino Rivadavia, Cs. Zoológicas* T. IV (1 y 2).

Cabrera, A., y J. Yepes
1960 *Mamíferos Sudamericanos.* 2da. edic. T. I y II. EDIAR, Buenos Aires.

Cabrera, A.L.
1971 Fitogeografía de la República Argentina. *Bolletin de la Sociedad Argentina de Botânica* XIV(1-2): 1-42.

Camacho, H.H.
1966 Invertebrados. *Paleontografía Bonaerense,* editado por A.V. Borrello, Fasc. III, Comisión de Investigaciones Científicas de la Provincia de Buenos Aires, La Plata.

Cattoi, N.
1966 Edentata. *Paleontografía Bonaerense,* editado por A.V. Borrello, Fasc. IV: 59-99. Comisión de Investigaciones Científicas de la Provincia de Buenos Aires, La Plata.

Contreras, J.
1973 La mastofauna de la zona de la Laguna Chasicó, Provincia de Buenos Aires. *Physis,* C, 32(84): 215-219.

Contreras, J., y O.A. Reig
1965 Datos sobre la distribución del género *Ctenomys* (Rodentia, Octodontinae) en la Zona Costera de la Província de Buenos Aires, comprendida entre Necochea y Bahía Blanca. *Physis* 25(69): 169-186.

Crawford, R.
1974 A través de la Pampa y los Andes. Editora EUDEBA, Buenos Aires.

Daino, L.
1979 Exégesis histórica de los hallazgos arqueológicos de la Costa Atlántica. *Prehistoria Bonaerense,* pp. 95-195. Municipalidad de Olavarría. Prov. de Bs. As.

Fidalgo, F.
1979 Upper Pleistocene-Recent marine deposits in northeastern Buenos Aires Province (Argentine). *International Symposium on Coastal Evolution in the Quaternary,* pp. 384-404. São Paulo.

Fidalgo, F., y E. Tonni
1981 Sedimentos eólicos del Pleistoceno tardío y Reciente en el área interserrana Bonaerense. *VIII Congreso Geológico Argentino* 3: 33-39.

Fidalgo, F., F. de Francesco, y U. Colado
1973 Geología superficial en las hojas Castelli, J.M. Cobo y Monasterio (Pcia. de Buenos Aires). *V Congreso Geológico Argentino, Actas* 4: 27-39.

Fidalgo, F., R. Pascual, y F. de Francesco
1975 Geología superficial de la llanura bonaerense (Argentina). *VI Congreso Geológico de Argentina, Relatorio* 103-138.

Figini, A., R. Huarte, y G. Gomez
1977 Datación por radiocarbono con contador proporcional. *Obra del centenario del Museo de La Plata, Geología,* T. IV: 289-296.

Frenguelli, J.
1931 Observaciones geográficas y geológicas en la región de Sayape (Pcia. de San Luis). *Escuela Normal Superior "José María Torres".* Paraná, Entre Ríos.

Gallardo, J.M.
1977 *Reptiles de los alrededores de Buenos Aires.* EUDEBA, Buenos Aires.

Greslebin, H.
1924 Fisiografía y noticia preliminar sobre arqueología de la región de Sayape (Pcia. de San Luis). Apéndice de Lucas Kraglievich. Bs. As.

Harris, G.
1964 Climatic changes since 1860 affecting European birds. *Weather* 19(3): 70-79.

Hassan, A., J. Termine, and C.V. Haynes, Jr.
1977 Mineralogical studies in bone apatite and their implications for radiocarbon dating. *Radiocarbon* 19:364-374.

Hoffstetter, R.
1950 Algunas observaciones sobre los caballos fósiles de la América del Sur, *Amerhippus* Gen. Nov. *Boletin Informativo Científico Nacional* III (26 y 27): 426-404. Editorial Casa de la Cultura Ecuatoriana. Quito.

1952 Les mammifères Pléistocènes de la Republique de L'Equateur. *Mémoires de la Société Géologique de France,* T. 31, Mémoire 66.

Longin, R.
1971 New method of collagen extraction for radiocarbon dating. *Nature* 230:241-242.

Madrazo, G.
1968 Hacia una revisión de la prehistoria bonaerense. *Etnía* 7: 1-12. Olavarría.

1972 Arqueología de Lobería y Salliqueló (Provincia de Buenos Aires). *Etnía* 15: 1-18. Olavarría.

1973 Síntesis de arqueología pampeana. *Etnía* 17: 13-25. Olavarría.

Moreau, R.E.
1969 Climatic changes and the distribution of forest vertebrates in West Africa. *Journal of Zoology* 158: 39-61.

Orquera, L.
1981 Arqueología y etnografía histórica de las regiones pampeanas. In *"Toponimia y arqueología del siglo XIX en la Pampa"*, edited by L. Piana, pp. 21-59 EUDEBA, Buenos Aires.

Ortega, H.E.
1970 Evolución de las comunidades, cambios faunísticos e integraciones biocenóticas de los vertebrados del Cenozoico de Bolivia. *IV Congreso Latinoamericano de Zoología, Actas* II: 985-990.

Palanca, F., y G. Politis
1979 Los cazadores de fauna extinguida de la Provincia de Buenos Aires. *Prehistoria Bonaerense*, pp. 70-92. Municipalidad de Olavarría, Pcia. Bs. As.

Palanca, F., L. Daino, y E. Benbassat
1972 Yacimiento "Estancia La Moderna" (Pdo. de Azul, Pcia. de Buenos Aires). Nuevas perspectivas para la arqueología de la Pampa Bonaerense. *Etnía* 15: 19-27. Olavarría.

Pascual, R.
1966 Rodentia. R. Pascual (dir.): *Paleontografía Bonaerense*, Fasc. IV: 100-139.

Pascual, R., E.J. Ortega Hinojosa, D. Gondar, y E. Tonni
1965 Las edades del Cenozoico mamalífero de la Argentina, con especial atención a aquellas del territorio bonaerense. *Anales de la Comisión de Investigaciones Científicas de la Provincia de Buenos Aires*, 6: 165-193. La Plata.

Paula Couto, C. de
1970 Evolução de comunidades, modificações faunísticas e integrações biocenóticas dos vertebrados cenozóicos do Brasil. *IV Congreso Latinoamericano de Zoología, Actas* II: 907-930, Caracas.

1979 *Tratado de paleomastozoología*. Academia Brasileira de Ciencias, Rio de Janeiro.

Raffino, R., E. Tonni, y A. Cione
1977 Recursos alimentarios y economía de la Quebrada del Toro. *Relaciones de la Sociedad Argentina de Antropología* XI: 9-30.

Reig, O.
1964 Roedores y marsupiales del Partido de General Pueyrredón y regiones adyacentes (Prov. de Buenos Aires, Argentina).*Publicaciones del Museo Municipal de Ciencias Naturales y Tradiciones*. Mar del Plata 1(6): 203-224.

1965 Datos sobre la comunidad de pequeños Mamíferos de la región costera del Partido de Gral. Pueyrredón y de los partidos limítrofes (Prov. Buenos Aires). *Physis* XXV(69): 205-211.

Ringuelet, R.A.
1955 Panorama zoogeográfico de la Provincia de Buenos Aires. *Notas, Museo de La Plata. Zoología* 18(156): 1-45.

1961 Rasgos fundamentales de la zoogeografía de la Argentina. *Physis* XXII(63): 151-170.

Ringuelet, R.A., y R.H. Arámburu
1957 Enumeración sistemática de los vertebrados de la Provincia de Buenos Aires. *Ministerio de Asuntos Agrarios, Buenos Aires, Publicación* 119.

Sanguinetti de Bormida, A.
1966 Las industrias líticas de Trenque Lauquen (Prov. de Buenos Aires). *Acta Prehistórica* V-VII: 72-94.

1970 La neolitización de las areas marginales de la América del Sur. *Relaciones de la Sociedad Argentina de Antropología* V(1): 9-23.

Schultz, C.B.
1972 Holocene interglacial migrations of mammals and other vertebrates. *Quaternary Research* 2: 337-340.

Tonni, E.P.
1980 The present state of knowledge of the Cenozoic birds of Argentina. *Natural History Museum of Los Angeles County, Scientific Contribution* 330: 105-114.

Tonni, E.P., y J.H. Laza
1976 Paleoetnozoología del área de la Quebrada del Toro. *Relaciones de la Sociedad Argentina de Antropología* X(n.s.): 131-140.

Tonni, E.P., y F. Fidalgo
1978 Consideraciones sobre los cambios climáticos durante el Pleistoceno Tardío - Reciente en la Provincia de Buenos Aires. Aspectos ecológicos y zoogeográficos relacionados. *Ameghiniana* XV(1-2): 235-253.

Tonni, E.P., y G. Politis
1980 La distribución del guanaco (Mammalia, Camelidae) en la Provincia de Buenos Aires durante el Pleistoceno Tardío y Holoceno. Los factores climáticos como causas de su retracción. *Ameghiniana* XVII(1): 53-66.

Tonni, E.P., G. Politis, y L. Meo Guzman
1982 La presencia de *Megatherium* en un sitio arqueológico de la Pampa Bonaerense (Rep. Argentina). Su relación con la problemática de las extinciones pleistocénicas. *VII Congreso Nacional de Arqueología,* Colonia, Uruguay, 1980.

Van Gelder, R.G.
1978 A review of Canid classification. *American Museum Novitates* 2646, New York.

Zetti, J.
1964 El hallazgo de un Megatheriidae en el "Médano Invasor" al SW de Toay, Provincia de La Pampa. *Ameghiniana* III(9): 257-265.

Zetti, J., E.P. Tonni, y F. Fidalgo
1972 Algunos rasgos de la geología superficial en las cabeceras del Arroyo del Azul (Provincia de Buenos Aires). *Etnía* 15. Olavarría.

Patagonian Prehistory: Early Exploitation of Faunal Resources (13,500 - 8500 B.P.)

GUILLERMO L. MENGONI GOÑALONS
Instituto de Ciencias Antropológicas
Universidad de Buenos Aires
25 de Mayo 217
1002 Buenos Aires
ARGENTINA

Abstract

Sufficient archaeofaunal evidence exists from contintental southern Patagonia to postulate the human adaptive strategies used throughout the known time of human occupation. Only the earliest period (13,500 - 8500 B.P.) is discussed here. Early human adaptive behavior appears to have been a generalized foraging economy. Utilized resources were not restricted to extinct fauna; guanaco, rodents, rhea, waterfowl and probably carnivores were also procured. Extinction of megafauna was gradual and did not affect the overall adaptive strategy. Economic status was assigned to each of the resources according to its predictability and abundance. The generalized strategy and the low expectation of finding prey corresponds to a young ecosystem, which is characteristic of the southern Patagonian biota.

The faunal information comes from the following sites: Mylodon Cave, Las Buitreras Cave, Fell's Cave, Palli Aike Cave, Los Toldos Cave 3, Cueva de las Manos and Cueva Grande del Arroyo Feo (see Figure 1)[1].

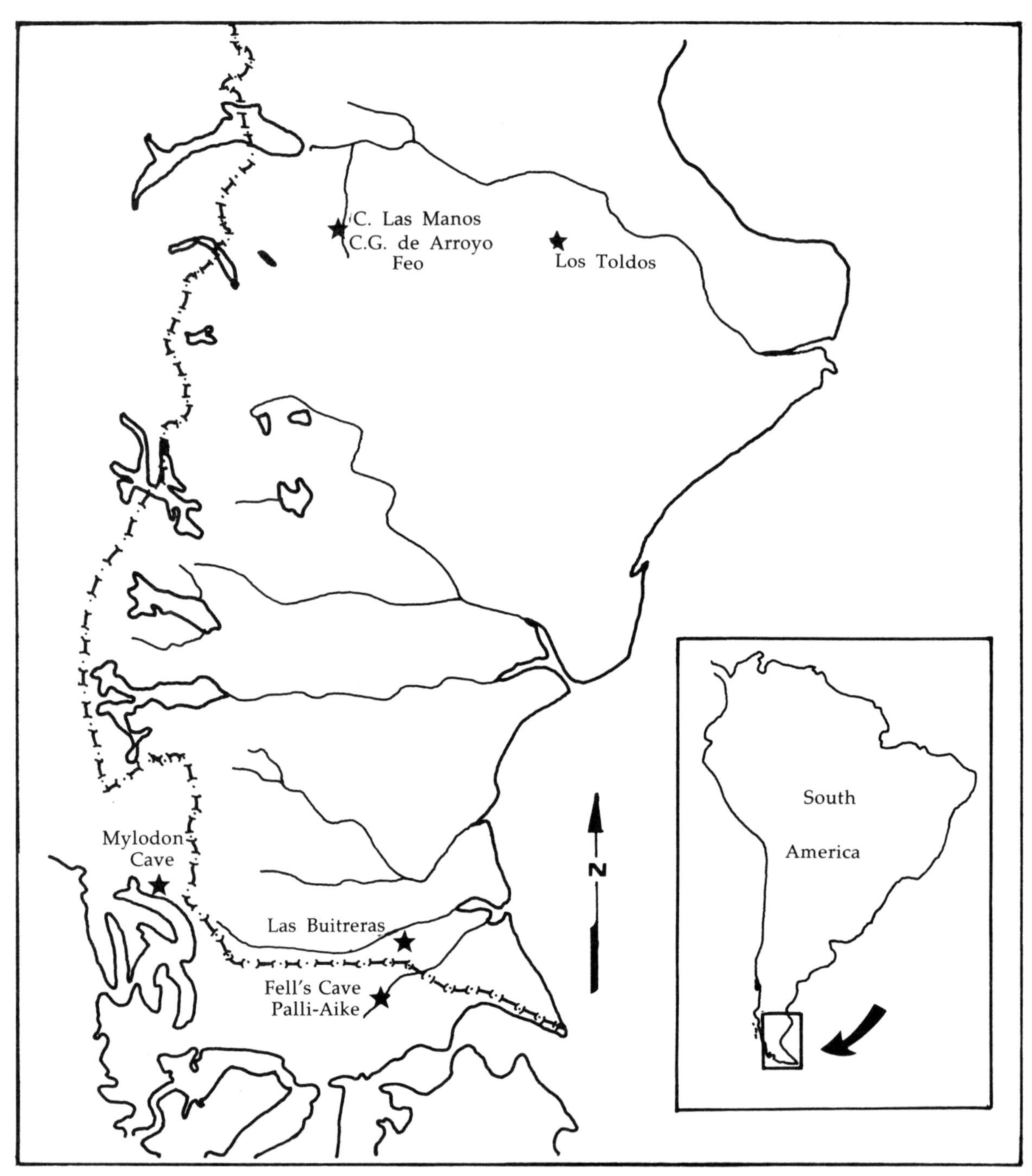

Figure 1. Archaeological sites in southern Patagonia.

MYLODON CAVE

Mylodon Cave was excavated by Nordenskjöld in 1899, by Hauthal in the same year, by the Emperaire's in 1953 and by Saxon in 1976. The best reports are those of Nordenskjöld (1900) and Saxon (1976, 1979).

Radiocarbon dates were obtained from Saxon's excavations for the human occupations of:

layer 7 BM-1201 5366±55 B.P. (charcoal sample)
layer 7 BM-1201A 5395±58 B.P. (recount of BM-1201)
layer 7 BM-1204 5684±52 B.P. (charcoal sample)
layer 7 BM-1204B 5643±60 B.P. (recount of BM-1204)
layer 8 BM-1207 7785±747 B.P. (burnt guanaco bone sample)

In these two layers only guanaco bones and stone tools were recovered. Sloth remains were found in the overlying layers 2 to 6 and in the underlying layers 9 to 10 (Saxon 1976). This sequence raised the possibility of the survival of sloth until ca. 5500 B.P. but without association with man.

Nordenskjöld (1900) found three layers:

A- shell midden with guanaco and huemul (*Hippocamelus* sp.) bones, later identified and dated by Saxon. The radiocarbon date of a charcoal sample was 2556±45 B.P. (BM-1202) (Moore 1978);

B- composed mainly of guanaco, horse and sloth bones. Nordenskjöld (1900) considered the latter to be intrusive from the underlying layer C;

C- contained sloth, horse, extinct large felid, canid and human bones. Nordenskjöld did not accept the association of man and sloth because the human bones seemed to him to be intrusive from above.

Nordenskjöld's layer B can hypothetically be correlated with the human occupations of layers 7 and 8 as identified and dated by Saxon; though no horse bones were found by Saxon.

It seems that Hauthal (1899) did not identify two bone awls he recovered as coming from a specific cultural layer, but simply from the lower part of the dung layer. This point leads to the conclusion that he treated the whole cave deposit as one unit and therefore his information adds nothing to the stratigraphic question discussed here. However it should be kept in mind that Lehmann-Nitsche (1902) illustrated another awl from Hauthal's excavations made on a rudimentary metacarpal of an equid.

The radiocarbon dates available for the older layers cluster between about 10,000 and 13,500 B.P.:

The vertebral fragment used in sample LU-794 came from the 1899 excavation of Nordenskjöld (Hakansson 1976), and probably dates the lower part of his layer C.

Neither Nordenskjöld's, the Emperaire's, nor Saxon's excavations have provided evidence for the association between *Mylodon* and man at this site. However, we have had the opportunity to examine part of the Hauthal collection at the La Plata Museum by courtesy of Dr. Rosendo Pascual in 1976 and 1980. On both occasions we found numerous cut marks on guanaco, horse and sloth bone. Although further research is needed it is clear that the question of man-giant sloth association at this site is not yet resolved.

One important fact is that we know with certainty that 13,500 years ago the area of southwestern Patagonia was inhabited by a variety of animals, including a high availability of *Mylodon* (Borrero, this volume), horse, guanaco and other taxa[3].

Table 1 Radiocarbon dates from Mylodon Cave.

Radiocarbon date B.P.	Laboratory Sample No.	Reference	Type of sample
10,200±400	Sa-49	Rivera 1978	sloth dung
10,400±330	A-1391	Long and Martin 1974	sloth hide
10,800±570	C-484	Bird 1952	sloth dung
10,864±720	C-484	Bird 1952	sloth dung
11,810±229	BM-1210	Burleigh, personal communication 1982	sloth dung
12,308±288	BM-1210B	Moore 1978	recount
12,496±148	BM-1209	Moore 1978	sloth dung
12,552±128	BM-1375	Saxon, personal communication 1982	sloth dung
12,984± 76	BM-728	Burleigh et al. 1977	sloth bone
13,183±202	BM-1208	Burleigh, personal communication 1982	sloth bone
13,260±115	LU-794	Hakansson 1976	sloth bone
13,500±470	NZ-1680	Rivera 1978	sloth hide and hair
13,560±180	A-1390	Long and Martin 1974	sloth dung[2]

LOS TOLDOS CAVE 3

As Los Toldos Cave 3 (see Figure 1) has no published faunal report, information must be gleaned from several publications (Cardich et al. 1973; Cardich 1977, 1978). Layer 11, dated 12,600±600 B.P., contained guanaco, horse and probably an extinct camelid (*Lama gracilis*) (see Table 1). We have no definite date for the extinction of horse, which is present up to layer 10; the radiocarbon date of 8750±480 B.P. comes from the overlying layer 9. The last two layers have greater faunal diversity (see Table 1) than layer 11. This fact probably anticipates the human adaptive strategy that was adopted after the megafauna (mainly herbivores) finally disappeared in this area.

FELL'S CAVE

At Fell's Cave (Figure 1) Bird (1938) found fireplaces which contained broken and burnt native horse, sloth and guanaco bones in the earliest occupation refuse.

Faunal remains from Bird's 1970 excavation have been studied by Saxon (1976, 1979). Mylodon and horse bones were again identified but according to Saxon the sloth bones did not show traces of burning or butchering. I believe that this material must be reexamined for cut marks, chop marks, fractures, or presence of expediency tools. This evidence and studies of disarticulation rate will provide further information in order to discuss this question.

The faunal information presented by the Mission Archaeologique Française from their 1959 excavations (Poulain-Josien 1963) shows that horse is present not only in layers corresponding to Bird Period I (layers XII and XI of their nomenclature) but also from layers X and IX, which have no apparent correspondence with the Bird periods. The only data they provide is that these two layers yielded only a few stone tools and no lithic or bone projectile points. Horse is absent after layer IX. Emperaire's layers VII and VI were tentatively correlated with Bird's Period II (Emperaire et al. 1963).

LAS BUITRERAS CAVE

An association of sloth, horse, extinct canid, guanaco, dolphin, rodents and unidentified birds (Table 1) was found at this site (Figure 1). This assemblage is valid for the faunal contents of layers 7 and 8, both dated before 9100 B.P. (Caviglia et al., this volume; Sanguinetti y Borrero 1977). These authors have shown:

1- sloth, guanaco and man in an interactive context, as seen by the amount of disarticulation of these animals, butchering techniques, and spatial distributions of the body parts (Caviglia et al., this volume);

2- association of the above with a human coprolite (Figuerero Torres 1982);

3- horse and dolphin are represented only by two molars in the first case and a single caudal vertebra in the second; therefore, it is difficult to assign economic status;

4- the presence of an extinct canid (*Dusicyon avus*) with dental morphological characteristics similar to some members of the genus *Canis* (Caviglia 1978).

According to Caviglia et al. (this volume; see also Caviglia 1978) we could safely assign the *Canis* cf. *familiaris* mentioned by Saxon (1979) as *D. avus*; this is the position adopted in this paper (see Table 1).

CUEVA DE LAS MANOS AND CUEVA GRANDE DEL ARROYO FEO

The earliest occupations of both these caves (Figure 1) have yielded only modern fauna (Mengoni Goñalons y Silveira 1976; Silveira 1979) (Table 1) about 9300 B.P. These sites are close to existing waterways, thus explaining the presence of freshwater fish, snails and waterfowl.

Although the evidence comes from only two sites located in an area of "cañadones y pampas" near the Andes, it is interesting to note the absence of extinct fauna in these early layers. Future excavation in this area will supply more information on the geographical distribution of extinct and modern faunal species during this period.

The importance of guanaco as the main resource seems to have been complemented by tucu-tucu (*Ctenomys* sp.) and probably rhea (Rheidae). Grey fox and puma are represented only by a few bone specimens and just one individual has been tabulated for each. The specimen identified as belonging to grey fox is a fragment of a probable awl. This would indicate occasional procurement. Freshwater snails, evidenced by few individuals, have no economic status. Rhea eggshell has been recovered from layer 11 of Cueva Grande del Arroyo Feo; this would point to seasonal procurement and knowledge of a highly predictable resource.

Table 2. Faunal information available for the sites discussed in the text in chronological order.

Radiocarbon date B.P.	Laboratory Sample No.	Site name	Capa layer	References of radiocarbon dates	fish	mussel (*Aulacomya ater*)	freshwater snail (*Chilin* sp.)	birds indet.	coot (*Fulica* sp.)	grebe (cf. *Podiceps antarticus*)	upland goose (*Chloëphaga* sp.)	ducks	partridge (*Eudromia* sp.)	hawks (Falconidae)	rhea (Rheidae)	dolphin (Delphinidae)	small rodents (Cricetidae)	tucutucu (*Ctenomys* sp.)	chinchillón (*Lagidium* sp.)	cat indet. (*Felis* sp.)	puma (*Felis concolor*)	pampas cat (*Felis colocolo*)	fox indet.	red fox (*Dusicyon culpaeus*)	grey fox (*Dusicyon griseus*)	guanaco (*Lama glama guanicoe*)	extinct camelid (*L. gracilis*)	horse (Hippidion-Onohippidion)	sloth (*Mylodon* (?) *listai*)	*Macrauchenia*	extinct canid (*D. avus*)	extinct felid
8480 ± 135	I-5143	Fell' Cave	layer 12	Saxon 1976								x		x				x				x		x		x					x	
8639 ± 450	C-485	Palli Aike	over volcanic ash layer	Bird 1952																						x		x	x			
8750 ± 480	BVA-Arsenal	Los Toldos	capa 9	Cardich et al. 1973																x			x									
no date		Los Toldos	capa 10			x							x	x	x											x	?	x				
9100 ± 150	I-5144	Fell's Cave	layer 13	Saxon 1976														x														
no date		Fell's Cave	layer 14															x													x	
no date		Fell's Cave	layer 15															x														
no date		Fell's Cave	layer 16									x		x				x				x		x							x	
9030 ± 230	I-5145	Fell's Cave	layer 17	Saxon 1976						x				x	x			x				x		x		x					x	
9100 (min.)		Las Buitreras	capa 7	Auer pers. comm. (it dates volcanic ash layer=capa 6 in Sanguinetti and Borrero 1977				x								x	x	x								x		x	x			
		Las Buitreras	capa 8					x									x									x			x		x	
9300 ± 90	CSIC-385	Cueva de las Manos	capa 6 base	Aguerre 1977																												
9320 ± 90	CSIC-138	Cueva de las Manos	capa 6 media	Gradin et al. 1976	x		x		x						x		x		x		x				x	x						
9330 ± 80	CSIC-396	Cueva Arroyo Feo	capa 11 base	Gradin et al. 1979				x							x		x	x	x						x	x						
10,080 ± 160	I-5146	Fell's Cave	layer 18	Saxon 1976						x	x			x				x				x		x	x	x					x	
10,720 ± 300	W-915	Fell's Cave	layer 19	Bird 1970																								x	x			
11,000 ± 170	I-3988	Fell's Cave	first occupation	Bird 1970																								no fauna				
12,600 ± 600	BVA-Arsenal	Los Toldos	capa 11 b	Cardich et al. 1973													x									x	x	x				
ca. 10,500	(1)	Mylodon Cave																														
ca. 13,000	(2)	Mylodon Cave																								x		x	x	x	?	x

(1) average of 4 radiocarbon dates mentioned in text
(2) average of 8 radiocarbon dates mentioned in text

If we could prove that for this period and area other herbivores (mainly horse and sloth) were not available, the guanaco — tucu-tucu — rhea association would indicate an adaptative strategy of the same nature as that of layers 9 and 10 of Los Toldos, and layers 17 to 13 (Bird Period II) and 12 (first occupation of Fell's Cave in Bird Period III).

PALLI AIKE CAVE

Here Bird (1938, 1952) found burnt bones of sloth, horse and guanaco on the surface of a volcanic ash layer. The upper part of that layer yielded Bird II materials. Bones of the three species were used as a sample for dating (8639±450 B.P., C-485, Libby 1952).

At the same site beneath the ash layer Bird (1938) found at least seven skeletons of sloths which seemed to him to have died naturally and not by human agency. Associated cultural materials were rather scanty.

DISCUSSION

Mylodon was a cave-dweller, as demonstrated by considerable dung accumulation at Mylodon Cave. The lack of sloth dung at Palli Aike and Las Buitreras questions the habitation of these caves by sloth. In the case of Las Buitreras there are human and carnivore coprolites associated with sloth bones in layers 7 and 8 (Figuerero Torres 1982).

In hunting sloth one would expect a high probability that hunters cornered prey in sheltered den locations after scheduled hunting or occasional sighting, stalking and persecution. This appears to have been the case at Las Buitreras, and Palli Aike, and perhaps at Fell's Cave. The possibility of human scavenging of carcasses must also be considered.

The above panorama could be explained by postulating a generalized adaptive strategy. Early human adaptive behavior in southern Patagonia has pointed toward generalization. As we have seen, extinct herbivores, including horse and sloth and camelids were hunted when available. But utilized resources were not restricted to big game mammals. Rodents, rhea, waterfowl and other birds, and probably carnivores were also procured.

The change from a generalized strategy, which included extinct fauna, to a generalized strategy lacking extinct species was gradual, as was the disappearence of the megafauna. These two aspects of the same kind of adaptive behavior may also have regional significance.

Extinction of the large herbivores in a comparatively short lapse of time would not have

Table 3 Economic status of utilized resources.

resources	spatial predictability	temporal predictability	abundance	economic status
guanaco	+	+	+	M
horse and sloth	+	+	?	C to O
rodents	+	+	+	C
rhea	+	+	?	O to C
waterfowl	+	+	+	C
carnivores	-	-	+	O

+ = high/ - = low   M = main/ C = complementary/ O = occasional

affected the situation of a young ecosystem, which is characterized by low species diversity and low biomass. Pianka (1978) states that "a low expectation of finding prey, or a high mean search time per item, demands generalization." The generalization evidenced by our data and the low expectation corresponds to the existence of a young ecosystem, which was characteristic of southern Patagonia at this early date.

Analysis of the structure, distribution of ecosystems, and taxonomic composition of plant and animal communities of the Magellanic biota was carried through by Pisano (1975). Pisano concluded that following the retreat of Patagonian glaciers after 13,000 B.P. (Mercer 1976) the migration to the south by plants and animals was a "highly selective process." This can be seen at the present: 1- low diversity and low population density of the terrestrial mammals; 2- high percentage of migratory or non-resident birds who temporarily occupy many possible niches of other consumers; 3- low diversity and scarce endemism of plants. Finally, Pisano concludes that there exist a series of historically young ecosystems with a low degree of homeostasis.

From a palaeoeconomic point of view, to understand an adaptive strategy one has to assign economic status to each of the resources procured. In Table 3, main, complementary and occasional resources (Higgs 1975) are distinguished according to two variables: predictability and abundance (Dyson-Hudson and Smith 1978).

NOTES

1. The site of Englefield is not discussed here although radiocarbon dates exist of 9248±1500 and 8450±1500 B.P. (Sa-20c) (Rivera 1978). Recently, another radiocarbon date of 3915±75 (GrN-8573) was published by Ortiz-Troncoso (1978). This date suggests that mixing of at least two occupational episodes occurred. The stratigraphic situation at Englefield Island must be clarified by further excavation. Also the faunal information provided by Josien (Emperaire and Laming 1961) is inadequate. Similarly, the earliest occupation of Alero Marazzi (Tierra del Fuego) was dated 9590±210 B.P. (Gif-1034) (Rivera 1978). The only evidence of hypothetical utilization of a marine resource for this period comes from Las Buitreras Cave (see below). The absence of early man sites on the Patagonia coasts of the Atlantic and Magellan Channel is most likely due to the post-glacial rise of sea level.

2. Radiocarbon date NZ-1680 (Rivera 1978) seems to be the same as R-4299 published by Long and Martin (1974). The latter was processed at the laboratory of the Department of Scientific and Industrial Research of the Institute of Nuclear Sciences of New Zealand. The sample was sloth hide and hair. This information was supplied in a letter by R.W. Wellman to Dr. Mateo Martinic (Instituto de la Patagonia, Punta Arenas, Chile) (L.A. Borrero personal communication 1981). Rivera (1978) apparently arrived at the same conclusion.

3. Birds of Mylodon Cave are under study by Dr. P. Humphrey (personal communication 1981) at the University of Kansas.

ACKNOWLEDGEMENTS

To the staff of the Laboratorio de Tritio y Radiocarbono (LATYR) of La Plata for their invaluable help and precise counselling.

To Dr. Rosendo Pascual and other members of the División Vertabrados del Museo de la Plata for their generosity and their willingness to exchange information.

To Dr. Alan L. Bryan for encouraging early man studies and for his precise editing of this paper.

To my colleagues and friends L.A. Borrero, S.E. Caviglia and H.D. Yacobaccio for their constructive criticism.

To my dear wife Mary Jo for her help, specially in my command of the English language.

REFERENCES CITED

Aguerre, A.M.
1977 A propósito de un nuevo fechado radiocarbónico para la Cueva de las Manos, Alto Río Pinturas, Provincia de Santa Cruz. *Relaciones de la Sociedad Argentina de Antropología XI:* 129-142.

Bird, J.
1938 Antiquity and migrations of the early inhabitants of Patagonia. *The Geographical Review* 28(2): 250-275.

1952 Fechas del radiocarbono para Sudamérica. *Revista del Museo Nacional* 21. Lima.

1970 Paleo-Indian discoidal stones from southern South America. *American Antiquity* 35: 205-209.

Burleigh, R., A. Hewson, and N. Meeks
1977 British Museum natural radiocarbon measurements IX. *Radiocarbon* 19: 143-160.

Cardich, A.
1977 Las culturas pleistocénicas y postpleistocénicas de Los Toldos y un bosquejo de la prehistoria de Sudamérica. *Obra del Centenario del Museo de La Plata* 2: 149-172, La Plata.

1978 Recent excavation at Lauricocha (Central Andes) and Los Toldos (Patagonia). In Early man in America from a circum-Pacific perspective, edited by A.L. Bryan, pp. 296-300. *Department of Anthropology, University of Alberta, Occasional Papers* No. 1. Edmonton.

Cardich, A., L.A. Cardich, and A. Hajduk
1973 Secuencia arqueológica y cronología radiocarbónica de la Cueva 3 de Los Toldos (Santa Cruz, Argentina). *Relaciones de la Sociedad Argentina de Antropología* VII: 85-123, Buenos Aires.

Caviglia, S.E.
1978 La presencia de *Dusicyon avus* (Burm.), 1864 en la capa 8 de la Cueva de Las Buitreras (Patagonia meridional). Paper presented at the *VI Congreso Nacional de Arqueología,* Salto, Uruguay.

Dyson-Hudson, R., and E.A. Smith
1978 Human territoriality: an ecological reassessment. *American Anthropologist* 80: 21-39.

Emperaire, J., and A. Laming
1961 Les gisements des iles Englefield et Vivian dans la mer d'Otway (Patagonie australe). *Journal de la Sociêtê des Amêricanistes* 50: 7-77.

Emperaire, J., A. Laming-Emperaire, and H. Reichlen
1963 La Grotte Fell et autres sites de la région volcanique de la Patagonie chilienne. *Journal de la Sociêtê des Amêricanistes* 52: 169-254.

Figuerero Torres, M.J.
1982 Análisis de coprolitos: el caso de Cueva Las Buitreras. *VII Congresso Nacional de Arqueología* 46-49. Colonia del Sacramento, Uruguay, diciembre 1980.

Gradin, C., C. Aschero, and A. Aguerre
1976 Investigaciones arqueológicas en la Cueva de las Manos, Alto Río Pinturas, Santa Cruz. *Relacione de la Sociedad Argentina de Antropología* X: 201-250, Buenos Aires.

1979 Arqueología del area Río Pinturas, Provincia de Santa Cruz. *Relaciones de la Sociedad Argentina de Antropología* XIII: 183-227.

Hakansson, S.
1976 University of Lund radiocarbon dates IX. *Radiocarbon* 18: 290-320.

Hauthal, R.
1899 Reseña de los hallazgos en las cavernas de Ultima Esperanza. *Revista del Museo de La Plata* 9: 411-420.

Higgs, E.S. (editor)
1975 *Palaeoeconomy*, Cambridge University Press, London.

Lehmann-Nitsche, R.
1902 Nuevos objetos de industria humana encontrados en la Caverna Eberhardt en Ultima Esperanza. *Revista del Museo de La Plata* 11: 55-67.

Libby, W.F.
1952 *Radiocarbon dating*. University of Chicago Press, Chicago.

Long, A., and P.S. Martin
1974 Death of american ground sloth. *Science* 186: 638-640.

Mengoni Goñalons, G.L., y M.J. Silveira
1976 Análisis e interpretación de los restos faunísticos de la Cueva de las Manos, Estancia Alto Río Pinturas (Prov. Santa Cruz). *Relaciones de la Sociedad Argentina de Antropología* X: 261-270.

Mercer, J.H.
1976 Glacial history of southernmost South America. *Quaternary Research* 6: 125-166.

Moore, D.M.
1978 Post-glacial vegetation in the South Patagonia territory of the giant sloth, *Mylodon*. *Botanical Journal of the Linnean Society* 77: 177-202.

Nordenskjöld, E.
1900 Jakttagelser och fyndi grottor vid Ultima Esperanza i Sydvestra Patagonien. *Kongliga Svenska Vetenkaps-Akademiens Handlingar* 33(3). Stockholm. English translation in possession of the author.

Ortiz-Troncoso, O.R.
1978 Nuevo fechado radiocarbónico para la isla Englefield (seno Otway, Patagonia Austral). *Relaciones de la Sociedad Argentina de Antropología* XII: 243-244.

Pianka, E.R.
1978 *Evolutionary Ecology*. Second edition. Harper & Row, New York.

Pisano, V.E.
1975 Características de la biota magallánic derivadas de factores especiales. *Anales del Instituto de la Patagonia* 6(1-2): 123-137. Punta Arenas.

Poulain-Josien, T.
1963 La Grotte Fell. Etude de la Faune. Appendice en "La Grotte Fell et autres sites de la région volcanique de la Patagonie chilienne" par J. Emperaire, A. Laming-Emperaire and H. Reichlen. *Journal de la Sociêtê des Amêricanistes* 52: 230-254.

Rivera, M.A.
1978 Cronología absoluta y periodificación en la arqueología chilena. *Boletín del Museo Arqueológico de La Serena* 16:13-41.

Sanguinetti, A.C., y L.A. Borrero
1977 Los niveles con fauna extinta de la Cueva Las Buitreras, Río Gallegos, Pcia. Santa Cruz. *Relaciones de la Sociedad Argentina de Antropología* X: 271-292.

Saxon, E.C.
1976 La prehistoria de Fuego-Patagonia: colonización de un habitat marginal. *Anales del Instituto de la Patagonia* 7: 63-73. Punta Arenas.

1979 Natural Prehistory: the archaeology of Fuego-Patagonian ecology. *Quaternaria* 21: 329-356.

Silveira, M.J.
1979 Análisis e interpretación de los restos faunísticos de la Cueva Grande del Arroyo Feo (Pcia. de Santa Cruz) Parte I. *Relaciones de la Sociedad Argentina de Antropología* XIII: 229-253.

Cazadores de *Mylodon* en la Patagonia Austral

LUIS ALBERTO BORRERO
Instituto de Antropología
Universidad de Buenos Aires
25 de Mayo 217
Buenos Aires
ARGENTINA

Abstract

The hypothesis is presented that before about 9000 B.P. Las Buitreras Cave (southern Argentina) was a ground sloth kill-site, as well as used for the consumption of guanacos hunted and butchered elsewhere. Recent examination of ground sloth bones recovered from early excavation in Mylodon Cave (nearby in southern Chile), show clear butchering marks. Based on clustered radiocarbon dates of ground sloth remains, sloth occupation of Mylodon Cave appears to have been between 13,000 and 12,300 B.P., although possibly also between 10,800 and 10,200 B.P. It is suggested that Mylodon Cave was also used by man as a place to hunt and butcher sloths. Palli Aike may also have served as a sloth trap.

INTRODUCCION

Las expediciones dirigidas por Junius Bird a la Patagonia austral, entre 1932 y 1937, mostraron la asociación de hombre y megafauna en las cuevas Fell y Palli Aike (Magallanes, Chile). Este fue sólo uno de los muchos resultados obtenidos en esas expediciones, pero uno especialmente importante, ya que llamó la atención hacia la probable antigüedad de los primeros pobladores de Sudamérica. Sobre la base de esos hallazgos (Bird 1938, 1946) y de los fechados radiocarbónicos realizados años después para los niveles inferiores de esos sitios (I-3988 = 11.000±170 B.P.: W-915 = 10.720±300 B.P.; I-5146 = 10.080±160 B.P.) (Saxon 1976), debió resultar obvio que el poblamiento de América necesariamente debía ser anterior a los clásicos 11.200 B.P. de la mayoría de los sitios Clovis de las Grandes Llanuras norteamericanas. Sin embargo no ocurrió asi. Parte de la explicación acerca de porqué esta información no fue bien utilizada en la evaluación del poblamiento americano, radica en que las excavaciones complementarias de investigadores franceses en la Cueva Fell (Emperaire et al. 1963) no pudieron comprobar la relación entre *Mylodon* y hombre. El mismo resultado habían arrojado las excavaciones que el mismo equipo había realizado en la Cueva del Mylodon (Ultima Esperanza, Magallanes, Chile) (Emperaire y Laming 1954). En los últimos años se ha vuelto a afirmar que los primeros pobladores de la Patagonia no cazaban *Mylodon* (Saxon 1976, 1979).

El trabajo desarrollado en un nuevo sitio, la cueva Las Buitreras (Santa Cruz, Argentina), mostró que, en caso, *Mylodon* (?) *listai* era indiscutiblemente cazado por el hombre. Sobre la base de esa evidencia, presentada principalmente en Sanguinetti (1976, 1981), Sanguinetti y Borrero (1977) y Caviglia et al. (1981), presentó la hipótesis de que Las Buitreras fue un sitio de matanza de *Mylodon*. Posteriormente haré consideraciones respecto a los otros sitios mencionados arriba. La hipótesis, como mostraré más abajo, es la que mayor sustento teorético tiene. Sin embargo, no toda la evidencia encaja perfectamente en esa explicación. De todas maneras es la hipótesis remanente al descartar varias hipótesis alternativas:

1. *Mylodon* no fue cazado, su presencia en el sitio se debe a factores naturales (ver refutación en Caviglia et al. 1981).
2. *Mylodon* fue cazado fuera del sitio, sus partes fueron transportadas allí, y luego fue consumido (ver refutación en Borrero 1980).
3. *Mylodon* fue consumido en el sitio, pero no fue cazado (cf. trozamiento y consumo de un animal muerto naturalmente o cazado por otros depredadores) (ver refutación en Borrero 1980).

EL CASO DE LAS BUITRERAS

La información sobre este sitio está presentada en otros trabajos, enumerados más arriba, por lo que no repetiré más que los datos esenciales para este análisis. En el Figura 1 sintetizo las principales variables conocidas que dan cuenta del proceso de formación de las capas VI-VII-VIII de este sitio. El objetivo de esta sección es el de presentar la evidencia pertinente a la hipótesis que ha sido retenida para explicar la función del sitio (sitio de matanza de *Mylodon*). La discusión estará dirigida hacia una comparación de los hallazgos de *Mylodon* (?) *listai* y de *Lama glama guanicoe* (las dos especies más representadas en los niveles inferiores). Comenzaré por entregar una síntesis de los patrones de conducta y de utilización del espacio por parte de ambas especies. Resulta difícil comparar una especie moderna (*Lama g.g.*) y una especie extinta (*Mylodon* (?) *l.*), ya que es muy despareja la calidad de la información. Sin embargo, a través del conocimiento de los patrones de conducta de especies relacionadas es posible realizar acercamientos útiles.

Existe evidencia para discutir la posibilidad de que los Edentata fósiles hayan contado con la capacidad para nadar. *Megalonyx* pudo quizá alcanzar Centroamérica durante el Terciario, es decir en tiempos en que estaba presente el geosinclinal Bolivar, como inmigrante extraviado probablemente azaroso (Patterson y Pascual 1972: 254). Al mismo tiempo alcanza las Antillas y probablemente en la misma forma (Patterson y Pascual 1972:269). Pero más allá del hecho de que efectuaran estas primeras migraciones nadando o como "navegantes" azarosos, subsiste el hecho de que mostraron una notable predisposición al movimiento. Se debe recordar que seis de las siete familias de edentados sudamericanos migran a Norteamérica (Patterson y Pascual 1982:292). Al mismo tiempo que la retracción de los Tardigrada hacia el norte muestra esa tendencia al movimiento, está indicando sus preferencias por paleoambientes tropicales y sub-tropicales (Scillato Yané 1976:311). La presencia de la subfamilia Mylodontinae en el sur del continente durante el Pleistoceno habla de adaptación al frío, obtenida, de acuerdo con Scillato Yané, por un mejoramiento de su termorregulación. Todo esto avala una notable movilidad al considerarlos como familias o subfamilias y en una escala temporal amplia, pero otro es el cuadro al considerar la conducta de individuos dentro de un ecosistema dado. En ese caso hay que destacar su lentitud y su hábito probablemente solitario (Mengoni Goñalons 1980). Su lentitud es una función de su peso y tamaño corporal (ver Paula Couto 1979). La evidencia, tanto patagónica como extra-patagónica, muestra una tendencia a habitar en cuevas, o por lo menos a utilizarlas con asiduidad como lugar para defecar. Efectivamente las concentraciones de excrementos han sido registradas como altamente localizadas (ver Martin et al. 1961; Martin 1975; Nordenskjöld 1900). Todo esto lleva a postular que seguramente el territorio diario de obtención de alimento de los edentados, era de un diámetro inferior al de cualquier herbívoro ágil y rápido. La adaptación al frío de *Mylodon* incluye una musculatura masticatoria más perfecta que sus ancestros terciarios, destinada a desmenuzar mejor los alimentos y obtener un mejor rendimiento energético (Scillato Yané 1980). Postular un territorio de alimentación pequeño es coherente con esa necesidad de mejor aprovechamiento energético creada por el frío Pleistoceno.

En el Figura 2 presento la distribución de *Mylodon* en el Pleistoceno Superior-Holoceno Temprano en sitios arqueológicos del sur de Sudamérica. Se observa una notable dispersión, lo que puede interpretarse como una abundancia relativamente grande de estos animales. Todo esto

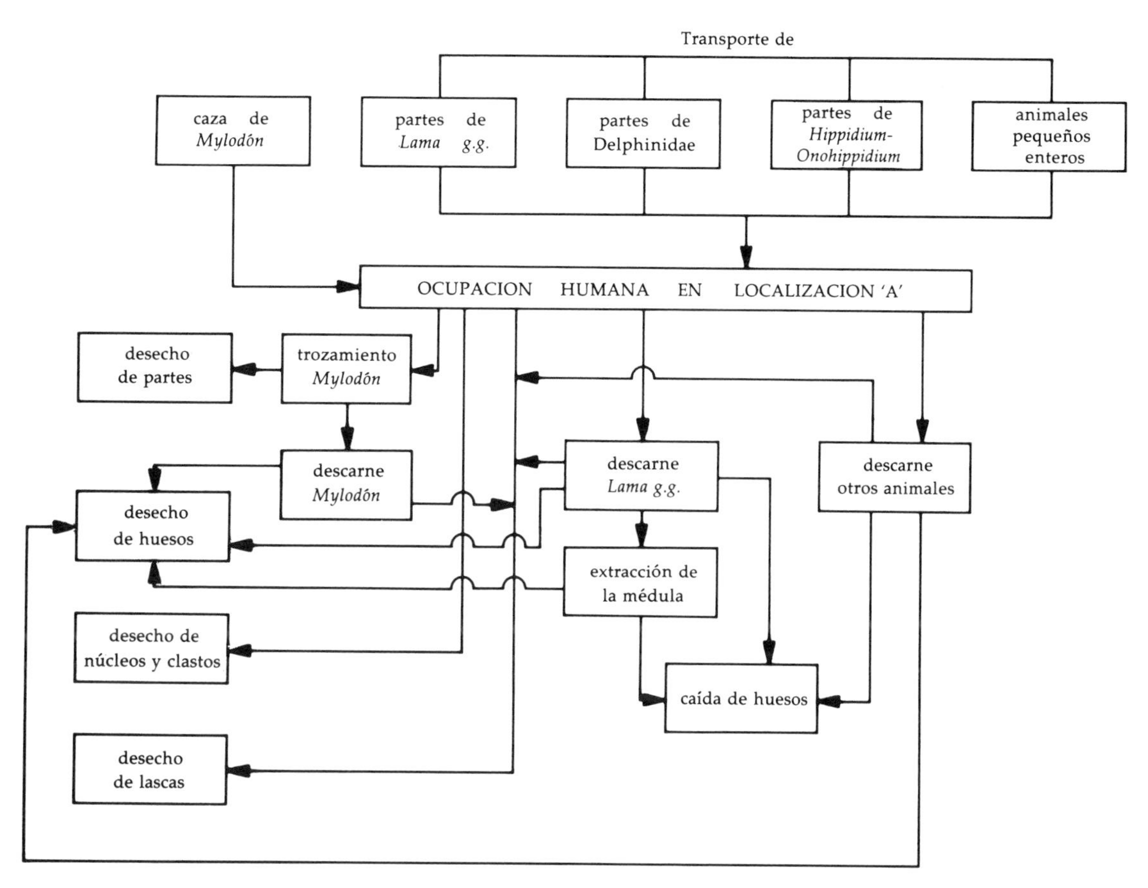

Figura 1a. Cueva Las Buitreras. Factores culturales que operaron en el proceso de formación de las capas VII-VIII.

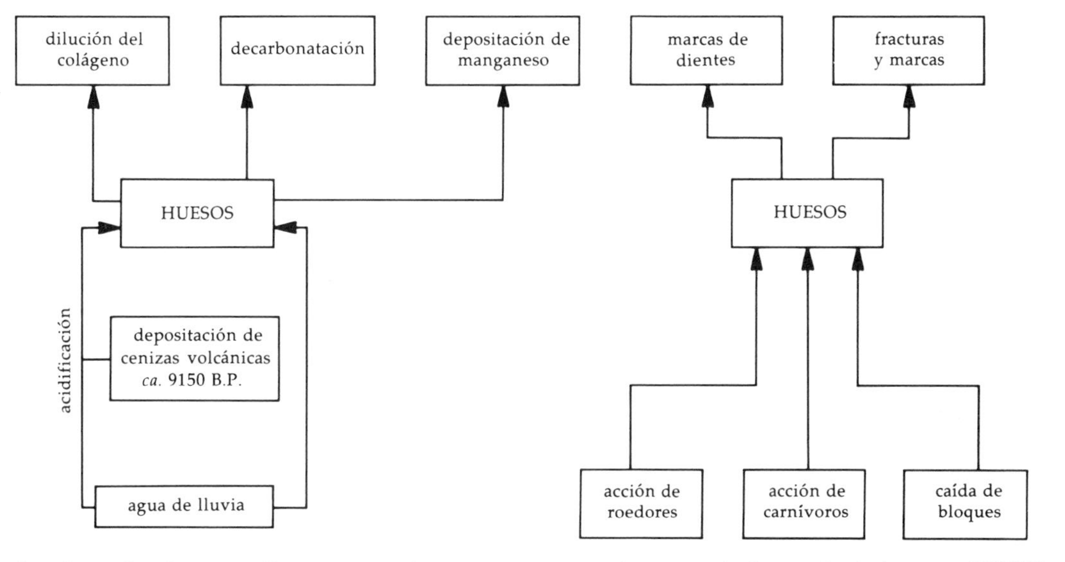

Figura 1b. Cueva Las Buitreras. Factores naturales que operaron en el proceso de formación de las capas VII-VIII.

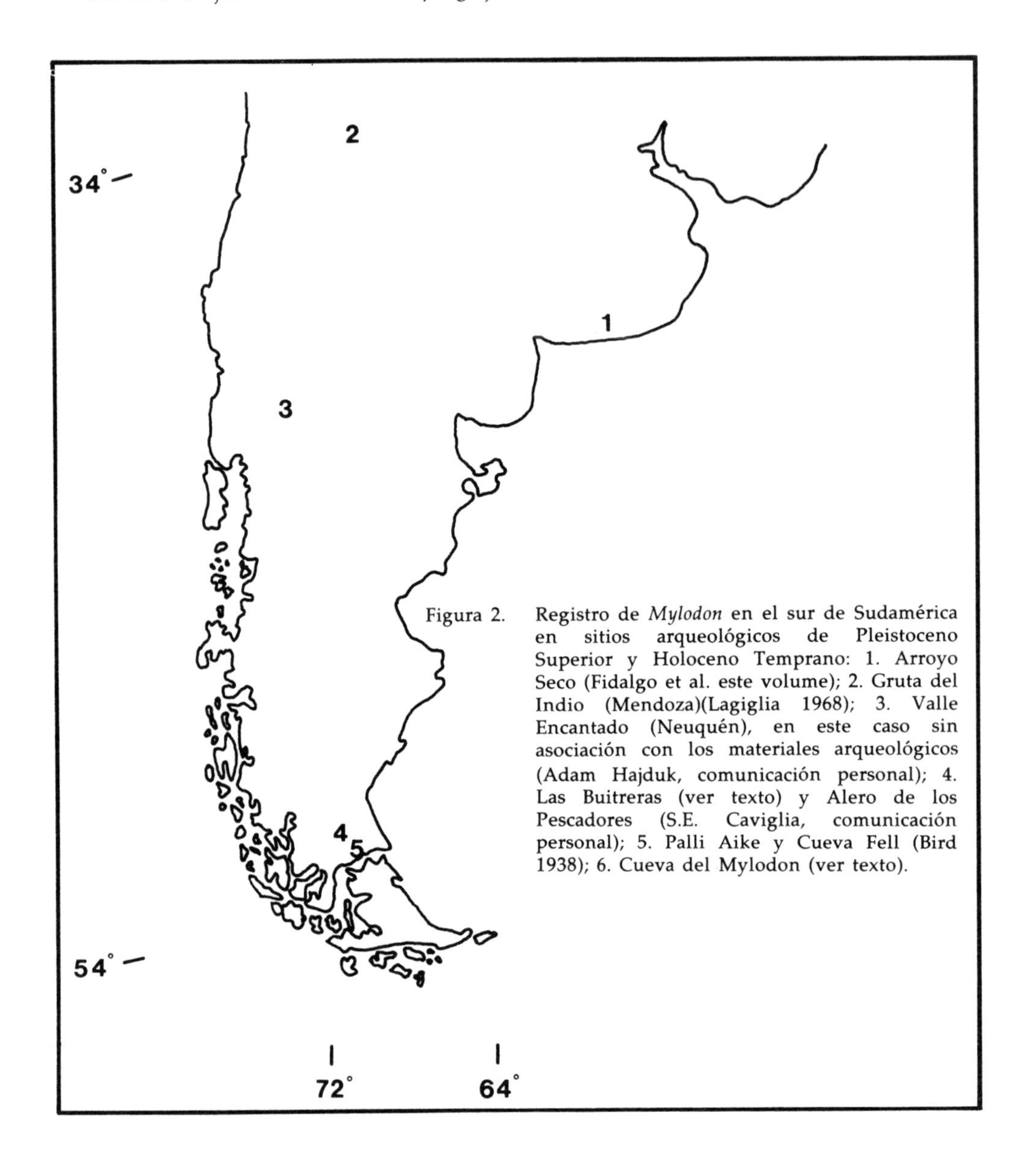

Figura 2. Registro de *Mylodon* en el sur de Sudamérica en sitios arqueológicos de Pleistoceno Superior y Holoceno Temprano: 1. Arroyo Seco (Fidalgo et al. este volume); 2. Gruta del Indio (Mendoza)(Lagiglia 1968); 3. Valle Encantado (Neuquén), en este caso sin asociación con los materiales arqueológicos (Adam Hajduk, comunicación personal); 4. Las Buitreras (ver texto) y Alero de los Pescadores (S.E. Caviglia, comunicación personal); 5. Palli Aike y Cueva Fell (Bird 1938); 6. Cueva del Mylodon (ver texto).

muestra que *Mylodon* podía estar disponible en cualquier sector del espacio al sur de los 38° de latitud Sur. Con una dispersión de este tipo no puedo sostener que *Mylodon* solamente vivía en cuevas. Probablemente, como cualquier otro habitante de las pampas, utilizaba las cuevas como refugio, y fue en estas donde se conservó el mejor registro de su existencia, debido a las mejores condiciones de sedimentación. En general se puede utilizar el registro terciario de Mylodontinae para confirmar la variabilidad de ambientes que utilizaba (ver Scillato Yané 1981:33).

En el caso específico de Las Buitreras se han hallado cuatro excrementos de carnívoro-omnívoro, atribuíbles a Canidae, y un excremento humano. El estado de conservación era bueno y por ello se consideró que la ausencia de excrementos de *Mylodon* en los niveles VII-VIII se debe a que este no habitó en la cueva (Figuerero Torres 1980). Sin embargo, entiendo que esa ausencia simplemente significa que probablemente *Mylodon* no defecó en la cueva. Existe evidencia de edentados que utilizan el mismo lugar para defecar, sin que este sea su lugar de habitación usual (ver Martin et al. 1961:123).

El caso de *Lama glama guanicoe* es muy diferente. Los restos obtenidos en sitios arqueológicos patagónicos muestran que era cazado en todos los ambientes: cañadones, cuencas lacustres, valles abiertos, llanuras, bosques, litoral marítimo (Mengoni Goñalons 1980). Esta información avala su notable capacidad adaptativa, que ha sido explicada por su euritermismo (Morrison 1966). Vallenas agrega otros mecanismos fisiológicos que explican la flexibilidad ambiental de los camélidos sudamericanos en general (Vallenas 1970:69). Sin embargo se ha señalado que su área de dispersión no incluye zonas bajas y pantanosas (Tonni y Politis 1980). A pesar de ello posee capacidad natatoria que le permite cruzar ríos o aun regulares cuerpos de agua de mar (así ha de haber efectuado el poblamiento de la isla Navarino, Tierra del Fuego, Chile). Prefiere habitar en espacios abiertos (llanuras o valles abiertos); aunque al norte del lago Fagnano (Tierra del Fuego) lo hace en bosque relativamente abierto (observaciones personales en los veranos de 1978 y 1980 y en el invierno de 1980) y al norte del canal de Beagle (Tierra del Fuego) en bosque muy cerrado (Luis Sosa, comunicación personal y observaciones personales en el invierno de 1980).

Para la ocupación del espacio exhibe una marcada territorialidad; así cada grupo familiar (por lo general 7 o más individuos) posee un territorio fijo de un diámetro aproximado de 30 kilómetros (Cajal 1979; Garrido et al. 1979). Este territorio es de utilización anual. Existen además tropas de machos, integradas por jóvenes que aun no han formado su grupo familiar y por viejos que han sido desplazados de los suyos. Estas tropas pueden incluir bastante más que 100 individuos. En el estudio de su conducta se ha propuesto la existencia de zonas de acceso libre (por ejemplo las aguadas), independientes del territorio familiar (Garrido et al. 1979). Evidentemente que, de ser verificada esta hipótesis, esos sectores de acceso libre ocuparían situaciones nodales respecto a varios territorios. Se debe agregar que son animales rápidos, capaces de moverse con facilidad en terrenos sinuosos. Todo esto muestra, al considerar cada grupo de "guanacos" en un ecosistema dado, una ocupación efectiva basada en la rapidez de movimientos y en la exclusividad en el uso de sectores del espacio. En áreas de recursos alimenticios sumamente escasos (Catamarca y La Rioja, en el noroeste de Argentina) los grupos familiares de "guanacos" pueden moverse distancias mayores que los treinta kilómetros anotados (Mengoni Goñalons, comunicación personal).

Quiero destacar, entonces, que estoy comparando un animal gregario, ágil y de tamaño mediano, con un animal seguramente solitario en sus desplazamientos, lento, torpe y de tamaño grande. En la consideración de ambas especies postulo que el territorio de alimentación de cualquier grupo familiar de *Lama glama guanicoe* era mayor que el de *Mylodon*. En la Figura 3 he dibujado la ubicación del sitio Las Buitreras y, asumiendo que fue habitada por *Mylodon* la he considerado el centro de un territorio de alimentación de radio inferior a 10 km; al mismo tiempo he dibujado el máximo teórico de territorios de alimentación de *Lama g.g.* inmediatamente accesibles desde ese sitio. Queda claro que, según las características en el uso del espacio que he anotado más arriba, se puede decir que ambas especies tenían localizaciones predecibles: *Lama glama guanicoe* con un rango máximo de unos 30 km y *Mylodon* en localizaciones específicas (cuevas) o sus inmediaciones. Me interesa destacar que los hallazgos de *Mylodon* (?) *listai*, *Lama glama guanicoe* y otras especies (ver Figura 1) se realizan en una cueva que se abre sobre una colada basáltica y que enfrenta una regular planicie graminosa (Sanguinetti 1976, 1981).

En los Cuadros 1 y 2 he anotado las partes presentes de ambas especies en Las Buitreras. Se observa una bajísima frecuencia de huesos del esqueleto axial de *Lama glama guanicoe*. Todos los restos de esa especie son astillas (excepto algunas falanges y un calcáneo fracturado longitudinalmente). Las frecuencias anotadas, entonces, no se refieren a huesos enteros prácticamente en ningún caso (para *Lama glama guanicoe*). En la Figura 4 he ilustrado las partes representadas en los niveles inferiores de Las Buitreras. El avanzado estado de

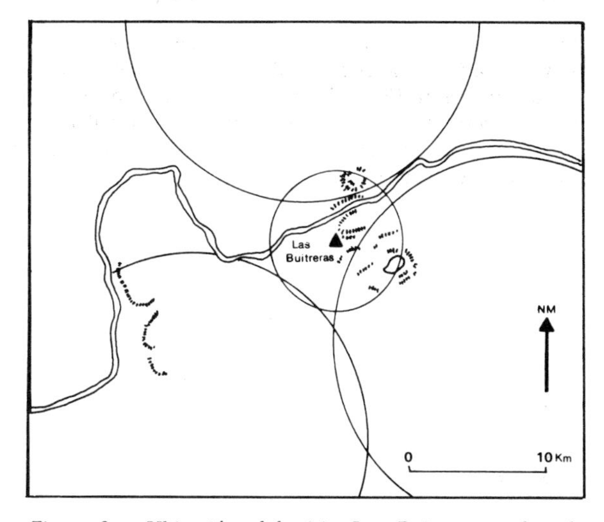

Figura 3. Ubicación del sitio Las Buitreras, sobre la margen derecha del río Gallegos, en su curso medio. El círculo central delimita el hipotético territorio de alimentación de *Mylodon*. Los círculos mayores delimitan hipotéticos territorios de alimentación de *Lama glama guanicoe*.

desarticulación, unido a las características que ya presenté más arriba, me llevan a decir que *Lama glama guanicoe* fue cazado lejos de la cueva. Las frecuencias relativamente altas de partes inferiores de los miembros (que son de bajo rendimiento alimenticio) son difíciles de interpretar. Una interpretación posible es que fueron transportados a la cueva para consumir la médula. Este problema aun no ha sido solucionado. Las partes superiores de miembros de *Lama g.g.* tienen buen rendimiento alimenticio (en carne y en médula disponibles).

Por otro lado el cuadro de restos de *Mylodon* informa que prácticamente todas las partes del cuerpo están representadas. Este hecho es más importante aun si se considera que la mayor parte de los huesos de *Mylodon* son indeterminables, debido al pésimo estado de conservación (ver Figura 1a). Además, se efectuaron varios hallazgos de huesos articulados, o concentraciones de huesos (desechados en un basural, según mi interpretación, ver Borrero 1980 y Caviglia et al. 1981) que seguramente estuvieron articulados en tiempos de su depositación. Es por eso que sostengo que *Mylodon* fue cazado muy acerca de la cueva. A partir de lo que ya informé sobre las costumbres de estos edentados digo que el hecho ocurrió en la cueva misma. Evidentemente esta explicación es coherente con la expectativa de transporte que presentan los restos de uno y otro animal: *Lama glama guanicoe* más liviano y con sus huesos portadores de médula suficiente como para justificar su extracción, es apropriado para transporte a grandes distancias; *Mylodon* mucho más

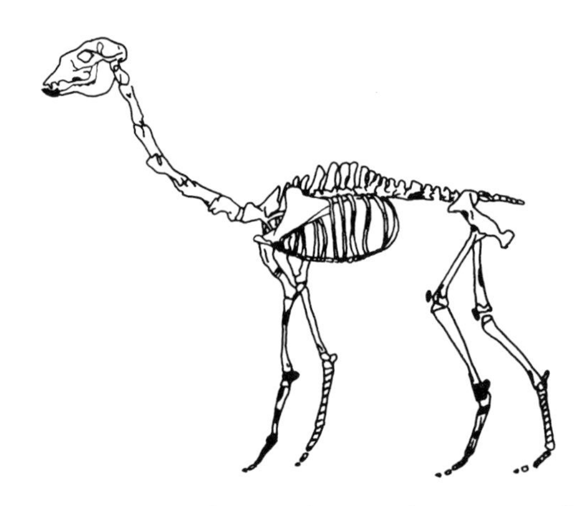

Figura 4. Partes de *Lama glama guanicoe* representadas en Capas VII-VIII de Las Buitreras. Rayado: fragmentos muy pequeños. Negro: fragmentos mayores (Caviglia y Figuerero Torres 1976; Sanguinetti y Borrero 1977; Sanguinetti 1981; Caviglia 1981).

Cuadro 1. Distribución de los óseos de *Lama g.g.* en sitios varios.

Material óseo de *Lama glama guanicoe* de capas VII-VIII de Las Buitreras (Caviglia y Figuerero Torres 1976; Sanguinetti y Borrero 1977; Sanguinetti 1981; Caviglia 1981) y de capa XIIIa de Fell (Poulain-Josien 1963). El material de Cueva del Mylodon no tiene procedencia estratigráfica clara y fue obtenido en el primer viaje de Hauthal (Roth 1899), se deben agregar fragmentos de diferentes miembros no identificados en la publicación. V=varios, todos los números indican cantidad de fragmentos.

	Las Buitreras	Fell	Cueva del Mylodon
Cráneo		1	1
Maxilar superior	1		V
Mandíbula	1		V
Vértebra	2	1	V
Pelvis	1		
Costilla	3	1	V
Escápula	1		2
Húmero	2		2
Radio-cúbito	4	1	
Metacarpo	4		
Fémur	1		
Tibia	1		
Calcáneo	1		
Metatarso	11		
Falange	14		
Sesamoideo	1	1	

Cuadro 2. Distribución de los óseos de *Mylodon* en sitios varios.

Material óseo de *Mylodon*. L.B. = Buitreras (Scillato Yané 1976; Caviglia 1981), F(1) = Fell (Emperaire et al. 1963; Poulain-Josien 1963), F(2) = Fell (Saxon 1976, 1979), C.M. (3) = Cueva del Mylodon (Roth 1899), C.M. (4) = Cueva del Mylodon (Roth 1904), C.M. (5) = Cueva del Mylodon (Nordenskjöld 1900), C.M. (6) = Cueva del Mylodon (Emperaire y Laming 1954), V = Varios, + = Existen más piezas que las informadas. El material de Las Buitreras proviene de capas VII-VIII, e de Fell(1) de capas XIII a y b, el de Fell(2) del Período 1 (apud Bird 1946), el de Cueva del Mylodon (3) y (4) no tiene procedencia estratigráfica, el de Cueva del Mylodon (5) del Nivel C, el de Cueva del Mylodon (6) de "niveles antiguos" y, en este último caso, no fue informado todo el material.

	L.B.	F(1)	F(2)	C.M.(3)	C.M.(4)	C.M.(5)	C.M.(6)
Cráneo	1	1	1	4	2+		1
Maxilar sup.				5+	1	3	2
Mandíbula			1	3+	2	11	
Molariformes	3			8			
Atlas		1					
Axis				1		1	
Vértebra	6	9		V	3	4+	
Pelvis	2			V			
Costilla	10	3		V	V		
Escápula	1	2		V	1+	9	
Clavícula				1	1		
Húmero	1			2		1	
Cúbito	1				1		
Carpo	1						
Metacarpo					1		
Fémur	1			V			1
Tibia	2	1		3			1
Peroné	1			2			
Tarso			2	3			
Astrágalo	2						
Calcáneo	1					V	
Metatarso				1			
Falange		2		5	V	V	
Uñas				16		V	
H. dérmicos	V	5		V		V	

pesado, con huesos sin médula suficiente para justificar su extracción, ofrece *a priori* menos atractivos para pensar en su transporte. Debo aclarar que la ausencia de excrementos de *Mylodon* en Las Buitreras es un argumento en contra de mi explicación. Sin embargo resulta insostenible decir que esos restos fueron acarreados al sitio, ya sea por el hombre o por otros depredadores.

Todo el sustento teorético de cazadores antiguos patagónicos presenta, hasta el momento, un modo de vida centrado en la caza de *Lama glama guanicoe*. Sobre esa base se puede discutir que la caza y consumo de esa especie fue programada desde algún tipo de sitio diferente a Las Buitreras. La caza de *Mylodon*, en cambio, no exigía más que una marcha hacia ciertos *loci* específicos (las cantidades de *Mylodon* disminuían día a día según avanzaba el Holoceno). El instrumental lítico hallado junto a los huesos en Las Buitreras, puede considerarse resultado de una tecnología expeditiva y no se ha hallado ningún instrumento que pueda interpretarse como especializado para la caza de *Lama glama guanicoe* (en general, el instrumental especializado, es apto para operaciones de caza a distancia). Lo que espero, entonces, es hallar ese instrumental en una Localidad de Actividades Generalizadas (*apud* Wilmsen 1970), con la salvedad de que una fracción del mismo debería encontrarse en sitios de matanza a cielo abierto. Estos sitios aun no han sido encontrados. Interpreto Las Buitreras como una Localidad de Actividades Limitadas (*apud* Wilmsen 1970), específicamente un sitio de matanza de *Mylodon*, donde además se consumieron otros animales obtenidos lejos del sitio (ver Figura 1).

EL CASO DE LA CUEVA DEL MYLODON

Todo el conocimiento paleontológico de la Cueva del Mylodon muestra que es un sitio caracterizado por notables concentraciones de *Mylodon darwinii*. Se han publicado muchos informes sobre este sitio y, en general, se ha tendido a interpretar que esos restos se depositaron naturalmente. A la luz de la evidencia de Las Buitreras puedo reinterpretar el sitio. He podido revisar huesos de *Mylodon* de la Colección Hauthal (publicados por Roth 1899 y 1904, y conservados en el Departemento de Paleontología de Vertebrados del Museo de La Plata) y he encontrado muy abundantes huellas de corte (confirmando observaciones publicadas por Lehmann-Nitsche 1899). Estas observaciones las realicé junto a S.E. Caviglia, G. Mengoni Goñalons, H.D. Yacobaccio y otros colegas. Hay un trabajo en preparación sobre estas huellas.

El estado de desmembramiento presentado por los restos de *Mylodon* encontrados en la cueva fue destacado por todos los autores que visitaron el sitio. Este desmembramiento puede deberse a muchos factores:

1) Pisoteo: Debo considerar que es grande la posibilidad de que el pisoteo tenga que ver con el estado de desarticulación observado. En la Cueva del Mylodon existe una capa de excrementos que, en algunos sectores (ver Figura 5), tiene más de un metro de espesor. Esto significa que, en cualquier caso, hayan vivido allí los animales o hayan concurrido a la cueva para defecar, el movimiento de los pesados *Mylodon* debió ser constante. Es preferible reconocer que el pisoteo puede dar cuenta de por lo menos parte del registro de huesos desarticulados; sobre todo si se recuerda que

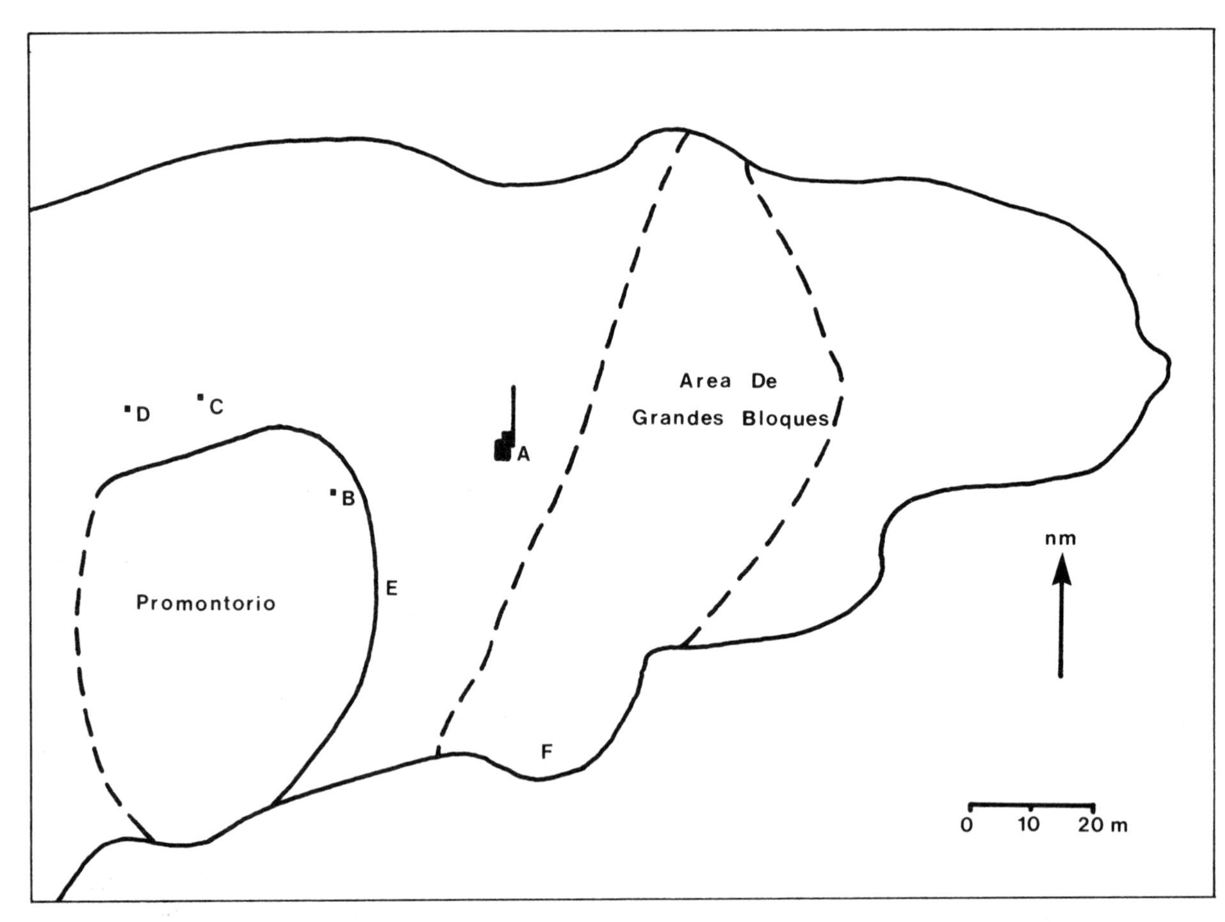

Figura 5. Planta de Cueva del Mylodon (reformada de Saxon 1979). A: Trinchera 2/7 (Expedición Saxon), en esta trinchera se efectuaron los hallazgos informados en el texto; B: Trinchera 3 (Expedición Saxon), entregó una secuencia paleoclimática (macroplantas); C: Trinchera 5 (Expedición Saxon); D: Trinchera A (Expedición Saxon), es un basural de canoeros (*ca.* 2500 B.P.); E: principal concentración de excrementos de *Mylodon*; F: esqueleto humano hallado en 1895 (Hauthal 1899).

el pisoteo puede inclusive provocar fracturas en espiral (Myers et al. 1980; Binford 1981:58). Resulta difícil estudiar una población de macizos huesos de *Mylodon* para determinar si hubo pisoteo, pues en ellos no cabe esperar la fractura columnar identificada por Gifford (ver Shipman 1981: 173-174).

2) Acción de carnívoros: El registro paleontológico de Cueva del Mylodon incluye restos de Canidae y de *Felis onca* (*apud* Nordenskjöld 1900:15 y 16); aunque puede discutirse la determinación específica en el segundo caso, no parece haber problemas con la genérica (ibid.). Posteriormente Roth asignó la mayoría de los restos de félidos de Cueva del Mylodon a *Felis listai,* al que describió como ". . . un gato mucho más poderoso /que *Felis onca*/" (Roth 1904:42). *Felis concolor* también está representado (Roth 1899:443). Una acumulación de esqueletos como la que ofrecía la Cueva del Mylodon seguramente debió atraer carnívoros (ver Binford 1981:40 s.); por lo que se debe considerar la posibilidad de que sus huellas estén representadas en el conjunto óseo.

3) Factores diagenéticos: La posibilidad de fracturas por compactación de sedimentos parece poco importante, debido al poco espesor y a las características de estos (formados principalmente por roca pulverizada de grano muy fino) (Wellman 1972). En la planta de la Cueva del Mylodon (Figura 5) se destaca un sector denominado "Area de grandes bloques"; aparentemente este sector está claramente diferenciado de la acumulación de excrementos y restos óseos de *Mylodon,* por lo que no cabe esperar una relación directa entre la caída de los bloques y el estado de desarticulación.

4) Actividad humana: El hombre perturbó los sedimentos de la cueva a través de múltiples excavaciones que llegaron, inclusive, a utilizar dinamita (sector centro-sur de la planta, entre el promontorio y el "Area de grandes bloques"). Esa actividad humana moderna ha producido todo tipo de huellas, fracturas y distorsiones en los huesos. El área revuelta por la dinamita no ha sido utilizada para extraer huesos; afortunadamente la mayor parte de la Colección Hauthal proviene de ese sector de la cueva y fue recogida antes de que se produjera esa perturbación. La otra forma de actividad humana que me interesa considerar es la debida a los antiguos cazadores que poblaron la Patagonia; me referiré a esta posibilidad en los párrafos siguientes.

Durante las excavaciones de la expedición Saxon en 1976, los hallazgos estratigráficos mostraron la presencia de una mandíbula (nivel 3), fragmentos varios entre niveles 4 y 6, el nivel 7 contenía solo restos de *Lama glama guanicoe* y, por debajo, se hallaron tres vértebras articuladas y un fragmento de escápula (todos estos materiales, salvo los indicados, corresponden a *Mylodon darwinii*). De manera que estos hallazgos avalan el estado de desarticulación a que hice referencia más arriba. Los niveles 7 y 8 contenían instrumental lítico, lascas con utilización y/o retoque y núcleos amorfos, en asociación con *Lama g.g.* y *Mytilus.* Por arriba y por debajo de esos niveles hay restos de *Mylodon.* Esta situación estratigráfica ya había sido reconocida anteriormente; efectivamente, Hauthal informó el hallazgo de instrumentos realizados sobre hueso de *Hippidium-Onohippidium s.l.* debajo de la capa de excrementos de *Mylodon* (Hauthal 1899). En 1947 se encontraron instrumentos líticos y óseos también por debajo de la capa de excrementos (Hammerly Dupuy 1948). Por todo esto es que sostengo que el hombre coexistió con *Mylodon* en esta cueva. Sobre la base de las huellas de corte identificadas en los materiales de la Colección Hauthal sostengo, además, que *Mylodon* fue cazado por el hombre.

Quiero destacar que la Colección Hauthal también incluye un molar de *Hippidium-Onohippidium s.l.* parcialmente quemado y varios fragmentos de *Mytilus;* este material se asocia indiscutiblemente con el Nivel B de Nordenskjöld (1900) (los hallazgos de *Mytilus* por debajo de niveles con *Mylodon* se repitieron durante la expedición Saxon). Resulta claro que la asociación "*Mylodon-Hippidium-Onohippidium s.l.-Mytilus*" en una cueva, difícilmente puede ser natural y obliga a agregar un cuarto término (el hombre) que le dé sentido. La presencia de *Mytilus* en una cueva situada a más de cinco kilómetros de distancia de las aguas del Océano Pacífico es difícil de explicar por factores naturales. Se sabe que algunas aves pueden transportar moluscos a grandes distancias de su habitat, pero ese no parece ser el caso de Cueva del Mylodon, fundamentalmente debido a los contextos estratigráficos en que aparece.

¿En qué momento ubico esa utilización del sitio por cazadores de *Mylodon*? Sobre la base del conocimiento actual de la cronología del poblamiento humano de la Patagonia, con fechas de ca. 12.500 B.P. para el curso medio del río Deseado (Cardich et al. 1973), unido a la evidencia estratigráfica que ya he presentado, sostengo que uno de los momentos de mayor disponibilidad de *Mylodon* en el área (cuyos restos han sido utilizados para realizar fechados radiocarbónicos) es precisamente el de la utilización humana. Los numerosos fechados radiocarbónicos realizados para los niveles inferiores del sitio se agrupan claramente en dos conjuntos:

1. Diez fechados radiocarbónicos que fluctúan entre 13.569 y 12.308 B.P. (detalle en Mengoni Goñalons, en este mismo volumen).
2. Tres fechados radiocarbónicos que fluctúan entre 10.832 y 10.200 B.P. (detalle en Mengoni Goñalons, en este mismo volumen).

A estos se pueden agregar siete fechados más, que fluctúan entre 7803 y 2556 B.P., pero que no están relacionados directamente con *Mylodon*.

El estado actual del conocimiento hace pensar que el segundo conjunto de fechas es el más adecuado para la aparición del hombre en el área. Sin embargo la evidencia de ninguna manera contradice, sino que por el contrario sostiene una hipótesis que mantenga al primer grupo de fechas. Evidentemente se necesita más trabajo de campo para solucionar esta disyuntiva. En cualquier caso debe tratarse de ocupaciones esporádicas por parte de grupos humanos pequeños. La presencia ya anotada de instrumentos mezclados con la capa de excrementos, o por debajo de la misma, es un argumento importante para sostener que no existe una utilización intensiva de la cueva.

En el Cuadro 1 presento la escasa información disponible para restos de *Lama glama guanicoe* de este sitio. Quiero destacar dos cosas:

1. El registro de *Lama g.g.* es mínimo, aseverando que la ocupación principal del sitio fue destinada a la explotación de *Mylodon* (si se puede mantener mi hipótesis).
2. Esos restos de *Lama g.g.*, si bien están desarticulados, incluyen abundantes huesos del esqueleto axial, lo que permite postular una cercanía efectiva del lugar de caza de "guanacos". Por otra parte los hallazgos son, principalmente, de huesos completos o fragmentos (no de astillas, como en el caso de Las Buitreras).

Por lo que dije más arriba, es muy difícil separar la acción de diferentes factores de desarticulación sobre el conjunto óseo acumulado en el sitio. No puede haber dudas que los depredadores de la región debieron estar activos sobre las carcasas de las varias decenas de animales representados en la cueva (quiero remarcar que Lehmann-Nitsche había destacado, en su publicación de 1904, que la colección de Cueva del Mylodon incluía "... un premaxilar fragmentado de *Onohippidium saldiasi* No. 1554/que/ presenta, en una extremidad, señales de que ha sido mordido por un carnívoro" (Lehmann-Nitsche 1904a:62). Sostengo que el hombre también acudió a la cueva, y que lo hizo para cazar *Mylodon*. Resulta obvio que, en este estado de la investigación, numerosas hipótesis alternativas pueden dar cuenta de los mismos hechos. Postulo que la Cueva del Mylodon es una Localidad de Actividades Limitadas, específicamente un sitio de matanza de *Mylodon*, y que su ubicación dentro del sistema de explotación de los recursos era cercano a los lugares de matanza de *Lama glama guanicoe* y, además, más cercano que Las Buitreras a una hipotética Localidad de Actividades Generalizadas. Si la explotación de recursos en un área dada es efectuada racionalmente, es lógico pensar que para una región como Ultima Esperanza, con disponibilidad de variedad de recursos (marítimos y terrestres), se acorten las distancias entre los distintos sitios que integraron el sistema de explotación humana.

OTROS SITIOS

Resulta imposible discutir en el mismo grado otros sitios sureños. Sólo quiero llamar la atención hacia la evidencia de la Cueva Fell que, por presentar abundantes puntas de proyectil junto a restos de *Mylodon darwinii, Hippidium-Onohippidium s.l.* y *Lama glama guanicoe*, puede quizá presentarse como un sitio de armamento. Para discutir esta hipótesis sería necesario conocer el índice de fragmentación de las puntas de proyectil, dato que no poseo. Por otra parte los restos de *Mylodon* (Cuadro 2) parecen apoyar la hipótesis derivada del caso de Las Buitreras y de la conducta de *Mylodon*, de que el sitio fue de matanza de edentados. No he podido revisar este material y, si es que debo guiarme por la información publicada recientemente (Emperaire et al. 1963; Saxon 1976, 1979) tengo que decir que esos restos han sido depositados naturalmente. Prefiero postular, en cambio, que esos animales han sido cazados por el hombre. Con ello quiero volver a la interpretación original de Junius Bird, quién afirmó por lo menos desde 1938, que el *Mylodon* era presa del hombre. Entiendo que también la colección de Palli Aike debe ser reestudiada (Bird 1938), pues las asociaciones y situaciones etratigráficas prácticamente repiten lo anotado para Las Buitreras y lo conocido de Fell. La situación topográfica de Palli Aike dentro de un volcán extinguido, muy alejado del agua potable, puede ser importante en la discusión de los factores de localización. En una situación topográfica semejante, en una caverna dentro de un volcán, ubicado en Markatsh Aike, Hauthal halló restos de industria arqueológica asociados a un molar de *Hippidium-Onohippidium s.l.* (Lehmann-Nitsche 1904b). Un equipo del Insituto de Ciencias Antropológicas de la Universidad de Buenos Aires, dirigido por la Profesora A.C. Sanguinetti está trabajando actualmente en sitios ubicados en el interior de volcanes en la misma área. Esos estudios junto con los que el mismo Bird y Mauricio Massone, del Instituto de la Patagonia de Punta Arenas, están realizando en el área de Palli Aike, seguramente ampliarán el conocimiento de los cazadores patagónicos de *Mylodon*.

AGRADECIMIENTOS

A Sergio E. Caviglia por la lectura crítica de una primera versión de este trabajo y por la preparación del Gráfico No. 3. A la Profesora A.C. Sanguinetti por su apoyo para la realización de este trabajo. Las investigaciones en Las Buitreras fueron realizadas con subsidios del CONICET a la Profesora Sanguinetti. A mis alumnas Marta Chiesa, Margarita Iaccarino, Ana María González y María Elisa Ponde quienes, en una monografía preparada para la materia Prehistoria Americana y Argentina 1, concluyeron que el carácter de la ocupación humana de Fell, Las Buitreras y Cueva del Mylodon era diferencial y probablemente explicable en términos funcionales. A los Doctores Anibal J. Figini, y Jorge E. Carbonari, Gabriel J. Gómez y Roberto A. Huarte del Laboratorio de Tritio Y Radiocarbono de la Universidad Nacional de La Plata, por sugerirme, mientras trataron infructuosamente de obtener suficiente colágeno de huesos de *Mylodon (?) listai* provenientes de Las Buitreras, la acción de aguas ácidas sobre esos huesos.

REFERENCIAS CITADAS

Binford, L.R.
1981 *Bones, ancient men and modern myths.* Academic Press, New York.

Bird, J.
1938 Antiquity and migrations of the early inhabitants of Patagonia. *Geographical Review* 28: 250-275.

1946 The archaeology of Patagonia. Handbook of South American Indians I. *Bureau of American Ethnology, Bulletin* 143(1): 17-24.

Borrero, L.A.
1980 La economía prehistórica de cazadores y cazadores-recolectores. MS, Informe al CONICET, Buenos Aires.

Cajal, J.
1979 Estructura social y área de acción del guanaco (*Lama guanicoe*) en la reserva de San Guillermo (Pcia. de San Juan). Trabajo presentado al III Congreso Internacional de Camélidos Sudamericanos (Noviembre 1979), Viedma.

Cardich, A., L.A. Cardich, y A. Hajduk
1973 Secuencia arqueológica y cronológica radiocarbónica de la cueva 3 de Los Toldos (Santa Cruz). *Relaciones de la Sociedad Argentina de Antropologia* (N.S.) VII: 85-123.

Caviglia, S.E.
1981 Informe sobre el material faunístico de capas 7 y 8 de Las Buitreras (Campaña 1981). MS en posesión del autor.

Caviglia, S.E., y M.J. Figuerero Torres
1976 Material faunístico de la Cueva Las Buitreras (Dto. Güer Aike, Santa Cruz). *Relaciones de la Sociedad Argentina de Antropologia* (N.S.) X: 315-319.

Caviglia, S.E., H.D. Yacobaccio, y L.A. Borrero
1981 Los niveles con megafauna de Las Buitreras: componentes culturales y faunísticos. El Poblamiento de América (Coloquio sobre "Evidencia arqueológica de ocupación humana en América anterior a 11.500 años B.P.", (A.L. Bryan Ed.), Comision XII, *X Congreso Unión Internacional de Ciencias Prehistóricas y Protohistóricas*, pp. 68-89, México.

Emperaire, J., y A. Laming
1954 La grotte de Mylodon (Patagonie occidentale). *Journal de la Société des Américanistes* 43: 173-206.

Emperaire, J., A. Laming-Emperaire, y H. Reichlen
1963 La grotte Fell et autres sites de la région volcanique de la Patagonie chilienne. *Journal de la Société des Américanistes* 52: 169-229.

Figuerero Torres, M.J.
1980 Análisis de coprolitos: el caso de Cueva Las Buitreras. Trabajo presentado en el VII Congreso Nacional de Arqueología, Colonia, Uruguay.

Garrido, J.L., J. Amaya, y Z. Kovacs
1979 Territorialidad, comportamiento individual y actividad diaria de una población de guanacos en la reserva faunística provincial de Cabo Dos Bahías. Trabajo presentado al III Congreso Internacional sobre Camélidos Sudamericanos (Noviembre 1979), Viedma.

Hammerly Dupuy, D.
1948 Importancia antropológica de la Patagonia occidental: nuevos hallazgos en la "Caverna Grande" de Ultima Esperanza. *Runa* 1:258-262.

Hauthal, R.
1899 El mamífero misterioso de la Patagonia *"Grypotherium domesticum"*. Reseña de los hallazgos en las cavernas de Ultima Esperanza. *Revista del Museo de La Plata* IX: 409-420.

Lagiglia, H.
1968 Nuevos aportes a los fechados de radiocarbón de la Argentina. *Notas del Museo* 8:1-8. San Rafael.

Lehmann-Nitsche, R.
1899 El mamífero misterioso de la Patagonia *"Grypotherium domesticum"*. III. Coexistencia del hombre con gran desdentado y equino en las cavernas patagónicas. *Revista del Museo de La Plata* IX: 455-472.

1904a Nuevos objetos de industria humana encontrados en la caverna Eberhardt en Ultima Esperanza. *Revista del Museo de La Plata* XI: 56-69.

1904b Hallazgos antropológicos de la caverna Markatsh Aiken. *Revista del Museo de La Plata* XI: 173-176.

Martin, P.S.
1975 Sloth droppings. *Natural History*, August-September, pp. 74-77.

Martin, P.S., B.E. Sabels, y R. Shutler Jr.
1961 Rampart Cave coprolite and ecology of the Shasta ground sloth. *American Journal of Science* 259: 102-127.

Mengoni Goñalons, G.
1980 El aprovechamiento de los recursos faunísticos en el interior de Patagonia meridional/continental: hipótesis y modelos. Trabajo presentado al VI Congreso Nacional de Arqueológia del Uruguay (Salto, 1978) (corregido en 1980).

Morrison, P.
1966 Insulative flexibility in the guanaco. *Journal of Mammalogy* 47: 18-23.

Myers, T.P., M.R. Voorhies, and R.G. Corner
1980 Spiral fractures and bone pseudotools at paleontological sites. *American Antiquity* 45:483-490.

Nordenskjöld, E.
1900 Jackttagelser och fynd i grottor vid Ultima Esperanza i Sydvestra Patagonien. *Konglinga Svenska Vetenskaps-Akademiens Handlingar* 33(3): 1-24.

Patterson, B., y R. Pascual
1972 The fossil mammal fauna of South America. In *Evolution, mammals and southern continents*, edited by A. Keast, F.C. Erk and B. Glass, pp. 247-309. State University of New York Press, Albany.

Paula Couto, C. de
1979 *Tratado de paleomastozoología*. Rio de Janeiro. Academia Brasileira de Ciencias.

Poulain-Josien, T.
1963 La grotte Fell. Etude de la faune. *Journal de la Sociêtê des Amêricanistes* 52: 230-255.

Roth, S.
1899 El mamífero misterioso de la Patagonia *"Grypotherium domesticum"*. II. Descripción de los restos encontrados en la caverna de Ultima Esperanza. *Revista del Museo de La Plata* IX: 421-453.

1904 Nuevos restos de mamíferos de la caverna Eberhardt en Ultima Espernaza. *Revista del Museo de La Plata* XI: 38-53.

Sanguinetti, A.C.
1976 Excavaciones prehistóricas en la cueva Las Buitreras (Santa Cruz). *Relaciones de la Sociedad Argentina de Antropología* (N.S.) X: 271-292.

1981 La cueva de Las Buitreras y el problema del estadio de caza temprano en el extremo austral de América. Trabajo presentado al Simposio sobre el Paleoindio en América, Smithsonian Institution, Washington, D.C.

Sanguinetti, A.C., y L.A. Borrero
1977 Los niveles con fauna extinta de la Cueva Las Buitreras (Río Gallegos, Santa Cruz). *Relaciones de la Sociedad Argentina de Antropología* (N.S.)XI: 167-175.

Saxon, E.C.
1976 La prehistoria de Fuego-Patagonia: colonización de un habitat marginal. *Anales del Instituto de la Patagonia* VII: 63-73. Punta Arenas.

1979 Natural Prehistory: the archaeology of Fuego-Patagonian ecology. *Quaternaria* XXI: 329-356.

Scillato Yané, G.
1976 Sobre algunos restos de *Mylodon* (?) *listai* (Edentata, Tardigrada) procedentes de la cueva Las Buitreras, provincia de Santa Cruz. *Relaciones de la Sociedad Argentina de Antropología* (N.S.) X: 309-312.

1980 Comentario a "Nueva especie de *Neocnus* (Edentata: Megalonychidae de Cuba) y consideraciones sobre la evolución, edad y paleoecología de las especies de este género", de N.A. Mayo. *Actas, II Congreso Argentino de Paleontología y Bioestratigrafía y I Congreso Latinoamericano de Paleontología* III: 234.

1981 Nuevo Mylodontinae (Edentata, Tardigrada) del "Mesopotamiense" (Mioceno tardío-Plioceno) de la Provincia de Entre Ríos. *Ameghiniana* XVIII(1-2): 29-34.

Shipman, P.
1981 *Life history of a fossil*. Harvard University Press, Cambridge.

Tonni, E.P., y G. Politis
1980 La distribución del guanaco (Mammalia, Camelidae) en la provincia de Buenos Aires durante el Pleistoceno tardío y Holoceno. Los factores climáticos como causas de su retracción. *Ameghiniana* XVIII(1): 53-66.

Vallenas Pantigoso, A.
1970 Fisiología de la digestión de los auquénidos. *Anales de la Primera Convención sobre Camélidos Sudamericanos*. pp. 69-78. Universidad Nacional Técnica del Altiplano. Puno.

Wellman, R.W.
1972 Origen de la Cueva del Milodón en Ultima Esperanza. *Anales del Instituto de la Patagonia* III(1-2): 97-101. Punta Arenas.

Wilmsen, E.N.
1970 Lithic analysis and cultural inference. A Paleoindian case. *University of Arizona Anthropological Papers* 16: 1-87. Tucson.

Las Buitreras: Convivencia del Hombre con Fauna Extinta en Patagonia Meridional

SERGIO ESTEBAN CAVIGLIA,
HUGO DANIEL YACOBACCIO, y
LUIS ALBERTO BORRERO
Instituto de Antropología
Universidad de Buenos Aires
25 de Mayo 217
Buenos Aires
ARGENTINA

Abstract

A summary of the results of the research at Buitreras Cave (southern Argentina) since 1974 is given. Although human occupation was limited, the lower levels yielded an association of extinct species (ground sloth, horse, fox), modern fauna, human and canid coprolites, and 16 lithic artifacts (principally scrapers). Many of the bone remains show clear signs of human activity in the form of impact and butchering marks. The early levels lack radiocarbon dates, being tentatively considered as older than 10,000 B.P. Taphonomical and bone alteration aspects are considered. Experimental studies were conducted on bones fed to several carnivores and rodents (particularly at the Buenos Aires Zoo), which allowed clear differentiation of animal and human activity on green bones. Microwear studies suggest that probably none of the lithic artifacts was used to work bone. The identification of canids from Patagonian sites is discussed. The human coprolite is constituted predominantly of plant material. Ground sloth, guanaco, and extinct fox were certainly part of the early inhabitants' diet, but the consumption of horse is not clear. Given the absence of ground sloth coprolites, it seems that this species did not live in the cave, and the presence of their bones in the site is of cultural origin.

> "Resumiendo los indicios que nos presentan los huesos cortados vemos que todas las partes del animal (*Mylodon*) han sido separadas y descarnadas al acaso sin ningún cuidado.
>
> No se puede saber con seguridad, por los restos, de qué manera ha sido muerto el animal. Este ser indefenso, y pesado, con sus molares inofensivos, probablemente fue muerto a golpes de maza en la cabeza. Una vez sacado el cuero, el cadáver, ha sido desmembrado; las partes mayores fueron cortadas en pedazos pequeños y comidas con placer. No dejaron nada más que las inserciones de los músculos y los tendones duros ...
>
> ...En el festín, la carne ha sido arrancada de los huesos con los dientes o quizás ayudándose con cuchillo. No es seguro que haya sido asada; su sabor ha sido al de un herbívoro; las astillas de hueso de animales jóvenes nos demuestran que aquellos glotones supieron apreciar muy bien la carne tierna".
>
> R. Lehmann-Nitsche (1899)

INTRODUCCION

La cueva Las Buitreras se halla ubicada sobre el curso medio del río Gallegos, en la Provincia de Santa Cruz (Patagonia Argentina). Está formada en una colada basáltica a 25 m sobre el nivel de base local y tiene una orientación N-S. La entrada tiene 6 m de ancho y la altura de la bóveda es de aproximadamente 4 m (ver Figuras 1 y 2). A su pie hay un antiguo meandro, hoy seco, del río Gallegos.

Los trabajos dirigidos por la Prof. A.C. Sanguinetti de Bórmida, comenzaron en 1974; en 1976; Caviglia y Figuerero Torres 1980; Scillato los contextos recuperados en ellas (Sanguinetti de Bórmida 1976), junto a una serie de artículos particulares referentes a la fauna asociada (Caviglia 1976; Caviglia y Figuerero Torres 1976; Scillato Yané 1976) y al contexto óseo (Curzio 1976). Posteriormente se realizaron estudios específicos referidos a los niveles con fauna extinta (Caviglia 1978, 1980; Borrero 1980; Figuerero Torres 1980; Sanguinetti de Bórmida 1980a, 1980b; Sanguinetti de Bórmida y Borrero 1977; Yacobaccio 1977, 1980).

Se desea aclarar que el material aquí estudiado fue rescatado de tres niveles (capa VII, VIII cúspide y VIII, ver Figura 4). Esta división es mantenida en algunos momentos con fines puramente prácticos, pero se les considera sólo un componente. Una de las razones que nos llevó a realizar esta reducción fue el notable efecto de telescopización (incremento del número mínimo de individuos) en los restos faunísticos (Grayson 1973).

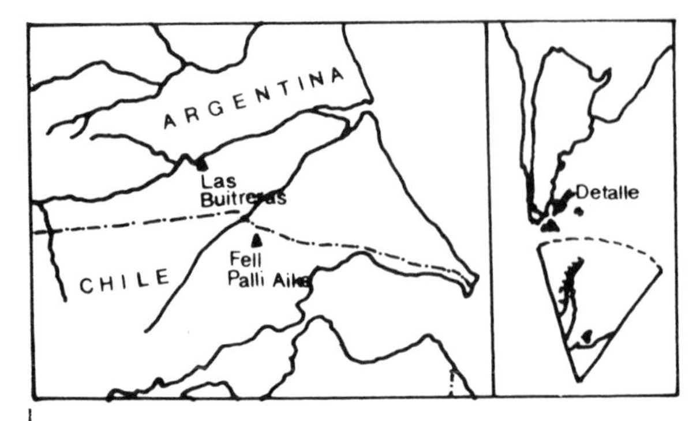

Figura 1. Mapa de Patagonia meridional.

Figura 2. Vista de la cueva Las Buitreras.

Los niveles estudiados presentan evidencias de asociación de especies extinguidas y fauna moderna, cuyos restos óseos muestran signos de

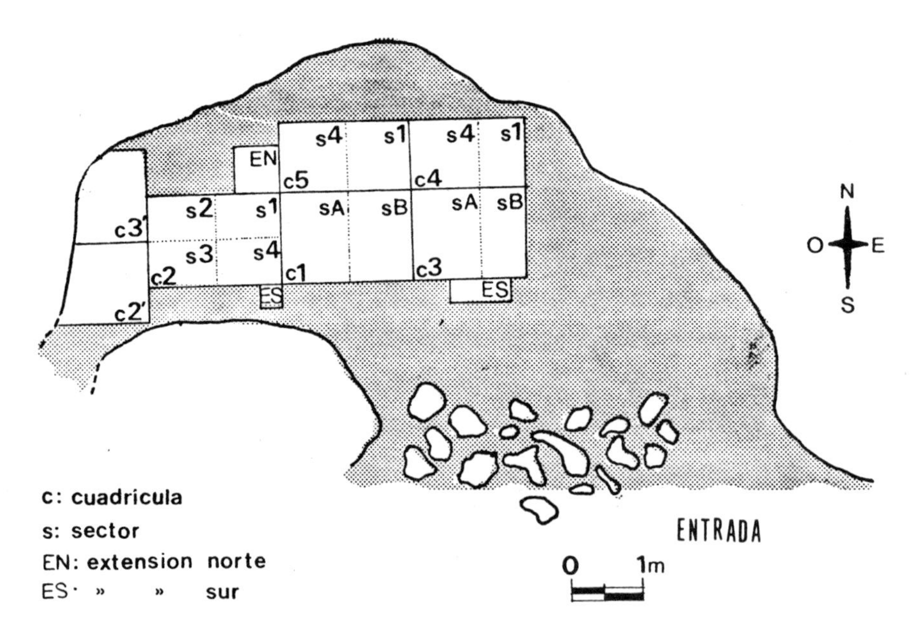

Figura 3. Mapa de las excavaciones de la cueva Las Buitreras.

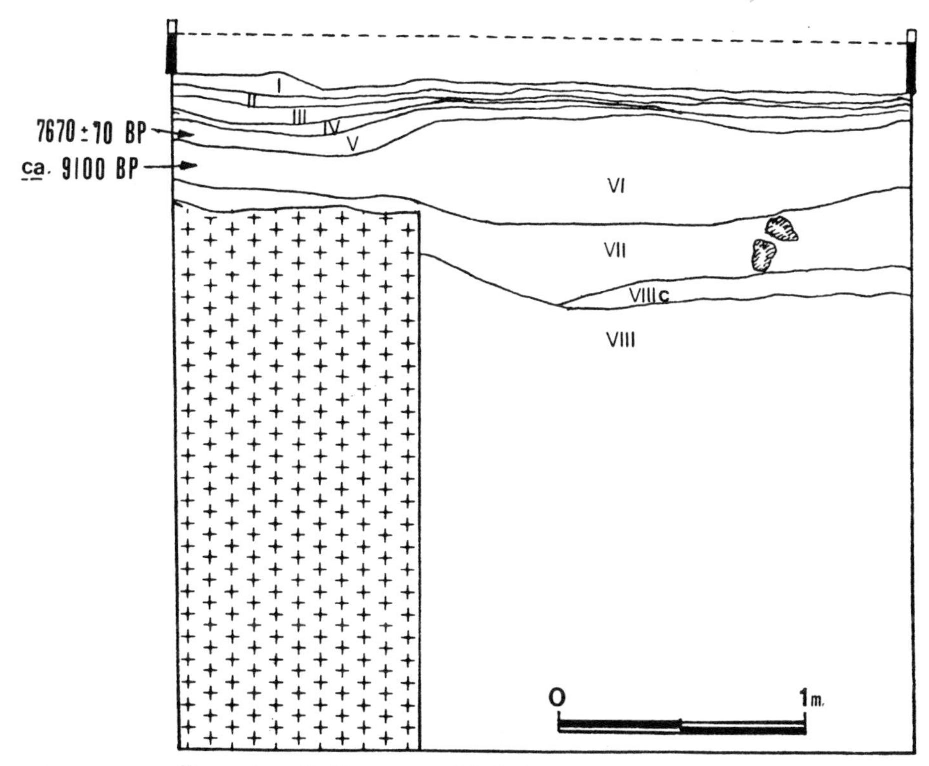

Figura 4. Perfil norte-cuadrícula 1.

actividad cultural que se suman a la presencia de artefactos líticos y coprolitos humanos y de cánidos.

El propósito de este trabajo es analizar por diferentes vías las características de la ocupación humana en esta cueva en una fecha cercana al 10.000 B.P.

CRONOLOGIA

No existe cronología absoluta para las capas VII y VIII; pero la siguiente aproximación es posible sobre la base de la evidencia disponible. Dado que:

a. La base de la capa V fue fechada en 7670±70 B.P. (CSIC-372).

b. La capa VI conformada por cenizas volcánicas, debe coresponder a la Erupción I (*apud* Auer 1974) fechada en el área, no en el sitio, en 9100 B.P. (idem).

c. La línea de bloques caídos sobre la cumbre de la capa VII puede sincronizarse con un episodio similar registrado en la cueva Fell (ubicada 50 km al sudeste) fechado entre 9030 y 10.600 B.P. (Saxon 1976).

Se puede especular que la ocupación de la capa VII ocurrió en algún momento anterior a los 10.000 B.P. (Sanguinetti de Bórmida y Borrero 1977).

ARTEFACTOS LITICOS

El conjunto de artefactos líticos recuperados asciende a 16 piezas. Todas provienen de la localización "A" de la cueva (ver más adelante). La tipología morfológica sumaria es como sigue:

Capa VII

Pieza No. 1 Lasca angular oblícua. 3,2 x 2,6 x 1 cm. Talón liso. Angulo de talón: 120°. Silice.

Pieza No. 2 Lasca angular recta. 3,0 x 5,4 x 0,8 cm. Talón liso. Angulo de talón: 115°. Basalto (Figura 5).

Pieza No. 5 Lasca de reactivación directa. 2,0 x 1,8 x 0,6. Sílice.

Pieza No. 6 Lasca con retoque marginal en bisel oblícuo. Retoque escamoso regular inverso en filo derecho. 2,1 x 1,9 x 0,5 cm. Filo cóncavo. Angulo de bisel: 60 - 70°. Sílice. (Figura 5).

Pieza No. 7 Nucleiforme de basalto (con tres extracciones en cara dorsal) 4,9 x 5,1 x 2,0 cm.

Pieza No. 8 Núcleo poliédrico agotado con extracciones multidireccionales. 4,7 x 3,1 x 2,9 cm. Basalto.

Pieza No. 9 Lasca angular recta. Talón liso. Angulo de talón: 115°. 2,2 x 1,6 x 0,6 cm. Basalto.

Lascas de desecho de talla y esquirlas: 6.

Pieza No. 10 Raedera lateral doble, + raspador frontal de filo restringido. El filo izquierdo es

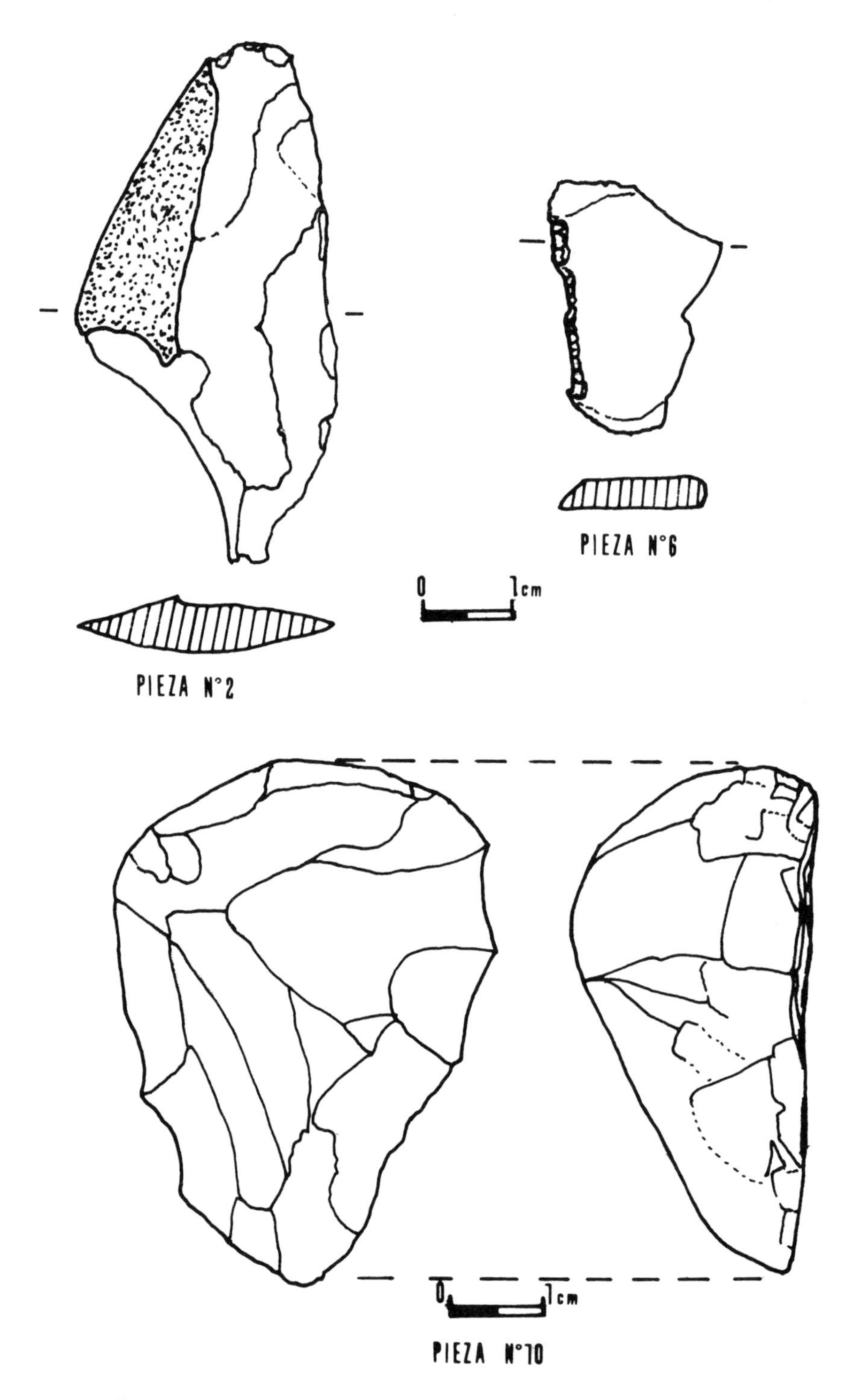

Figura 5. Artefactos líticos.

convexo y el derecho levemente denticulado. Angulo de filo: 65° (izq.), 65° - 85° (der.), 85° (frontal). 5,0 x 3,9 x 1,8 cm. Sobre lasca de guijarro (arenisca silicificada) (Figuras 5 y 6)

Capa VIII cumbre.

Pieza No. 3 Lasca angular recta. 2,5 x 3,0 x 0,9 cm. Talón liso. angulo de talón: 70°. Basalto.

Pieza No. 4 Lasca de desecho de talla. Sílice.

Aunque la muestra es muy pequeña permite algunas consideraciones. Las lascas son secundarias en su totalidad, por lo tanto se descarta la actividad primaria de talla en el sitio (ausencia de núcleos de sílice). La presencia del núcleo y nucleiforme de basalto, junto a lascas de esta misma materia prima, puede indicar cierta actividad de talla sobre basalto. El núcleo, no denota una dirección única para la extracción de lascas. Los talones lisos representan una preparación de la plataforma de percusión mediante la formación de una superficie plana. Los ángulos de los mismos sugiere una técnica de talla por percusión. La talla por presión no está presente, dado que las dos piezas retocadas lo fueron por percusión (retoque corto, con alto índice de ''step fractures'').

Cinco piezas están confeccionadas en basalto negro de grano fino (lascas, núcleo y nucleiformes). Cuatro en sílices de distintos colores, mayormente calcedonia, (lascas y lascas con retoque) y una arenisca silicificada (raedera). Todas estas rocas son alóctonas a la cueva. Las fuentes de aprovisionamiento de sílices se ubican a más de 50 km del sitio.

Figura 6. Vista lateral derecha del instrumento N° 10.

RESTOS FAUNISTICOS

El material faunístico recuperado no es significativo cuantitativamente, pero sí lo es cualitativamente. Por ello no presentamos número de individuos sino sólo valores nominales (presencia-ausencia) y su discusión (ver Cuadro 1).

Roedores

Entre los animales pequeños encontramos tres especies de roedores: *Reithrodon physodes* (n.v. "rata

CUADRO 1. Material Faunístico proveniente de los niveles con Megafauna de Las Buitreras (Campañas 1974-1981).

MATERIAL FAUNISTICO	Capa VII	Capa VIII Cúsp	Capa VIII
Ave (gen. et sp. indet).		X	
Reithrodon physodes	X		X
Euneomys chinchilloides	X		
Ctenomys sp.	X		
Lama glama guanicoe	X	X	X
Delphinidae (gen. et sp. indet).	X		
Dusicyon avus			X
Hippidium-Onohippidium (s.l.)	X		
Mylodon (?) *listai*	X	X	X
Coprolito cánido	X		X

conejo de Patagonia"); *Euneomys chinchilloides* (n.v. "rata austral con pelo de chinchilla") y *Ctenomys* sp. (n.v. "tucu tucu").

Reithrodon physodes es un cricétido que usualmente no habita en cuevas. *Euneomys chinchilloides* es una especie considerada poco frecuente (Hershkovitz 1962). Pero la evidencia proveniente de sitios del SE de la Provincia de Santa Cruz y Tierra del Fuego (Orejas de Burro; El Volcán 1; Güer Aike; Bloque Errático 1, Cabeza de León y Punta Maria 2), parecen mostrar que se trata de un roedor relativamente abundante en esta zona. *Ctenomys* sp. es probablemente la especie más abundante de mamíferos pequeños en sitios arqueológicos de Patagonia. Lamentablemente la discusión de problemas taxonómicos sobre la base de restos óseos no es aun posible. De cualquier manera, no queremos dejar de destacar la notable diferencia de tamaño de los restos hallados en los niveles inferiores de Las Buitreras con los actuales. La media de los ejemplares de estos niveles supera holgadamente a los mayores ejemplares recientes.[1]

Bond et al. (1981:101-104) discuten los controles para demostrar depositación natural o artificial (por el hombre) de roedores en sitios patagónicos. En Las Buitreras nos resulta difícial discernir, con la evidencia que se posee, acerca de uno u otro tipo de depositación. Debido a ello, consideramos que por el momento sólo se puede hablar de asociación *s.s.* sin ningún otro tipo de implicaciones.

Guanaco

Los restos de *Lama glama guanicoe* no son muy abundantes. Fueron determinados 40 fragmentos y existen unas 80 astillas sin determinar anatómicamente, pero que atribuímos a esta especie.

La mayoría de los huesos están muy fragmentados y algunos presentan sus bordes "pulidos" (sin que esto implique actividad cultural) y otros están en muy mal estado de conservación. Prácticamente todos los restos presentan claras evidencias de fractura traumática intencional sobre hueso fresco ("green bone").[2]

La preponderancia de elementos apendiculares por sobre esqueleto axial es notable; de este último sólo hay 1 fragmento de cuerpo vertebral, 2 fragmentos de costillas y piezas dentarias con fragmentos de maxilar (ver Figura 7).

Se pudo establecer la presencia de restos provenientes de por lo menos tres individuos. Estos fueron determinados sobre la base de: estado de fusión de epífisis de metapodios; tamaño deferencial de calcáneos, rótulas y de radio-cúbito. Uno de

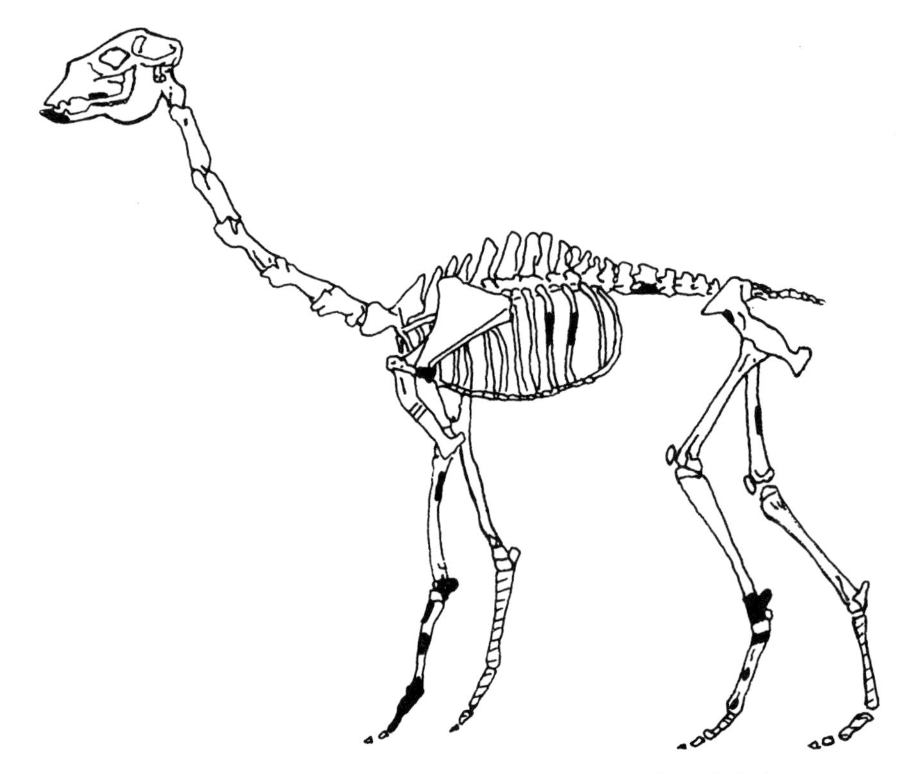

Figura 7. Ubicación en el esqueleto de *Lama glama guanicoe* de los fragmentos óseos halladas.

los restos pertenece a un individuo nonato o neonato.

Presentaremos ahora algunos de los restos de *Lama g.g.* que exhiben claros indicios de actividad cultural:

-Fragmento de epífisis distal de falange con parte de diáfisis. Presenta una fractura traumática intencional longitudinal con muesca. Presenta huellas de corte anteriores a la fractura (Figuras 8, 9, y 10).

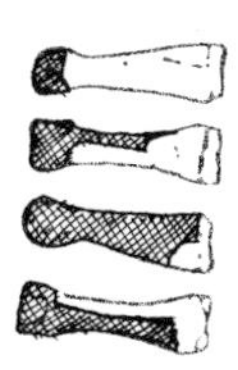

Figura 9. Falange completo.

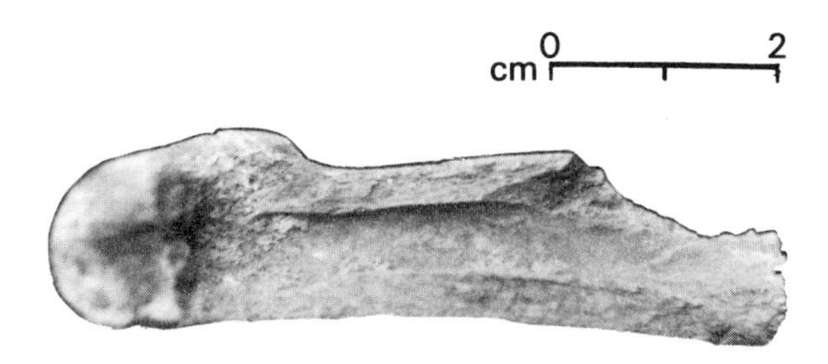

Figura 8. Falange con fractura.

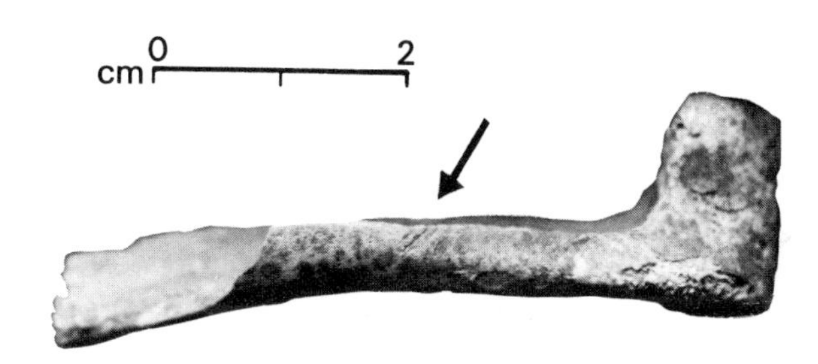

Figura 10. Falange con corte.

Un fragmento de calcáneo partido longitudinalmente. Presenta una muesca notable y otro punto de impacto anterior fallido (Figuras 11, 12, 13 y 14).

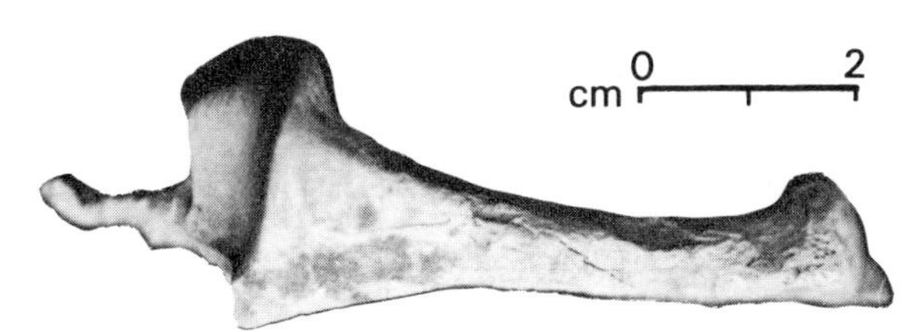

Figura 11. Calcáneo fracturado.

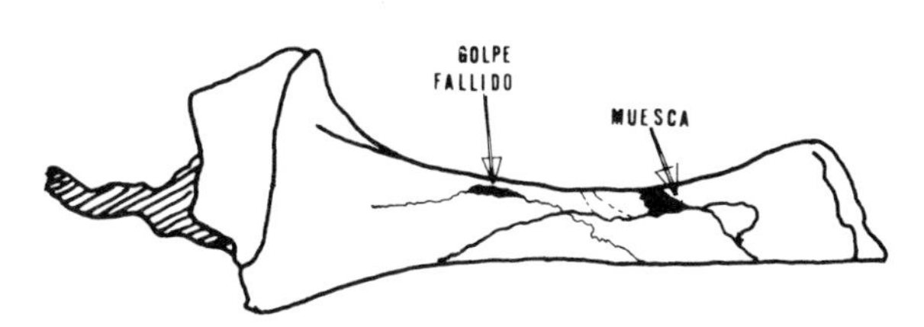

Figura 13. Calcáneo con puntas de impactos.

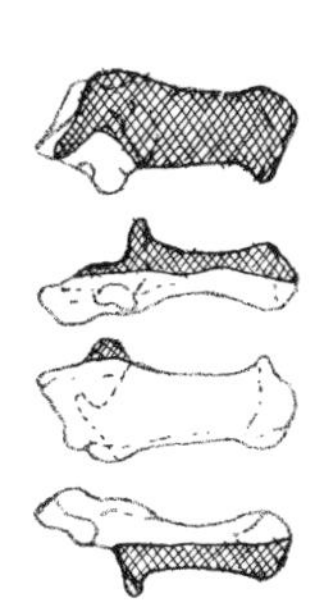

Figura 12. Calcáneo completo.

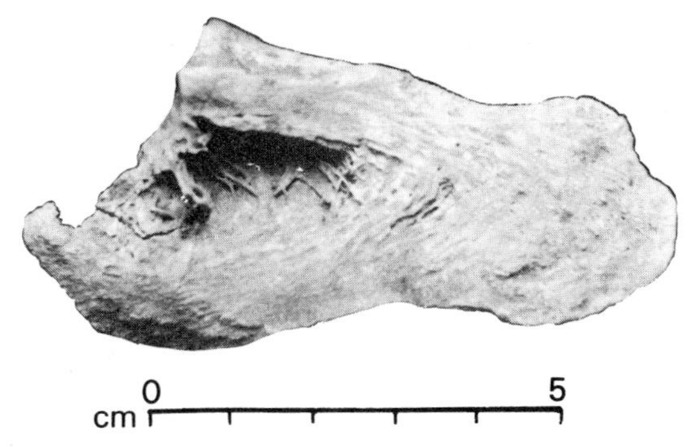

Figura 14. Calcáneo fracturado.

Un fragmento posterior de diáfisis de fémur que presenta fractura en espiral y con claras huellas de descarne. Hay dos tipos de huellas, unas (A) leves superficiales y cortas y otras (B) con líneas finas y paralelas dentro de un surco principal. Dos de estas últimas son anteriores a la fractura (Figuras 15, 16, y 17).

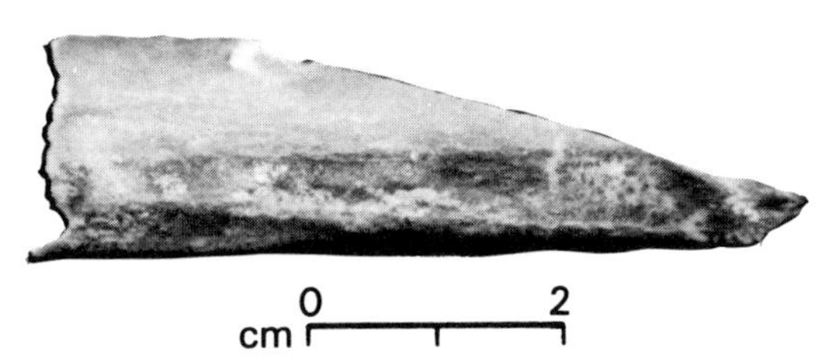

Figura 15. Fémur con fractura espiral.

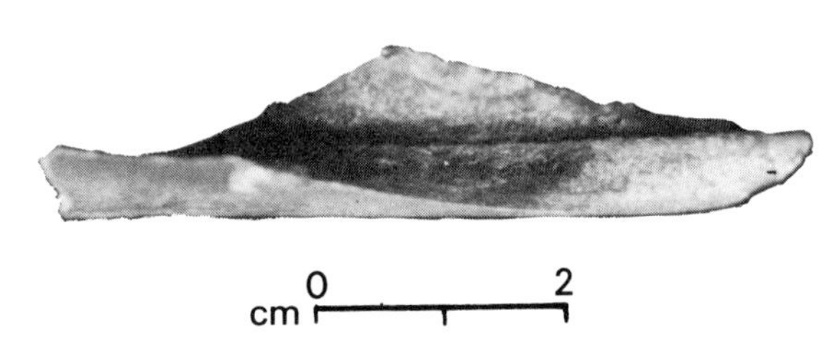

Figura 16. Fémur con fractura espiral.

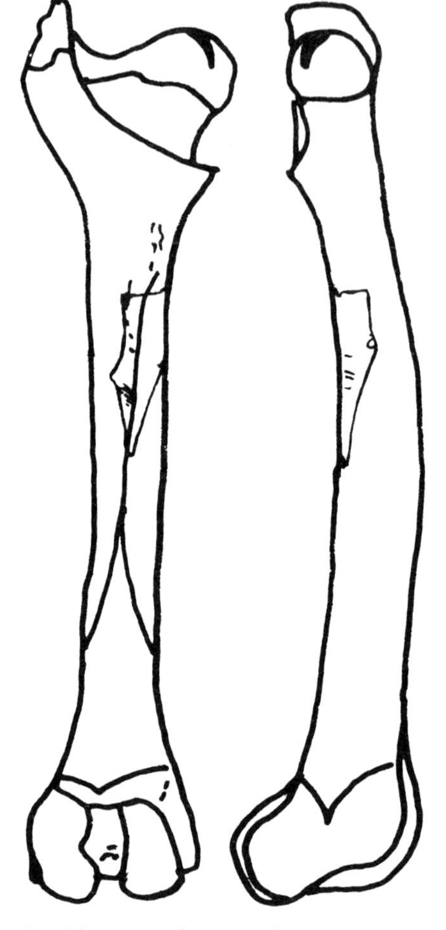

Figura 17. Fémur con fracturas de impacto.

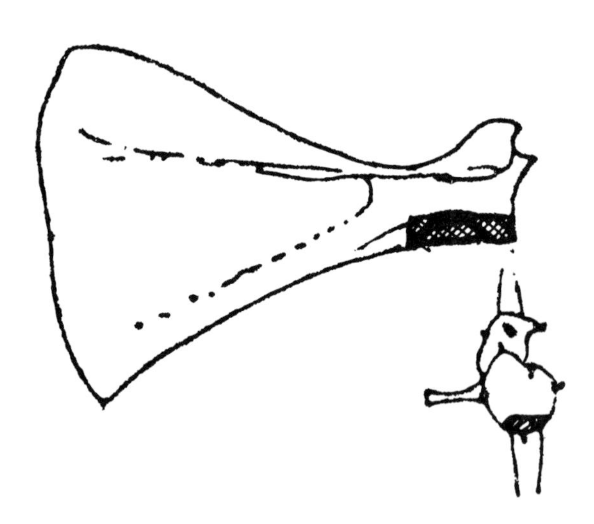

Figura 18. Un fragmento de epífisis proximal de escápula derecha con huellas de descarne anteriores a la depositación de óxido de manganeso.

Un fragmento de calcáneo izquierdo con fractura traumática intencional (Figuras 19 y 20).

Fragmento lateral posterior izquierdo de radio derecho de un ejemplar nonato o neonato. Este fragmento presenta huellas de descarne notablemente claras, probablemente reflejo de una actividad semejante al "filleting" (*apud* Binford

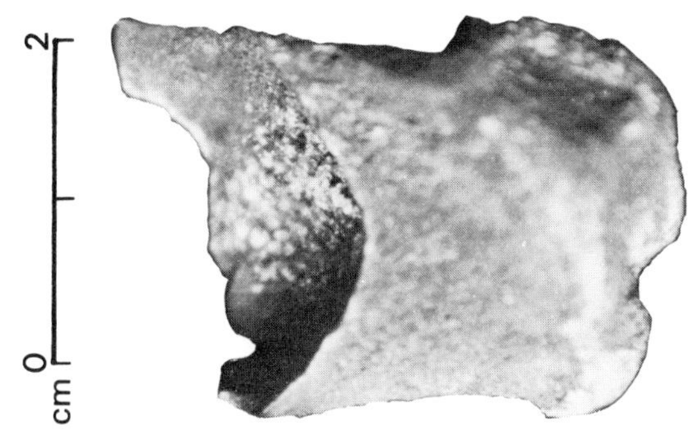

Figura 19. Calcáneo con fractura.

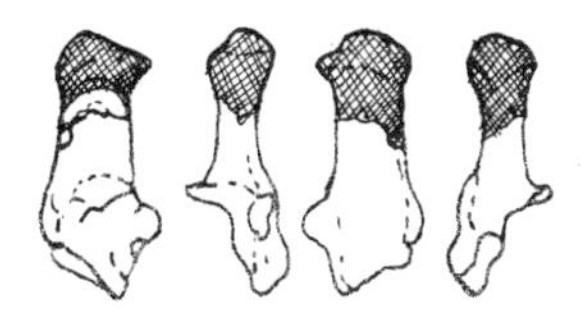

Figura 20. Calcáneo con fractura.

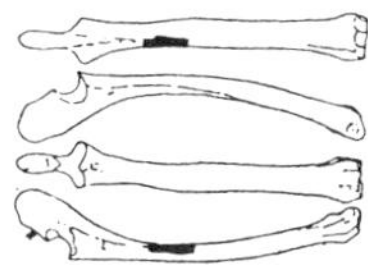

Figura 21. Radioulna completo.

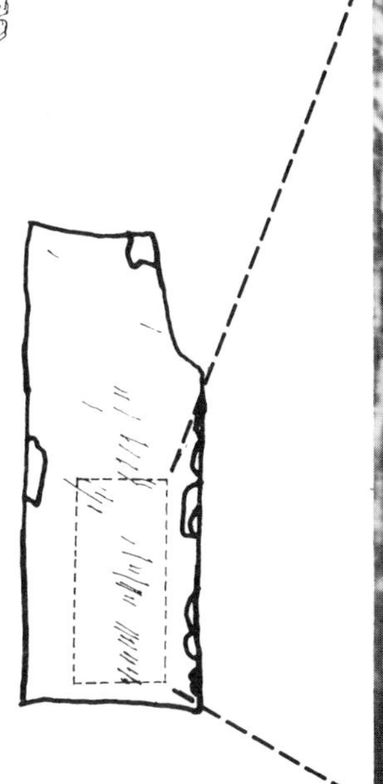

Figura 22a. Fragmento con huellas.

Figura 23. Metapodio con huella de corte.

1981:126 ss.), lamentablemente este es demasiado pequeño y es el único ejemplar con este tipo de marcas, como para hacer consideraciones al respecto (Figuras 21, 22a y 22b). La longitud de las huellas oscila entre 0,19 y 0,40 cm.

Un fragmento de diáfisis de metapodio (?) con una clara huella de corte con estrías más finas y paralelas dentro del surco principal (Figura 23).

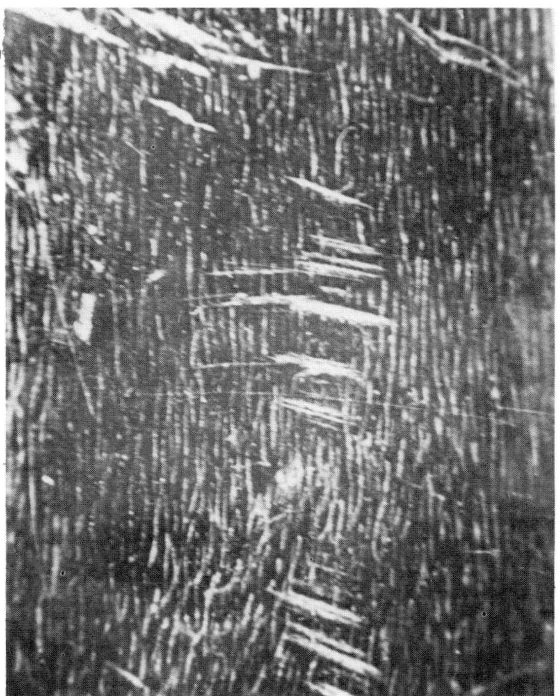

Figura 22b. Fragmento con huellas.

Caballo Americano (Figura 25)

El significado de la presencia de *Hippidium-Onohippidium* es discutido en el punto VI. Sólo fueron hallados dos molariformes (Figura 24).

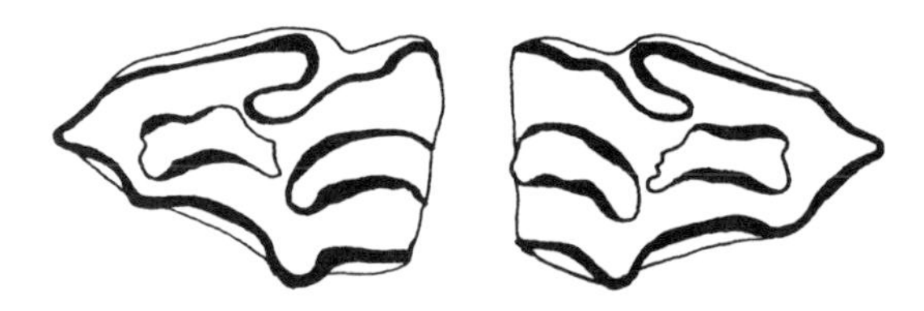

Figura 24. Vista oclusal de dos muelas.

Figura 25. Caballo Americano Nativo (Modificado de Bird 1967:537).

Perezoso Gigante de Patagonia (Figura 26)

La distribución de los restos de *Mylodon* (?) *listai* es igualmente discutida en el punto VI. Nos interesa mencionar aquí otros datos referentes al mismo material. Los huesos - de dos individuos juveniles, posiblemente completos - están en general en muy mal estado de conservación (usualmente se desintegran) o en caso contrario se encuentran "quemados" por una muy fuerte oxidación de manganeso. En los casos que (a pesar del manganeso) se pueden identificar huellas de descarne, la cubierta ha sido útil para controlar si las mismas han sido anteriores a su depositación.

Figura 26. *Mylodon* sp.

Entre las excepciones se hallaron cuatro fragmentos de costillas (Figuras 37 y 40, Costillas 1, 2 y 3) en buen estado de conservación y con huellas que atribuímos a razones culturales como se discutirá más adelante. En los huesos de *Mylodon* no se observó ningún tipo de fractura de origen cultural.

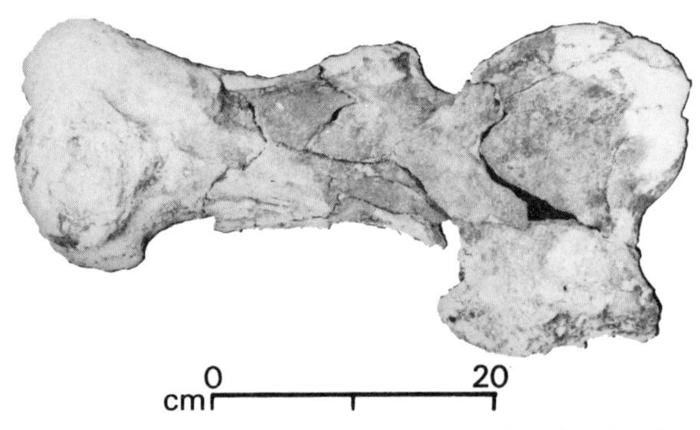

Figura 27. Húmero de *Mylodon* (?) *listai* (Escala 10 cm).

Zorro Extinguido

La presencia de *Dusicyon avus* fue detallada en trabajos anteriores (Caviglia 1978, 1980). En este se consultó el material de cánidos pertenecientes al Museo Argentino de Ciencias Naturales "Bernardino Rivadavia" (donde se halla el Holotipo y la mayoría del material de *D.a.*), del Museo de La Plata; el material proveniente de Los Toldos y ejemplares de sitios arqueológicos depositados en el Instituto de Ciencias Antropológicas (UBA).

Sobre esta base se discutió el status actual de *Dusicyon australis, D. culpaeus, D. lycoides* y *D. avus.* Asimismo se estudió el problema del "Perro Yámana" y se examinó el registro de *Canis* sp. en sitios arqueológicos tempranos de Sudamérica.

Para ello hemos trabajado con diagramas logarítmicos que incluyen variación específica y consideran 27 medidas craneanas. Cuando fue necesario se utilizaron caracteres morfológicos más específicos. Los resultados de este análisis permitieron conocer aquellos rasgos diagnósticos para cada especie, como así también su variación.

En Las Buitreras fue hallado un fragmento de maxilar derecho con sus M^1 y M^2, determinados como *Dusicyon avus* (Figura 28).

La importancia de una correcta determinación de estos cánidos (*D.a.*) radica en el valor cultural que se le ha atribuído en sitios arqueológicos de

Figura 28. Fragmento de maxilar.

Patagonia meridional, dado que por su gran tamaño pueden ser confundidos con especies de *Canis s.s.*

Sabemos que las especies de *Canis* viven en grupos familiares basados en jerarquías dominantes, por ello el troquelado (imprinting) hacia el hombre puede ser extendido durante la vida adulta. En cambio, los cánidos de hábitos solitarios como las especies de *Dusicyon s.l.* no poseen una estructura social tan elaborada (o compleja), por lo tanto el troquelado en vida adulta como su reproducción en cautiverio es dificultosa (Clutton-Brock 1977:1340).

La presencia de *D. avus* en Las Buitreras implicó una revisión de los materiales de otros sitios patagónicos en donde se mencionaban restos atribuídos a *Canis* cf. *familiaris* (Fell's Cave, en Saxon 1979)[3]; *Canis familiaris* (Porción mandibular de Los Toldos, en Cardich et al. 1977) en relación con niveles culturales. Como resultado de este análisis, se atribuyen a *D. avus* ciertos restos de Fell's Cave (ejemplar 20421) y a *D. aff. avus* la porción mandibular de Los Toldos. A ellos se suman los restos determinados por Roth (1902) para la "Cueva del Mylodon" (Ultima Esperanza, Chile) de los que se desconoce con seguridad el contexto del cual provienen. Bird y Bird (1937) mencionan la presencia de *Dusicyon* (*Pseudalopex*) *avus* para los niveles correspondientes al Período II de la secuencia del Estrecho de Magallanes, cuyos restos fueron determinados por G.G. Simpson.

La importancia en determinar correctamente las especies de estos géneros radica entonces en las posibilidades cinegéticas diferenciales que cada uno brindaría. Cardich et al. (1977:116) consideran la presencia de perros en niveles Casapedrenses como una posibilidad alternativa para explicar la práctica de caza: "Hoy al encontrar huesos de perros (*Canis familiaris*) surge otra posibilidad, de que estos animales, en calidad de domesticados, pudieran haber contribuído también en la caza".

Restos de *D. culpaeus* y *D. griseus* son relativamente frecuentes en sitios patagónicos y seguramente por sus hábitos deben haber sido presa ocasional (ver Mengoni-Goñalons 1978) y por ende un recurso de baja predictibilidad y abundancia (Mengoni, este volume). *D. avus* seguramente poseía hábitos semejantes a *D. culpaeus* y *D. griseus*, por lo tanto podríamos considerarlo también como presa ocasional.[4]

Aves

Sólo fue hallado un coracoides izquierdo que presenta surcos muy profundos. Estos llamativamente se hallan en una arista interna y parecen ser posteriores a algunas líneas de fisura y/o rajadura. Estas razones nos llevan a no considerar estas huellas como de actividad de descarne, sin que por el momento podamos ofrecer otra alternativa más que las evidentes huellas. (Figura 29). La longitud de las huellas es de 0,25 cm.

COPROLITOS

Fueron hallados 5 especímenes que fueron estudiados por Figuerero Torres (1980), quién realizó un minucioso exámen de los mismos, de donde tomamos esta información. Cuatro de ellos fueron determinados como pertenecientes a un carnívoro/omnívoro (no humano) *cf.* zorro gris. El restante fue determinado como humano por su morfología, reacciones, características (olor levemente fecaloide) y contenido. Su contenido es casi totalmente vegetal, con un clasto de basalto (10 mm) y otros más pequeños de cuarzo, fragmentos de hueso sumamente pequeños y carbones.

DISTRIBUCION DE LOS RESTOS DENTRO DE LA PLANTA

En cuanto a la distribución de los restos dentro de la planta de la cueva recordemos que Sanguinetti había diferenciado, sobre la base de la evidencia conocida hasta 1975, dos áreas:

1) Parte central de la cueva; abarca las cuadrículas 3A, 5 (sectores 1 y 4); porción Este de la cuadrícula 1.

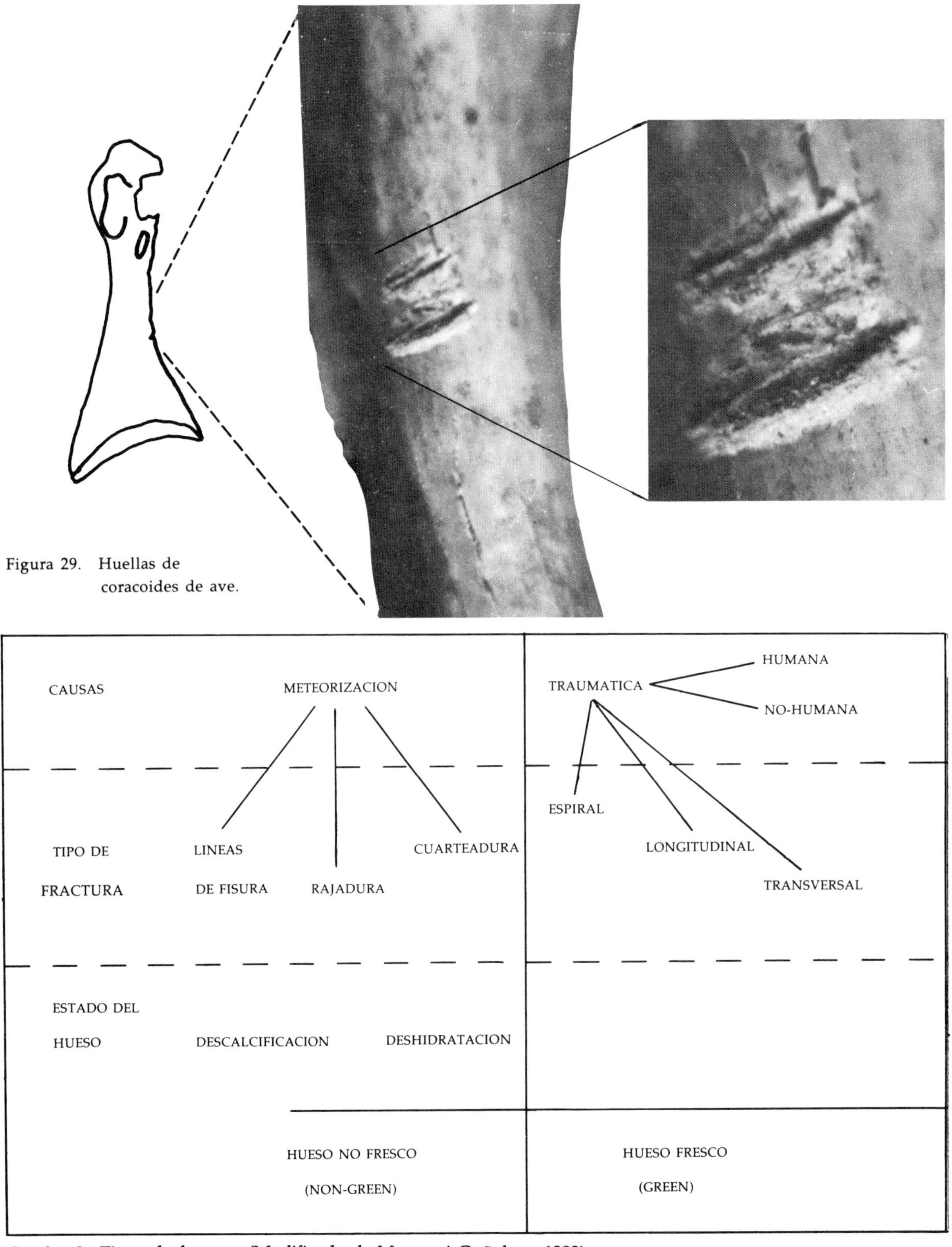

Figura 29. Huellas de coracoides de ave.

Cuadro 2. Tipos de fractura (Modificado de Mengoni-Goñalons 1982).

	NUMERO DE INDIVIDUOS	COMEN HABITUALMENTE CARNE CON HUESO	HUESOS COMIDOS RECOGIDOS EN LA MUESTRA	DEJAN CARNE EN EL HUESO	HUESOS LIMPIOS HASTA LA CUBIERTA TENDINOSA	HUESOS LIMPIOS TOTALMENTE	HEUSOS CON HUELLAS
HURON (Mustelidae)	5		9	x	x		
ZORRINO (*Conepatus* sp.)	4		5		x	x	
ZORRO GRIS (*Dusicyon griseus*)	9	x	37	x			1
CHACAL (*Canis aureus*)	1	x	11	x	x		
LOBO (*Canis lupus*)	2	x	9	x			
PUMA (*Felis concolor*)	3	x	16	x	x	x	
GATO MONTES (*Felis colocolo*)	6		15	x	x	x	2
TOTALES	30		102				3

Cuadro 3.

2) Comprende el sector Oeste, ocupado por la cuadrícula 1 (porción Oeste), cuadrícula 2 y 2′ y sus respectivas prolongaciones (Sanguinetti de Bórmida 1976:280).

La prosecución de los trabajos nos permite precisar algo más estas subdivisiones. Distinguimos cuatro localizaciones en la distribución de los huesos y materiales líticos en la cueva Las Buitreras:

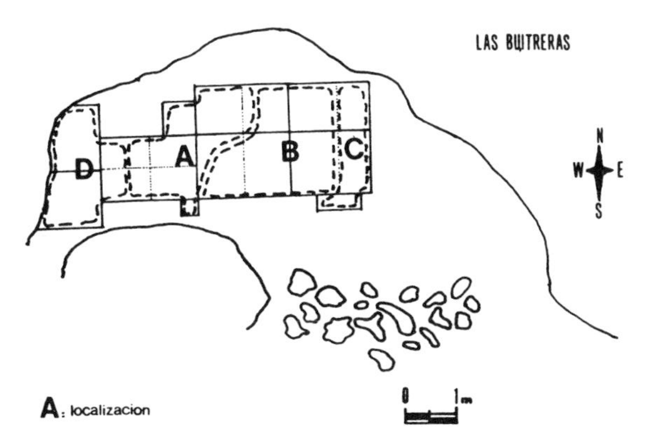

Figura 30. Mapa de los sectores de la cueva.

a. Cuadrícula 2 (especialmente sectores 4 y 1 y mitades orientales de sectores 2 y 3), cuadrícula 5 (sector 4). En capa VII los huesos se encuentran en regular estado de conservación, pero su distribución es horizontal. Esto último quedó avalado principalmente por la aparición de tres vértebras de *Mylodon* articuladas.

b. Cuadrícula 1 (sectores A y B), cuadrícula 5 (sector 1), cuadrícula 3 (sector A) y cuadrícula 4 (sector 4). La distribución de los restos está claramente concentrada en la capa VIII y es vertical. El estado de conservación es muy malo. La mayoría de los huesos lleva en su superficie una capa de manganeso. La notable concentración de los huesecillos dérmicos en esta localización obliga a pensar que allí estuvo por lo menos parte de los cueros de *Mylodon*. Aquí aparecen los únicos restos atribuíbles al cráneo de este animal.

c. Cuadrícula 3 (Extensión sur y sector B), Cuadrícula 4 (Extensión NE y sector 1). Depositación vertical de los huesos. Presencia de manganeso en algunos de los huesos. El estado de conservación es bueno, en algunos casos muy bueno. Se observó un caso de vértebras articuladas en situación vertical en la cuadrícula 3 (Extensión sur) y dos escápulas superpuestas de un mismo individuo en este mismo sector.

d. Cuadrícula 2′ y 3′. En sectores de esta localización, la capa VII aparece algo alterada por un basural atribuído a la ocupación de la capa V. La capa VIII aparece en buena situación estratigráfica en toda la localización. Los huesos presentan muy buen estado de conservación. Se trata de la localización más interna, por ende la menos expuesta a los procesos de meteorización.

El estado de desmembramiento anatómico de los huesos de *Mylodon* observables en "a", y ante la sugerencia ya efectuada de que su depositación es cultural, permite hipotetizar una cercanía relativa del sitio de matanza (*apud* Binford 1978: 64). Por lo arriba dicho la discusión no se puede realizar incluyendo "b" y "c". La información de "d" asimismo, conviene utilizarla tan sólo como control de lo que sucede en "a". La evidencia de "c" avala la premisa de que el trozamiento es una tarea de desmembramiento de conjuntos de huesos articulados y no de huesos individuales. El estado de desmembramiento muy avanzado de los huesos de *Lama glama guanicoe*, en cambio, permite hipotetizar que el sitio de matanza está muy alejado del *locus* final de esos huesos. Estas situaciones son coherentes con lo que constituye otra premisa entre pueblos cazadores: cuando mayor es el tamaño de la presa, más cerca del lugar de caza se lo consume. El pesado *Mylodon* es cazado cerca y el relativamente liviano guanaco es cazado más lejos. La premisa mencionada puede ser leída de dos maneras: 1) si se planifica la caza de un animal de gran porte, se escoge un área cercana; 2) si se caza un animal de gran porte en el transcurso de una partida de caza indiferenciada, su consumo debe hacerse en una localización cercana a la del hecho de caza. En el segundo caso debe hacerse una salvedad: si en lugar de consumirlo en las cercanías el animal es procesado allí y luego transportado por partes a sitios lejanos, entonces lo que se localizará será el área de faenamiento de ese animal en las cercanías del hecho de caza. Esta última circunstancia es la que podríamos pensar para Las Buitreras, si no fuera porque los restos de *Mylodon*, además de casi completos, no son los únicos encontrados. Por el contrario la asociación entre esos huesos y los de *Lama glama guanicoe* fue lo suficientemente estrecha como para que pensemos que nos encontramos ante la evidencia de un episodio único. El hecho de que la localización sea una cueva apoya la idea de que se trata de un lugar de consumo, cualquiera sea la distancia existente al sitio de matanza.

Así vista, la depositación final de los huesos es, en nuestro entendimiento:

a. Cultural
b. Natural
c. Natural
d. Cultural (con las restricciones formuladas arriba)

El carácter cultural de "a" ortoga también ese rótulo a las localizaciones "b" y "c". Pero en esos casos habría seguridad respecto a la actuación de agentes postdepositacionales que alteraron la situación original. En conclusión: "a" se mantuvo *in situ*, "b" y "c" fueron removidas y "d" pudo haber sido removida o no.

Nuestra hipótesis para explicar la diferencia de conservación y de profundidad existentes entre el material de "a" y "b" es que "b" funcionó como basural en tiempos de la ocupación de "a" o de "a-d". Luego, pensamos que el hombre ocupó la cueva, en la que consumió restos de diversos animales. Los restos de *Mylodon* sufrieron muy poco transporte humano, probablemente ninguno; los restos de *Lama glama guanicoe*, en cambio, fueron transportados de algún lugar relativamente alejado de la cueva. La falta de excrementos de *Mylodon* versus la buena conservación de excrementos de carnívoros-omnívoros (incluído el hombre) autorizan a pensar que, aunque se lo pudo cazar allí, *Mylodon* no vivió en el sitio en tiempos de la ocupación de la capa VII (Figuerero Torres 1980).

Merece destacarse el hallazgo de sólo dos molariformes de *Hippidium-Onohippidium* (*s.l.*) pertenecientes sin lugar a dudas a un mismo individuo (Figura 24). Estos fueron encontrados en el extremo oriental de la cuadrícula 2′. Resulta muy difícil explicar su presencia por causas naturales debido a las siguientes razones:

a. los únicos restos hallados fueron estos dos molariformes aislados;
b. en caso de conservación diferencial (huesos-dientes), ¿porqué no están presentes los dientes restantes?
c. igualmente en este sector fueron hallados restos de otros animales, razón que invalidaría en parte el problema de conservación diferencial;
d. pensando en depredadores (carnívoros, por ejemplo) es difícil explicar el hecho de que se hallan trasladado sólo dos piezas dentarias (molariformes opuestos), siendo lo más lógico hallar dos dientes contiguos o un fragmento no tan específico como los hallados.

Creemos que estas razones son ya suficientes como para considerar que la única explicación posible para la presencia de estos molares es de índole cultural. Esta afirmación no implica su presencia debido a razones de consumo.

ALTERACIONES DE INDOLE NATURAL Y ACTIVIDAD CULTURAL SOBRE EL MATERIAL OSEO

Fracturas

Mengoni-Goñalons (1982) sistematiza los diferentes tipos de fracturas de la siguiente manera[5] (ver Cuadro 2).

Las líneas de fisura están usualmente orientadas a lo largo del axis de los huesos largos y son distintas de las rajaduras; pero se diferencian de las traumáticas en que estas últimas pueden ser perpendiculares al axis (Tappen y Peske 1970).

El estudio de fracturas por meteorización es de gran utilidad para establecer las condiciones bajo las cuales fueron depositados los especímenes. Este nos da una información inicial acerca de las condiciones postmortem que han afectado los huesos (Tappen y Peske 1970; Behrensmeyer 1978). Los resultados obtenidos por Behrensmeyer y otros investigadores, nos indicarían que las fracturas por meteorización seguirían patrones similares aún en diferentes ambientes.

Con respecto a la fractura traumática intencional longitudinal sólo consideraremos aquella con muesca que no deja lugar a dudas sobre su origen cultural (esto es sólo válido para cuevas relativamente pequeñas como Las Buitreras).

La fractura en espiral es considerada por Bonnichsen (1978, 1979), Morlan (1978, 1979) y Stanford (1979) como de origen cultural. Sin embargo Myers et al. (1980) presentan restos óseos de sitios paleontológicos a cielo abierto del Mioceno y Plioceno de USA, en donde hay gran cantidad de huesos con fractura espiral. Consideran que la explicación más económica es el pisoteo (trampling) por parte de otros animales. Por ello demuestran que este tipo de fractura no es exclusivo de la actividad humana, pero que mediante una adecuada contrastación esto puede ser correctamente determinado. Haynes (1980: 348-349) menciona el caso de grandes carnívoros (osos por ejemplo) capaces de fracturar huesos compactos de grandes ungulados, sin que posteriormente se observen huellas de dientes en los fragmentos.

En el punto IV fueron mencionados determinados tipos de fracturas halladas en huesos de *Lama glama guanicoe* que nos indican una clara actividad cultural sobre los mismos. Las muescas y los tipos de fractura (longitudinal, espiral y transversal) sobre los restos, encontrados en Las Buitreras, no podrían ser explicados por pisoteo, dado que fueron hallados en el interior de la cueva. Morlan igualmente sugiere la posibilidad de fractura espiral por caída de bloques dentro de una cueva, en tal caso los huesos se hallarían debajo de los mismos, cosa que por lo general no ocurre en este sitio. Aunque en determinados sectores fueron hallados huesos de *Mylodon* debajo de bloques, no presentan fracturas dada la característica de la muestra.

Otra alternativa sería el pisoteo por parte de *Mylodon*, pero esto no explicaría las marcas y variedad de fracturas en los huesos de *Lama glama guanicoe*.

Alteraciones

Dentro de este tema se prestó especial importancia a las huellas producidas por dientes. Nuestro objetivo era ver la posibilidad de establecer controles adecuados para discriminar huellas de origen cultural (descarne, corte, etc.) de las producidas por animales. Para ello se trabajó sobre la base de investigaciones específicas sobre el tema y con datos experimentales. Los trabajos experimentales fueron realizados en el Zoológico de la Municipalidad de Buenos Aires. Se efectuaron estudios de huesos comidos por 30 individuos pertenecientes a 3 familias distintas (Mustélidos, Cánidos y Félidos) y siete especies, durante tres días consecutivos. El siguiente cuadro resume los resultados:

Igualmente fue utilizado material obtenido de animales domésticos (perros, cabayos y hamsters) y material proveniente de sitios arqueológicos en los cuales las huellas fueron de indudable origen.

Nuestros resultados nos muestran un muy bajo índice de presencia de huellas. Estos coinciden con los datos brindados por Haynes (1980) y Kent (1981). Por esta razón, para determinar la actividad de carnívoros se debe también recurrir a otro tipo de indicios.

Haynes (1980) considera que los carnívoros mastican los huesos siguiendo patrones característicos. Zapfe (1939), Sutcliffe (1970), Bonnichsen (1979), Haynes (1980) y Binford (1981) también realizan exhaustivos análisis acerca de los diferentes patrones de alteración producidos por carnívoros a carcasas o conjuntos de huesos. La experimentación llevada a cabo por nosotros es con huesos desmembrados, al igual que la realizada por Kent (1981).

Sutcliffe (1970, 1973), Brain (1970), Sutcliffe y Collings (1971) y Bouchud (1974) presentan huellas de dientes de carnívoros, roedores y cérvidos, haciendo referencia a sus características distintivas.[6] Potts y Shipman (1981) en un trabajo reciente utilizan scanning electron microscopy (SEM) para el análisis de dichas huellas; estos presentan claramente las características de huellas de carnívoros y roedores comparadas con otras de índole cultural (corte, raspado y machacado) y marcas producidas durante la preparación del material (preparation scratches). Consignan asimismos excelentes fotografías de cada uno de los tipos.

También se han desarrollado técnicas que permiten extraer moldes de siliconas de las huellas. Esta técnica permite establecer la sección transversal de los surcos para determinar su origen (Walker y Long 1977; Bunn 1981). Esta técnica fue aplicada a las encontradas en *Mylodon* con resultados negativos, debido a la presencia de óxido de manganeso en el interior de los surcos.

Ilustramos ahora huellas producidas por zorro y cricétidos (Figuras 31; y 32-33 respectivamente).

Podemos entonces sintetizar en dos grandes categorías las huellas que aquí nos interesan:

1- Huellas de roedores: dejan dos surcos paralelos, profundos, de sección cuadrangular y separado por una pequeña cresta. Estas están usualmente producidas por un cincelado muy limpio y claro. Existen variaciones respecto a la manera de actuar según la forma y característica de los huesos, pero el rasgo definitorio (dos surcos y una cresta) se conserva como unidad elemental.

2- Huellas de carnívoros: surcos de sección cóncava con sus bordes usualmente astillados. Este astillamiento es producido por un diente de cúspides romas (exceptuando micromamíferos) cuya acción se manifiesta más en la presión por arrastre que en el corte. Los carnívoros suelen también destruir las epífisis de los huesos, en donde dejan surcos como los ya mencionados.

En una de las escápulas de *Mylodon* fueron halladas "huellas" que macroscópicamente se asemejan a las de corte, pero que resultaron ser canales arteriales o venosos. En la fotografía se puede observar cuando estas penetran en el hueso (Figuras 34 y 35).

Las huellas halladas en *Lama glama guanicoe* fueron presentadas en el punto IVb y son el resultado de actividad cultural.

En *Mylodon* sólo fueron halladas huellas de corte en 4 costillas y sobre su cara interna. Borrero (1980:22) y Caviglia et al. (1976:87 Lamina IV) ilustran algunas huellas de la costilla No.1, de claro origen humano y seguramente producto de la actividad de descarne.

Las huellas presentes en estas costillas son oblícuas o perpendiculares al eje longitudinal del hueso. Se hallan usualmente aisladas y son en general poco profundas (salvo una excepción). Su longitud varía entre 0,3 y 1,2 cm. (ver Figuras 36, 37, 38 y 39).

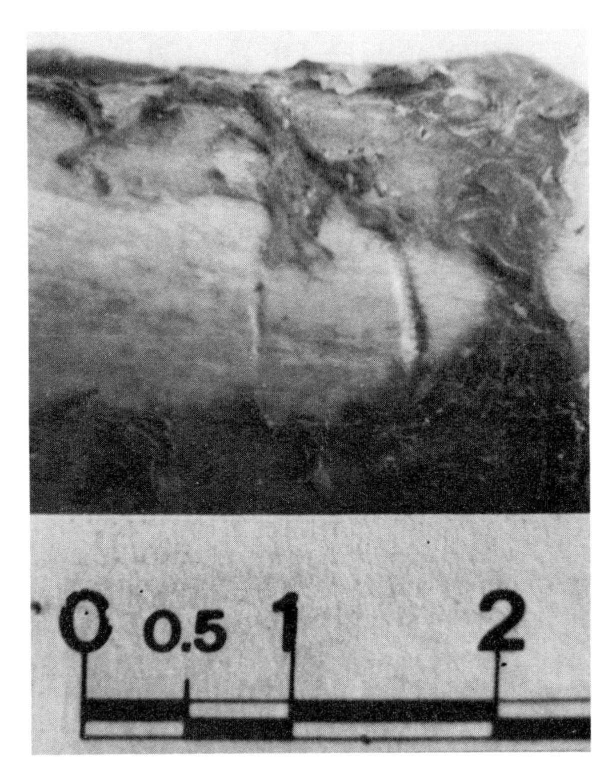

Figura 31. Huellas producidas por zorro.

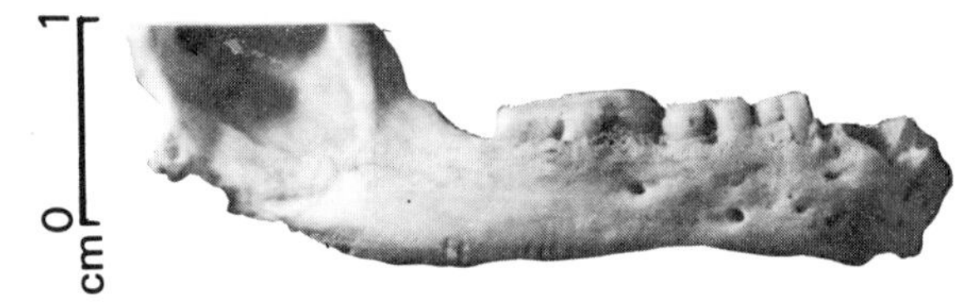

Figura 32. Huellas producidas por cricétidos.

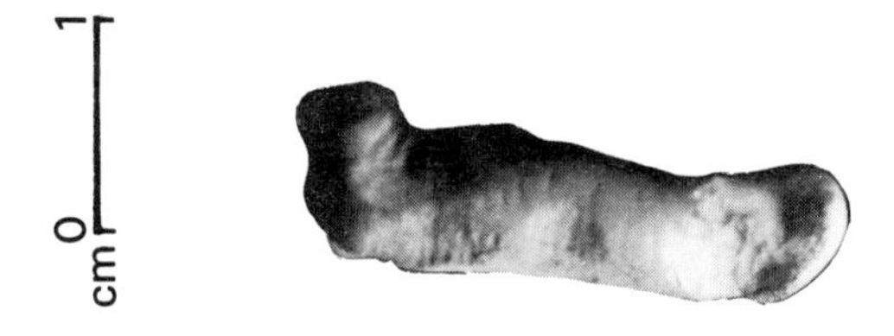

Figura 33. Huellas producidas por cricétidos.

Figura 34. Huellas naturales.

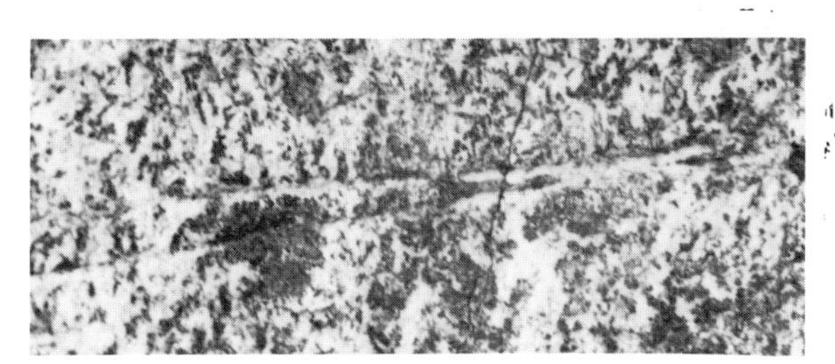

Figura 35. Huellas naturales.

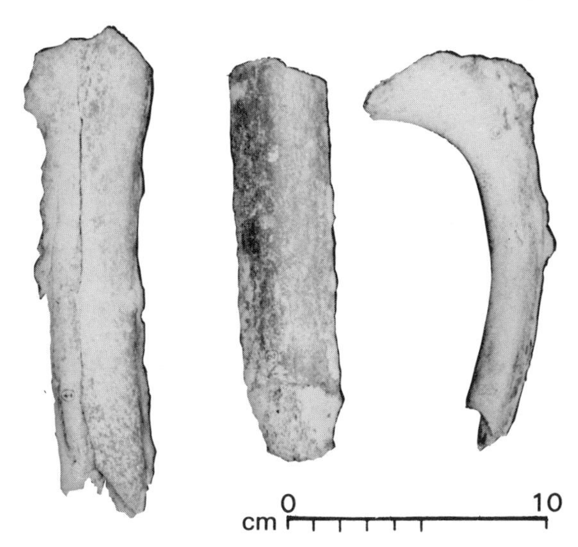

Figura 36. Costillas de *Mylodon* N° 1, 2 y 3.

Figura 37. Huellas de corte en costilla N° 2.

Figura 38. Huellas de corte en costilla N° 1.

Figura 39. Huellas de corte en costilla N° 3.

ANALISIS FUNCIONAL

Las piezas provenientes de la localización "a" (referida en el punto VI) fueron sometidas a un estudio microscópico de huellas de uso de acuerdo a una metodología aplicada por uno de nosotros y que se explica en otra parte (Yacobaccio 1982). De este análisis se obtuvo:

- La lasca con retoque en bisel oblícuo fue utilizada para raspado (pieza No. 6).
- Una lasca de basalto (pieza No. 2) fue utilizada en actividades de corte.
- La raedera/raspador fue utilizada: (a) los filos derechos e izquierdo para raspado de substancia dura, (b) el filo frontal para raspado de substancia blanda.

En base a este análisis se puede establecer que sólo la pieza número 2 (lasca de basalto) podría ser la causante de las huellas de descarne observadas. Sin embargo, el grado de desgaste del filo es leve. Si hubiera cortado substancia dura (hueso) con el ángulo de filo que posee (45°), aquel debería presentar un desgaste mucho mayor, como lo demuestran los experimentos llevados a cabo. Además, las microfracturas presentan terminaciones normales y comienzos curvados. Si hubieran trabajado sobre substancia dura presentarían un porcentaje de terminaciones angulares y en ángulo agudo y comienzos con puntos de iniciación.

En cuanto a la lasca con retoque en bisel oblícuo (pieza No. 2) observamos que el desgaste es leve, presentando sólo pequeñas fracturas (menores que 0,5 mm) con terminaciones normales sobre la cara dorsal. No podemos determinar la substancia trabajada, pero se asemeja al tipo de desgaste registrado por Hayden - definido como ''minute crushing'' (1979:220 y Figure 20), aunque raramente, en raspadores para piel de colecciones etnográficas de los esquimales de Alaska y Arapaho.

La raedera/raspador tampoco fue utilizada directamente sobre hueso, de acuerdo a la muestra ósea que manejamos (ningún hueso evidencia signos de raspado). Los filos laterales que rasparon substancia dura pudieron ser destinados a madera.

De este modo, pensamos, que no se puede establecer una relación directa de actividad entre el material lítico y el óseo. Ambos se presentan como evidencias independientes de diversas actividades.

CONCLUSIONES

La evidencia recuperada en la excavación de los niveles VII y VIII de la cueva Las Buitreras nos muestra:

- En cuanto a los roedores sólo se puede hablar de asociación *sensu stricto* sin ningún otro tipo de implicaciones.
- En los restos de guanaco se pudo determinar la presencia de fracturas traumáticas intencionales, como así también huellas de corte. Todos los fragmentos son pequeños y provienen de restos de por lo menos tres individuos, uno de ellos nonato o neonato.
- Los dientes de caballo americano como así también la vértebra de delfínido, no son indicadores de consumo de estos animales en el sitio.
- *Mylodon* presentó claros elementos que permiten suponer su consumo. Esto se ve avalado por el aprovechamiento de esta especie en otros sitios de Patagonia Meridional (Mengoni Goñalons 1982 y en este volumen; Borrero, en este volumen). Los restos pertenecen a dos individuos juveniles.
- *Dusicyon avus* (zorro extinguido) seguramente ha sido una presa ocasional.
- El coprolito humano muestra un alto contenido de vegetales, aunque también pequeñas astillas de hueso. La escaséz de esta evidencia no permite hacer generalizaciones sobre la dieta.
- Los instrumentos están confeccionados en materia prima alóctona a la cueva y son en general de tamaño pequeño. Tanto la técnica de talla como la de retoque fueron por percusión. El análisis de microdesgaste por uso demuestra su participación en diversas actividades (Corte de substancia blanda y raspado sobre substancia blanda y dura).
- La distribución en planta nos muestra la existencia de cuatro sectores. Dos de ellos han sido afectados por procesos naturales, uno con depositación *in situ* y el restante interpretado como de origen cultural a partir de este último. Tanto las muestras óseas mayormente analizadas como el material lítico en su totalidad, provienen de la localización "a" de origen cultural.
- El estado de desmembramiento de *Lama glama guanicoe* hace pensar en un *locus* de matanza alejado de la cueva. Asimismo las actividades fundamentales de descarne no fueron realizadas en el sitio sino que las astillas presentes fueron seguramente llevadas al sitio con los trozos de carne a consumir.
- Aunque se piensa que *Mylodon* no vivió en el sitio (ausencia de excrementos), el registro óseo nos permite pensar que fue cazado y desmembrado dentro de la cueva.
- Como se ha visto, el conjunto de artefactos líticos no ha trabajado directamente sobre hueso. Es posible suponer que las marcas de corte han sido efectuadas con instrumentos que fueron llevados del sitio por sus ocupantes al abandonarlo.[7]
- Es dable pensar que la ocupación del sitio estuvo vinculada al consumo de carne de guanaco y de *Mylodon* y a la realización de pocas tareas complementarias.

- Se debe suponer que la ocupación fue corta, probablemente un solo evento, y por un núcleo reducido de personas implicado en la baja demografía de Patagonia Meridional hacia el 10.000 B.P. (*vide* Borrero 1980).
- Estas características nos permiten clasificar el sitio como una "Localidad de actividades limitadas" (*apud* Wilmsen 1970:75).

NOTAS

1. Esto mismo ocurre con ejemplares obtenidos en el "Volcán Orejas de Burro" (ca. 15 km al este de Pali-Aike).

2. La nomenclatura para fracturas y huellas es aclarada en el punto VII.

3. Estos restos fueron determinados por J. Clutton Brock, quien ante la evidencia de la presencia de *Dusicyon avus* en Patagonia, decidió dejar incierto su status taxonómico, considerando necesaria una revisión de estos restos (Clutton Brock, comunicación personal).

4. Hinde (1970: 657-675) realiza una excelente discusión acerca de pautas de comportamiento y su relación con problemas taxonómicos.

5. En este trabajo se realiza una excelente síntesis y sistematización de los estudios de fracturas realizadas hasta ese momento, sumándose a ello trabajo experimental. Por esta razón sólo realizaremos unas breves consideraciones y remitimos a éste para cualquier ampliación.

6. Igualmente Bang y Dahlström (1975) dan abundante información acerca de huellas y señales de mamíferos y aves europeas.

7. Esta característica, bajo índice de artefactos líticos en relación a los restos faunísticos, se repite en diversos sitios de matanza en las planicies de USA.

AGRADECIMIENTOS

A la Profesora A.C. Sanguinetti que nos ofreció el material para este estudio.

Al Dr. C. Irigoyen, del Museo Argentino de Ciencias Naturales "Bernardino Rivadavia", que nos permitió publicar el dibujo de *Mylodon*.

A la Dra. M. Mantecón y autoridades del Jardín Zoológico de la Municipalidad de Buenos Aires por su colaboración en la etapa experimental de nuestro trabajo.

Al Lic. G.L. Mengoni Goñalons por facilitarnos su trabajo inédito y por el fructífero intercambio de ideas.

A M.J. Figuerero Torres por seleccionar la cita que introduce este trabajo y las charlas sobre coprolitos.

No queremos dejar de mencionar "Lizzie" y "Gwendolyn" (perras), "Pippin" (hamster), "Gigi" (cobayo) y los demás animales del zoológico, que tan gentilmente han mordido los huesos que les brindamos.

REFERENCIAS CITADAS

Auer, V.
1974 The isorhythmicity subsequent to the Fuego-Patagonian and Fenoscandian ocean level transgression and regressions of the latest glaciation. *Annales Academiae Scientiarum Fennicae.* Serie A III Geologica-Geografica 133: 1-88. Helsinki.

Bang, P., and P. Dahlström
1975 *Huellas y señales de los animales de Europa.* Editorial Omega, Barcelona.

Behrensmeyer, A.K.
1978 Taphonomic and ecologic information from bone weathering. *Paleobiology* 4: 150-162.

Binford, L.R.
1978 *Nunamiut ethnoarchaeology.* Academic Press, New York.

1981 *Bones. Ancient men and modern myths.* Academic Press, New York.

Bird, J.
1967 In Chilean caves forlorn. In *Conquistadors without swords,* edited by L. Devel, pp. 528-539. Schocken Books. New York.

Bird, J., and M. Bird
1937 Human artifacts in association with horse and sloth bones in southern South America. *Science* 36(2219): 36-37.

Bond, M., S.E. Caviglia, y L.A. Borrero
1981 Paleoetnozoologia del Alero de los Sauces (Neuquen, Argentina); con especial referencia a la problemática presentada por los roedores en sitios Patagónicos. Instituto de Ciencias Antroplógicas, Faculdad de Filosofia y Letras, Universidad de Buenos Aires, *Trabajos de Prehistoria* 1: 95-111.

Bonnichsen, R.
1978 Critical arguments for Pleistocene artifacts from the Old Crow Basin, Yukon: a preliminary statement. In Early man in America from a circum-Pacific perspective, edited by A.L. Bryan, pp. 102-118. *University of Alberta, Department of Anthropology, Occasional Papers* No. 1. Edmonton.

1979 Pleistocene bone technology in the Beringian Refugium. National Museum of Man, *Archaeological Survey of Canada, Mercury Series* 89. Ottawa.

Borrero, L.A.
1980 La relación entre los primeros cazadores Americanos y la fauna Pleistocénica: consideraciones demográficas. *Segundo Congreso Argentino de Paleontología y Bioestratigrafía y Primero Congreso Latinoamericano de Paleontología, Actas III*: 211-221. Buenos Aires.

Bouchud, J.
1974 Les traces de l'activité humaine sur les os fossiles. *Premier Colloque International sur l'industrie de l'os dans la Prêhistoire,* pp. 27-33. Editorial Universitaire, Provence.

Brain, C.K.
1970 New finds at the Swartkrans Autralopithecine site. *Nature* 225: 112-119.

Bunn, H.T.
1981 Archaeological evidence for meat-eating by Plio-Pleistocene hominids from Koobi Fora and Olduvai Gorge. *Nature* 291: 574-577.

Cardich, A., E.P. Tonni, y N. Kriscautzky
1977 Presencia de *Canis familiaris* en restos arqueológicos de Los Toldos (Prov. de Santa Cruz, Argentina). *Relaciones de la Sociedad Argentina de Antropología* XI(N.S.): 115-119.

Caviglia, S.E.
1976 Sobre la presencia de un cetáceo en asociación con *Hippidium-Onohippidium* (*s.l.*) y *Mylodon* en la Cueva Las Buitreras. *Relaciones de la Sociedad Argentina de Antropología* X(N.S.): 313-314.

1978 La Presencia de *Dusicyon avus* (Burm.), 1864, en la capa VIII de la Cueva Las Buitreras (Patagonia, Argentina): su relación con otros hallazgos de Patagonia Meridional. Trabajo Presentado en VI Congreso de Arqueología de Uruguay. Salto.

1980 Ibid. (Resumen). *Runa* XIII(1-2): 31-33.

Caviglia S.E., y M.J. Figuerero Torres
1980 Material faunístico de la Cueva Las Buitreras. *Relaciones de la Sociedad Argentina de Antropología* X(N.S.): 315-319.

Caviglia, S.E., H.D. Yacobaccio, y L.A. Borrero
1976 Los niveles con megafauna de Las Buitreras: componentes culturales y faunísticos. Comisión XII, *X Congreso Uniôn Internacional de Ciencias Prehistôricas*, pp. 68-89. México.

Clutton-Brock, J.
1977 Man-made dogs. *Science* 197: 1340-1342.

Curzio, D.E.
1976 Consideraciones tipológicas del contexto oseo de la Cueva Las Buitreras. *Relaciones de la Sociedad Argentina de Antropología* X(N.S.):293-307.

Figuerero Torres, M.J.
1980 Análisis de coprolitos: el caso de la Cueva Las Buitreras. *VII Congreso Nacional de Arqueología de Uruguay*, Colonia. pp. 46-49.

Grayson, D.K.
1973 On the methodology of faunal analysis. *American Antiquity* 39: 432-439.

Hayden, B.
1979 Snap, shatter, and superfractures: use wear of stone skin scrapers. In, *Lithic use-wear analysis*, edited by B. Hayden. Academic Press, New York.

Haynes, G.
1980 Evidence of carnivore gnawing on Pleistocene and Recent mammalian bones. *Paleobiology* 6: 341-351.

Hershkovitz, P.
1962 Evolution of neotropical cricetine rodents (Muridae). Zoology. *Fieldiana* 46: 1-524. Chicago.

Hinde, R.A.
1970 *Animal behaviour*. McGraw Hill Kogakusha Ltd., Tokyo.

Kent, S.
1981 The bone-gnawing experiment. *Journal of Field Archaeology* 8: 369-372.

Lehmann-Nitsche, R.
1899 El mamífero misterioso de la Patagonia *"Gryotherium domesticum"*. III. Coexistencia del hombre con gran desdentado y equino en las cavernas patagónicas. *Revista del Museo de La Plata* IX:455-472.

Mengoni-Goñalons, G.L.
1978 El aprovechamiento de los recursos faunísticos en el interior de Patagonia Meridional continental: hipótesis y modelos. Trabajo Presentado en VI Congreso Nacional de Arquelogía de Uruguay, Salto.

1982 Notas zooarqueológicas I: fracturas en huesos. *Actas VII Congreso de Arqueología de Uruguay*, pp. 87-91. Colonia.

Morlan, R.E.
1978 Early Man in northern Yukon Territory: perspectives as of 1977. In Early man in America from a circum-Pacific perspective, edited by A.L. Bryan, pp. 78-95. *University of Alberta, Department of Anthropology, Occasional Papers* No. 1. Edmonton.

1979 A Stratigraphic framework for Pleistocene artifacts from Old Crow River, northern Yukon Territory. In *Pre-Llano cultures of the Americas: paradoxes and possibilities*, edited by R.L. Humphrey and D. Stanford, pp. 125-145. Anthropological Society of Washington. Washington, D.C.

Myers, T.P., M.R. Voorhies, and R.G. Corner
1980 Spiral fractures and bone pseudotools at paleontological sites. *American Antiquity* 45: 483-490.

Potts, R., and P. Shipman
1981 Cutmarks made by stone tools on bones from Olduvai Gorge, Tanzania. *Nature* 291: 577-580.

Roth, S.
1902 Nuevos restos de mamíferos de la Caverna Eberhardt de Ultima Esperanza. *Revista del Museo de La Plata* XI: 37-50. La Plata.

Sanguinetti de Bórmida, A.C.
1976 Excavaciones prehistóricas en la cueva de Las Buitreras (Prov. de Santa Cruz). *Relaciones de la Sociedad Argentina de Antropología* X(N.S):271-292.

1980a El sitio de Las Buitreras como aporte de fuentes prehistóricas del temprano poblamiento sudamericano. *Runa* XIII(1-2): 11-20.

1980b La cueva de Las Buitreras y el problema del estadio de caza temprana en el extremo austral de América. Presentado en El Paleoindio en América del Sur. Seminario Internacional. Antofagasta.

Sanguinetti de Bórmida, A.C., y L.A. Borrero
1977 Los niveles con fauna extinta de la Cueva Las Buitreras. *Relaciones de la Sociedad Argentina de Antropología* XI(N.S.): 167-175.

Saxon, E.C.
1976 La prehistoria de Fuego-Patagonia: colonización de un hábitat marginal. *Anales del Instituto de la Patagonia* 7: 63-74. Punta Arenas.

1979 Natural prehistory: the archaeology of Fuego-Patagonian ecology. *Quaternaria* XXI:329-356.

Scillato Yané, G.
1976 Sobre algunos restos de *Mylodon* (?) *listai* (Edentata, Tardígrada) procedentes de la Cueva Las Buitreras. *Relaciones de la Sociedad Argentina de Antropología* X(N.S.): 309-312.

Stanford, D.
1979 The Selby and Dutton sites: evidence for a possible Pre-Clovis occupation on the High Plains. In *Pre-Llano cultures of the Americas: paradoxes and possibilities,* edited by R.L. Humprhrey and D. Stanford, pp. 101-123. Anthropological Society of Washington, Washington, D.C.

Sutcliffe, A.J.
1970 Spotted hyaena: crusher, gnawer, digester, and collector of bones. *Nature* 227: 1110-1113.

1973 Similarity of bones and antlers gnawed by deer to human artefacts. *Nature* 246: 428-430.

Sutcliffe, A.J., and H.D. Collings
1971 Gnawed bones from the Crag and Forest Bed deposits of East Anglia. *Suffolk Natural History* 15(6): 497-498.

Tappan, N.C., and G.R. Peske
1970 Weathering cracks and split-line patterns in archaeological bone. *American Antiquity* 35:383-386.

Walker, P.L., and J.C. Long
1977 An experimental study of the morphological characteristic of tool marks. *American Antiquity* 42: 605-616.

Wilmsen, E.N.
1970 Lithic analysis and cultural inference. A Paleoindian case. *University of Arizona Anthropological Papers* 16. Tucson.

Yacobaccio, H.D.
1977 Patrones de fractura y modos de uso en artefactos en la capa VII y capa VIII cumbre de Las Buitreras (Santa Cruz, Argentina). *Relaciones de la Sociedad Argentina de Antropología* XI(N.S.): 176-178.

1980 Aspectos tipológicos y funcionales de los artefactos líticos de los niveles con fauna extinta de Las Buitreras. *Runa* XIII(1-2): 27-30.

1982 Estudio de microdesgaste por uso en análisis lítico I: fracturas. *Actas VII Congreso de Arqueología de Uruguay,* pp. 162-166, 3 Láminas, Colonia.

Zapfe, H.
1939 Lebensspuren der eiszeitlichen Höhlenhyáne. *Paleobiologica* 7: 111-146.

The Cultural Relationships of Monte Verde: A Late Pleistocene Settlement Site in the Sub-Antarctic Forest of South-central Chile

TOM D. DILLEHAY
Department of Anthropology
University of Kentucky
Lexington, Kentucky 40506
U.S.A.

Abstract

Multidisciplinary research at Monte Verde, a Late Pleistocene occupation site located in south-central Chile, is reported. The research has yielded a minimally modified, stone tool industry and a wood industry in direct association with an architectural feature and the bone remains of at least five to six individual mastodons. Radiocarbon dates from the site show that the village was occupied about 13,000 years ago. A recently discovered earlier occupation, dated 33,000 B.P., is being excavated in 1985.

Cultural materials at the site are buried in the banks and adjacent sandy knolls of Chinchihuapi Creek, a southern tributary of the Rio Maullin. The stream drains a wet boggy area in a humid sub-antarctic forest that has existed there since Late Pleistocene times.

The stratigraphy consists of eight geologic strata. The lowest deposits are MV-8, gravel, and MV-7, gray heavy-grained sand with lenses of volcanic ash. Overlying MV-7 is MV-6, a narrow stream-bed deposit, consisting of medium-grained brown sand and gravels. Both these strata made up the surface at the time of the cultural event. Geologic studies show that at the time people occupied the site the ancient creek was flanked by low sandy knolls and shallow bogs. Later, the bogs, represented by stratum MV-5, expanded and covered the stream, creating a peat deposit that overlies and seals in the archaeological materials. All of the cultural materials are confined to the upper parts of MV-6 and MV-7 and the lower part of MV-5. Overlying these strata are more recent sterile sediments.

Most of the lithic collection consists of minimally modified pebble tools made of both local and exotic raw materials. There is no securely associated evidence for bifacial flaking, thus creating a problem in defining the type and form of lithic technology at the site. It is hypothesized that a process of selection of naturally occuring fractured pebbles, either for immediate use or minimal alteration prior to use, took place. Discussion of a method designed to compare the context, morphology and edge-wear characteristics of cultural stones, natural stones and replicative stones is discussed.

Wooden artifacts include two hafts made of hard wood on which are mounted stone tools, two modified branches with grooves and associated sub-triangular gouges, a modified log, and about 85 pieces with cuts, grooves, modified edges and abrasions. The data are discussed in light of the total context of the cultural locality and of the local natural setting.

Several osteological elements of the individual mastodons show clear evidence of cultural modification. In addition, clusters of bone elements occurred in association with lithics and wood in discrete cultural activity areas within the site.

A wooden wishbone-shaped architectural structure in direct association with two hearths and the modified log is described. The structure and several discrete clusters of bone, wood and lithics form a semi-circular activity pattern along the creek bank, suggesting a planned occupation.

A diversity of floral specimens were preserved at the site. Preliminary results of analyses conducted on these materials by specialists are presented in order to reconstruct the paleoenvironment of the region during the Late Pleistocene era.

The general cultural relationships and implications of these data and patterns are provided for insight into the technology, economy and activity organization of Early Man at Monte Verde.

INTRODUCTION

Monte Verde (X-1) is a Late Pleistocene village occupational site located 33 air km southwest of Puerto Montt in south-central Chile. The site is situated in the north bank of Chinchihuapi Creek, (a small tributary of the Maullin River, which lies 8.7 km north of the site) and 15 km northwest of the Pacific Ocean at an elevation of 50 m above sea level. The exact position of the site is long. 73° 15′ W, lat. 41° 30′ S (Figure 1).

Chinchihuapi Creek drains a wet boggy terrain in a humid subantarctic forest that has existed there at least since Late Pleistocene times. Radiocarbon dates from the site indicate that a single cultural episode took place there between 12,000 and 14,000 years ago.

Human activity at this site is revealed by the direct association of selected and fractured osteological remains of at least five to six individual mastodons, a minimally modified pebble tool technology, a well-preserved wood industry, a diversified array of ecofacts, and the structural foundation of an architectural feature.

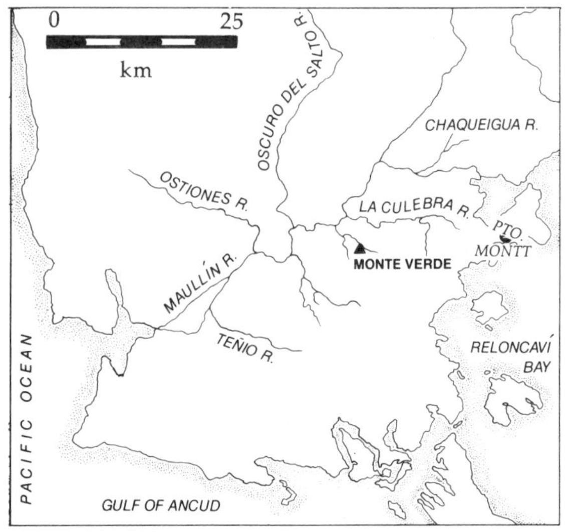

Figure 1. Location of the Monte Verde site.

A brief history of research at the site and an overview of the archaeological, geological and ecological evidence for early human activity at Monte Verde is presented here. Specifically, I will concentrate on the types of archaeological data recovered for documentation of the cultural event, on the research problems and on the logistics of data acquisition. No specific conclusive interpretations will be provided since further fieldwork is required at the site and the multidisciplinary research team has not yet completed analyses of the materials recovered to date.

BACKGROUND TO ARCHAEOLOGICAL RESEARCH AT MONTE VERDE

In October, 1976, the author and students of the Universidad Austral de Chile (U.A. Ch.), Valdivia, conducted a cursory surface inspection of Chinchihuapi Creek, locating fractured bones, one biface, and modified pieces of wood eroding from the bank of a stream. In total, 112 bone specimens, the biface, and three wood fragments were collected from the surface of the modern creek bed. Subsequent analysis of these materials at the archaeological laboratory (U.A. Ch.) suggested that human intervention might account for the observed alterations.

In January, 1978, a grid pattern was established at the site, consisting of one m square units, and a subsurface testing program was conducted in the area where materials were eroding from the bank. In addition, extensive surface inspection was carried out along other exposed vertical cuts of the creek bank and on the crest of adjacent sandy ridges.

This work defined: 1) the buried horizontal

and vertical extent of the archaeological remains in the north bank of the creek and on the sandy ridge south of the creek; 2) the nature of the geological deposits; and 3) the direct and patterned association of artifacts and ecofacts.

Three distinct areas of the site were defined: Area A, along the north side of the creek, where the wood and bone materials were eroding from the bank; Area B, where the dental remains of an infant mastodon and two pebble "flakes" were exposed in an *in situ* position in the south bank of the creek about 300 m downstream from Area A; and Area C, along a low, sandy ridge on the south side of the creek, where modified pebbles and the remains of a scattered hearth were located (Figure 2). Altogether, some 42 m^2 were excavated in Area A and 5 m^2 in Areas B and C.

In Areas A and B, artifactual material was recovered at a depth between .95 to 1.10 cm below the surface in the ancient stream bed of Chinchihuapi Creek (see the following section of geology of the location). Discrete clusters of lithic, bone and wooden artifacts were concentrated in a 24 m^2 section of the excavation in Area A. In Area C, the hearth and lithics were were recorded at a depth of .95 cm below the surface. All areas of the site were excavated by natural strata. In total, 117 bone elements, 181 lithics, and 85 worked pieces of wood were excavated during the 1978 season.

Excavation of the cultural materials at the site raised two research problems. The first was a concern with preservation of the exposed floral remains. These remains were sealed and preserved by a super-imposed peat layer. A slow and tedious recovery technique, utilizing syringes, horse-hair brushes, soft bamboo splints and water pump and hoses, was required to minimize damage to these remains during excavation. These materials were soaked in appropriate chemicals for conservation and later transported to the archaeological laboratory in Valdivia for further treatment. (Description of the types of chemicals used and the procedural steps employed in this conservation treatment are beyond the scope of this paper).

The second problem was a methodological and theoretical one. Based on the direct contextual association (including clusters of cultural materials suggestive of distinct activity areas) of the bone, wood and lithic specimens in a relatively small, concentrated horizontal segment (4 m^2 of the

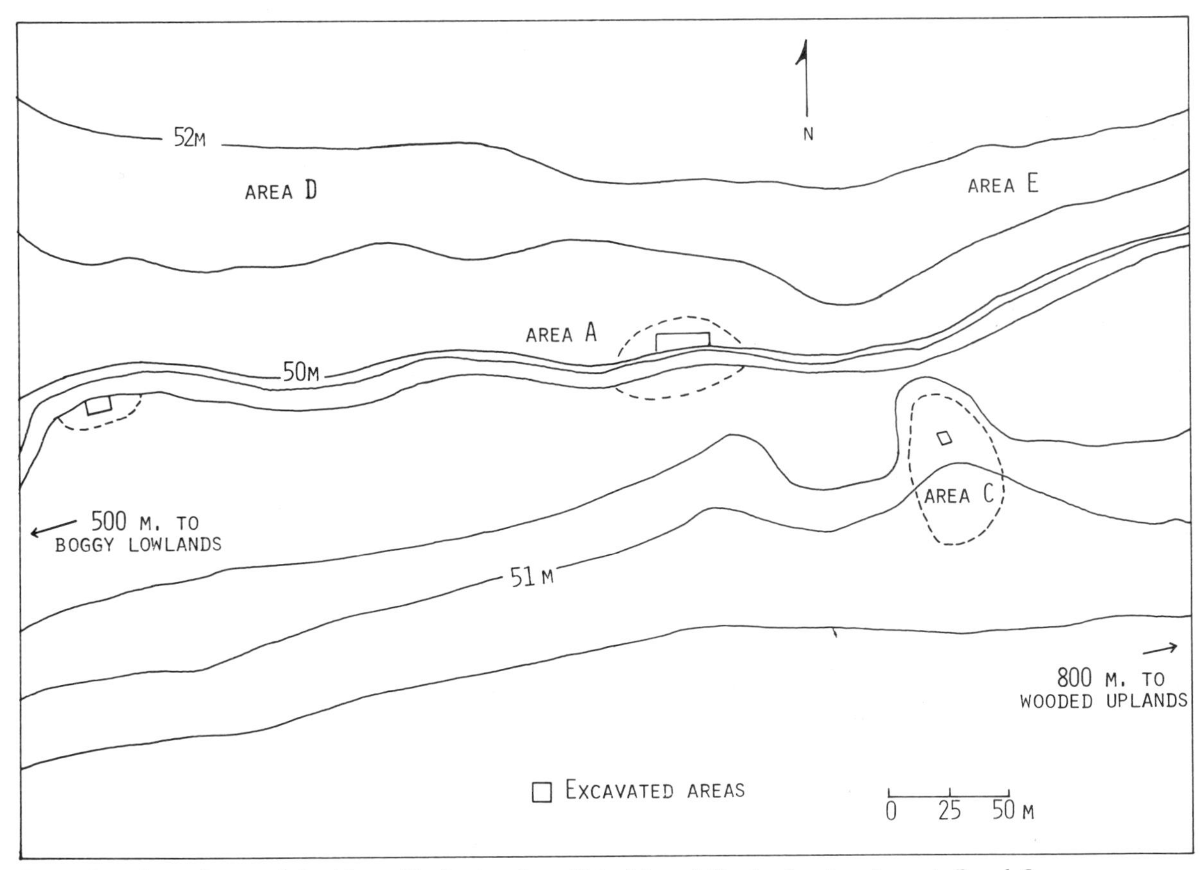

Figure 2. General map of the Monte Verde site along Chinchihuapi Creek, showing Areas A, B and C.

excavated 42 m^2 along the north bank of the stream) and in a 10 to 15 cm thick vertical deposit, we were convinced that the placement and modification of these different categories of materials were related to human agency (see Figure 3).

A cursory field inspection of the disarticulated and fractured nature of the excavated osteological remains revealed that only certain skeletal portions (i.e., skull, teeth, torso and limbs) of five to six individual mastodons were represented. Although a large majority of the bone elements exhibit little or no direct evidence of butchering activity, at least eight of the excavated bone elements show fractures or "cut" marks or are contextually associated with other materials which have resulted from human activity. Also, an ilium was attached by bitumen to a modified, vertically placed wooden branch stub, and the distal end of a femur housed eight fractured shell valves representing six individual mollusks. It was difficult to accept all of these characteristics as the result of natural forces in a low energy environment like Chinchihuapi Creek, which at its maximum extent in both the modern and the ancient creek channels was only 4 m wide and 20 cm deep (Pino n.d.).

The recovered pebble specimens appeared to have little morphological alteration, and thus little functional value. Percussion and pressure flaking were not apparent on the lithics, except for one percussion-flaked biface which had eroded from the creek bank in Area A. Furthermore, clusters of certain types of rock forms (i.e., spherical rocks, including two full-grooved stones, and fractured wedge-shaped specimens) were observed in and around concentrations of modified bone and wood elements.

For instance, seven almost perfectly spherical stones (see Collins and Dillehay, this volume), including one grooved specimen, were clustered within a 1.5 m^2 area in close proximity to two full-length grooved branches. Three fractured pebbles exhibiting battered edges were closely associated with a femur and two modified branches. Three proximal rib fragments, approxomately of the same size and length and showing similar alterations on their distal ends,

Figure 3. Lithic, bone and wood artifacts *in situ* in Area A. Note the full-length grooves in two branches located in the right center of the photograph. The grooved branch to the left has a subtriangular gouge-like stone *in situ* in one end.

were also in close association. Other clusters were characterized by numerous piles of modified branches, other organic matter, lithics and small bone fragments. As described in detail below, several bone, wood and lithic specimens were attached, forming composite implements.

All of these clusters were concentrated in a 24 m^2 zone of Area A. The tight horizontal concentration and the lens-like stratigraphic position of these materials, as well as the absence of any internally intrusive features, is suggestive of a single cultural episode.

No previously recorded early wood assemblage from the New World is available that would serve as a type collection for definition of cultural criteria in the analysis and classification of the Monte Verde wood materials. In addition to the rich organic matter recovered from the site, the Monte Verde artifact collection included several wooden specimens that had been tied together with a local *juncu* reed, and two specimens that possessed fractured, gouge-like stones resting in grooves or canals cut the full-length of the branches. In addition, two other fractured pebbles were hafted onto small branches. Several other wooden pieces showed flat surfaces, modified edges, cuts and unnaturally curved forms. (Wood technicians from the Universidad Austral were brought to the site to compare the type of modification and form of these specimens with naturally occurring branches in different environmental contexts of the vicinity. These specialists were unable to determine any natural mechanisms which might account for the supposed alterations, and suggested that the wooden pieces had probably been collected and modified by human activity [Diaz-Vaz n.d.]).

The direct contextual association of cultural and ecofactual materials at Monte Verde binds together a minority of lithic, bone and wood specimens that are morphologically of unquestionable human production. Current knowledge of early lithic technology in the New World has provided little aid in the definition and classification of the Monte Verde stone collection. Moreover, the problem with the total material remains from the site is amplified by the lack of a type collection for analysis of the modified Monte Verde woods. Collectively, these problems necessitated a better understanding of the natural mechanisms which might compete with the human processes as explanations for alterations observed on lithic, bone and wood remains.

Human intervention in the direct contextual association of these materials and in the modification of many individual lithic, bone and wood specimens can be verified and explained. However, since a majority of the lithic and wood specimens are little modified from their natural state, it is difficult to define the vague boundary between some natural forms and cultural forms. Thus, in essence, an overriding concern is: "What is the defining limit of material culture, particularily in regard to the altered stone, at the site?" In an effort to resolve these issues, much of the field and laboratory work was oriented toward examination and definition of the composition and distribution of inorganic and organic materials in both the buried cultural context and in active and buried natural contexts. Stratigraphic "natural" contexts along Chinchihuapi Creek and along the buried ancient creek bed (outside the cultural area) were excavated to define the type and distribution of stones and organic matter and the natural processes affecting these materials. In addition, analysis of the distribution of various extant tree species, the form and location of different sizes of branches in these trees, and the natural processes determining physical condition and contextual displacement of fallen branches were performed for purposes of obtaining a natural "type" collection for comparison with the excavated wood assemblage. Replicative experiments are also being performed on selected stones and woods from "natural" contexts in an attempt to duplicate the morphological characteristics resulting from human production. These collections will be compared to both the excavated archaeological materials and to the excavated natural specimens. Ultimately, these analyses will aid our effort to define the cultural-natural boundary, particularily for the majority of stone specimens, and to better understand any competing natural and cultural factors.

Since these analyses are incomplete and would require a detailed presentation of multiple data sets here, only brief reference to specific tactics and findings will be mentioned in the following appropriate disussions as recourses to solving problems.

During the 1979 field season, an additional 95 m^2 of the site area was excavated in Areas A, B, and C. Test pits were also placed on the slopes and tops of surrounding ridges in an attempt to discover other activity areas. No new cultural areas were discovered, but the work in these other areas aided in defining the local geology.

In addition, a group of multidisciplinary specialists were integrated into the project to study the geological and ecological context of the Monte Verde area. Professor Mario Pino (Instituto de GeoCiencias, U.A. Ch.) conducted the geological investigations. Local modern-day biotic zones were analyzed by Professor Claudio Briones (Instituto de Botánica, U.A. Ch.). Pollen analysis was performed by Dr. John Heusser (Department of Biology, S.U.N.Y., Tuxedo). The recovered faunal materials were examined by Dr. Rodolfo Casamiquela (Centro de Investigaciones Cientifica, Viedma, Argentina). Dr. Juan Diaz-Vaz O. (Director of the

Instituto Tecnológico de Madera, U.A. Ch.) studied the preserved wood specimens and other paleo-floral material from the site. In 1980, Dr. Michael B. Collins (Department of Anthropology, University of Kentucky) joined the project as a lithic specialist, concentrating primarily on the analysis of use-wear patterns with Dillehay.

In January, 1981, an additional 12 m² of the site were excavated. A wishbone-shaped foundation of an architectural feature, a modified log, two hearths, and four fractured pebbles were found in an excavation block in Area A about 1 m north of the concentration of materials recovered from the same area in 1978.

GEOLOGY OF THE REGION AND THE SITE

South-central Chile is defined to the west by a narrow littoral and a low coastal mountain range and to the east by the high Andean mountain chain. The Central Valley of Chile, lying longitudinally between these two ranges, has been formed by a graben filled with lacustrine and fluvio-glacial alluvium. Several westerly flowing rivers descend the Andean slopes, cross the Central Valley and empty into the Pacific Ocean. Numerous volcanoes rise along the western slope of the Andes and numerous lakes dot both the mountainous and Central Valley landscape. Llanquihue Lake, drained by the Rio Maullin, is situated some 25 km north of Puerto Montt. Glaciers emanating from the Andes have left extensive, poorly-sorted till deposits and moraines in this part of the Central Valley. In places, these have been reworked by the Maullin and other rivers and their tributaries, but the lakes, moraines, bogs and other effects of glaciation dominate much of the Central Valley terrain. The site of Monte Verde is located on the bank of an ancient tributary emplaced on the Pleistocene terrace of the Rio Maullin.

Although Pino's geological study of the location is still in progress, his preliminary conclusions follow. Two geologic formations and eight strata are distinguished in the immediate area of the Monte Verde site (Figure 4). Strata MV-7 and MV-8 were surface strata of the Salto Chico Unit, the high terrace formed by the Rio Maullin drainage system during Late Pleistocene times. This unit is made up of large-grained sands, and igneous and metamorphic stones; the presence of matrices of sandy silt with some volcanic ash is also a common characteristic of the unit. The age of the Salto Chico Unit has been estimated to be no younger than 19,500 years (Pino n.d.). This Unit is overlain by the Monte Verde Formation (strata MV-1 to MV-6). Radiocarbon dates for several of these strata are discussed below.

Overlying a narrow strip of MV-7, is MV-6, a 4 m wide ancient creek bed deposit lying unconformably on the eroded surface of strata MV-7. The fluvial materials that comprise the ancient stream bed are not the product of primary transport or sedimentation processes of the Rio Maullin drainage system, but rather are the result of secondary local selection of the smaller sized particles (e.g., sand, smaller gravels, etc.) from the upper levels (MV-7 and MV-8) of the Salto Chico Unit. A minority of the pebbles in the MV-6 stratum are fractured. The fractured nature of these stones are accounted for by sedimentary processes that occurred during the formation of MV-7 and

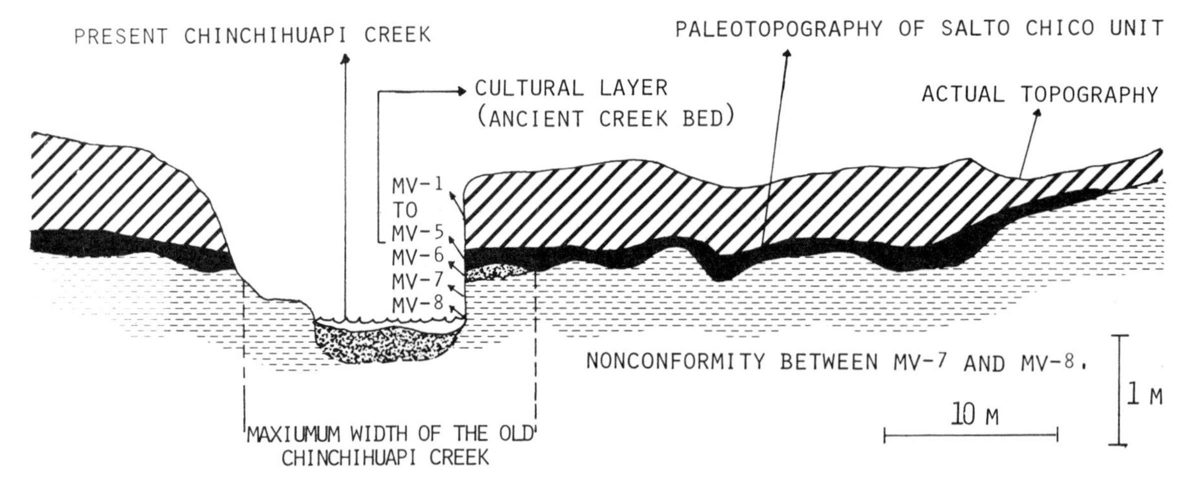

Figure 4. Stratigraphic section of the Monte Verde site, showing the location of the old and present creek channel.

MV-8 and not during the later development of MV-6 itself. Coarse gravels of granodiorites, trachytes, andesites and basalts and quartz-rich sands are the most common rock types in both the ancient and modern creek beds.

In the area of the site, the maximum extent of gravel fill of the ancient stream bed (MV-6) is 4 m wide and 3 to 20 cm thick. Along the edges of the ancient creek bed, including the area where part of the cultural debris is deposited in close association with other materials on the old surface of the creek bank (MV-7), the gravel lenses out to a scattered single layer of stones.

The present Chinchihuapi Creek has cut into and partially exposed the earlier, filled channel (MV-6) of the same creek. The cultural materials rest on the thin gravel fill of a gently inclined edge of the old channel and on the sandy banks, MV-7, of that channel (see Figure 14). Cultural materials recovered from test pits and adjacent sandy knolls (Area B) to the southeast were found in the buried surface of MV-7, while the dental remains found downstream (Area B) were lying on top of MV-6.

Strata MV-6 (creek bed) and MV-7 (sandy banks along the creek bed) are both superimposed by MV-5, a peat layer that encroached from the nearby bogs and covered and sealed the site. Above MV-5 are strata MV-1 through MV-4, are together about 85 cm thick and composed of small pebbles and sands.

Core drilling and test pits outside of the site area show that at the time stratum MV-6 was deposited the stream was flanked by low sandy knolls and small bogs, restricting access to the creek. Later the bogs, represented by stratum MV-5, expanded and covered the old channel, creating a peat deposit 1 to 25 cm thick peat deposit that sealed in the archaeological materials. All of the bone, lithic and wood materials are contained within a maximum 15 cm thick layer composed of the upper parts of strata MV-6 and MV-7 and the base of MV-5. (In some excavated units of Area A the organic artifacts are differentially preserved by the uneven thickness of the peat layer, which is explained by surface erosion in geological times.) The architectural feature, which lies on the surface of MV-7, adjacent to the cultural materials on top of MV-6, has a maximum height of 1.2 m and is unevenly covered by a heavily eroded MV-5. In some places the feature is overlain by stratum MV-4.

There has been no post-depositional disturbance (other than the erosion of MV-5) of strata MV-3 through MV-8, except that in recent times the modern stream has reincised a channel through them, exposing one edge of the ancient MV-6 creek bed that contains the cultural materials. Stream action as a possible agent for modification of stone, wood and bone has been negated by Pino (n.d.). Detailed geological analysis shows that the ancient creek banks of MV-6 were about 20 cm high and that the maximum depth of water passing through the site was about 15 cm. These data coupled with examination of the inclination of gravels in the stream and of sediments have determined that the ancient creek was a low-energy environment unlikely to have altered the stones, bone and wood. The low-energy flow of the creek is also evidenced in the excavated zone of area A where soft organic materials and culturally modified branches are in direct association with the lithics and bone remains. Any rapid creek flow after the cultural event took place would certainly have disturbed their context as well as the discrete artifactual clusters previously mentioned.

Alteration of materials due to frost fracture has also been dismissed since paleo-ecological studies (Hoganson and Ashworth 1982; Mercer 1972) have determined that the climate throughout at least the past 15,000 years or so has changed remarkably little. (The lowest annual temperature recorded in the area over the past 50 years has been -4° to -6° C.

Although Professors Briones, Pino and Diaz-Vaz's reconstruction of the paleo-environment is only beginning at this time, they have tentatively concluded that the cool and wet conditions similar to the present-day environment prevailed in the region some 12,000 to 16,000 years ago. Similar conditions for the same time span have also been postulated by the geologists Hoganson and Ashworth (1982), and Mercer (1972, 1982), who have worked independently in regions a few kilometers north of the Monte Verde site.

It should be noted, however, that palynologists (Heusser 1966, 1974; Heusser and Streeter 1980) have presented an alternative interpretation which views a gradual warming trend having occurred from about 13,000 to 11,300 years ago. Further collaborative studies on the floral, faunal and glacial geomorphological evidence from this area will hopefully resolve the conflicting interpretations of the late-glacial climatic conditions of southern Chile.

RADIOCARBON CHRONOLOGY

One sample of wood from the base of MV-4, four samples of wood and charcoal from MV-5, one bone sample from the top of MV-6, and a wood fragment from the modified log lying in close proximity to the architectural feature on the surface of MV-7, were submitted for radiocarbon analysis. Both the corrected and uncorrected dates are listed in Table 1 for samples from all strata.

As can be determined from Table 1 the sequence of the radiocarbon data is in congruence with the geological stratigraphy. The single cultural event at Monte Verde is estimated to date between 12,000 to 14,000 years ago. Two humanly altered objects from the cultural layers were dated. Samples on bone from stratum MV-7 place the cultural event between approximately 12,350 and 13,030 years ago. Previously it was reported that the Tx-3208 date of 13,965±250 was an older charcoal fragment deposited in the MV-5 peat layer (Dillehay et al. 1982:548). Subsequent geological work by Pino has shown that the gradual *in situ* development of MV-5 covered and sealed the cultural materials lying on top of MV-6. Thus, sample Tx-3208 originated in MV-6. The Tx-3760 date of 12,350±200 on bone may be younger than the cultural event. Additional radiocarbon dates on wood from MV-5 and on both bone and wood artifacts from MV-6 and MV-7 hopefully will resolve this problem. It may simply be a problem that different materials yield incompatible dates.

The geological evidence, as analyzed at the time of this writing, is compatible with the radiocarbon dates. The dating of these geological strata is in general agreement with Mercer (1972), Heusser's (1966, 1974) and Hoganson and Ashworth's (1982) radiocarbon dates for similar contexts in areas where they worked.

CULTURAL MATERIALS

Five categories of cultural materials have been excavated at Monte Verde. These include architecture, lithic, wood, bone and shell materials. Analyses of these materials are incomplete; thus, only preliminary comment on the findings are

Table 1. Radiocarbon Dates from Stratum MV-4, MV-5, MV-6 and MV-7 at the Monte Verde site.

Sample No.	Material	Total Carbon Counted	Counting Time	Age B.P., Half-life 5730
Non-Culture Bearing Deposits Lower Layer of MV-4				
Tx-4436	Wood	3.500 gm	2800 min	8270±130
Upper Layer of MV-5				
Tx-3207	Wood and Charcoal	2.409 gm	2700 min	11,155±130
Tx-3210	Wood	0.551 gm	5500 min	12,115±470
Lower Layer of MV-5				
Tx-3472	Wood	2.412 gm	3700 min	11,950±120
Culture Bearing Deposits Upper Layer of MV-6				
Tx-3760	Bone	1.682 gm	2700 min	12,350±200
Tx-3208	Charcoal	1.592 gm	5600 min	13,965±250
Upper Layer of MV-7				
Tx-4437	Wood	6.000 gm	2800 min	13,030±130

given here. I particularily will be brief with regard to the lithic assemblage since Collins and Dillehay's paper (this volume) specifically treats this category. With the exception of the biface exposed for surface collecting, only the excavated cultural materials from Area A will be treated here.

Distinct activity areas are also identified at Monte Verde, and although the analyses of these data are also incomplete, the spatial patterning of activity areas will be briefly demonstrated.

Architectural Artifacts

During the 1981 field season the foundation of a wishbone-shaped wooden structure was found in a matrix of compact gravels (from the nearby creek bed) and sand on the surface of MV-7, below stratum MV-5 (the peat layer). (The ancient creek bed, MV-6, is located about 3-5 m farther south.) The structure has a semi-rectangular platform protruding from the exterior side of its base (Figure 5). An entranceway opens to the southeast. Two hearths (as evidenced by the presence of burned clay and charcoal) were found on the floor of the structure. (Soil samples from these stains as well as from the foundation wall are being processed to determine soil type, chemical constituents and the identification of any preserved organic matter.) In addition, the remains of a 42 cm long "pole" was located inside the structure and centered in front of the inner face of the platform. The buried end of the vertical pole had been cut and pointed. The upper end was not preserved above the matrix of the overlying MV-5 peat layer. No modified lithics were recovered from the floor of the structure, although numerous "stains" were observed. Also, sequentially arranged vertical wooden "stubs" about 3 cm to 8 cm in length were recovered from the arms of the structure. These wooden stubs are interpreted as the remains of side walls made of branches. The compact sand and gravel base is the foundation support of these walls.

Positioned near the southwestern side of the structure was a 1.2 m long horizontal log which had been heavily modified by burning and scraping the exterior walls. The upper part and side walls of the log had been scraped to form a semi-retangular object (Figure 6A). The bottom side of the log had a 5 cm wide groove running down its full length. Midway down the groove was a bridge cut perpendicular to the groove (Figure 6B).

One end of the log was positioned in a heavily modified, V-shaped wooden implement, which possibly served as a "vise" to hold the log in place while it was being worked. The V-shaped, vise-like object measures about 90 cm in total length and 30 cm in width. Four wooden pegs were found placed in the sand against the east side of the log, presumably to further secure its position while being worked.

In combination these two features reveal a substantial labor commitment to the construction of a semi-permanent to permanent dwelling at the site. Although the function of the modified log is not completely understood at this time, its close proximity to the dwelling and its size and form suggests that it was an unfinished device probably related to architectural construction. Hopefully, further excavation in this area of the site will reveal more about the nature and function of these features and possibly others of that type.

Lithic artifacts

Most of the lithic collection from Monte Verde exhibits minimal evidence of cultural modification and use, thus creating an intriguing problem in

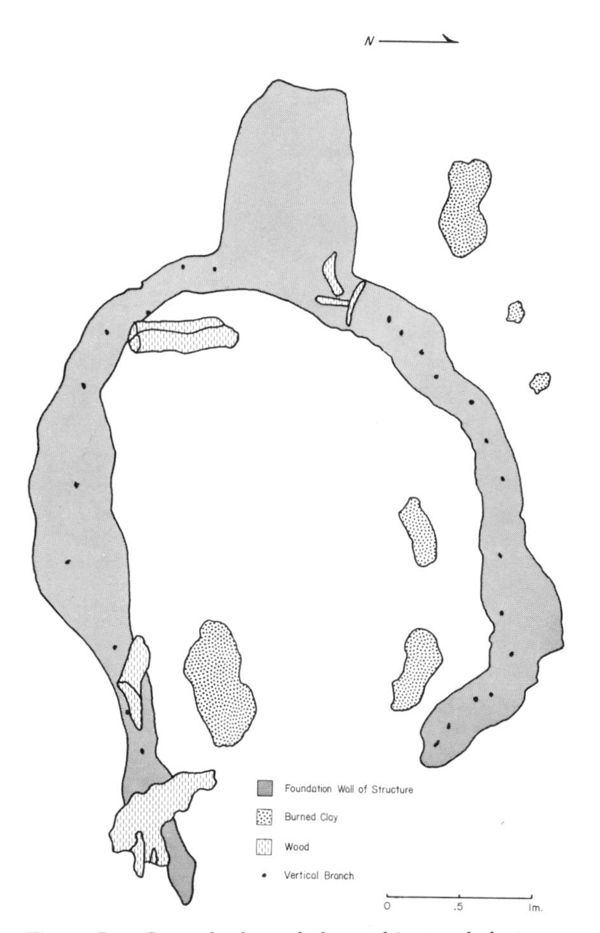

Figure 5. General plan of the architectural feature at Monte Verde site.

Figure 6. A. Modified log and associated "vice-like" implement *in situ* about 0.4 m south of the architectural feature. B. Underside of the log, showing the cut canal and perpendicularly cut "bridge."

defining the type and form of stone technology at the site. Although stones exist which can be morphologically identified as cultural implements, there many more pieces which to the naked eye do not show clear evidence of cultural use, but based on their form, edge-wear attributes and context, they appear to have been utilized by humans. This particular problem in regard to the lithics is discussed in greater detail in the paper presented by Collins and Dillehay; however, a few comments will be made here.

No bifacially flaked tools have been excavated at the site. We have attempted to recover micro-lithic debitage by passing the excavated portions of strata MV-5 and MV-6 through fine screens (1/6 in.). No chipping debris was discovered, suggesting that flaked tool types, including knives, scrapers or projectile points made from secondary and tertiary flakes were not manufactured. Of course, it is possible that such tool types were produced or retouched outside the main site area (A), for instance, on the sandy knoll to the southeast. If this were the case, however, we would expect that tools would have been utilized to process the plant materials and the hide and meat of the animals in Area A. The same reasoning applies to the absence of bifacially-flaked stone projectile points.

One notable artifact is a flaked bifacial chopper (Collins and Dillehay this volume: Figure 5c) made of foreign quartzite. This tool was found exposed at the edge of the modern creek bank. It is one of the most technologically "advanced" tools in the total lithic assemblage. Unfortunately, its surface context weakens its interpretation. The source material from which it is made is not found locally. Professor Pino, the project's geologist, has traced the nearest source of this type of quartzite to an area some 80 km to the north. Surface micro-cleavages in the quartzite contain both MV-6 sands and MV-5 peat, suggesting that the tool had eroded from the buried cultural site exposed in the MV-6 stratum of the modern creek bank.

The remaining lithic artifacts at Monte Verde belong to an industry which incorporates only minimal modification of the raw material. It is the process of *selection* of the natural form (either for immediate use or for minimal alteration prior to use), rather than bifacial manufacture which is the key to understanding the lithic artifact assemblage. The selection is believed to have been carried out at two levels; the first is the collection of specific natural forms which are later used, and the second is the selection of specific attributes of those forms either for direct use, or through some degree of manufactured modification. Thus, one of the hypotheses that we are testing is that the individuals who utilized the site of Monte Verde programmatically selected certain forms and types of rock for utilization as tools. The null hypothesis is that there is a random distribution of stone forms external to the cultural site area. The hypotheses are being tested in part by statistically comparing the frequency and distribution of the lithic forms and types between the area identified as the site (Areas A, B and C) and the naturally occurring gravels which have already been excavated and analyzed.

However, only culturally modified stones were recovered from the MV-7 stratum. The excavated MV-6 stratum of the Monte Verde site contained both altered artifacts and naturally occurring lithics. Because there is a gradation in the degree of alteration of original form between shaped tools and the minimally altered implements, it was necessary to construct an analytical methodology which would allow one to distinguish between the cultural artifacts and the "ecofacts." Criteria for making this assessment incorporate form, size, material type, patterned edge alterations, and context. These crtieria are being compared with the same set of characteristics for stones occurring outside the cultural context of Monte Verde.

Focusing on this problem, we have employed the following technique as one method to solve the problem of obtaining a sample of natural stones. Prior to continued excavation in the cultural areas of the site during the 1979 field season we excavated 12 one m^2 control units into the buried stratum MV-6 in distant "natural" areas. Each excavated stone in these non-cultural areas of excavation was counted. All fractured stones were catalogued and classified according to size, form, weight and type of stone. In total, 128,457 stones from the natural areas were counted, and of these some 1842 fractured ones were classifed and analyzed in terms of the above-mentioned characteristics. (For comparative purposes the same technique was applied to all stones recovered from excavation in the 42 one m^2 control units in the cultural Area (A). In total, 3834 stones from the cultural area were examined and 843 of these were fractured). Attribute analysis of these naturalstones is being performed for comparison with stones recovered from the excavated cultural areas.

Our preliminary analysis of these materials shows that about 18 different forms of fractured stones (Sanzana 1980) are available in the natural context of the ancient stratum MV-6. These 18 forms occur randomly in the natural areas, but tend to be clustered in specific locales of the cultural area where concentrations of wood and bone are found. Furthermore, 11 particular natural forms were found to be concentrated in and around the discrete concentrations of bone and wood materials. Statistical analysis of both the "natural" collection and the "cultural" assemblage will aid in our interpretation of cultural form and function.

During the 1979 field season, experiments on lithic replication were also performed on stones from the natural area in an attempt to establish associations between the edge characteristics of the different forms of naturally-occurring pebbles fractured when applied to different types of wood and to ox and horse bones. Pressure and angle of application, length of flake, motion and number of scraping and cutting strokes applied on experimental bone, wood and other stone material will be quantified to detect any correlations between the various forms of fractured pebbles and their edge-wear attributes, and specific tasks and material type. Data recovered from these experiments also will aid analysis of the lithic use wear described in more detail by Collins and Dillehay.

Wooden Artifacts

The wooden artifacts include two hafts which were mounted on fractured pebbles made of quartzite (Figure 7; also see Collins and Dillehay this volume, Figure 7), two modified branches with full-length grooves and associated sub-triangular gouge-like stones (Figure 8) and about 85 pieces of wood with cuts, grooves, modified edges and abrasions.

In addition, four branches of wood show worked and utilized areas. Each of these specimens exhibit cuts and utilized tips with similar edge angles (38 degrees to 45 degrees) and thinned, curved "handles" (Figure 9). These four artifacts were found in close association with the bone remains of the torso region of three individual mastodons. Three of the specimens averaged about 40 cm in length. The largest branch is about 1.3 m long.

Another artifact is of special interest here. A growth ring section of a luma branch had been cut into a lanceolate-like form with a concave base and ventral side. The distal point was diagonally cut to one side and had been burned (Figure 10). The specimen is 0.8 cm thick and 7.7 cm long.

Dr. Juan Diaz-Vaz (n.d.) has analyzed the cellular structure of most of the modified pieces of wood and has determined that these alterations were not caused by natural processes (such as stones being pressed or mashed into the wood, thus creating depressions), but that the modified sections of wood specimens definitely had been "cut" or "scraped."

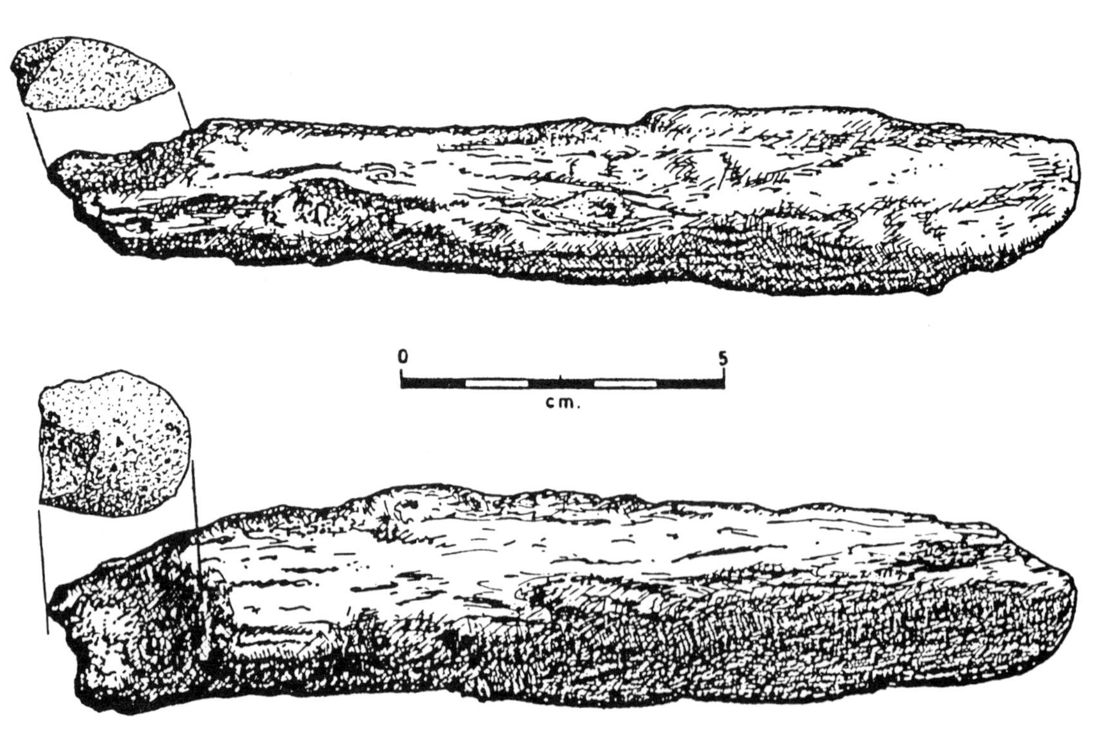

Figure 7. Wooden handle with fractured pebble tool mounted on one end.

Figure 8. Modified end of a tree branch. Arrows indicate cut surfaces and scars.

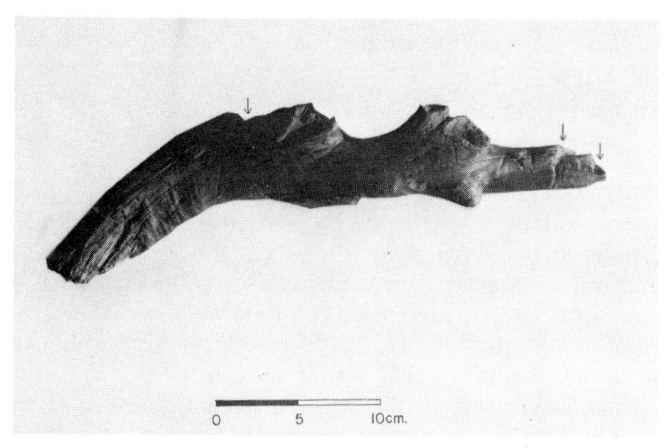

Figure 9. Modified end of a tree branch. Arrows indicate cut surfaces and scars.

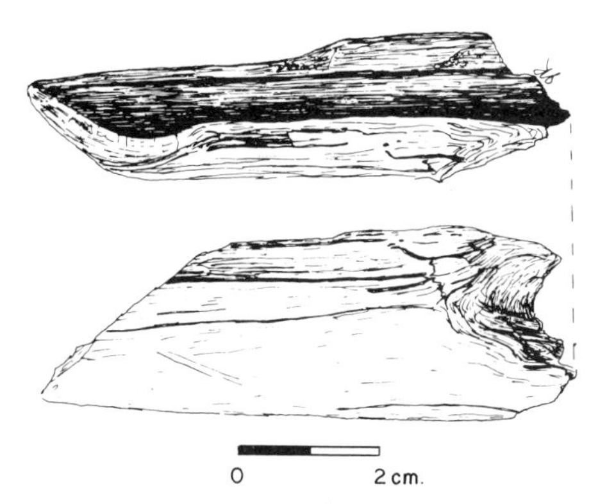

Figure 10. Lanceolate shaped wood fragment with a concave base and interior side.

Eight burned pieces of a plant material identified as juncu reed (*Juncus bufonius*, Briones n.d.) were concentrated within a m² area. These specimens measure 4.5 to 5 cm in length. Three of them exhibit conical-shaped tips.

Wood replicative experiments are also being performed on various fresh and dry branches of different tree species to determine the nature and type of cellular and structural modifications as a result of human intervention. In addition, the attribute studies are also being conducted on naturally occurring branches on land and in the creek. Collectively, these comparative data will provide detailed information on the various natural and cultural stages that can alter the morphological characteristics of wood specimens and thus help to define more precisely the nature and type of the wood technology employed by the inhabitants of the Monte Verde site.

Other Floral Materials

A varied array of plant specimens, including the wooden artifacts, were preserved by the overlying peat. The floral materials include leaves and branches of trees, seeds of fruit, flowers and pods, and pollen. Dr. Carlos Ramirez, Director of the Universidad Austral de Chile, has conducted a preliminary analysis of these materials. Most of the species have been identified and it has been determined that the paleo-environment at the time of the cultural event was similar to that of the present day; that is, a humid, sub-antarctic forest and cool, wet climatic conditions.

The most common plant remains recovered at Monte Verde are branches of the luma (*Amomyrtus luma*) tree and large quantities of fruits (*Aristotelia chilensis, Rubus contrictus,* and *Berberis busifolia*) and other miscellaneous shrubs (Briones n.d.).

The data currently at hand also show a distinction between the types of plants found in direct association with the cultural materials from the upper layer of strata MV-6 and MV-7 and those remains recovered from strata MV-6 and MV-5 peat in the "natural" areas. A more diversified and larger quantity of edible plants were found in the cultural area. These plants are also spatially patterned in that they are clustered in the activity areas of the site.

Osteological Remains

Mastodon bones of the genus *Cuvieronius* (Casamiquela n.d.) constitute all of the bone sample. The sample includes 112 surface collected bone elements, which had eroded out of the site, and another 54 excavated bone specimens. Based on analysis of both the surface collected and excavated molars from the site, Dr. Casamiquela has estimated that at least five to six individual animals are represented in the bone collection, including two young, one sub-adult and two old animals. (His estimation is based on the minimal number of bone elements representative of an individual.)

Several excavated bone elements exhibit modifications. One osteological element which has been identified as a cultural element is an ilium which has a modified branch "standing" in its center (Figure 11). The bone was recovered from the top of MV-7 and the base of MV-6 about 10 cm below the level which contained other cultural materials. However, the vertically arranged wood protruded up through the strata and thus was located on the same level as the cultural activity area. The wood was attached to the bone with bitumen. The bone exhibits numerous shallow, narrow grooves (ca. 3 mm wide) on the surface which holds the wood.

Figure 11. Vertically placed branch attached to ilium with bitumen.

Four other bone elements (2 femora, 1 skull fragment and 1 long bone fragment) display narrow grooves measuring 1.4 to 3.1 cm in length. Casamiquela has determined that these grooves were produced by human activity. He also noted that the tip of one long bone fragment from Area A exhibits several micro-cuts and depressions similar to modifications made on bone flakers recovered from the lower levels of the Tagua-Tagua site in central Chile (Montané 1968). One other unidentified bone element had been modified to form a sub-triangular object with weak shoulders and a narrow contracting stem (Figure 12).

In addition, two tusk fragments, exhibiting heavy polish and deeply cut, parallel microstriations on one end, were excavated during the 1981 season (Figure 13).

The partial skeletons of individual mastodons recovered from Area A are disarticulated and 95% of the bones are fractured. Only the limb, torso and head (skull, teeth and tusk fragments) portions of the animals are represented in the collection. The disturbed nature of these skeletons can be most easily explained by butchering and food processing techniques, although the possibility of natural processes accounting for some fracturing has not been ruled out. Collectively, the nature and patterning of the osteological remains and the absence of any typical lethal weaponry (e.g. projectile points) used in killing the animals (see Collins and Dillehay this volume) suggest that the location of

Figure 12. Modifed bone fragment with a pointed tip and contracting "stem."

Figure 13. Tusk fragment of mastodon, showing modifications along one edge. Brackets indicate heavy polish and diagonal striations.

the kill and initial butchering and selection of bone for later removal of marrow and for a source of raw material for tools is probably situated either outside the main site area or in a distant place.

Eight fractured mollusk shells, representing six different individuals, were also excavated at the site. These shells were found "packed" in clay (the source of which has not been located along Chinchihuapi Creek) inside one of the mastodon femur bones. As yet, biologists at the Universidad Austral de Chile have not been able to identify the species. The fractured nature and location of these shells at the site suggest that mollusks were also part of the diet of the Monte Verde people.

Activity Areas

Archaeological materials recovered from Areas A, B and C indicate the presence of different activity areas along a 400 m stretch of Chinchihuapi Creek. As all the cultural debris from these areas are contained in the upper part of MV-6 or on the surface of MV-7 and the artifactual assemblages from these areas are similar in form and type, the assumption can be made that these different activities are culturally related. Further archaeological work may reveal more concrete connections between Areas A, B and C as well as the presence of additional localities up and down the creek.

The spatial configuration of materials recovered from Area A takes on a partial semi-circular shape (Figure 14). Further excavation along the north side of this feature may reveal a more complete circular formation. The semi-circle is comprised of the architectural structure and four spatially distinct, but slightly overlapping, clusters of wood, bone and lithic materials. These clusters are tentatively interpreted as distinct work zones where tool manufacturing, food processing, and other tasks were performed. This configuration imparts a sense of internal planning of activities along the creek bank. (I might add that this semi-circular layout does not conform to any natural topographic formations along the ancient creek bank. The excavations have shown that the old creek bank is slightly sloped and free of any depressions or lateral drainages.)

Detailed presentation of the type, frequency and distribution of artifacts that compose each of

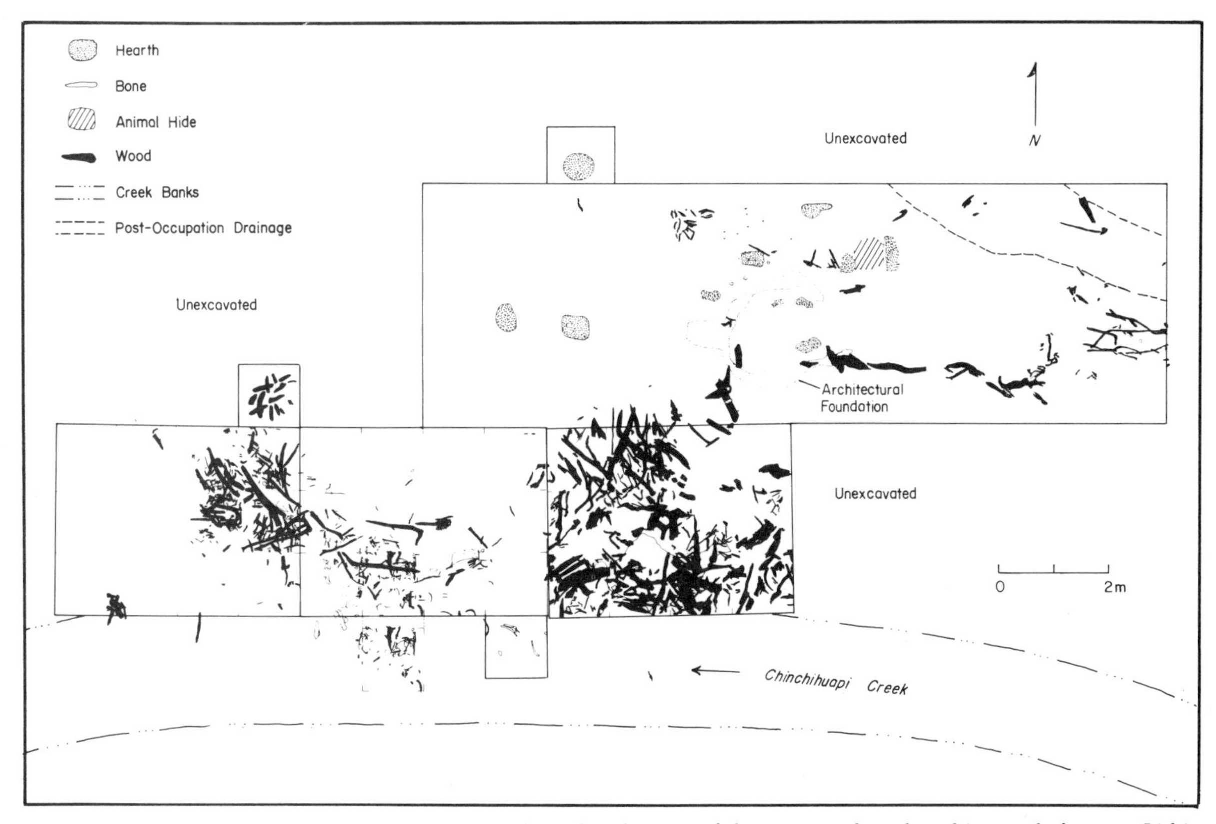

Figure 14. Area A excavation showing non-random distribution of bones, wood and architectural feature. Lithic materials are not shown here. Note the distinct clusters of materials forming a semi-circle. Areas outside of the block enclosures have not been excavated.

these clusters is beyond the scope of this brief report. Nevertheless, it is important to point out that these clusters are relatively homogeneous in their artifact content and type and in their spatial size and arrangement. Their pattern tentatively suggests that several similar, but spatially discrete tasks were being performed simultaneously in Area A, instead of different activities conducted sequentially. Once the analysis of these activity areas are complete and the remaining unexcavated portion of Area A is worked, a better understanding of the organization of labor and possibly the group size will be gained.

It needs to be noted that Figure 14 does not include the lithic assemblage as part of the artifact clusters. Since a small portion of the stone collection recovered from Area A has not been completely analyzed, and thus not definitely determined to be modified either by human agency or natural processes, this category is excluded from the figure. However, I will point out that neither the clearly identified cultural stones nor the suspected cultural stones are distributed randomly in Area A; rather, all are concentrated in and around the discrete clusters of bone and wood.

The type and structure of activity loci in Areas B and C are even less understood because excavation in these areas has been limited. The bone remains from stratum MV-6 in Area B have not yet produced any definitive evidence of human agency as the cause of deposition. The hearth and modified bones recovered from test pits in Area C were lying on the buried surface of MV-7, which is a 3 m high ridge on the south side of the ancient creek. The limited data from this area suggest that the ridge might have been utilized as another occupational loci at the site. No wooden or bone materials have been as yet recovered in Area C. Since the crest of this buried ridge was higher than the peat fill (MV-5), it is likely that these materials were not preserved in the area. Further work should elucidate the cultural relationship between Areas A, B and C, as well as other possible activity loci which, as yet, have not been identified.

Lastly, the current archaeological and geological evidence from the site reveals that a single cultural episode took place. However, it is not completely understood whether the site was continuously occupied over a prolonged period of time or whether it was visited by a small group of people over a short period of time. The selected and disarticulated nature of the bone collection, the presence of the architectural structure, and the tight spatial arrangement of the relatively homogenous activity areas in Area A imparts a sense of internal planning and of a lengthy and probably continuous stay at the site. It is also likely that the selective nature of the bone collection represents occasionsl distant kills or scavenging episodes. The meat obtained from several mastodons coupled with plant foraging in the ecologically rich forest would certainly have supplied an adequate food base for a prolonged occupation. On the other hand, Areas A, B and C may suggest that a small group may have repeatedly exercised similar economic and social activities. Whichever the case may be, the activity structure and the artifact assemblage show the same cultural group occupied the site.

CONCLUSION

Monte Verde is an archaeological site that provides significant cultural and ecological data to enable us to reconstruct and explain not only the paleoenvironmental conditions of south-central Chile some 12,000 to 14,000 years ago, but also how early people were adapted to them. Although not discussed in detail above, the data suggest that the minimally modified stone and wood industry reflect a diversified economy. The activity structure reflected in the composition and configuration of the cultural materials indicates that the site was a locus of intensive and extensive occupational activity.

Research conducted at Monte Verde has provided a unique opportunity to reveal, and hopefully to understand, the interplay between natural and cultural factors in the alteration of selected elements in one kind of paleoenvironment, as well as those aspects of nature that preserved the cultural by-products of human behavior. Further work at Monte Verde, and the final analysis and interpretation of the recovered cultural materials will undoubtedly amplify and change our thinking not only on the adaptation of Early Man to his environment, but also on the early selection of natural forms of stones and woods, which with little or no human modification, exhibit attributes of cultural use. Moreover, the implications of the type and form of activities performed at the site are very important for consideration of later socio-economic developments that took place in the Andes.

ADDENDUM

As a postscript to this paper, a brief comment on findings of the 1983 field season is necessary. Extensive excavation of areas around the architectural foundation recovered in Area A during the 1981 fieldwork revealed a wooden branch-lined plaza fronting the entranceway of the

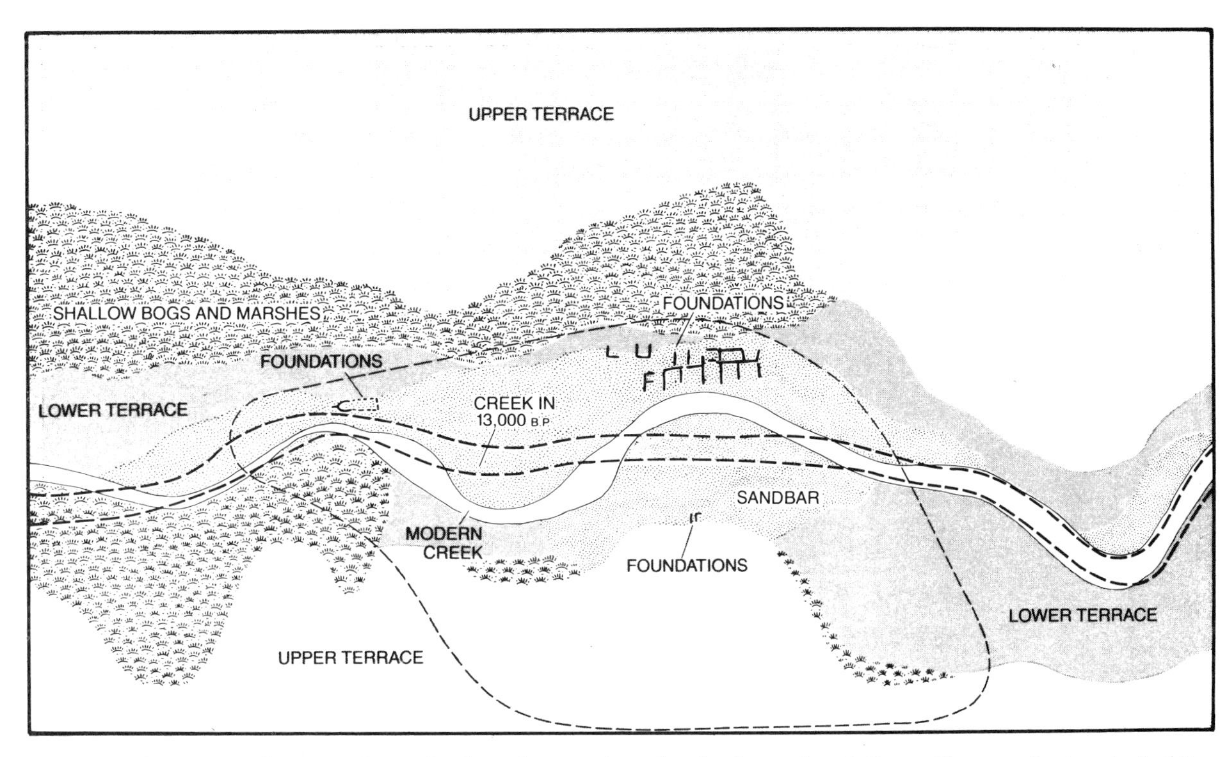

Figure 15. Reconstructed topography at the time of occupation, showing house foundations with approximate extent of the village encircled. The original creek bed is shown by parallel lines, and the modern creek by solid lines. (This figure is reproduced from "A Late Ice-Age Settlement in Southern Chile" by T.D. Dillehay. Copyright © 1984 by Scientific American, Inc.).

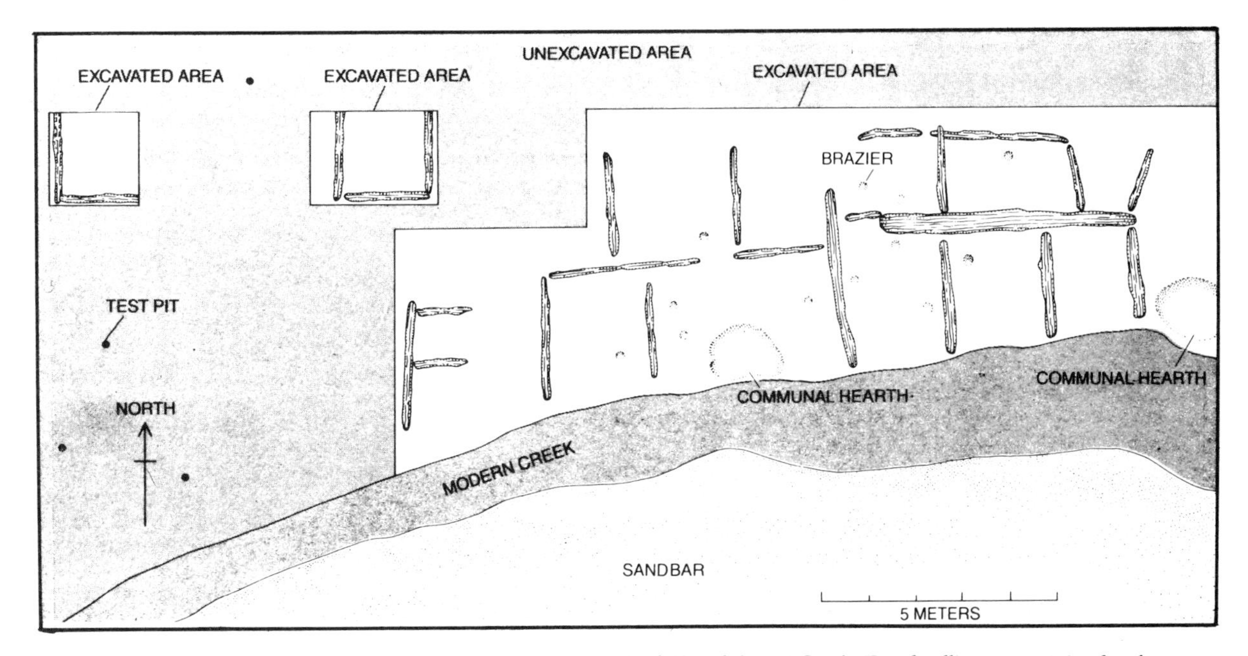

Figure 16. Excavated area exposing 12 dwellings on north bank of Chinchihuapi Creek. Ten dwellings were joined to form two rows parallel to the creek. Shallow clay-lined braziers are situated on the floors of several huts, and two large hearths are located adjacent to the modern creek. (This figure is reproduced from "A Late Ice-Age Settlement in Southern Chile" by T.D. Dillehay. Copyright © 1984 by Scientific American, Inc.).

structure. The plaza measured about 3 by 5 m and was associated with several hearths containing carbonized seeds and other plant remains. In addition, a separate 90 m[2] were excavated in Area D about 30 m east of Area A. This work recorded ten additional architectural foundations which were rectangular in form and agglutinated in a linear layout along the creek bank. Each of these foundations were characterized by wooden planks and logs that were held in position by wooden stakes. Resting on the occupational floors of these structural units were small, clay-lined hearths, wooden artifacts and lithic (including a few flaked stone tools) scatters. Several larger hearths and lithic and bone concentrations were found near the entranceway of these units (Dillehay 1984).

Of particular interest were three wooden mortars which contained a well-preserved assortment of floral remains, including seeds, fruits and stalks of various edible plants. Several grinding stones made of local basalt and quartzite pebbles were located in direct association with these mortars.

Additional test pits placed systematically across the site yielded a rich array of artifactual and ecofactual data which not only linked the various areas of the site but also provided new information on the overall size and complexity of the settlement. For instance, several test pits revealed the remains of other architectural features, showing that other discrete occupational zones remain intact at the site.

All of the above mentioned materials were recovered from the buried surface of the creek bank (MV-7) of the ancient creek (MV-6) and were preserved by the same peat layer (MV-5) that covered the previously excavated residue.

Besides these general findings, a wealth of multidisciplinary information was generated by specialists working primarily in ethnobotany, ecology, geology and experimental archaeology. A thorough study of the extant edible plant types and their distribution and seasonality in the surrounding indigenous forests and bogs will provide a comprehensive study of the year-round subsistence base, and a good comparative collection for the culturally associated floral remains. The other studies resolved a number of previously unanswered questions on local geology, environment and the technology of resource procurement.

Once the final work at the site and the subsequent analyses of all recovered materials are completed, we will learn that the sub-antarctic forest of south-central Chile contained sufficient essential resources for a Late Pleistocene human population with a mixed plant collecting and hunting economy to develop a semi-permanent to permanent and planned settlement.

POSTSCRIPT

Since this paper was written, additional data have come to light at Monte Verde, as yet unanalyzed and too recently discovered to add to the presentation above. A research strategy of the 1983 field season at the site was the systematic placement of test pits on the gently sloping terraces adjacent to the main occupational area. The results of this work were the recovery of five stone artifacts in direct association with light scatters of carbonized wood and charcoal. These materials were located at the base of MV-7 about 2 m below the cultural evidence described in the foregoing paper.

The lithic collection consists of three edge-battered stones, one percussion-split pebble, and one flake. Since the only gravel bearing deposit of the terrace is the younger MV-6 creek bed, it can be determined that the stone artifacts, which are made of basalt or andesite, were brought to the site probably from the Rio Maullin drainage. Two radiocarbon dates on wood and charcoal from the base of the MV-7 stratum at the site and at geological locations in the region have been determined at 33,370±530 (Beta-6754) and greater than 33,020 (Beta-7825). Additional archaeological and geological research will be performed in order to gain a fuller and more precise understanding of the nature and context of these materials and their cultural meaning.

ACKNOWLEDGEMENTS

The research described was supported by the Universidad Austral de Chile, Valdivia; the University of Kentucky, Lexington, Kentucky; the National Science Foundation; the National Geographic Society; and by the University Research Institute of the University of Texas, Austin, Texas for a grant to Dr. E. Mott Davis, who underwrote the radiometric measurements. I would like to particularly thank Professor Mauricio Van de Maele (Director, Museo de Historica y Antropológia, Universidad Austral de Chile, Valdivia), who during my period of tenure at that university, gave me his support and encouragement. I am also grateful to many Chilean colleagues for sharing their knowledge of Chilean prehistory with me. I am deeply indebted to my numerous Chilean friends and students who dedicated their time and talents to the successive field seasons at the site. These students were patient, bold fieldworkers who often had to work under very harsh conditions at the site. This paper is dedicated to the late Mr. Ivan von Leifner, a Hungarian-Chilean, who diligently worked with enthusiasm on the conservation of the wooden artifacts. Lastly, I thank Michael B. Collins, William Brown,

Lawrence Keeley, Eric Gibson, Richard Boisvert, Susan Graham, and David Pollack for their assistance in the analysis of cultural materials, and to my Andean colleagues, who, on numerous occasions have discussed the site with me, for their support. I am responsible, however, for the presentation and interpretation of the data, as well as for the inadequacies in this study.

REFERENCES CITED

Briones, C.
n.d. La vegetación de Monte Verde, Chile. Manuscript in possession of author.

Casamiquela, R.
n.d. Informe preliminar sobre los restos vertebratos de Monte Verde, Chile. Manuscript in possession of author.

Diaz-Vaz, J.
n.d. Identificación de Madera Pleistocénica de Monte Verde, Chile. Manuscript in possession of the author.

Dillehay, T.D.
1984 A late Ice-Age settlement in southern Chile. *Scientific American* 251(4): 106-117.

Dillehay, T.D. M. Pino Q., E.M. Davis, S. Valastro, Jr., A.G. Varela, and R. Casimiquela
1983 Monte Verde: radiocarbon dates from an Early Man site in south-central Chile. *Journal of Field Archaeology* 9:547-549.

Heusser, C.J.
1966 Late Pleistocene pollen diagrams from southern Chile. Manuscript in possession of the author.

1974 Vegetation and climate of the southern Chilean Lake District during and since the last interglaciation. *Quaternary Research* 4:290-315.

Heusser, C.J., and S.S. Streeter
1980 A temperature and precipitation record of the past 16,000 years in southern Chile. *Science* 210:1345-1347.

Hoganson, J.W., and A.C. Ashworth
1982 The Late Glacial climate of the Chilean lake region implied by fossil beetles. *Proceedings of the Third North American Paleontological Convention*, Vol. 1:251-256.

Mercer, J.H.
1972 Chilean glacial chronology 20,000 to 11,000 carbon-14 years ago: some global comparisons. *Science* 176:1118-1120.

1982 Holocene glacier variations in southern South America. In *Holocene Glaciers*, edited by W. Karlen, *Striae*. 18:35-40. Uppsala.

Montané, J.C.
1968 Paleo-Indian remains from Laguna de Tagua-Tagua, central Chile. *Science* 116:1137-1138.

Pino, Q.M.
n.d. La Geología de Monte Verde, Chile. Manuscript in possession of author.

Sanzana, P.
1980 Analisis Geo arqueologico en gravas fracturadas de Monte Verde I, X Region, Chile. Tesis de Licenciatura en Filosofía y Ciencas Sociales. Universidad Austral de Chile, Valdivia.

The Implications of the Lithic Assemblage from Monte Verde for Early Man Studies

MICHAEL B. COLLINS and TOM D. DILLEHAY
Department of Anthropology
University of Kentucky
Lexington, Kentucky 40506
U.S.A.

Abstract

The 12,000-14,000 year old deposits at Monte Verde, Chile, yielded 207 small stone objects in association with mastodon bones, features, and artifacts of bone and wood. A majority of these stones are little modified from their natural states. They are associated with a biface, a core, and wedge-shaped quartz pieces produced by direct percussion flaking. All of the pieces in the collection consist of igneous and metamorphic stones, with the majority being coarse-grained. Seven of the 207 specimens appear petrologically to be exotic, whereas the rest evidently came from gravel exposures at or near the site. Nearly spherical forms make up over a third of the collection, and include two intentionally grooved bolas stones. The remainder have one or more acute edges which, by experimental replication, are shown to be effective cutting and scraping tools; a significant number of these manifest nicking and microscopic polish and striations interpretable as use wear along their edges. This assemblage is interpretated as the product of an early, distinctive human technology characterized primarily by the selecting and using of naturally occurring stones, or minimally modifying such stones for use.

The paradigm of Early Man lithics analysis must be expanded conceptually as well as procedurally to deal with material culture that is little modified from its natural state.

INTRODUCTION

As Dillehay has reported in the preceding paper, the Monte Verde site offers a wealth of information destined to alter a number of our views on Early Man in the Americas, and the lithics are very much a part of this rethinking. In the following presentation the collection of 207 stones is described, the interpretive issues raised by this assemblage are discussed, and tentative interpretations are offered. Further work with this collection, and hopefully the documentation of other comparable lithic assemblages, will undoubtedly amplify and change the present interpretations.

This study of the lithics from Monte Verde cannot be conclusive for a number of reasons. The interpretative problems are unusual and complex. Other collections of comparable materials have not been documented. The extant paradigm of Early Man lithic technology in the New World provides very little guidance for the interpretation of such an assemblage. The microscopic examination of specimens for traces of use-wear is incomplete. Also, the petrology of the sundry rock outcrops and complex extensive gravel deposits in the region is neither readily available nor sufficiently detailed for unequivocal source-area identifications.

It must be emphasized that these limitations reside primarily in the interpretive milieu apart from the collection itself. The lithic specimens, except for one biface, were excavated from a buried deposit which not only has been reliably dated but contained features and artifacts of bone, wood, and other plant materials as well (Dillehay, this volume). Furthermore, the patterns in the lithic assemblage are clearly and strongly expressed. It is the meaning and significance of these patterns as well as the search for additional patterns that will benefit from advances in the efficacy of research. The quest to be initiated as a result of this study is not so much for an improved data base as for more powerful research procedures.

INTERPRETIVE ISSUES

To anticipate the major conclusion of this analysis, the crux of the interpretive problem presented by the lithic assemblage from Monte Verde is that the prevalent technology used by the occupants of Monte Verde seems to have been selecting and using naturally occurring stones or minimally modifying such stones for use (see Dillehay, this volume, for fuller discussion of this point). Thus, the primary task of distinguishing the three categories of unutilized natural stones, utilized natural stones, and minimally modified stones is sufficiently difficult that a definitive identification is probably impossible for much of this collection. This is because the site is on an old terrace of the Rio Maullin, a terrace composed of poorly sorted fluvioglacial sands, gravels, and cobbles of igneous and metamorphic materials. Since nearly all of the specimens from the site are pebble to small cobble-sized pieces of igneous and metamorphic rock, the cultural/natural boundary is indistinct.

Fortunately, there are at least three lithic specimens morphologically of unquestionable human production (a grooved stone, a biface, and a core) and at least six others which exhibit forms (a chopper, another grooved stone, a split cobble, and three edge-battered stones) much more easily explained as the result of human rather than natural processes. In addition, a small split pebble, itself not obviously an artifact, was found hafted to a wooden handle. These ten objects and some 16 more stones which are probably unnatural occurrences, all in buried association with the features, wooden artifacts, mastodon bones, and other evidences of human activity at the site, are the fortunate circumstance which brings attention to a lithic assemblage that would undoubtedly be overlooked under less favorable conditions. The

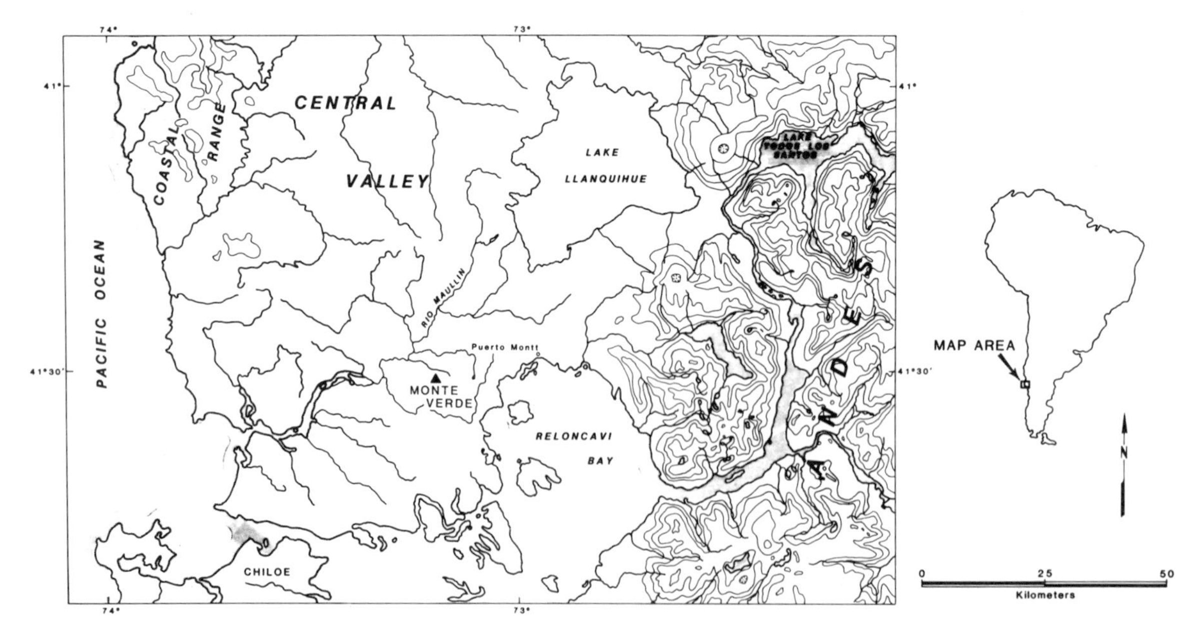

Figure 1. Location and geographic setting of Monte Verde, Chile.

conditions, however, are not entirely favorable. Had these stones been found more securely removed from their geologic source, perhaps in a limestone cave or in an aeolian dune, human agency would be more readily apparent. Instead, almost all of the specimens in question were found contiguous with, or in close proximity to, the geologic deposits from which they probably were derived.

Superficially, many of the Monte Verde specimens appear to have little functional value, They are small and have dull edges compared to those of, for example, chert flakes. These first appearances are faulty, as the following discussion will show. Finally, the search for vestiges of use wear on coarse crystalline rocks is a research domain in need of major improvements. Edge and surface characteristics of certain Monte Verde specimens are apparently the result of use; however, more work is needed to verify this interpretation.

GEOGRAPHIC AND GEOLOGIC SETTING OF MONTE VERDE

The southern Chilean landscape near Monte Verde consists of a narrow Pacific littoral and a low coastal mountain range (of metamorphosed Precambriam and Paleozoic rocks) some 30 km wide on the west, the Andes (volcanic and intrusive rocks) on the east, and in between a central longitudinal valley (a graben filled with lacustrine and fluvio-glacial alluvium) some 75 km wide. Several westerly flowing rivers rising in the Andes cross the Central Valley and breach the coastal range (Figure 1). A chain of volcanoes extends along the western flanks of the Andes, and a series of lakes dots the eastern margin of the Central Valley at the foot of the Andean slope. Llanquihue Lake is the southernmost of the large lakes in this series, and it drains to the west-southwest via the Rio Maullin. Glaciers emanating from the Andes have left poorly-sorted till deposits and moraines in this part of the Central Valley. In places, these have been reworked by the Maullin and other rivers and their tributaries; but the lakes, moraines, bogs, and other effects of glaciation dominate much of the Central Valley terrain. The site of Monte Verde is on a Pleistocene terrace of the Rio Maullin. Cut at this point by Chinchihuapi Creek, the terrace consists of an underlying gravel and sand formation (the Salto Chico) estimated to be no younger than 19,500 years (Pino n.d.) and an overlying Monte Verde Formation, in part radiocarbon dated to about 12,000 to 14,000 years ago.

The gravels of the Salto Chico and Monte Verde Formations are made up of igneous and metamorphic stones. At the site, the present Chinchihuapi Creek has cut into and partially exposed an earlier, filled channel of the same creek (Figure 2). Except for low, sandy ridges in the immediate vicinity of Monte Verde, boggy forested land bordered the creek until recently, as it evidently did when the earlier channel flowed. The cultural materials at Monte Verde rested on the gravel fill of the old channel of Chinchihuapi Creek and on the sandy banks of the channel. Peat encroaching from the nearby bogs covered and sealed the site.

Essential to this lithic analysis is an understanding of the nature and distribution of the naturally occurring stones at and near the site of Monte Verde. The geology of the Llanquihue region is known only in a general way (Figure 3). The fluvio-glacial fill of the Central Valley west of Puerto Montt consists primarily of coarse gravels of diorites, granodiorites, trachytes, andesites, basalts, and quartz-rich sands transported from Andean elevations to the east. Petrologically, these Andean source-areas are dominated by Cretaceous granites and diorites and Quaternary andesites and basalts with significant amounts of metamorphic rocks as

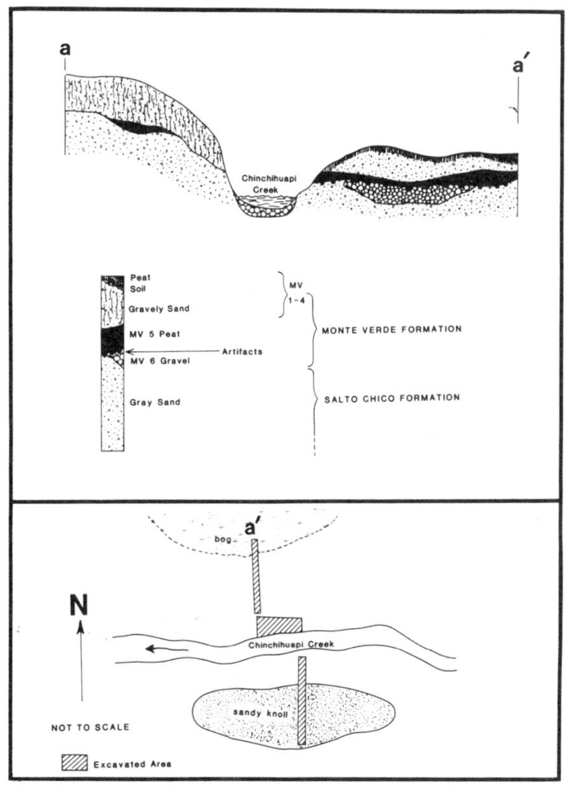

Figure 2. Simplified geologic section through Chinchihuapi Creek and the site of Monte Verde.

Table 1. Inferred source areas for lithics in the Monte Verde site.

Artifact Category	Probable Source Area			
	Maullín Gravels	Coastal Range	Unknown	Totals
1. Biface		1		1
2. Chopper			1	1
3. Core			1	1
4. Grooved stones	2			2
5. Spherical stones	14			14
6. Subspherical	22			22
7. Edge Battered	3			3
8. Single-faceted	86			86
9. Multi-faceted	27	2	1	30
10. Miscellaneous	1		1	2
11. Other	45			45
Totals	200	3	4	207

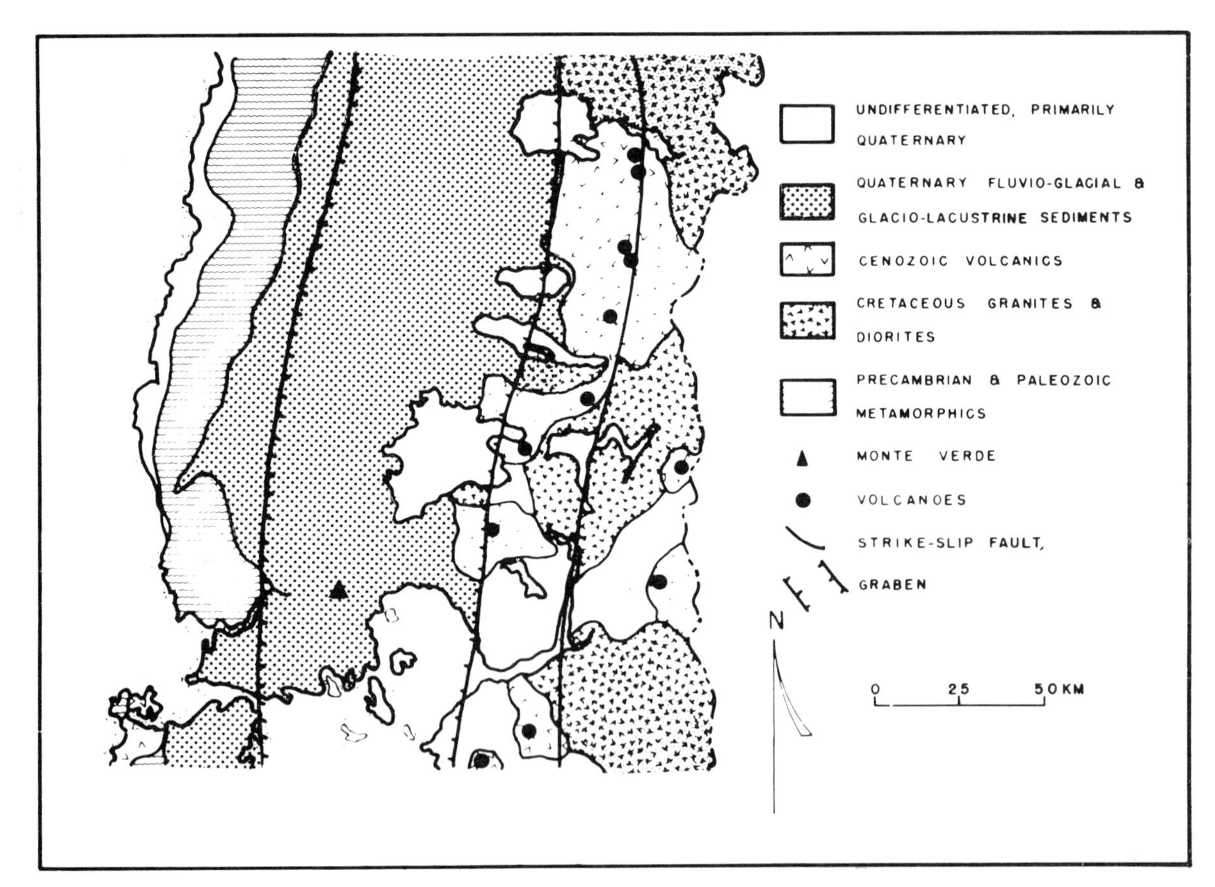

Figure 3. Geologic setting of Monte Verde, Chile.

well. Detailed mapping has not been done in these mountains, and exposures of as yet unreported kinds of igneous or metamorphic rocks may exist. Thus far, however, the extensive examinations of Maullin and Chinchihuapi gravels by Dillehay, Pino, and their associates have revealed no exceptions to the rock types listed above, a finding which is in total agreement with what we would expect from the known source-area lithologies. If exceptions do exist, they must be extremely rare. Morphologically, these gravels are composed of smooth, subrounded, rounded, and well-rounded pieces for the most part, with a notable minority exhibiting fractures along natural cleavages in the stone. High energy battering is virtually absent among gravel specimens, as is fracturing other than along planes of weakness.

From this background, then, two observations may be made with respect to the lithology of the Monte Verde collection (Table 1). First, 200 (or 96.6%) are of the rock types and forms occurring naturally in the fluvioglacial gravels of the Maullin Valley. However, compared to Chinchihuapi Creek gravels, a disproportionately high number of split pebbles and nearly spherical ones were recovered at Monte Verde. Second, seven pieces (3.3%) differ in significant ways from the documented gravels of Chinchihuapi Creek, including three specimens (1.4%) of quartzites distinctly different from anything reported in the Andean source areas or observed in the Maullin Valley gravels; these resemble quartzites reportedly outcropping in the coastal range. A core of basalt, a chopper of gray-pink granite, and a very large cobble of basalt (?) are, in general, lithologically consistent with the composition of Maullin terrace gravels in the vicinity of Monte Verde; but otherwise exhibit some differences which may indicate importation. Another item of uncertain origin is a fractured piece of grayish-white quartzite.

THE COLLECTION

The 207 stones here briefly described constitute the "collection" from the excavations at Monte Verde (Dillehay, this volume). The term "collection" is used to denote all of those pieces which, on contextual evidence, *could* have been employed culturally; that is, they were found among the modified wooden and bone pieces at the base of the peat deposit (MV-5) or, in the case of two pieces, in the sandy ridges adjacent to the creek at the base of the Monte Verde Formation. Within this collection are definitely identifiable artifacts, objects exhibiting contextual and/or morphological characteristics probably indicative of cultural intervention, and a majority whose cultural status is problematical. Eleven descriptive categories have been established upon morphological criteria; and and the objects in each category examined and described in terms of morphology, lithology, and evidence for use-wear. The latter evaluation has been aided by experimentally replicating some of the wooden artifacts, using stones from elsewhere in Chinchihuapi Creek, and the shaping of stone using replicas of the grooved wooden artifacts. The former two observations are made in comparision to extensive data collected on natural Chinchihuapi Creek gravels by Dillehay, Pino, and William Brown.

Of the 207 specimens, 204 were recovered from the peat and peat-gravel or sand-gravel contact in Area A (Figure 2). Also from Area A is the biface (category 1, below), but its association with the peat is inferential. Two specimens (the core, category 3; and one of the single-faceted stones, category 8) were found in Area C at the base of the sand, in a deposit stratigraphically correlated with the majority of the specimens from Area A.

After the obviously man-made biface, core, and grooved stones are removed, the stones in this collection are readily sorted into two major categories: unbroken smooth stones and fractured stones. Shape is the next sorting criterion. Beginning with the relationship of L>W>T among the three principal axes of each stone, the relative length of each axis, when plotted on a triangular coordinate graph, displays the basic geometry of the stone (Figure 4). In this display, the shapes of the nine major categories in this collection are discernible. These and the two residual categories are described below:

CATEGORY 1: BIFACE. A bifacially-flaked quartzite piece was recovered from the eroded face of Chinchihuapi Creek prior to the excavations (Figure 5c). It had been dislodged by erosion, but its imprint in the face of MV-5 and peat adhering to it strongly associate it with the remainder of the collection. It is 147 mm long, 62 mm wide, and 41 mm thick, superficially resembling an Abbevillian handaxe. It is almost symmetrically lanceolate, with rounded convergencies at both ends, strongly biconvex in cross section; and exhibits deep flake scars on both faces, with concomitantly sinuous edges. It is of banded medium grey to dark purplish grey quartzite. This material is unknown in the Central Valley but reportedly outcrops in the low metamorphic Coastal Range some 60 km northwest of Monte Verde. No use-wear has been discerned on this specimen.

CATEGORY 2: CHOPPER. A small, rounded cobble of dense pinkish-grey granite exhibits two intersecting fracture scars at one end (Figure 5a). One scar appears to be the result of a single force application, the other of two. It is 61 mm

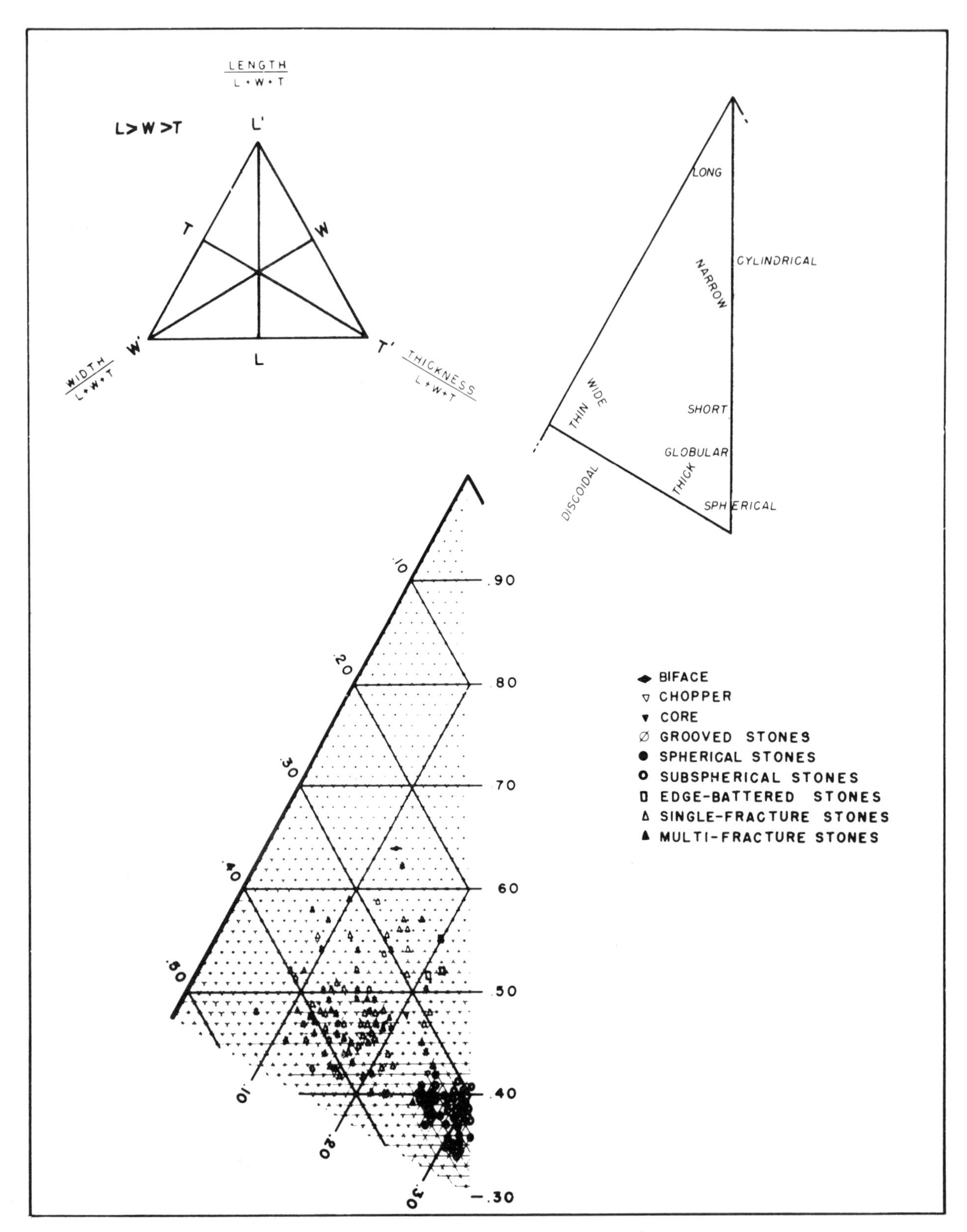

Figure 4. Triangular coordinate graph illustrating shapes of the nine major artifact categories.

long, 44 mm wide, and 35 mm thick. The pink color, density, and large size of this piece of granite are not duplicated in the Chinchihuapi gravels; and the force required to produce the fractures is not expectable in this natural environment. The acute edge of this specimen is battered and smoothed, but the cause of this attrition is unknown.

CATEGORY 3: CORE. This is a piece of black, very fine-grained basalt with a remnant of cobble cortex on one face (Figure 5b). The other two faces are a single-faceted flake scar which served as a striking platform and a convex face exhibiting at least ten flake scars and numerous smaller chip facets. The three largest flake scars were struck from the prepared platform. The core

Figure 5. Chipped stone artifacts from Monte Verde. a, chopper; b, core; c, biface.

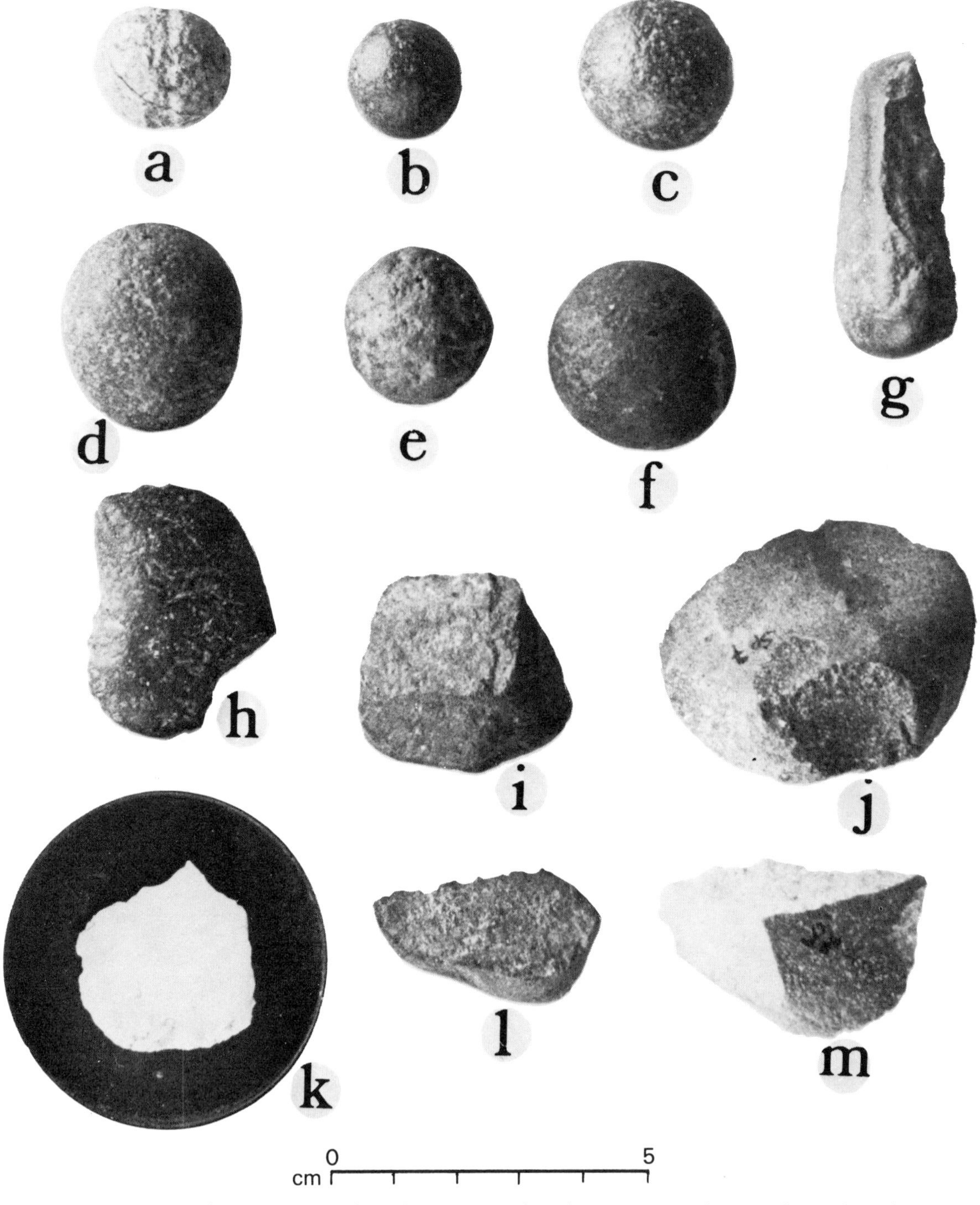

Figure 6. Stone objects from Monte Verde and experimental replicas: a, grooved stone; b-c, spherical stones; d-e subspherical stones; f, experimental replica of spherical stone; g, battered stone; h-i, single-faceted pieces; j, experimental replica of single-faceted stone; k-l, multi-faceted stone; all pieces except "k" are of locally obtainable materials.

is 96 mm long, 62 mm wide, and 40 mm thick. The fine-grained black basalt of which it is made is weathered to light brown cortex. The cortical layer is that of a moderately battered cobble. The original cobble (or boulder) represented by this piece was larger, more fine-grained, and more heavily stream battered than any other observed in Chinchihuapi gravels, though its lithology is entirely expectable in the Maullin terrace materials. It was recovered near the base of the sand in Area C.

CATEGORY 4: GROOVED STONES. These are two nearly spherical pieces, each with a well centered groove which girdles the mid point perpendicular to the longest axis of each stone (Figure 6a). In one case, the entire groove is artificial; and in the other it consists of an artificial extension of a natural fissure. These are 22 mm and 28 mm in maximum diameter, and have sphericity indices of .90 and .88. They are of locally obtainable material, and may be naturally spherical or shaped in part culturally. The surfaces of these two stones have a distinctive muted luster not unlike that of a matte finish on a photograph.

CATEGORY 5: SPHERICAL STONES. These nearly spherical pieces are 14 well-rounded smooth stones with indices of sphericity ranging from .90 to .96 (average .92). These objects, which are strikingly spherical in appearance, may have been selected from natural occurrences; or may have been modified in manufacture or in use (Figure 6b-c). Maximum diameters range from 18.8 mm to 51.0 mm and average 28.1 mm. An experiment, described below, was conducted which raises the possibility that at least a portion of these stones were partially shaped culturally. They exhibit the same muted luster as the grooved stones, and are all of locally obtainable kinds of stone.

CATEGORY 6: SUBSPHERICAL STONES. Twenty-two subspherical smooth well-rounded stones of locally occurring materials have indices of sphericity from .80 to .89 (average .85). Maximum diameters range from 20.7 mm to 63.1 mm and average 36.5 mm. As with the spherical stones, these probably occur naturally but have been selected and/or modified culturally (Figure 6d-e). Some of these stones have the muted lustrous surfaces mentioned above.

CATEGORY 7: EDGE-BATTERED STONES. Three small elongated stones exhibit battering and abrading scars along some, but not all, prominent ridges on their edges and face. Each has fracture scars originating from the battered ridges (Figure 6g). It is evident that repeated impacts of sufficient force to fracture these pieces were concentrated on two or three promontories while other equally exposed ridges retaining smooth, stream-polished surfaces, were not impacted. These are of locally obtainable material, and thus far have not been determined to have any use-wear.

CATEGORY 8: SINGLE-FACETED STONES. The most numerous category in the collection consists of 86 split pebbles and small cobbles (Figure 6h-i). The salient characteristics of these is one or more edges formed by the intersection at an acute angle of a rounded exterior with a nearly planar fracture surface. On one specimen (Figure 7a), the fracture surface is a prominent, fresh percussion scar cutting through the dense crystalline mass of the cobble. The remainder are less fresh, and the majority appear to follow planes of weakness in the material. Experiments, discussed below, have shown that similar stones, whether selected from nature or made by fracturing rounded stones, are extremely effective cutting tools. Nicking and dulling of edges is common in this category; and specimens exhibit microscopic features that may be the result of use wear, though further research is needed in this regard. All of these are of locally occurring kinds of stone.

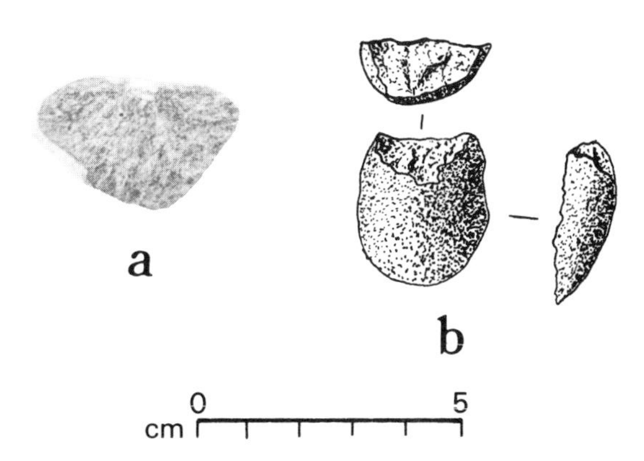

Figure 7. Faceted artifacts from Monte Verde: a, split cobble showing point of impact (top center) and radiating shatter lines; b, split pebble (multi-faceted) found seated on wooden haft.

CATEGORY 9: MULTI-FACETED STONES. These are thirty small pieces of stone exhibiting two or more fractured surfaces which intersect to form sharp edges (Figure 6k-l). One of these is a grey to white quartzite and two are of white quartz. The former is of unknown origin, whereas the latter two almost certainly derive from the coastal area. The remainder are locally available kinds of stone. One piece in this category is a split pebble of crystalline rock. One edge of the otherwise hemispherical piece is broken away. The specimen was found seated on a wooden haft with the cleavage plane down and the broken edge protruding slightly beyond the edge of the haft (Figure 7b). It is unknown if the ragged break was the result of use, or if it was intentionally broken to facilitate hafting. Two specimens, under magnification, show striations near the edges. These are apparently the result of use. Although still ambiguous at this stage of analysis, four more of these pieces exhibit edge modifications which may be the result of use.

CATEGORY 10: MISCELLANEOUS. Two miscellaneous stones have possible cultural significance (Figure 8). The first is a porous, medium-grained basalt cobble, roughly disc-shaped, with most of the perimeter exhibiting a pitted cobble surface (Figure 8a). The obverse face is nearly planar, though pitted, with the higher surfaces of the topography exhibiting a moderately lustrous polish; the reverse is an irregular fracture plane. Under magnification, the polished surface is seen to be heavily worn with randomly oriented, mostly short striations. The higher points in the microtopography are worn down and polished, extending part way down into the pits. The other stone in this category is a large (126 x 76 x 70 mm) piece of what appears to be fine-grained basalt (Figure 8b). It is a smooth, well-rounded cobble weighing some 730 grams. Its size and weight so greatly exceed anything else known to occur in Chinchihuapi gravels that the possibility of human transport must be considered.

CATEGORY 11: OTHER. This residual category contains 45 fractured stones that are quite similar to those found generally in Chinchihuapi Creek gravels; they have no morphological or other characteristics to suggest cultural utility or human modification.

EXPERIMENTATION

In order to establish reliable criteria for identifying use-wear on Monte Verde specimens, naturally fractured pebbles from Chinchihuapi Creek were

Figure 8. Miscellaneous pieces from Monte Verde: a, coarse-grained cobble exhibiting polish on prominent surfaces of obverse face; b, large cobble of fine-grained basalt.

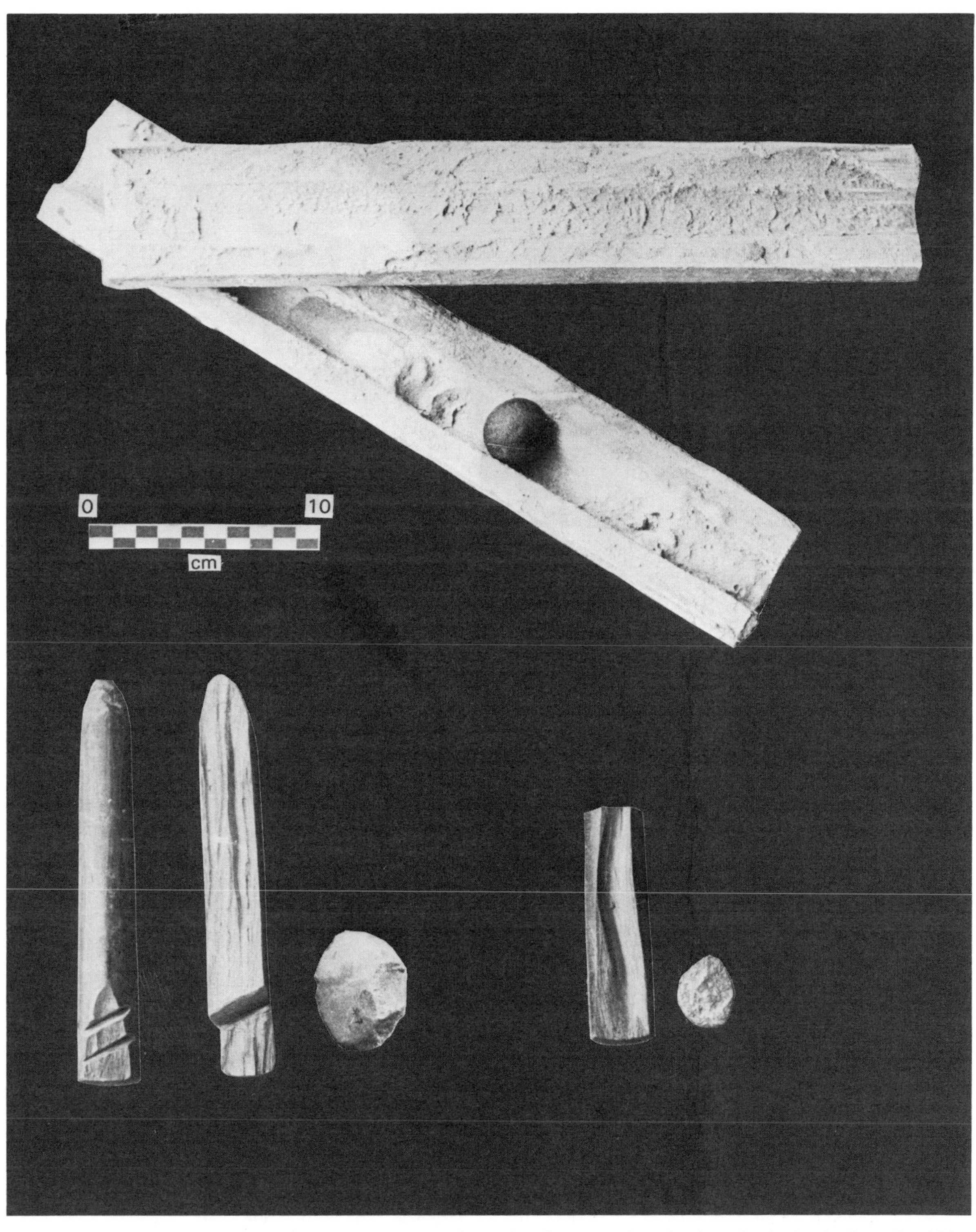

Figure 9. Replicative experimental pieces. Top, grooved wooden device used to abrade and round subspherical pebble; lower left, split cobble used in sawing and typical kerfs cut in hard wood; lower right, split cobble used in scraping and typical groove scraped in hard wood; kerfs and groove replicate closely those found at Monte Verde.

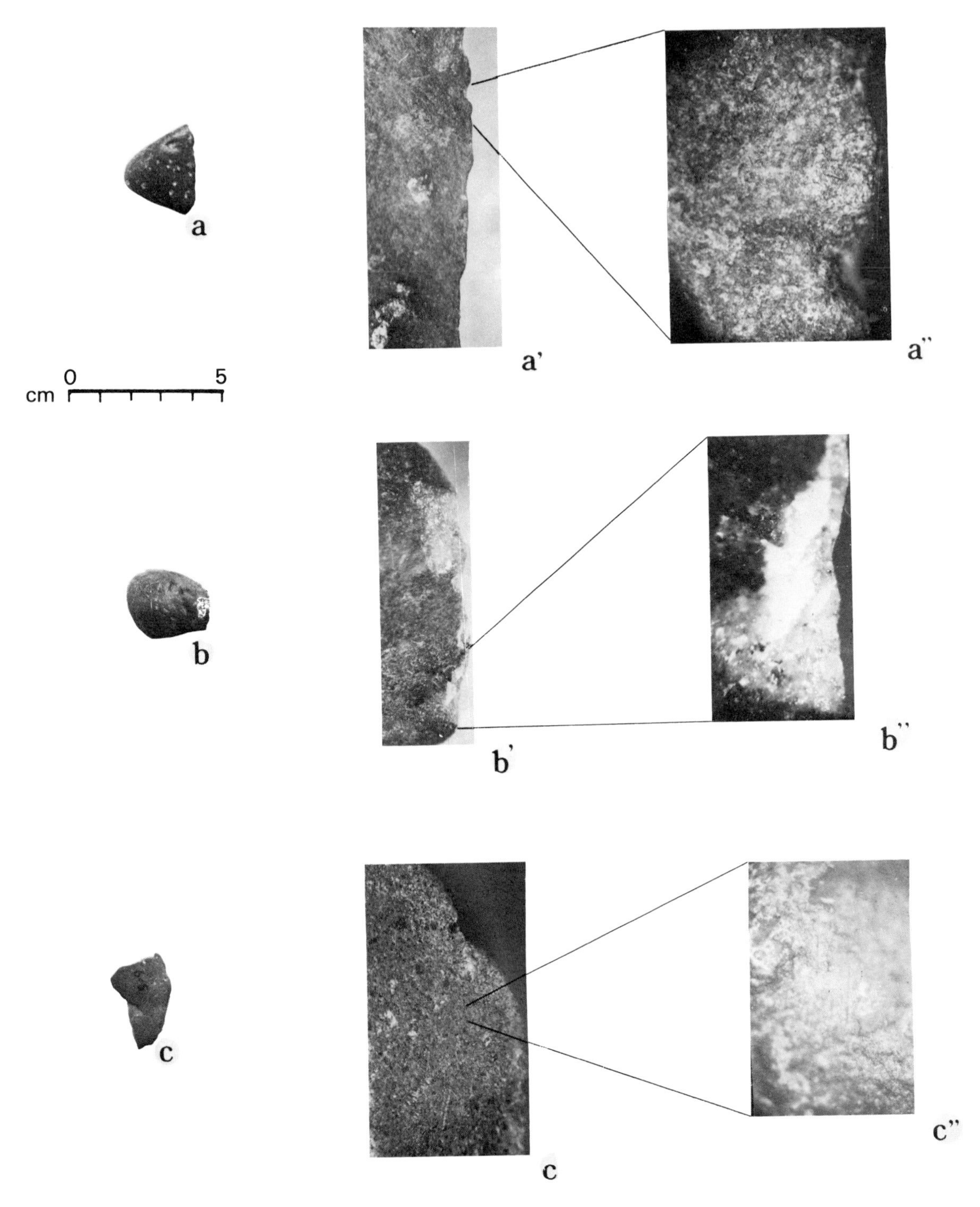

Figure 10. Examples of use-wear on Monte Verde and experimental pieces: a, prehistoric, single-faceted piece; a′, piece a magnified 7 times showing edge nicking and striations near the edge; a″, piece a magnified 60 times showing striations parallel and perpendicular to edge; b, experimental single-faceted piece showing edge nicking; b′, b magnified 7.2 times; b″, b magnified 25 times to illustrate angular crystalline mass exposed by attrition after 1000 strokes of scraping hard wood; c, experimental single-faceted piece used to saw soft wood; c′, piece c magnified 10 times showing numerous fine striations parallel to edge; c″, piece c magnified 220 times showing minute striations.

selected, and rounded ones were fractured for use-experiments. In all, 22 stones were utilized in the cutting and scraping of wood (pine and mahogany). There were little, if any, discernible differences in the effectiveness of naturally occurring versus experimentally fractured stones. All were dense, crystalline masses with ample amounts of angular quartz crystals in their composition. These proved to be moderately effective cutting and scraping tools when initially used on the pine, but soon became inoperative when wood residue adhering to the stone reduced the contact area to a minimum. However, when used on mahogany they remained highly effective cutting and scraping tools (Figures 9 and 10). Channels or grooves could be easily scraped into the surfaces of the mahogany parallel to the grain, and the wood could be sawed across the grain (Figure 9). The exposed quartz crystals cut readily into the hard wood and did not become clogged with wood residue, as had been the case with the pine. In several instances, the stone fabric gradually broke away, exposing fresh cutting (or more accurately, abrading) surfaces (Figure 10). The stones used in these experiments were examined under magnifications ranging from 10x to 300x before, repeatedly during the operations, and after, using procedures adopted from those of Keeley (1980).

Once cleared of wood residue, three categories of change could be discerned on the utilized pieces: attrition, nicking, and striations. On virtually all, there was reduction or removal of weathered surfaces followed by attrition of the crystalline mass. This was proportional to the duration of use, and was greatest on the high points of contact on the microtopographies. Attrition occurred primarily as individual crystals broke away from the mass. On some specimens, nicking along the edges exposed fresh surfaces (Figure 10). Least commonly observed were striations resulting from dislodged crystals being dragged across intact surfaces of stone (e.g., Figure 10). Although clear striations have thus far been recorded for only two experimental specimens, refining the procedures for microscopic examination is expected to increase this number substantially. In spite of the weathered condition of the Monte Verde specimens, examples of what appear to be all three of these use characteristics were documented.

In another experiment, a subspherical stone was selected from Chinchihuapi gravels to see if it could be made more spherical using wooden channels like those found in the site of Monte Verde. The stone initially was well rounded, with principal axes of 32, 29, and 26 mm. Two pieces of wood with matching longitudinal channels 29 mm across were prepared. These were then liberally coated with fine-grained sand and water. The stone was placed between the boards, in the grooves, and vigorously rolled back and forth for seven minutes (Figure 9). When cleaned, the stone exhibited two changes (Figure 6f). First, the weathered cobble surface was totally removed and fresh stone exposed over its entire exterior. This surface is much fresher than, but otherwise similar to, the dull matte finishes observed on the grooved stones, spherical stones, and some of the subspherical stones from Monte Verde. Second, the major axis of the stone was shortened by almost a millimeter to 31 mm. An additional 12 minutes of this same treatment produced no further observable or measurable changes in the stone. Evidently, the abrasion had greater effect on the weathered zone than on the interior of this material.

These experiments are entirely preliminary, but do establish the feasibility of using these stones to work wood; and of shaping stone using abrasives and wooden objects like those found in the site. A great deal more experimentation is needed before conclusive use-wear analyses can be conducted with the Monte Verde collection.

The authors have somewhat independently examined portions of the Monte Verde collection of 207 pieces for evidence of use wear. Collins first examined 80 specimens; and, with his results as the point of departure, Dillehay has examined 77 specimens. Because the emphases, approaches, and results of these two studies differ, each is reported separately in the next section of this paper.

EVIDENCE OF USE-WEAR ON MONTE VERDE SPECIMENS

Experimenting with crystalline rocks like those from Monte Verde established that coarse abrasive surfaces of angular quartz crystals held in matrices of various other crystals were effective in cutting and scraping hard woods. The most characteristic form of "wear" in these experiments was attrition by dislodging crystals from the stone fabric. Under magnification, the irregular surfaces of often bright, multi-colored crystals provide unfavorable conditions for observing use-wear. Among the 207 stones in the Monte Verde collection, preliminary microscopic examination by Collins has been conducted on 80 specimens. The results of these examinations are summarized in Table 2 and discussed below.

Nineteen Monte Verde specimens exhibit features like those created on stones during the replicative experiments. Striations (Figure 10a) are considered to be the best indicator of wear, particularly when occurring in multiple groups located and oriented in a manner consistent with a functional application of a particular specimen. These occur on seven pieces. Contact between stones in natural gravels is more likely to result in attrition and nicks than striations, although the coincidence of two or more of these on and near optimum use edges is suggestive of human agency. Among the 12 single-faceted pieces showing evidence of use-wear (Table 2), one shows only attrition, four show only nicking, two exhibit attrition and nicking, two exhibit nicking and striations, and three manifest all three. Among the six multi-faceted pieces, one exhibits only nicking, three exhibit attrition and nicking, and two exhibit all three.

Orientation of individual striations was observed as a basis for distinguishing between operations such as sawing or scraping. One sawing pattern was observed; the other six were mixed.

No basis currently exists for distinguishing hafted from prehended pieces, nor for determining what materials were worked with the apparently utilized stones in this collection. The preservation of a wooden haft and of sawn and scraped (gouged?) wooden objects in the site provided the partial, and only available, answers to these questions.

Dillehay has re-examined approximately half of these 80 specimens as well as a number of additional ones (a total of 77) to investigate more fully the nature of the striations observed by Collins. It was obvious from Collins' work that striae held the greatest promise for interpretation if our concepts and procedures of examination could be refined to map the patterns of these occurences on specimen surfaces. While manipulating the light angle and polarization in the metallurgical microscope to accentuate striations, improved control was gained of another variable, polish. It is difficult to identify polish because the prominent quartz crystals under examination appear very bright on fresh as well as weathered surfaces. However, after first carefully distinguishing among control and experimentally used pieces, it is possible to recognize true polish and its direct association with striations on the Monte Verde specimens.

This effort enhanced a key element in our view of the stone collection from the site. Preliminary microscopic examinations of the composition and topography of coarse-grained crystal aggregates exposed on the fractured surfaces of the Monte Verde stones, and the location and type of modification evidenced on these surfaces, has led to an expanded conceptualization of use-wear analysis. The crystalline structure and topography of fractured stone surfaces are characterized by irregularly distributed promontories individually composed of one or a few large, slightly elevated quartz crystals surrounded by a lower-lying matrix of smaller crystals and particles. Use-wear appears almost exclusively on edge-specific promontories in the coarse-grained stone collection.

Table 2. Summary of Collins' preliminary use-wear observations, Monte Verde collection.

Artifact Category	Number Examined	Number with Use Evidenced	Use Wear Observed*			
			Attrition	Nicking	Striation	None
1. Biface	1	0				1
2. Chopper	1	0				1
3. Core	0	0				0
4. Grooved	1	0				1
5. Spherical	3	0				3
6. Subspherical	3	0				3
7. Edge Battered	1	0				1
8. Single-faceted	44	12	6	11	5	32
9. Multi-faceted	20	6	5	6	2	14
10. Miscellaneous	1	1	1			0
11. Other	5	0				5
Totals	80	19	12	17	7	61

*Column totals each represent "minimum numbers" because if any doubt, specimens were not counted; use-wear totals exceed number of utilized specimens when more than one kind of wear was noted on a single specimen.

The implications of these observations is that the specific use characteristics and spatial distribution of individual promontories collectively define the overall use-wear pattern of a specimen. When viewed from this perspective, the edge surface is considered as a micro-topography more heavily impacted in areas where prominent crystal aggregates occur. This is not to say that lower-lying surfaces are not impacted; but either because of more rapid attrition or less contact pressure, less record of wear occurs there. Thus, each promontory is first taken as a microcosm with its own specific history of use as reflected in its distinct pattern of alteration (polish and/or striations). Multiple promontories on a single surface can then be considered collectively for overall patterning. Because of extensive attrition which occurs in these materials during use, the location of any given promontory with respect to inward migrating edges will change, as will its prominence in the immediate microtopography.

Friction pressure on fine-grained, homogeneous edges frequently results in a more uniform use-wear pattern along a broader edge topography. Scarring impact has thus been conceptualized as a continuously flowing pattern, occasionally broken by unaltered surface areas (e.g., Semenov 1964; Keeley 1978; Odell and Odell-Vereeken 1980). In contrast, the kinds of materials from Monte Verde are seen as small patches of scarring isolated by large areas of seemingly unaltered surfaces.

Recognition of the analytical and interpretive significance of the aggregate promontories and of the micro-wear on them came about through an integrated study of the three different stone collections retrieved from, in, and near the site: the non-cultural (or natural) stones, the experimental stones, and the stone collection from the excavated site. Each of these stone categories is discussed briefly in terms of cumulative stages of analysis, and of the complementary information ascertained from each.

During the course of fieldwork, naturally occurring stones were systematically collected from survey and excavation work along culturally sterile areas of Chinchihuapi Creek. These stones comprised a random collection of non-fractured and fractured materials. Fractured pebbles were examined under the microscope to determine the topography and composition of the crystalline matrices; and to define the type, distribution and degree of alteration exposed on surfaces of these as a result of natural forces such as solifluction, stress fracture, and steam rolling. Damage in the form of polish or striations is sparse, poorly developed, random; and may occur anywhere on a surface. Examination of these stones provided knowledge on the nature, location, and range of modifications caused by natural agents.

As described in a foregoing section on experimentation, various replicative activities were performed using the full range of non-cultural stone types. Unused control samples of the fresh surface topographies of these stones were inspected under the microscope. No discernible modification (striations or polish) was visible on these fresh surfaces. The acute edges subjected to use were then examined. The microwear patterns on these edges are characterized by a series of narrow, well-defined striations and by an associated moderate polish that occurs almost exclusively on the prominent crystal aggregates along and near utilized edges. Comparision of the edges on the experimental specimens and on the naturally impacted, non-cultural stones established a controlled set of physical criteria for distinguishing between the location, type, and range of modification produced either by human forces or by natural forces.

Collins' general comments on micro-wear patterning on the specimens from the excavated stones have already been given. Thus the brief statements provided here pertain only to that portion of the collection which has been studied in light of the more specific research problem; that is, the nature of striations and polish on individual promontories, and the relationships among them.

Micro-wear patterns in the form of deep, narrow striations and bright polish appear on surfaces near acute edges of 33 of 77 specimens out of the collection of 207. Eighty percent of these patterns occur on the irregularly spaced promontories located along specimen edges. These patterns were particularly well developed on the quartz crystals and less pronounced as they grade onto the lower lying aggregate of adjacent crystals. Topographically lower areas between these promontories revealed little or no modification. This study has shown that there are clear and predictable microwear patterns on selected prominent areas of specimen edges.

Results of the integrative analysis of the three stone categories tend to confirm the pointed contact and use wear evidence on edge specific crystalline formations. Use wear as a result of human action can be consistently distinguished from alteration from natural action. As this line of analysis proceeds in the future, it is hoped that the promontory specific contact approach employed here will be even better developed so that we may detect and classify the subtle range of variation in the numerous micro-wear patterns observed along an edge in order to explain the use-wear history of a specimen.

Unfortunately, at this writing, it has not been possible to integrate Dillehay's results with those of Collins; and precise numbers of specimens with various kinds of use wear have not been determined. Approximately half of the collection has been microscopically examined by at least one of us, and some 40 specimens have been observed to have at least some evidence of wear.

INTERPRETATIONS

Not all of the stone objects from Monte Verde may be unquestionably identified as human products. The biface and core are familiar kinds of chipped stone objects, one certainly and the other probably of exotic stone. As no flakes of either material were recovered, it must be inferred that the knapping of these two pieces occurred outside of the sampled area of the site or at other sites. Technologically, direct percussion is indicated by these artifacts. The two nearly spherical, grooved stones clearly are humanly made of local materials, perhaps being natural stones requiring only the grooving or perhaps first being abraded to their spherical form from less spherical natural forms and then grooved.

The hafted split pebble is another established artifact; although in the absence of the haft, its form is little different from many of the objects whose statuses as artifacts are less certain.

At the next level of interpretation are the objects which may be natural but have possible cultural attributes, or were found under conditions strongly suggesting cultural intervention. These include the three edge-battered stones; the two multi-faceted pieces of white quartz; the granitic chopper; the split cobble; the spherical stones; and most, if not all, of the subspherical stones. The utility of the wedge-shaped edges on these crystalline rocks has been established experimentally in wood cutting and scraping tasks replicative of those found on the wooden artifacts at the site, and at least 40 of these stone pieces show evidence of use. The original question of whether any of these were artifacts no longer exists; and we are left with determining more precisely which were and which were not actually used. With use wear criteria established and a measure of proficiency developed with these materials, such determinations are largely a matter of time. It has been further suggested experimentally that the spherical stones and the grooved stones may represent subspherical pebbles with artifically enhanced sphericity. The polished pieces of porous basalt and the 730 gram cobble also seem to be artifacts, though such an agency as moving ice cannot be entirely ruled out as the explanation for their presence and form.

Thus far, nothing in the preliminary study of the stones from Monte Verde links them to the *Mastodon* bones. (None of these would seem to be a lethal weapon used in bringing down a large mammal, though sling stones might be used in driving or in effectively harassing a trapped animal). Nor is there any evidence that any of these were butchering or meat processing implements. In fact, the working of wood, and the possible use of channeled wooden objects for finishing spherical stones, are the only activities tentatively suggested by this study.

The overriding interpretation that emerges is that selection of naturally occurring stones was the dominant theme of the lithic technology represented by this assemblage. Pecking-abrading and fracturing of the dense crystalline materials as well as flaking of the brittle, finer-grained samples seem to have been equally well developed stone-working techniques.

Beyond these interpretations, this study identifies three primary research needs: 1) increased efficacy of research with materials of this kind; 2) documentation of sites with comparable assemblages; and 3) much more thorough examination of this collection, particularly for evidences of use-wear. Currently, statistical comparision of these stones with a random sample of Chinchihuapi Creek gravels is underway, more experimental applications are planned, and more exhaustive microscopic examination of the collection is in progress. In the absence of well-documented additional collections, however, no study of the Monte Verde stones is likely to provide a conclusive answer to the primary question of what is cultural and what is natural for all of the pieces in the collection. It may well be that as the search for Early Man in the New World continues, the interpretive effort will of necessity focus more sharply on material culture that is little modified from its natural state. Research of this kind must begin with data from unimpeachable geologic contexts and proceed with a thorough understanding of the natural environmental factors which compete with human agency as explanations for the phenomena under study. This research will require fresh approaches, conceptually as well as procedurally.

ACKNOWLEDGMENTS

The invaluable assistance of Boyce N. Driskell, Susan Graham, David Pollack, William Brown, Bruno Hanson, Eric Gibson, Richard A. Boisvert, and Lawrence Keeley is gratefully acknowledged without burdening them with any responsibilities for inadequacies in these studies.

REFERENCES CITED

Keeley, L.H.
1978 Note on the edge damage of flakes from the Lower Paleolithic sites at Caddington. In *Paleoecology and archaeology of an Acheulian site at Caddington, England,* edited by C.C. Sampson, Southern Methodist University Press, Dallas.

1980 *Experimental determination of stone tool use: a microwear analysis.* University of Chicago Press, Chicago.

Odell, G.H., and F. Odell-Vereeken
1980 Verifying the reliability of lithic use wear assessments of "blind tests": the low power approach. *Journal of Field Archaeology* 7:87-120.

Pino, Q.M.
n.d. Investigaciones geologicas en la zona de Monte Verde, Puerto Montt, Sur la Chile. Manuscript in author's possession.

Semenov, S.A.
1964 *Prehistoric technology.* Translated by M.W. Thompson. Cory, Adams and MacKay, London.

Organizing Research on the Peopling of the Americas

ROBSON BONNICHSEN and
MARCELLA H. SORG
Center for the Study of Early Man
Institute for Quaternary Studies
University of Maine
Orono, Maine 04469
U.S.A.

INTRODUCTION

When and how the Americas were peopled is a problem of broad significance to both the humanities and the sciences. At the most general level, it is central to an understanding of how our own species was able to colonize and settle most of the earth's continents. As such, this subject cuts across national boundaries and provides a unifying focus on our Pleistocene cultural heritage.

There have been a number of distinguished and interesting attempts in past decades to review theories and findings on human origins in the New World. Writers of these reviews have had the difficult task of synthesizing uncoordinated research efforts with often contradictory results. Our task here is not to criticize their efforts, but rather to look at some of the underlying factors that have led to this lack of coordination and difficulty of synthesis.

Research on the Pleistocene peopling of the Americas has been moving slowly. Investigations have been underfunded; many discoveries have been poorly reported, and some have not been reported at all. Linguistic and geographic barriers between scientists have reduced communication. Findings of general interest to the public have not been shared in an organized or systematic manner. This in turn leads to inadequate public involvement and financial support. But whatever the reasons for this limited progress, lack of interest in the topic is not one of them.

From a specifically anthropological perspective, this problem is of major consequence to three major subfields. For physical anthropology, assumptions about the origins and chronology of early human populations underlie hypotheses about the original genetic stock and the spread of people through the Americas. These are important in turn for explaining rates of microevolutionary change and degrees of genetic variation among and between present-day American Indian, Aleut and Eskimo populations.

Resolving the question of human occupation of the Americas is obviously of considerable concern to prehistoric archaeology. The study of early sites is important for understanding how people in American Pleistocene environments used their tool-making and tool-using repertoires, as well as how they applied their knowledge of local resources in adapting to specific settings. Understanding these adaptive patterns will help to explain how the American continents were originally colonized. Conversely, this understanding can also illuminate the problem of how and why peoples of later periods developed certain strategies for living in their particular ecological context and social milieu.

For social and cultural anthropology, the subject of the earliest American populations is more problematic. Obviously, this is a topic of consequence to such subdisciplines as cultural evolution and cultural ecology. These, in turn, are the areas of anthropological theory to which archaeologists and physical anthropologists have looked for direction. Unfortunately, little guidance has been forthcoming from this quarter. This lack of theoretical guidance seems to us both symptom and cause of more basic problems.

Anthropology, the discipline in which most American archaeologists and physical anthropologists are trained, has experienced several decades of increasing specialization. Along with the resulting fragmentation of knowledge has come a shift in emphasis from culture history to "process" studies. With these developments, anthropology's ability to provide what students of early human life in the Americas need most — a holistic, integrative theoretical framework — has decreased. Consequently, knowledge of early human history in the New World has developed in a piecemeal, fragmented manner.

On the other hand, interdisciplinary research efforts provide an integrative structure for organizing scholarly work research. Quaternary sciences, in particular, now provide a holistic and unifying framework for placing the study of human occupation of the Americas in the context of local, regional and global environmental conditions. However, this kind of multidisciplinary research is very expensive and requires leadership, cooperation, and institutional support that has not been generally forthcoming.

This lack of an organizational structure has been as important as the lack of theoretical frameworks in retarding progress in this field. Specialists interested in the Pleistocene people of the Americas lack both good institutions for exchanging information and ideas with one another, and means for making significant research results available to the public. This absence of established mechanisms for exchange and dissemination of information often results in poorly coordinated research efforts. The global scale of the field of research, differences in intellectual traditions within and between nations, and variable funding levels in different countries all exacerbate the problem.

HISTORICAL BACKGROUND

Prior to World War II, a unifying theme in American anthropology was a focus on historical origins. Archaeologists were largely concerned with construction of local and regional cultural sequences; physical anthropologists were concerned with developing physical and racial typologies from which to infer the origins and relationships of human populations; and cultural and linguistic anthropologists concerned themselves with the distribution and relationships between language and culture groups.

Following World War II, anthropology followed a cross-disciplinary trend of increasing specialization. Rather than attempt holistic syntheses, cultural anthropologists chose to investigate how specific social, political, economic, and religious systems worked. Interest in material culture, although of obvious importance to archaeologists, was by and large eliminated from the purview of cultural anthropology. This problem was partially rectified by renewed interest in ethno-archaeological and experimental research. However, the research overlap between cultural anthropology and archaeology remains very thin, particularly on this topic.

Anthropological linguistics followed a similar course in assuming independence as a subdiscipline. Rather than continue to study historical phenomena, linguists addressed new questions. Subfields arose within linguistics to emphasize the study of the generative nature of grammar and the social and psychological parameters of the acquisition and use of language.

Physical anthropology followed similar trends beginning in the 1950s. Advances in genetics and the development of the synthetic theory of evolution stimulated the growth of population genetics as a new subfield within physical anthropology. A parallel development of primatology as a subfield also occurred at this time. Human osteology continued as a third subfield, usually connected with studies of growth and development or paleoanthropology.

Although evolutionary theory was incorporated as a unifying paradigm within physical anthropology, the discipline became somewhat fragmented as analysts matured and a new generation of specialists was trained. As the field became more closely linked with biology, debates even arose among practitioners about whether physical anthropology (or "human biology") should continue to be part of anthropology. These trends are all symptomatic of the compartmentalization within anthropology generally which has reduced its tendency to focus on "big picture" problems.

Following World War II, archaeologists began to realize that archaeological remains could not be fully explained by simply considering cultural historical factors. A new emphasis was placed on understanding the dynamic relationship betweeen culture and environment in specific contexts through the use of process models. Numerous archaeologists have contributed to the development of what is called the environmental archaeological approach.

With the objective of reconstructing human adaptive systems, an emphasis was placed on the collection and analysis of geological and biological remains rather than archaeological remains alone. Because this new framework provided a means for integrating state educational and museum programs working with regional problems, environmental archaeology received widespread acceptance. New data have been taken into account in developing a new and more sophisticated understanding of how people adapted to environments by creating settlement and subsistence patterns.

Environmental archaeological reconstruction emphasizes an understanding of group behavior dynamics. The overriding emphasis by archaeologists on the reconstruction of settlement and subsistence patterns has overshadowed development of more micro- and macroscopic approaches which are important for producing a holistic understanding of how and when the Americas were peopled.

Little systematic attention has been devoted to the basic issues surrounding the question of how to differentiate objects modified by nature from those modified by humans. Approaches useful for understanding relationships among cognition, behavior and material products are only now being considered in a systematic manner.

Likewise, theory in which human behavior is related to global-scale phenomena such as climate change, which is important for understanding the peopling of the Americas, has been slow to develop. For example, a number of models have been proposed for considering how and when the Americas were peopled, but few have been advanced with predictive consequences — as tested against the literature or by additional fieldwork.

It is our belief that the development of a coherent understanding of human occupation of the Americas will involve more than the collection of data. The increased specialization and fragmentation of the major subdisciplines of anthropology must be overcome. New approaches will be required to integrate all levels of anthropological theory. Anthropological theory alone will not be sufficient.

THE MULTIDISCIPLINARY SOLUTION

If increased specialization is part of the problem, multidisciplinary research is part of the solution. For example, developments in the Quaternary sciences have provided a powerful approach to reconstructing the physical and paleoecological settings for prehistoric human activity, including global-scale processes that influenced the inter-continental spread and exchange of plants and animals.

The multidisciplinary approach often results in the development of new synthetic concepts with strong interpretive potential. For instance, paleoecological concepts such as the "arctic-steppe biome" bring back into focus the need for examining the complex and interdependent relationships among prehistoric environments and cultural systems. Only with the systematic development of broad and complex concepts will we be able to understand the relationships between regional and local data sets.

Most researchers now recognize the importance and power of the multidiscipinary approach. However, there are well-recognized problems in its application: it is expensive of time and funds; it requires cooperation among a team of individual scientists; and it requires leadership from people who have made the intellectual investment in acquiring a working knowledge in several areas.

The first generation of American anthropologists were trained as generalists. With the proliferation of specializations in the past several decades, it has become clear that specialists need each other in order to solve large-scale problems and to provide the best technical expertise modern science has to offer. The team approach, however, requires some scholars who are also generalists: they must be people who can appreciate the problems of coworkers so that all the research may provide evidence to solve a common problem.

Specialists interested in the peopling of the Americas have been constrained by a lack of adequate funding for high-quality multidisciplinary research. This is related, in turn, to problems of inadequate communication and information flow within and between disciplines involved, as well as to low public and scientific visibility for significant research on the topic. As a result, progress in the field has been sluggish and generally unfocused during past decades. These problems point to the need for an institutional mechanism for exchanging information and ideas and for publicizing research results.

CENTER FOR THE STUDY OF EARLY MAN

The Center for the Study of Early Man at the University of Maine at Orono was created specifically to deal with some of these problems. The Center was established in July 1981 with the assistance of seed funds from Mr. Bingham's Trust for Charity. Funding was predicated on the assumption that it was more important to address the issues of disciplinary fragmentation, limited flow of information flow, and low public visibility than simply to finance further archaeological excavations. The stated mission of the Center is to promote research, education, and dissemination of information. Its programs are designed to coordinate research and communication between scientists, humanists, and the public in North and South America. Its central premise is that scholars working on the problem of human life in the American Pleistocene must have access to each other, to information, and to the public.

In addition to ongoing research at a number of sites, three programs are now in place: preparation of an international bibliography, organization of conferences, and academic training.

Center staff are preparing an international bibliography on early sites and related research in the Americas. Initially focusing on North America, the bibliography includes both retrospective and current components. Work has begun to locate funding for a parallel effort to cover South American literature.

Beginning in 1984 the Center will be organizing and seeking funds for conferences which focus on significant problems in New World Pleistocene archaeology. Our intent is to facilitate specialists' development of discipline-wide goals and multidisciplinary research efforts. Our attempt is not to advocate any one approach to the problem, but to encourage organization toward collectively-developed goals which include a variety of approaches.

The Center, as a subunit of the Institute for Quaternary Studies, will continue to support graduate training in Quaternary archaeology. Our goal is to model the multidisciplinary approach to archaeological research and education.

Central to the development of research on early peoples in the New World is public support. One of the most direct ways to assure long-term support of research is to involve the public in the fascinating story of America's earliest cultural heritage. The staff at the Center for the Study of Early Man is implementing the "Peopling of the Americas" publication program of books and serials to communicate research results to both the scientific community and the interested public.

The major long-range goal of the Center is to stimulate and facilitate the development of knowledge about Pleistocene peoples of the Americas and human origins in general. In order to accomplish this, we have addressed our programs to the problems discussed here: increased specialization, the need for multidisciplinary research, and the need for an organizational format for exchanging ideas and research results. We invite comments and suggestions from our colleagues as we move forward with these efforts.

ACKNOWLEDGEMENTS

Editorial suggestions by George L. Jacobson, Jr., and Lisa Feldman led to substantial improvements in former versions of the manuscript. Karen Hudgins typed the manuscript. For their efforts we extend sincere thanks. The authors are responsible for all errors, omissions and interpretations.

Index